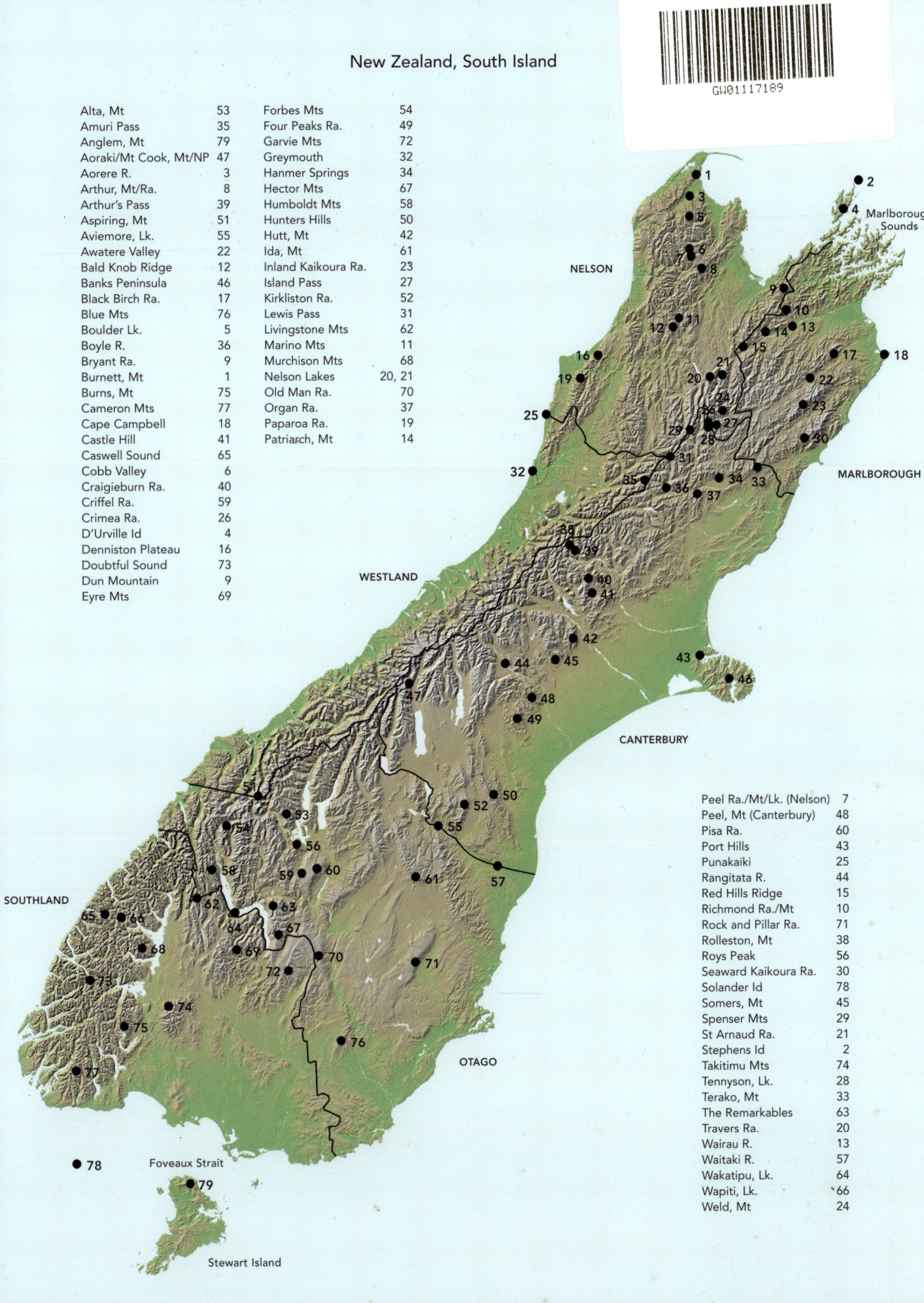

An Illustrated Guide to New Zealand Hebes

An Illustrated Guide to
New Zealand Hebes

M. J. Bayly and A. V. Kellow

With introductory chapters by
Peter J. de Lange, Phil J. Garnock-Jones and Kenneth R. Markham
Photography by W. M. Malcolm

First published in New Zealand in 2006 by
Te Papa Press, P.O. Box 467, Wellington, New Zealand.

© Text and images the Museum of New Zealand Te Papa Tongarewa, 2006;
   Fig. 103A Auckland War Memorial Museum.

This book is copyright. Apart from any fair dealing for the purpose of private study, research, criticism, or review, as permitted under the Copyright Act, no part of this book may be reproduced by any process, stored in a retrieval system, or transmitted in any form, without prior permission of the Museum of New Zealand Te Papa Tongarewa.

TE PAPA® is the trademark of the Museum of New Zealand Te Papa Tongarewa.
Te Papa Press is an imprint of the Museum of New Zealand Te Papa Tongarewa.

National Library of New Zealand Cataloguing-in-Publication Data
Bayly, Michael, 1970-
An illustrated guide to New Zealand hebes / by Michael Bayly and Alison Kellow.
Includes bibliographical references and index.
ISBN-13: 978-0-909010-12-6
ISBN-10: 0-909010-12-9
1. Hebe (Plants)—New Zealand. 2. Hebe (Plants)—New Zealand—
Pictorial works. I. Kellow, Alison, 1969- II. Title.
635.93395—dc 22

Cover design: Mission Hall Creative
Internal text design and typesetting: Afineline
Printed by: Everbest Printing Co, China

Contents

Acknowledgements viii

List of Symbols and Abbreviations ix

Scope of the Book xii

Part A *General Chapters*

Introduction 3

Classification and Evolution 7
  Taxonomic History at Species, Subspecies and Variety Level   *8*
  History of Generic and Infrageneric Classifications   *11*
  Relationships of the *Hebe* Complex Inferred from DNA Sequences   *13*
  Defining the Limits of *Hebe* and *Veronica*   *16*
  Constructing an Infrageneric Classification for *Hebe*   *17*

Distribution, Habitats and Biogeographic History 19
  Distribution and Habitats of *Hebe* and *Leonohebe*   *19*
  Patterns of Species Richness   *21*
  Patterns of Endemism   *23*
  Biogeographic History of *Hebe* and *Leonohebe*   *24*

Morphology (P. J. Garnock-Jones) 27
  Vegetative Morphology   *27*
  Reproductive Morphology   *32*
  Developmental Morphology   *37*

Flavonoid Biochemistry (K. R. Markham) 38
  Introduction to the Flavonoids   *38*
  Flavonoids in *Hebe* and *Leonohebe*   *38*
  Screening for Flavonoids   *40*
  Taxonomic Application of Flavonoid Characters   *41*

Chromosomes 45

Reproductive Biology (P. J. Garnock-Jones) 61
  Flower Biology   *61*
  Flower Visitors and Pollinators   *63*
  Gender   *64*
  Fruit and Seed Biology   *66*

Conservation (P. J. de Lange) 67
    Conservation Status *67*
    Conservation Management *72*

Cultivation 73

Part B *Identification and Description of Species*

Materials and Methods 77
    Scope of Survey *77*
    Concepts of Species and Infraspecific Taxa *77*
    Morphological Descriptions *79*
    Arrangement of Species *79*
    Treatment of Hybrids *79*
    Distribution Maps *79*
    Flowering Times *80*
    Photography and Imaging *80*

Taxonomic Treatment 88
    Synopsis *88*
    Implicit Character States *88*
    Other Notes on Descriptions *88*

Hebe 90
    "Flagriformes" (the Whipcords) *91*
    "Connatae" *113*
    "Subcarnosae" *129*
    "Occlusae" *144*
    "Buxifoliatae" *215*
    Small-leaved "Apertae" *224*
    Large-leaved "Apertae" *266*
    "Grandiflorae" *286*
    "Pauciflorae" *288*

Leonohebe 290
    Sect. *Leonohebe* (the Semiwhipcords) *291*
    Sect. *Aromaticae* (*Leonohebe cupressoides*) *300*

Nomenclature 302
    Names of Genera *302*
    Names of Subgenera *302*
    Names of Sections *302*
    Names of Series *303*
    Subsectional Names Without a Clear Indication of Rank *303*
    Names of Species and Infraspecific Taxa in *Hebe* *303*

    Names of Species in *Leonohebe*   *323*
    Incertae Sedis   *324*
    Names of Possible Wild Hybrids   *325*
    Names of Horticultural Forms   *327*
    Excluded Names   *328*

Common and Māori Names 330
    English Common Names   *330*
    Māori Names   *331*
    Names from French Polynesia   *331*

Part C *Indices*

Appendices 335
    Appendix 1: Informal Names used by Druce (1980, 1993) and Eagle (1982)   *335*
    Appendix 2: Variation in *Hebe hectorii*   *336*
    Appendix 3: Variation in *Hebe lycopodioides*   *338*
    Appendix 4: Examples of Leaf Outlines for Selected Species   *340*
    Appendix 5: Details of Voucher Specimens for Photographs and Illustrations   *352*

Glossary 357

References 368

Picture Credits 380

Picture Credits 380

Index 381

Acknowledgements

Many individuals and organisations assisted with work toward this book. For help with fieldwork and/or for providing plant material we thank Tim Adam, Rebecca Ansell, Ross Beever, Ilse Breitwieser, Barbara Brown, Pat Brownsey, Kate Bull, Shannel Courtney, Peter de Lange, Arnold Dench, Alison Evans, Allan Fife, Kerry Ford, Lisa Forester, Tim Galloway, David Glenny, Michael Heads, Illona Keenan, Sue Lake, Julia Lee, Arlene McDowell, Bianca Maich, Bill and Nancy Malcolm, Kevin Mitchell, Wendy Nelson, David Orlovich, Nan Pullman, Leon Perrie, Barbara Polly, Karen Riddell, Jenny Steven, Alan Tennyson, Mick Thomas, Santha Traill, Stephanus Venter, Anthony Vadala, Rainer Vogt, Aaron Wilton and Debra Wotton.

Curators at Percy Reserve, Petone (Robyn Smith, Jill Broome), at Otari-Wilton's Bush (Anita Benbrook, Jane Wright, Robyn Smith) and at the Landcare Research Gardens, Lincoln (Stuart Oliver), provided access to plants in cultivated collections. Many landowners allowed access to, or across, their properties. The Department of Conservation (DoC) granted collecting permits for reserves under their control, and DoC staff throughout the country provided advice or assistance with fieldwork.

Successive research assistants (Rebecca Ansell, Gina Williams, Illona Keenan, Debra Wotton, Barry Sneddon) assisted with the collection, mounting and databasing of specimens; measurements used in descriptions and morphometric analyses; digital imaging; and compiling seed descriptions (Barry Sneddon). The detailed research by Kevin Mitchell and Ken Markham into leaf flavonoids provided data invaluable to this revision. Padmini Ekanayake-Carlson dealt with a large number of interloan requests for nomenclatural publications. Neville Walsh provided Latin translations for descriptions of many new taxa.

Various Australian Botanical Liaison Officers (Don Foreman, Bob Chinnock, Peter Bostock, Annette Wilson) provided information on specimen and library holdings at K, or assisted with our visit there. Directors and staff of the herbaria of AK, BAB, BISH, BM, C, CANU, CHR (particularly Kerry Ford and Mary Korver), K, MEL, NZFRI, OTA, P, SGO, WAIK, WELT and WELTU provided loan material and/or access to collections.

Tim Galloway helped with fieldwork, provided many line drawings and did a sterling job of laying out the plates. This book would not be anywhere near as useful without the excellent photographs of Bill Malcolm (Micro-Optics Ltd, Nelson). We also thank all others who provided photographs (listed individually in the picture credits on page 380).

The authors greatly appreciate the erudite contributions from Peter de Lange (Science and Research Unit, Department of Conservation, Auckland), Phil Garnock-Jones (Victoria University of Wellington) and Ken Markham (Industrial Research Ltd, Lower Hutt). Anne Bayly provided much needed childcare while we finished work on this book. Ilse Breitwieser, Ewen Cameron, Murray Dawson, John Dawson, Peter de Lange, Phil Garnock-Jones, David Glenny, Peter Heenan, Jill Kellow, Kevin Mitchell, Leon Perrie, Barry Sneddon and Steve Wagstaff made valuable comments on the manuscript. All stages of this project, from drafting the initial grant application to editing of the final copy, benefited from the supervision, advice, support, friendship, attention to detail and hard work of Patrick Brownsey. This work was funded by the Foundation for Research Science and Technology (MNZX0201).

List of Symbols and Abbreviations

adj.	adjective
AK	herbarium of the Auckland War Memorial Museum
AKU	herbarium of the University of Auckland
APG	Angiosperm Phylogeny Group
Apr	April
Art.	Article (of ICBN)
asl	above sea-level
Aug	August
BAB	herbarium of the National Institute of Farming and Livestock Raising Technology, Buenos Aires
BISH	Herbarium Pacificum, Bishop Museum, Honolulu
BM	herbarium of The Natural History Museum, London
C	herbarium of the University of Copenhagen
c.	*circa*; about, approximately
CANTY	herbarium of the Canterbury Museum [now at CHR]
CANU	herbarium of the University of Canterbury, Christchurch
cf.	*confer*; compare
CHBG	herbarium of Christchurch Botanic Gardens
CHR	Allan Herbarium, Landcare Research, Lincoln
Ck	Creek
cm	centimetre(s)
dbh	diameter at breast height
Dec	December
DNA	deoxyribonucleic acid
e.g.	*exempli gratia*; for example
egl.	eglandular
esp.	especially
et al.	*et alii*; and others
Ex.	Example (of ICBN)
f.	forma
Feb	February
Fig., Figs	figure, figures
FRST	Foundation for Research, Science and Technology
Gk	Greek or Greek-derived
gl.	glandular
Herb.	herbarium (or herbarium of)
ICBN	International Code of Botanical Nomenclature
Id, Ids	Island, Islands
i.e.	*id est*; that is
ITS	internal transcribed spacer, a region of nuclear ribosomal DNA
Jan	January

Jul	July
Jun	June
K	herbarium of the Royal Botanic Gardens, Kew
km	kilometre(s)
L.	Latin or Latin-derived
Lk., Lks	Lake, Lakes
m	metre(s)
Mar	March
MEL	National Herbarium of Victoria, Melbourne
mg	milligram(s)
mm	millimetre(s)
MR	micropylar rim
Mt, Mtn, Mts	Mount, Mountain, Mountains (usually place names)
Mya	million years ago
Myr	million years
n	haploid chromosome number
NI	North Island
nom.	*nomen*; name
nom. illeg.	*nomen illegitimum*; illegitimate name
nom. nud.	*nomen nudum*; a name published without an accompanying description
Nov	November
NP	National Park
n.v.	*non vidi*; not seen (by the authors)
NZ	New Zealand
NZFRI	herbarium of New Zealand Forest Research Institute, Rotorua
NZMS	New Zealand Map Series
Oct	October
OTA	herbarium of the University of Otago, Dunedin
P	herbarium of the Muséum National d'Histoire Naturelle, Paris
pers. comm.	personal communication
pl.	plural
p.p.	*pro parte*; in part
R.	River
Ra.	Range
*rbc*L	chloroplast gene encoding large subunit of the photosynthetic enzyme ribulose-1,5-bisphospate carboxylase/oxygenase
s. lat.	*sensu lato*; in a broad sense
s. str.	*sensu stricto*; in a strict sense
sect.	section
Sep	September
ser.	series
SGO	herbarium of the Museo Nacional de Historia Natural, Santiago
SI	South Island
sp., spp.	species (singular), species (plural)
Stm	Stream
sts	sometimes
subg.	subgenus
subsp.	subspecies

TS	transverse section
us.	usually
Va.	Valley
var., vars	variety, varieties
WAIK	Waikato Herbarium, University of Waikato, Hamilton
WELT	herbarium of the Museum of New Zealand Te Papa Tongarewa, Wellington
yr	year
WELTU	H. D. Gordon Herbarium, Victoria University of Wellington
μm	micrometre(s) (1 μm = 0.001 mm)
<	less than
>	greater than
≤	less than or equal to
≥	greater than or equal to
±	more or less
≈	approximately equal to
=	taxonomic (heterotypic) synonym (where used in list of synonyms)
≡	nomenclatural (homotypic) synonym
!	indicates specimens seen by the authors (when following a specimen number)
♂	male
♀	female
⚥	hermaphrodite
2n	diploid chromosome number

Scope of the Book

This book deals with the identification, classification and biology of the flowering plant genera *Hebe* and *Leonohebe* (family Plantaginaceae; formerly included in Scrophulariaceae). These plants occur primarily in the New Zealand region, although *Hebe* includes two species that also naturally occur in South America, and one that is endemic to Rapa in French Polynesia. The main aim of the book is to provide a tool for the identification of species, subspecies and varieties of *Hebe* and *Leonohebe*, and to serve as a reference on the known variation, distribution and classification of these groups. Keys to identification are provided. Detailed descriptions of species, subspecies and varieties are accompanied by colour photographs and distribution maps, as well as information on recognition, variation, flowering time, chromosome number and the etymology of names. Introductory chapters provide information on classification, evolution, biogeography, ecology, conservation, reproductive biology and flavonoid chemistry. Toward the end of the book, information is included on matters of botanical nomenclature (listing synonymous names, type specimens, excluded names and names of uncertain application) and common and Māori names, and there is also a glossary of technical terms and a list of references.

PART A

General Chapters

Introduction

Hebe is New Zealand's largest genus of flowering plants, comprising eighty-eight species as defined in this book. It and related plants are conspicuous components of the natural landscape of New Zealand, particularly in subalpine to alpine areas, and are equally conspicuous in urban settings, being commonly grown in gardens, both locally and overseas.

Hebes are remarkable for their ecological and morphological diversity. Different species occur in habitats that range from coastal margins to alpine rocks, up to *c.* 2800 m above sea-level, about the highest altitude attained by any flowering plant in New Zealand. Likewise, plants range from large-leaved shrubs (e.g. *H. speciosa*) or small trees (e.g. *H. parviflora*) to "whipcord" forms, with small scale-like leaves (e.g. *H. hectorii*, *H. salicornioides*, *H. lycopodioides*). This striking diversity in appearance provides a range of forms for use in horticulture, and a range of questions to those interested in the evolution of plant diversity.

Hebes have a reputation for being a "difficult" group to study. The large number of species (by New Zealand standards), the general similarity of many species to each other, and the extent of variation within some species, have all contributed to this perception. Another contributing factor has probably been the widespread notion that hybridisation between species is rife, a problem that, at least among wild plants, seems to have been overemphasised. Some groups of species, and occasional specimens with unusual combinations of features, are difficult to identify. However, most species can be identified if attention is paid to detail, often necessitating the use of a hand lens. Those interested in learning more about the group should not be daunted by its reputation.

The last complete taxonomic revision of *Hebe* was published in *Flora of New Zealand Vol. 1* by Moore & Ashwin (in Allan 1961). This recognised seventy-nine species, divided into ten informal subdivisions. It provided a substantial advance over previous classifications and has, for more than forty years, been the standard and most complete account of the genus. A range of guidebooks published since the *Flora*, built largely on the information it contained, have provided useful tools for the identification of some groups of hebes, and have been used widely by amateur and professional botanists alike. However, with the exception of the recent, largely horticultural book of Hutchins (1997), and that of Chalk (1988), none has focused solely on *Hebe* and none has attempted taxonomic reappraisal of the group.

In the years since the *Flora* was published, continued botanical exploration of New Zealand has suggested the existence of additional species, subspecies and varieties of *Hebe* (some of which are rare and/or threatened), as well as raising questions about the validity or limits of some previously recognised species. In *Hebe*, as in other groups of New Zealand plants, taxonomic research has not kept pace with the new ideas and taxa proposed by field botanists. Absence of an up-to-date, comprehensive revision of *Hebe* has resulted in the widespread use of informal names, or letter/number codes, for proposed new taxa among field botanists, in herbarium databases, in Department of Conservation (DoC) reports,

TABLE 1 Differences, at species and infraspecific ranks, from the classification in *Flora of New Zealand Vol. 1* (Moore, in Allan 1961). Names accepted here are given in roman type, those considered synonyms are in italics (= indicates taxonomic synonyms; ≡ indicates nomenclatural synonyms).

Name (in *Flora*)	Taxonomic change
Hebe adamsii	reinstated
H. allanii	here ≡ H. amplexicaulis f. hirta
H. amplexicaulis var. *erecta*	here = H. amplexicaulis s. str.
H. biggarii	reinstated
H. carnosula	reinstated
H. coarctata	here ≡ H. hectorii subsp. coarctata
H. colensoi var. *hillii*	here = H. colensoi s. str.
H. dilatata	reinstated
H. elliptica var. *crassifolia*	here = H. elliptica s. str.
H. fruticeti	here = H. subalpina
H. gracillima	here = H. leiophylla
H. haastii var. *humilis*	here ≡ H. macrocalyx var. humilis
H. haastii var. *macrocalyx*	here ≡ H. macrocalyx var. macrocalyx
H. hectorii var. *demissa*	here ≡ H. hectorii subsp. demissa
H. laingii	here = H. hectorii subsp. hectorii
H. lycopodioides var. *patula*	here = H. lycopodioides s. str.
H. macrocarpa var. *brevifolia*	here ≡ H. brevifolia
H. macrocarpa var. *latisepala*	here = H. macrocarpa s. str.
H. matthewsii	placed *incertae sedis*
H. parviflora var. *angustifolia*	here ≡ H. stenophylla
H. parviflora var. *arborea*	here = H. parviflora
H. pauciramosa var. *masoniae*	here ≡ H. masoniae
H. petriei var. *murrellii*	here ≡ H. murrellii
H. poppelwellii	here = H. imbricata
H. recurva	here = H. albicans
H. subsimilis	here ≡ H. tetragona subsp. subsimilis
H. subsimilis var. *astonii*	here = H. tetragona subsp. subsimilis s. str.
H. treadwellii	reinstated
Veronica ×bishopiana	here ≡ H. bishopiana
V. salicifolia var. *angustissima*	here ≡ H. angustissima
V. salicifolia var. *paludosa*	here ≡ H. paludosa
—	H. arganthera (described since *Flora*)
—	H. calcicola (described since *Flora*)
—	H. crenulata (described since *Flora*)
—	H. cryptomorpha (described since *Flora*)
—	H. flavida (described since *Flora*)
—	H. mooreae (described since *Flora*)
—	H. pareora (described since *Flora*)
—	H. perbella (described since *Flora*)
—	H. pimeleoides subsp. faucicola (described since *Flora*)
—	H. pubescens subsp. rehuarum (described since *Flora*)
—	H. pubescens subsp. sejuncta (described since *Flora*)
—	H. rigidula var. sulcata (described since *Flora*)
—	H. scopulorum (described since *Flora*)
—	H. societatis (described since *Flora*)
—	H. stenophylla var. hesperia (described since *Flora*)
—	H. stenophylla var. oliveri (described since *Flora*)
—	H. tairawhiti (described since *Flora*)

in checklists of New Zealand plants (in particular those of Druce 1980, 1993) and in some popular books (most notably that of Eagle 1982). Although informal names may meet a short-term need, their extensive use creates confusion. Such names are not subject to the formal rules of botanical nomenclature. They are easily created, regardless of whether or not they are soundly based, and are often not clearly defined. They are commonly used in different senses by different workers, as evidenced by the labelling of herbarium specimens, and it is not unusual for multiple names to be applied to the same biological entities (Appendix 1). Resolution of the resulting confusion and correct application of formal botanical names is necessary for proper assessment of conservation risks and management strategies, and to provide the basis for better understanding of ecology, phylogenetic relationships and biogeographic patterns.

Against this background, Pat Brownsey from Te Papa (the Museum of New Zealand Te Papa Tongarewa), in collaboration with Phil Garnock-Jones (Victoria University of Wellington), Ken Markham (Industrial Research Ltd) and Bill Malcolm (Micro-Optics Ltd), successfully applied for funding from the New Zealand Foundation for Research, Science and Technology (FRST) to undertake a biosystematic revision of *Hebe*. Mike Bayly and, later, Alison Kellow were appointed research scientists at Te Papa to work on this project. Primary aims of this research were to: reassess the classification of *Hebe*, particularly the classification and distribution of species, subspecies and varieties; test hypotheses about origins and evolution of the group; and provide user-friendly tools for identification. The purpose was to provide scientific support for the conservation, management and utilisation of hebes as a natural resource. Some research findings from the project have been published as scientific papers (Mitchell et al. 1999, 2001; Bayly et al. 2000, 2001, 2002, 2003, 2004; Garnock-Jones et al. 2000; Wagstaff et al. 2002; Kellow et al. 2003*a*, 2003*b*, 2005; Bayly & Kellow 2004*a*, 2004*b*; Markham et al. 2005), and this book is the culmination of the project.

Research for the book has focused on all groups of *Hebe* treated in the *Flora of New Zealand*, except for the subdivision "Paniculatae", which has been described as a distinct genus, *Heliohebe*, and for which a detailed account is provided by Garnock-Jones (1993*b*). It was initially assumed – for example, from the morphological study of Garnock-Jones (1993*a*) – that this book would focus on a "natural" or "monophyletic" group. However, it is now apparent, from recent studies of DNA sequences (e.g. Wagstaff & Garnock-Jones 1998; Wagstaff et al. 2002), that the included species comprise two distinct evolutionary lineages that are probably not each others' closest relatives. These lineages are treated here as distinct genera. *Hebe* is the largest of the two genera. The smaller genus, *Leonohebe*, comprises just five species (four "semiwhipcord" species, and the distinctive *L. cupressoides*). *Leonohebe* was first described by Heads (1987) with a much broader circumscription than that adopted here. *Hebe* and *Leonohebe* are treated together in this book because of the general historical association of their members, and because detailed information on the species of both groups should be of use to those interested in the identification of New Zealand plants. Not all botanists will agree with the generic boundaries accepted here, and the merits of some alternative classifications are discussed in the next chapter.

This book deals only with naturally occurring ("wild") hebes and leonohebes, and not with the extensive range of horticultural forms. Some plants commonly used in horticulture are indistinguishable from wild species, being vegetatively propagated from, and therefore genetically identical to, original wild collections. Many garden hebes, however, are either highly aberrant forms, or hybrids. These hybrids may have been deliberately bred for desirable characteristics, or have arisen by chance, as often happens when many hebes are grown together in gardens, and are, as a result, of uncertain parentage. This book should be used with caution if trying to identify garden-grown plants; many commonly grown plants are not included. That is not to say, however, that it is of no use to those interested in hebes in horticulture. Knowledge of the attributes of wild species can assist in determining appropriate growing conditions for garden plants, whether or not a plant matches a wild species, along with possible parentage of cultivated hybrids and sources of useful characteristics for inclusion in future breeding programmes. Books that cover horticultural forms in more detail are those of Hutchins (1997), including detailed descriptions of many cultivars, especially those grown in Britain; Metcalf (2001), a register of all *Hebe* cultivars, including brief descriptive notes on many; and Wheeler & Wheeler (2002), dealing mostly with the uses of hebes in horticulture, but with descriptions and photographs of a range of cultivars. A history of *Hebe* as a horticultural subject has been written by Heenan (2001).

In preparing this account of *Hebe* and *Leonohebe*, certain aspects of the classification have been resolved and are advancements on earlier works. However, as is often the case in this type of work, there are still aspects of the classification that are unsatisfactory. In particular, the limits of some species are still imperfectly understood or delineated (this is generally noted in the entries for these species), and the arrangement of species into natural groups remains problematic. Undoubtedly there will be botanists who disagree with some of the taxonomic decisions in this work, chiefly because of different perceptions of morphological traits, or because of different interpretations of what constitute genera, species, subspecies and varieties. Disagreement of this sort, though inevitable, can be a driving force behind continued botanical investigation.

The introductory chapters aim to present background information on classification, evolution, biogeography, conservation, reproductive biology and flavonoid chemistry. These are mostly reviews of previously published works, but do include some new information. Some of the material is, by its nature, quite technical. We have tried to present these topics in a manner accessible to the general reader, but have included references to relevant scientific publications for those with deeper interest in the subject matter. There are, however, certain areas of discussion in which a basic understanding of plant classification and biology has, for the purpose of brevity, been assumed.

Substantial differences between the treatment presented here and that of the *Flora of New Zealand* include (apart from generic limits, and the arrangement of species into informal groups) the recognition of eleven additional species, three varieties and three subspecies described since 1961, and the reinstatement of eight species that were either not recognised, placed *incertae sedis* or treated as hybrids in the *Flora* (**Table 1**). Five species and six varieties recognised in the *Flora* are not recognised here at any rank. Other changes relate mostly to the ranks at which taxa are recognised, with five recognised here as species rather than varieties; two recognised here as subspecies, and one as a form, rather than species; and one recognised as a subspecies rather than a variety. Most of these changes are not proposed here for the first time, but have been suggested, either by ourselves or others, in books, papers and checklists published since 1961.

The account presented here is, we hope, a reasonable and up-to-date summary of what is known about *Hebe* and *Leonohebe*, in a format that suits the needs of a wide range of users. It should be useful for those needing, or wanting, to identify hebes and leonohebes, and provide both the stimulus for, and background to, further investigation of these plants and their evolution, classification, conservation and uses.

Classification and Evolution

Hebe and *Leonohebe* have traditionally been placed in the flowering plant family Scrophulariaceae (tribe Veroniceae in many treatments). However, recent studies of DNA sequences (Olmstead & Reeves 1995; Olmstead et al. 2001) suggest that traditional classifications of that family are not well founded, and that a revised classification is warranted. On this basis, recent publications (e.g. Judd et al. 1999, 2002; APG 2003; Heads 2003; Kellow et al. 2003b, 2005; Albach & Chase 2004; Bayly et al. 2004; Garnock-Jones & Lloyd 2004) generally include *Hebe* and its allies in a revised circumscription of the family Plantaginaceae.

Hebe and *Leonohebe* are segregates of the larger, mostly northern hemisphere genus *Veronica*. They are associated with a group of Australasian genera (**Fig. 1**)

FIG. 1 Examples of genera related to *Hebe* and *Leonohebe*.
A *Veronica plebeia*;
B *V. serpyllifolia*;
C *Derwentia perfoliata*;
D *Parahebe hookeriana*;
E *Heliohebe hulkeana*;
F *Chionohebe glabra*.

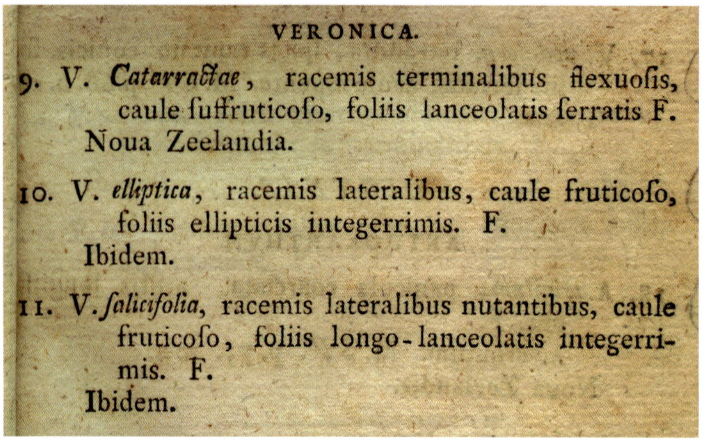

VERONICA.

9. V. *Catarractae*, racemis terminalibus flexuoſis, caule ſuffruticoſo, foliis lanceolatis ſerratis F. Noua Zeelandia.

10. V. *elliptica*, racemis lateralibus, caule fruticoſo, foliis ellipticis integerrimis. F. Ibidem.

11. V. *ſalicifolia*, racemis lateralibus nutantibus, caule fruticoſo, foliis longo-lanceolatis integerrimis. F. Ibidem.

FIG. 2 The first published descriptions of New Zealand *Hebe* species (as *Veronica elliptica* and *V. salicifolia*). Reproduced from Forster (1786).

that, at least in recent local floras and associated works, have also been distinguished from *Veronica*. This group includes the genera *Parahebe* (Oliver 1944; Garnock-Jones & Lloyd 2004), *Chionohebe* (Briggs & Ehrendorfer 1976; treated as *Pygmea* in Allan 1961 and some earlier works), *Heliohebe* (Garnock-Jones 1993*b*), *Derwentia* (Briggs & Ehrendorfer 1992) and *Detzneria* (Diels 1929; Van Royen 1983). For convenience these Australasian segregates are referred to here as the "*Hebe* complex", a term also used by Heads (1992, 1994*a*, 1994*b*), Garnock-Jones (1993*a*), and Wagstaff & Garnock-Jones (1998).

The following sections summarise the taxonomic history of *Hebe* and *Leonohebe*, evidence for their evolutionary relationships, and some current issues regarding their classification at both generic and subgeneric levels. The first section deals with the history of taxa at the ranks of species, subspecies and variety, and presents data for all taxa included in *Hebe* and *Leonohebe* as defined here (in other words, regardless of the genus in which they were originally described), but ignores taxa that have been included in wider circumscriptions of *Hebe*. The sections on generic classification and evolutionary relationships necessarily involve further consideration of *Veronica* and other members of the *Hebe* complex. These groups are not the main focus of this work, but the cited references provide further information about them.

TAXONOMIC HISTORY AT SPECIES, SUBSPECIES AND VARIETY LEVEL

The first formal descriptions of hebes from New Zealand (as *Veronica elliptica* and *V. salicifolia*) were published by Georg Forster (Forster 1786), who, together with his father, Johann Reinhold Forster, was a botanist on James Cook's second voyage to New Zealand in 1773. Forster's descriptions are brief (**Fig. 2**), but sufficient for formal publication of botanical names. More detailed descriptions of five New Zealand hebes were prepared by Daniel Solander, one of the botanists on Cook's first voyage to New Zealand, 1769–70, but they, and accompanying copperplate engravings, were not published.

Prior to the publication of Forster's descriptions, Conrad Moench had formally described one hebe (as *V. decussata*) from live specimens originally collected on the Falkland Islands (Moench 1785). That description was based on the same species as Forster's *V. elliptica*. The name *V. decussata*, being published first, would normally be accepted as the correct name for the species under the rules of the International Code of Botanical Nomenclature (ICBN). But, since Moench's description has been overlooked by botanists for more than 200 years, a proposal, under Art. 56 of the ICBN, to retain the use of the well-known name *V. elliptica* has recently been made (Bayly & Kellow 2004*a*).

An outline of the history of description of species, subspecies and varieties of *Hebe* and *Leonohebe*, from 1785 to the present, is shown in **Fig. 3**. Descriptions immediately following those of Moench and Forster were made by Aiton (1789) and Gmelin (1791), probably both of these from Falkland Islands material, and both of the same species previously described by Forster as *V. elliptica*. The next descriptions were of *V. macrocarpa* and *V. parviflora* by Martin Vahl (1794), based

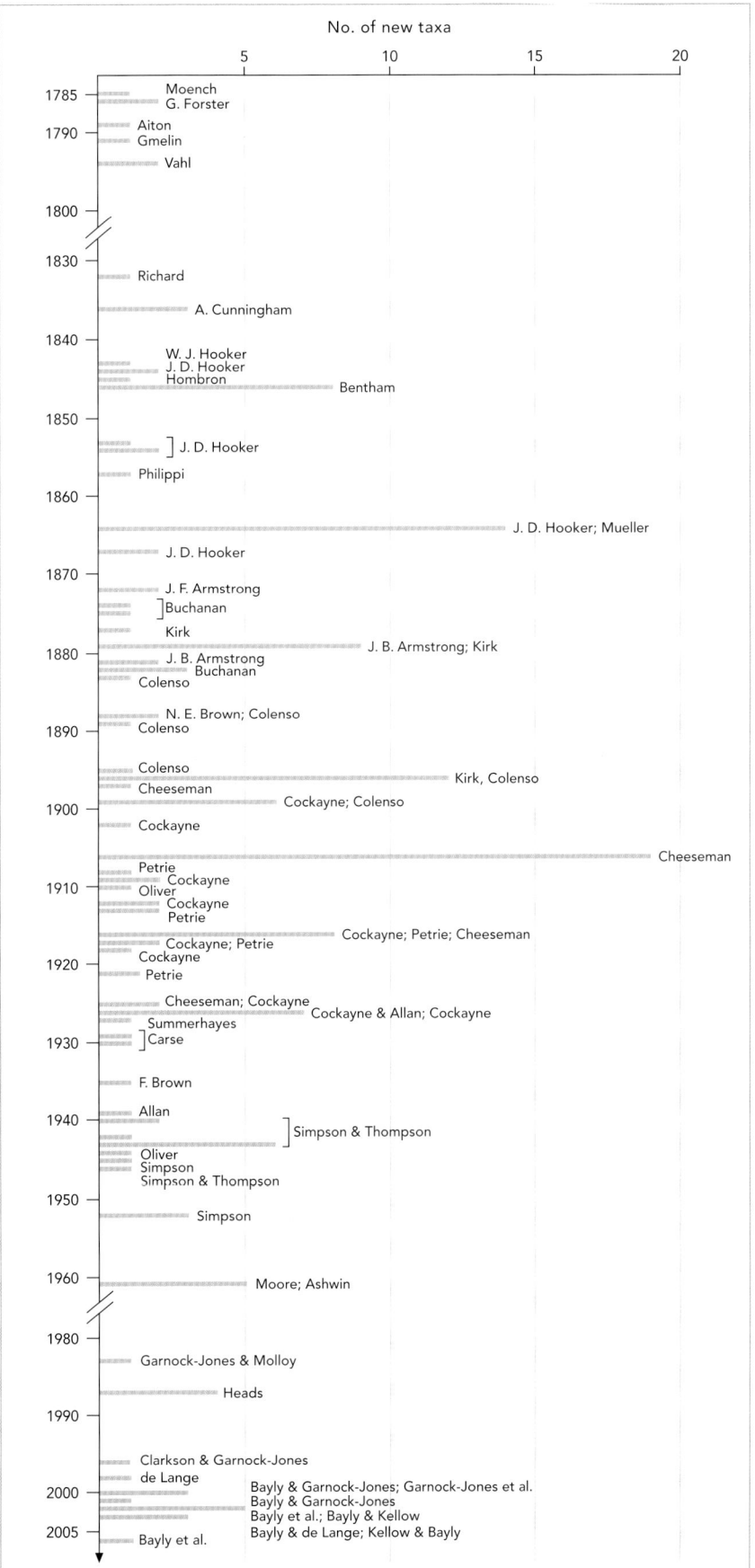

FIG. 3 Graph showing the numbers of new descriptions of *Hebe* and *Leonohebe* species and infraspecific taxa published every year between 1785 and the present. This includes new descriptions and replacement names, but not new combinations based on previously published names; authors are named beside each bar.

on specimens most likely collected on Cook's first voyage to New Zealand (there is no evidence that these species were collected by the Forsters, who did not visit any localities where *H. macrocarpa* is known to occur). After Vahl's descriptions there was a gap of thirty-eight years before another new species (*V. angustifolia*) was described by Achille Richard (Richard 1832), from specimens collected on Dumont d'Urville's voyage to New Zealand in 1827.

From 1832 until the publication of volume 1 of *Flora of New Zealand* (Allan 1961), the description of new taxa continued with little interruption (**Fig. 3**) as the New Zealand flora became better known. Authors during that time commonly described small numbers of taxa, mostly fewer than ten, as they came to light. A few notable works describing larger numbers of taxa are those of Joseph Hooker, in his *Handbook of the New Zealand Flora* (Hooker 1864); Thomas Kirk (Kirk 1896) who, prior to his death in 1898, was working toward a new flora of New Zealand, published posthumously in part (Kirk 1899); and Thomas Cheeseman, in the first edition of his *Manual of the New Zealand Flora* (Cheeseman 1906).

A substantial deviation from the trend of describing new taxa was made by Mueller (1864), who interpreted twenty previously named taxa to represent "forms of one species", to which he gave the illegitimate name *V. forsteri*. He considered that the forms "though in their extremes habitually so dissimilar, are linked together by an uninterrupted chain of graduations". This extreme view on the classification of the group has not been adopted by any subsequent authors.

Until 1867 all work on the classification of *Veronica/Hebe* was published overseas by European or Australian authors. That situation changed with the description of *V. anomala* (here treated as a synonym of *H. odora*) by J. F. Armstrong (1872), a botanist resident in New Zealand. Since that year the vast majority of work on the species-level classification of *Hebe* and *Leonohebe* has been published by local botanists. Many of the notable figures of New Zealand botany have taken an interest in the group at some stage, and the list of authors in **Fig. 3** is almost a *Who's Who* of local flowering-plant botanists.

After publication of the *Flora of New Zealand* (Allan 1961), there was a gap of twenty-two years before the publication of any new names in *Hebe* or *Leonohebe* (Garnock-Jones & Molloy 1983*a*). Since then the most significant numbers of new descriptions have been provided by Heads (1987) and in work leading to the revision presented here (Bayly et al. 2000, 2001, 2002, 2003; Garnock-Jones et al. 2000; Kellow et al. 2003*a*).

Many of the taxa described since 1961 are not "new" discoveries. Most were represented in herbaria and appeared on informal checklists of indigenous New Zealand vascular plants (e.g. Druce 1980, 1993) for one to several decades prior to formal description. Although some genuinely new discoveries have been made recently – for example, *H. societatis* (Bayly et al. 2002) – and others may remain to be found, hebes are a well-known and well-collected group of plants.

As the number of named species, subspecies and varieties has increased there has been further consideration of the limits of these taxa, and of their relative placement and most appropriate rank. The number of names identified as synonymous has generally increased in successive revisions of the group (**Fig. 4**), and there has been intermittent tinkering with the ranks of taxa (**Fig. 5**), as well as their relative placement, since the work of Hooker (1853).

Different choices of rank reflect different interpretations of both the plants themselves (e.g. their morphological variation, distinctiveness, affinities and distribution), and of how the ranks of species, subspecies and variety should be used.

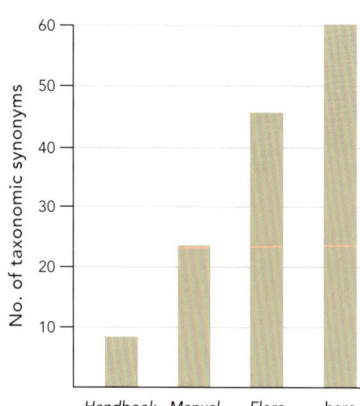

FIG. 4 Graph showing the number of synonymous names at species and infraspecific ranks identified in substantial accounts of *Hebe* and *Leonohebe*, namely Hooker (1864), *Handbook of the New Zealand Flora*; Cheeseman (1906), *Manual of the New Zealand Flora*; Moore & Ashwin (in Allan 1961), *Flora of New Zealand*; and this current treatment. These numbers represent groups of taxonomic synonyms only; the total numbers of synonyms, including all nomenclatural synonyms, is much larger.

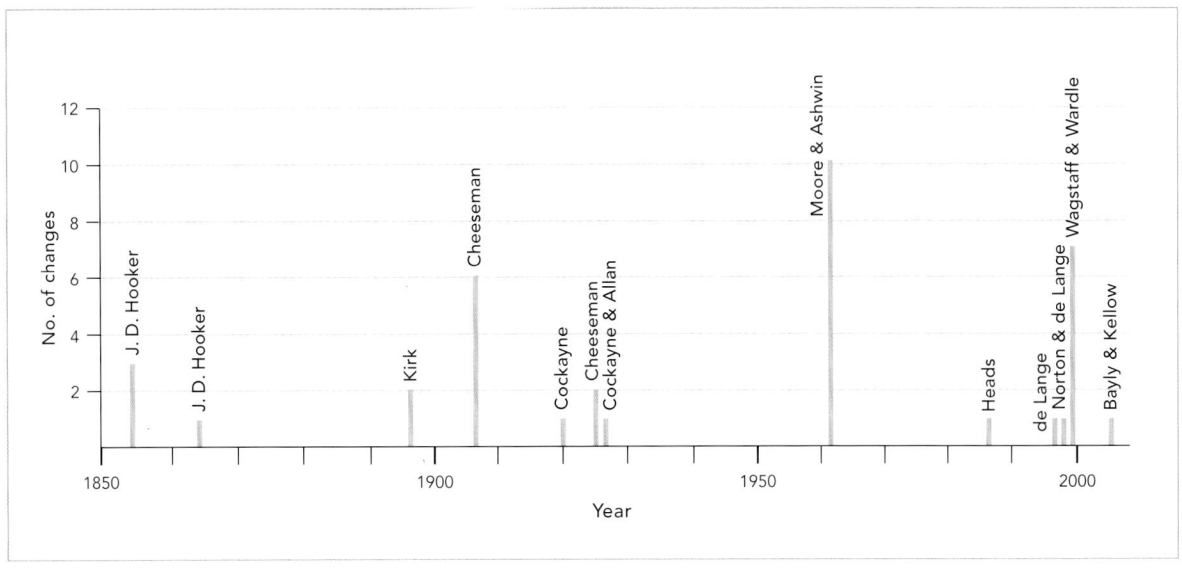

FIG. 5 Graph showing the numbers of newly proposed changes in ranks of species and infraspecific taxa of *Hebe* and *Leonohebe* from 1853 to the present. Authors are named beside each bar.

In New Zealand botany in general, much rearrangement has involved the movement of taxa from variety to species rank, and vice versa. Traditionally, infraspecific taxa in New Zealand were described mostly at the rank of variety – for example, this was the only rank used by Allan (1961). More recent works have increasingly used the rank of subspecies, often as the only infraspecific rank, but frequently without providing explanation of this choice (e.g. Wagstaff & Wardle 1999). In this revision the infraspecific ranks of subspecies, variety, and even forma, are all used. Discussion of choices, regarding both specific and infraspecific recognition, is provided in the section "Materials and Methods".

HISTORY OF GENERIC AND INFRAGENERIC CLASSIFICATIONS

The genus *Hebe* was first proposed by Jussieu (1789), and the first species, *H. magellanica*, evidently based on South American material of *H. elliptica*, was described by Gmelin (1791). The genus name was not generally used until reinstated by Pennell (1921) in a treatment of South American species. The distinguishing features he discussed were: the nature of capsule dehiscence in which the carpels part, sagittally splitting the septum, followed by formation of a distal median split through the septal wall of each carpel; a shrub or tree habit; flowers borne in specialised axillary racemes; "an exceedingly baffling tendency to form local races, a habit in contrast with that of the other 'Veronicas'"; and austral distribution, which "suggests genetic remoteness" and "emphasises *Hebe*'s claim to recognition as a genus".

Pennell's arguments were soon taken on board by New Zealand botanists; the genus *Hebe* was accepted for New Zealand by Oliver (1925), and the formal transfer of species was begun by Andersen (1926) and Cockayne & Allan (1926a, 1926b, 1926c). Initially, the circumscription of *Hebe* in New Zealand included those species still placed in the genus, together with those now included in *Leonohebe* and *Heliohebe*. Allan (1939) also included species treated by Cheeseman (1906, 1925) under *Veronica* division Euveronica, but these were later transferred by Oliver (1944) to his new genus *Parahebe*, which is the treatment followed in this book.

The genus *Leonohebe* was first described by Heads (1987), and fuller details of species, their distributions and historical biogeography were provided later

TABLE 2 Comparison of the present classification of *Hebe* and *Leonohebe* with those in other significant treatments (adapted from Wagstaff et al. 2002). The classifications of Cheeseman (1925) and Moore & Ashwin in Allan (1961) included further subdivision of some groups (not practical to display here).

Bentham (1846)	Hooker (1864)	Armstrong (1881)	Wettstein (1891)	Cheeseman (1925)	Moore & Ashwin in Allan (1961)	Heads (1987, 1994b)	Garnock-Jones (1993a, 1993b)	Albach et al. (2004a)	Present treatment
Veronica p.p.	*Veronica* p.p.	*Veronica* p.p.	*Veronica* p.p.	*Veronica* p.p.				*Veronica* p.p.	
sect. *Hebe* p.p.	sect. 6 p.p.	subg. *Koromika* p.p.	sect. *Hebe* p.p.	division *Hebe* p.p.	*Hebe* p.p.		*Hebe*	Subg. "*Hebe*" p.p.[1]	*Hebe*
		sect. 1 p.p.	subsect. *Serratae* p.p.	subdivision B p.p.	"Grandiflorae"	*Parahebe* p.p.		"	"Grandiflorae"
	sect. 1, 2 & 3	sect. 6, 4 p.p., 5 p.p. & 7 p.p.	subsect. *Integrae* p.p.	subdivision A p.p.		*Hebe*		"	
Decussatae p.p.[2]	"	"	"	"	"Subdistichae"	sect. *Subdistichae*	sect. *Subdistichae*	"	"Apertae"[3]
Decussatae p.p.; *Speciosae* p.p.[2]	"	"	"	"		sect. *Hebe*	ser. *Hebe*	"	"
"	"	"	"	"	"Apertae"	ser. *Hebe*		"	"
"	"	"	"	"	"Occlusae"	ser. *Occlusae*	ser. *Occlusae*	"	"Occlusae"[4]
"	"	sect. 2	"	"	"Subcarnosae"	sect. *Glaucae*	sect. *Glaucae*	"	"Subcarnosae"[4]
						Leonohebe p.p.			
Decussatae p.p.[2]	"	"	"	"	"Buxifoliatae"	sect. *Buxifoliatae*	"Buxifoliatae"	"	"Buxifoliatae"; "Pauciflorae"
"	sect. 5	sect. 3	"	"	"Connatae"	sect. *Connatae*	"Connatae"	"	"Connatae"
"	"	sect. 1 p.p.	subsect. *Serratae* p.p.	subdivision B p.p.	"	sect. *Apiti*	"	"	"
"	sect. 4	subg. *Pseudoveronica*	subsect. *Integrae* p.p.	subdivision A p.p.	"Flagriformes"	sect. *Flagriformes*	"Flagriformes"	"	"Flagriformes"
			"	"		sect. *Salicornioides*	"	"	
			"	"			"	"	*Leonohebe*
			"	"		sect. *Aromaticae*	"	"	sect. *Aromaticae*
			"	"	"Semiflagriformes"	sect. *Leonohebe*	"Semiflagriformes"	"	sect. *Leonohebe*

1 Albach et al. (2004a), informally, use the name subg. *Hebe* for this group, but the correct name is either subg. *Koromika* or subg. *Pseudoveronica*.
2 These are subsectional names without a clear indication of rank.
3 Includes small- and large-leaved "Apertae".
4 Circumscriptions differ slightly from those of Moore (in Allan 1961).

12   AN ILLUSTRATED GUIDE TO NEW ZEALAND HEBES

(Heads 1992, 1994a, 1994c). As originally described, the genus had a much broader circumscription than that adopted here (**Table 2**).

Successive treatments of *Veronica* and/or the *Hebe* complex have presented infrageneric classifications of varying complexity (**Table 2**). Some have used formal ranks of subgenus, section or series, while others have used entirely informal groupings. All classifications under *Veronica* have treated *Hebe*, or the *Hebe* complex, as a distinct group of some kind. Within *Hebe* and *Leonohebe* there is partial agreement between the infrageneric groups recognised in some treatments, but also some significant points of difference.

The most recent work on generic classification of the *Hebe* complex (Albach et al. 2004a) suggests that all genera in this group should, once again, be placed in a broad circumscription of *Veronica*. That classification (not adopted here) is based mostly on evidence from DNA sequences, and is discussed in the following sections.

RELATIONSHIPS OF THE *HEBE* COMPLEX INFERRED FROM DNA SEQUENCES

Assessments of relationships of *Hebe* and *Leonohebe* have followed a progression seen in the study of many plant groups. They started with traditional classifications based on morphology and distribution – for example, infrageneric classifications in the spirit of Hooker (1864). As additional sources of information on cytology (Frankel & Hair 1937; Frankel 1941; Hair 1967) and secondary chemistry (Grayer-Barkmeijer 1979) became available, these were also used to formulate ideas. Discursive accounts of relationships (e.g. Yamazaki 1957; Moore 1967; Heads 1993) have been increasingly replaced by, or supplemented with, more rigorous phylogenetic analyses employing a range of data types, including morphology (Hong 1984; Garnock-Jones 1993a; Kampny & Dengler 1997) and, most recently, DNA sequences (Wagstaff & Garnock-Jones 1998, 2000; Wagstaff & Wardle 1999; Albach & Chase 2001, 2004; Wagstaff et al. 2002; Albach et al. 2004b, 2004c). Such analyses, mostly using cladistic methods, developed from those proposed by Hennig (1966), produce branching diagrams (trees) whose structure is an hypothesis of evolutionary relationships, directly inferred from data on character variation. Some of the terms used to describe relationships shown on these trees (discussed below) may not be familiar to the general reader, and are defined in the glossary.

Useful information on the relationship of the *Hebe* complex to *Veronica* comes from analyses of the internal transcribed spacer region (ITS) of nuclear ribosomal DNA (nrDNA; Albach & Chase 2001; Wagstaff et al. 2002) and several chloroplast markers – the *rbc*L gene (Wagstaff et al. 2002), the *trnL* intron and *trnL-F* spacer (Albach & Chase 2004; Albach et al. 2004b) and the *rps16* intron (Albach et al. 2004c). Analyses based on all of these sources support the notion that the *Hebe* complex is derived from within *Veronica* as it is presently circumscribed – in other words, that *Veronica* is paraphyletic, with some members more closely related to the *Hebe* complex than they are to other members of *Veronica* (**Fig. 6**). Analyses of ITS sequences also suggest the *Hebe* complex is part of a monophyletic Australasian group; analyses of chloroplast sequences do not include sufficient taxa or sequence variation to assess this relationship properly.

Within the *Hebe* complex, the most comprehensive DNA dataset for assessing relationships is that based on ITS sequences presented by Wagstaff et al. (2002). That study included sixty-five samples from the *Hebe* complex, representing most genera (except for the monotypic New Guinean genus *Detzneria*), and all the infrageneric groups recognised by Moore & Ashwin (in Allan 1961).

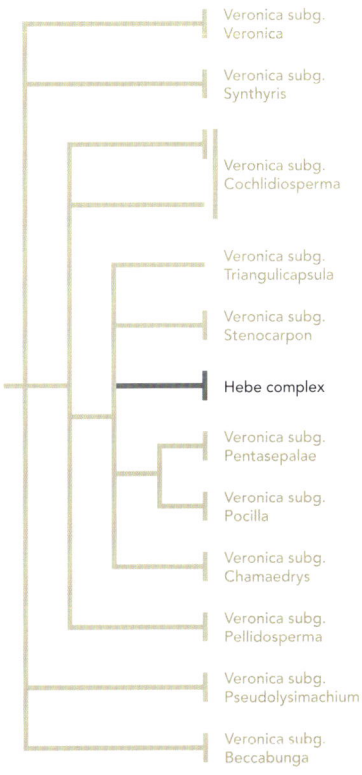

FIG. 6 Phylogenetic tree showing hypothesised relationships of the *Hebe* complex to *Veronica*. This is a summary of the trees presented by Albach et al. (2004a, 2004b) from analysis of ITS and chloroplast DNA sequences.

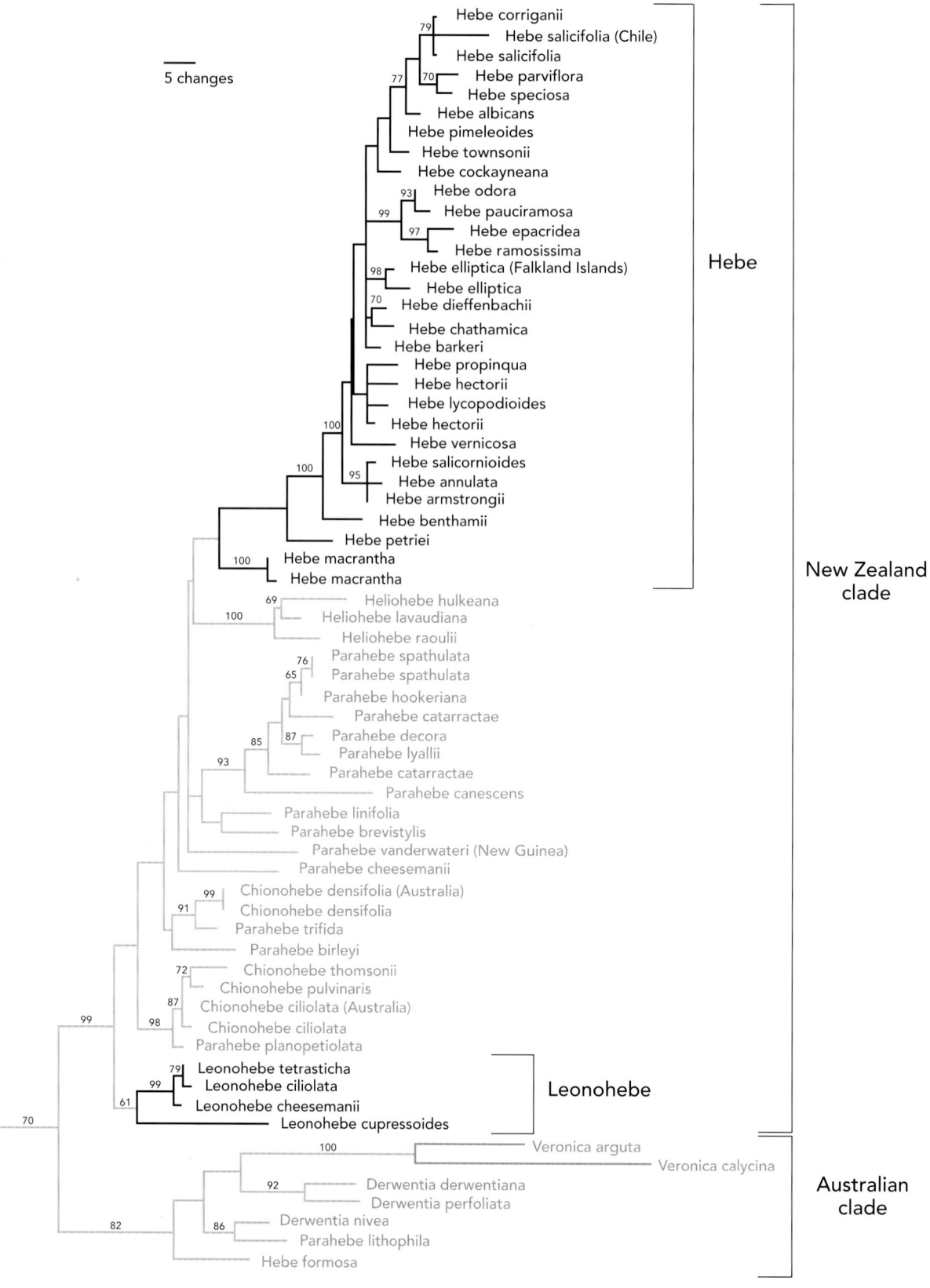

The phylogenetic trees presented by Wagstaff et al. (2002; **Fig. 7**) show *Hebe* and *Leonohebe* as part of a large, monophyletic group that occurs primarily in New Zealand (with a few notable exceptions), and includes members of *Parahebe*, *Chionohebe* and *Heliohebe*. Both *Hebe* and *Leonohebe*, as defined here, are shown as monophyletic, but are not each other's closest relatives. *Heliohebe* is also monophyletic, but *Parahebe* and *Chionohebe* are both polyphyletic, and part of a grade below *Hebe* and *Heliohebe*. Some groups within this large, mostly New Zealand, group are well supported (**Fig. 7**), but relationships between them – for example, of *Heliohebe* to *Hebe*, and within the *Chionohebe-Parahebe* grade – are generally weakly supported.

Most closely related to the largely New Zealand group is a clade of Australian taxa. This heterogeneous group includes species recently placed in *Derwentia*, *Veronica*, *Parahebe* and *Hebe* (Briggs & Ehrendorfer 1992; Briggs & Makinson 1992; Briggs et al. 1992; Garnock-Jones 1993*a*). The group highlights some clear errors in previous classifications, and it would be sensible for all of these taxa to be included in the one genus (*Derwentia* under the classification adopted here).

Relationships within *Hebe*, as defined here, are not well resolved or supported by analyses of ITS sequences. This is mostly because there is limited ITS sequence variation within the genus, providing few useful characters for determining relationships. Several relationships within *Hebe* are, however, worthy of note. These are:

1. The majority of *Hebe* species form a well-supported monophyletic group (**Fig. 7**), with weaker support for *H. petriei*, and then *H. macrantha*, being next most closely related. *H. macrantha* has long been placed in an isolated or ambiguous position. It lacks some of the potentially derived morphological features common to many other members of *Hebe* (e.g. large leaf buds, entire leaf margins and latiseptate capsules), and it was placed by Moore (in Allan 1961) in its own grouping, "Grandiflorae", and was included by Heads (1987, 1994*b*) in *Parahebe*.
2. The group "Connatae" appears to be polyphyletic, with multiple origins, since *H. petriei* and *H. benthamii* are placed near the base of the New Zealand group (and not most closely related), whereas *H. epacridea* and *H. ramosissima* are higher in the tree, and most closely related to the two representatives of "Buxifoliatae".
3. A close relationship of the whipcord hebes to "Buxifoliatae", as proposed by Moore (1967, in Allan 1961), is not supported.
4. The whipcords are not resolved as monophyletic, although three species – *H. annulata*, *H. salicornioides* and *H. armstrongii*, which share possession of fused anterior calyx lobes and a chromosome number based on $n = 21$ – form a well-supported clade (the monophyly of the whipcords is discussed in more detail in the notes for *H. tetragona*).
5. The other infrageneric groups of Moore (in Allan 1961), namely "Occlusae", "Apertae", "Subdistichae" and "Subcarnosae", are also not resolved as monophyletic.

Within *Leonohebe*, analysis of ITS sequences strongly supports the monophyly of the semiwhipcords (sect. *Leonohebe*). *L. cupressoides* (the only species of sect. *Aromaticae*) is more weakly supported as sister to that group. Only one species of the genus, *L. tumida* (sect. *Leonohebe*), was not included in the ITS dataset.

opposite
FIG. 7 Phylogenetic tree showing hypothesised relationships among species of the *Hebe* complex. This is derived from analysis of ITS sequences (Wagstaff et al. 2002), and is one of the equally best supported trees using parsimony analysis. Branch lengths are proportional to the number of inferred changes in DNA sequences. Jackknife values, which provide a statistical assessment of the support for branches, are shown when > 60 per cent (the closer to 100 per cent, the greater the support for a branch). Species of the "New Zealand Clade" are endemic to New Zealand except where indicated.

DEFINING THE LIMITS OF *HEBE* AND *VERONICA*

There are no hard and fast rules about the way in which genera (and other taxa) should be defined. A range of criteria is used when defining generic limits. These include ease of recognition, patterns of character state distribution, breeding compatibility, geographic distribution, number of species, nomenclatural history, needs of the various users of a classification, and phylogenetic relationships. Classifications are usually a compromise between competing goals, and they are always subjective.

Since the development of explicit methods for analysing phylogenetic relationships, the notion that classifications should reflect those relationships, and that genera should be monophyletic, has been widely, although not universally (e.g. Meacham & Duncan 1987; Kinman 1994; Brummitt 2002), adopted. Classifications that reflect monophyletic groups are appealing to many biologists because they are potentially more information-rich and predictive (Nelson & Platnick 1981; Wiley 1981; Judd et al. 1999).

Accepting that recognition of monophyletic genera is desirable, the classification of *Veronica* and the *Hebe* complex is problematic. Even though relationships among taxa are imperfectly known, it seems evident that *Veronica*, and some genera of the *Hebe* complex (*Derwentia, Chionohebe, Parahebe*) are not monophyletic as usually circumscribed (**Fig. 7**).

For *Veronica*, there are two basic ways in which these problems could be resolved. One solution, proposed by Albach et al. (2004*a*), is to expand the circumscription of *Veronica* to include the *Hebe* complex and other segregate genera; that is, returning to the generic classification widely used before 1926, requiring different names to be used for all taxa treated here under *Hebe* and *Leonohebe*. The alternative solution is to divide *Veronica* into a series of smaller, putatively monophyletic genera; Albach et al. (2004*a*) identify nine groups within *Veronica* that could potentially be recognised as genera under such a scheme.

If *Veronica* is subdivided into smaller genera, a range of possible taxonomic solutions could be used to recognise only monophyletic genera within the *Hebe* complex. One solution would be to include all Australasian taxa under one genus, *Hebe*. Another would be to recognise one Australian genus (*Derwentia*) and one large genus (*Hebe*) including all New Zealand taxa and at least some from New Guinea. Another would be to recognise *Derwentia* (as above), and a series of smaller New Zealand genera – in other words, similar to the current classification but with some revision of generic limits (which would be assisted by further data on relationships of *Chionohebe* and *Parahebe*).

Since a number of different classifications could satisfy the desire to recognise monophyletic genera, choosing the most appropriate requires assessment using other criteria, many of which are discussed by Albach et al. (2004*a*). Weighing up the different options is no simple task because scientific names for *Veronica* and the *Hebe* complex are required throughout much of the world by people with differing needs (e.g. plant taxonomists, ecologists, professional horticulturists, gardeners, farmers, weed managers, biosecurity agencies and conservation managers). Lumping all taxa into *Veronica* would create a large genus (*c.* 450 species), including plants as strikingly different as *V. plebeia* (**Fig. 1A**), *Chionohebe glabra* (**Fig. 1F**) and *Hebe barkeri* (**Fig. 109**). On the other hand, splitting up *Veronica* could place European species that are confusingly similar in morphology, but genetically distinct, into different genera, many of which would not retain the traditional name *Veronica* (Albach et al. 2004*a*).

The impacts of revised generic classifications would be different in different parts of the world. Including the *Hebe* complex in *Veronica* would have greatest impact on southern hemisphere users, but relatively little effect on those in the northern hemisphere, apart from gardeners and those in the nursery trade (where the common name "hebe" could probably still be used). Splitting *Veronica* into a series of segregates would, conversely, have little effect in the southern hemisphere, but a very substantial one in the north.

The classification used here retains the genera *Hebe* and *Leonohebe* as distinct from one another and from *Veronica*. This is partly for pragmatic reasons associated with the preparation of this book. More than this, however, we take the view that much effort has gone into understanding the diversity and relationships of the *Hebe* complex, and that lumping all members into a large and variable genus (especially *Veronica*, rather than a purely Australasian genus) will create a classification that is information-poor, obscuring obvious diversity and some clear relationships.

Ultimately, it is the users of a classification who will decide what generic limits become generally accepted. Work in progress by P. J. Garnock-Jones, D. Albach and B. Briggs (P. J. Garnock-Jones pers. comm. 2005) will provide names in *Veronica* for all members of the *Hebe* complex. Once these names are available, users can adopt whichever classification suits their needs, and time will tell which scheme prevails.

CONSTRUCTING AN INFRAGENERIC CLASSIFICATION FOR *HEBE*

The infrageneric classification used here for *Hebe* is based on that of Moore (in Allan 1961), with some modification (discussed below). It does not use formal taxonomic names – for example, of subgenera, sections or series – covered by the ICBN. Instead, it uses a series of informal, rankless names; quotation marks are used to indicate the informal nature of these names, as in "Connatae" and "Occlusae". This classification is intended chiefly to aid identification. Informal groups are mostly defined by combinations of easily discernible features, so that users can quickly ascertain the group to which a specimen belongs and then proceed to the key for that group, or browse the photographs and notes on species.

While some groups in this classification may be "natural", or monophyletic, others are almost certainly "artificial" in that they are not based on the evolutionary relationships of species. Potentially monophyletic groups are "Buxifoliatae" and "Flagriformes" (discussed in the notes under *H. tetragona*, pages 92 and 94). All other groups, with the exception of the monotypic "Grandiflorae" and "Pauciflorae", are probably artificial. Each will certainly include some groups of closely related species, but some species may have their closest relatives in other groups (see notes under individual species; **Fig.** 7).

A classification that reflects the evolutionary relationships of species would be preferable (as discussed above for genera), but current data on species-level relationships are insufficient to build such a system confidently. The DNA sequences so far examined (Wagstaff & Garnock-Jones 1998; Wagstaff et al. 2002) do not show sufficient variation, or taxon sampling is insufficient, to resolve many relationships. Nor have there been any species-level cladistic analyses of morphological data; the morphological analysis of Garnock-Jones (1993*a*) made a priori assumptions about the monophyly of some groups that may not be supported when further data are considered (Wagstaff & Garnock-Jones 1998; Wagstaff et al. 2002).

The lack of clear evidence on species-level relationships is another reason for adopting an informal infrageneric classification. Even though some formal names, at the ranks of section and series, are provided by Heads (1987), these are not sufficient to encompass all groups included here (since Heads takes a narrower view of *Hebe* than we do). As such, a formal classification would require publication of new names. We would prefer to have greater confidence in the circumscriptions of groups before proposing such names.

Major differences between the infrageneric classification adopted here, and that of Moore (in Allan 1961) are as follows:

1. Moore's group "Subdistichae" is included here in "Apertae", which is divided into two subgroups based on leaf length. This has been done primarily to simplify the process of identification, since some of the features used to define "Subdistichae" were not clear-cut.
2. *H. pauciflora* is removed from "Buxifoliatae" and placed in a new monotypic group, "Pauciflorae".
3. "Flagriformes" no longer includes *L. cupressoides* (= *L.* sect. *Aromaticae*).
4. Some species have been moved to different groups – for example, *H. breviracemosa* to large-leaved "Apertae", because fresh specimens have shown it to have a leaf bud sinus; and *H. albicans* and *H. decumbens* to small-leaved "Occlusae", which they resemble and are thought to be closely related to (Kellow et al. 2005; Mitchell et al. in prep.).
5. Most of Moore's subgroups (denoted by letter codes in her synopsis) are not explicitly recognised, but are often represented in the order of arrangement of species in the "Taxonomic Treatment" section.

Distribution, Habitats and Biogeographic History

DISTRIBUTION AND HABITATS OF *HEBE* AND *LEONOHEBE*

Hebe occurs primarily in New Zealand (**Fig. 8**), with only three species occurring naturally elsewhere. One of these species, *H. rapensis*, is endemic to Rapa in French Polynesia. The other two, *H. elliptica* and *H. salicifolia*, occur both in South America and on South Island, New Zealand (*H. elliptica* is also common on the New Zealand subantarctic islands). Within New Zealand, *Hebe* occurs in most major land areas, from the Kermadec Islands in the north, throughout the three main islands, to the Chatham Islands in the east, and to Auckland and Campbell islands in the south.

The genus has a wide ecological range; some examples of habitats in which species occur are listed in **Table 3**. Although this range includes most types of terrestrial ecosystems, the majority of species occur in open situations (most abundant in areas above tree-line), and the genus is poorly represented in forests.

Leonohebe occurs only on South Island, New Zealand (**Fig. 9**). When compared with *Hebe*, it shows less habitat diversity. The four "semiwhipcord" species of sect. *Leonohebe* all occur in similar, exposed rocky places above the tree-line. *L. cupressoides* occurs in grey scrub (defined by Meurk et al. 1987 and Wardle 1991) on rock outcrops, bouldery moraine and slips, usually near rivers and lakes in dry montane basins.

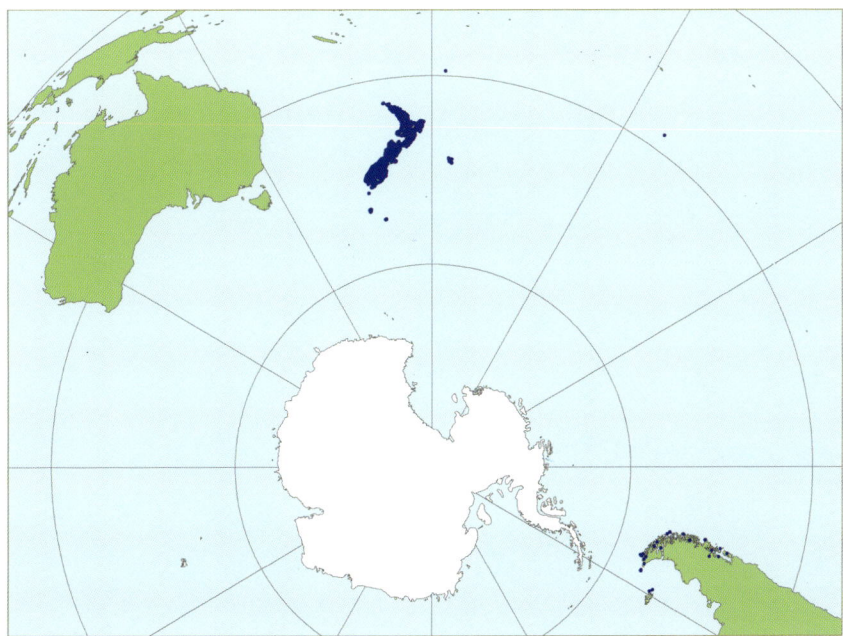

FIG. 8 Distribution of *Hebe*.

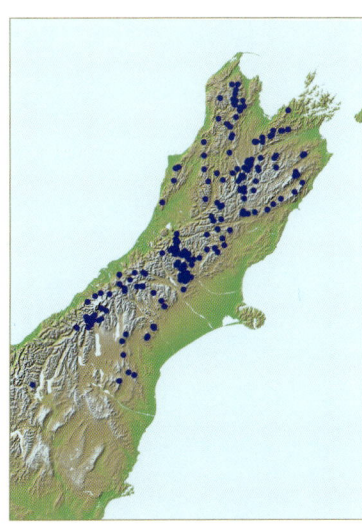

FIG. 9 Distribution of *Leonohebe*.

TABLE 3 Examples of habitats occupied by *Hebe*. This list is not meant to be exhaustive. The taxa listed are those that occur primarily (although not always exclusively) in the given habitat, but are not necessarily the only ones that might be found in these situations.

Coastal rocks, beach margins or exposed coastal sites
  H. elliptica, H. chathamica, H. insularis, H. obtusata

Near-coastal forest, bluffs or scrub (often further above the water-line and/or in less exposed sites than the above), on:
(a) various substrates
  H. speciosa, H. ligustrifolia, H. pubescens, H. bollonsii
(b) ultramafic rock
  H. brevifolia

Lowland mesotrophic wetland
  H. paludosa

Riverbanks (not necessarily rocky)
  H. acutiflora

Rocky walls of river gorges
  H. colensoi, H. rupicola, H. angustissima, H. pimeleoides subsp. *faucicola, H. pareora*

Cloud forest (northern North Island)
  H. flavida

Beech forest in montane areas (often in small light gaps or at forest margins)
  H. vernicosa

Lake margins/river terraces (usually in dry inland basins)
  H. pimeleoides subsp. *pimeleoides*

Subalpine shrubland (with woody vegetation 1–2 m tall), on:
(a) various substrates
  H. crenulata, H. cryptomorpha, H. venustula, H. brachysiphon, H. topiaria, H. rakaiensis, H. evenosa, H. glaucophylla, H. truncatula
(b) marble rocks
  H. calcicola

Penalpine grassland (with woody plants < 1 m tall), on:
(a) various substrates
  H. macrantha, H. salicornioides, H. masoniae, H. pauciramosa, H. societatis
(b) marble rocks
  H. ochracea
(c) ultramafic rocks
  H. carnosula

Rock in subalpine to alpine areas, on:
(a) various rock substrates
  H. pinguifolia, H. amplexicaulis, H. gibbsii, H. epacridea
(b) mostly solid rock outcrops
  H. biggarii
(c) boulder fields and coarse rock rubble
  H. ramosissima, H. petriei, H. murrellii, H. annulata
(d) coarse scree
  H. haastii, H. macrocalyx

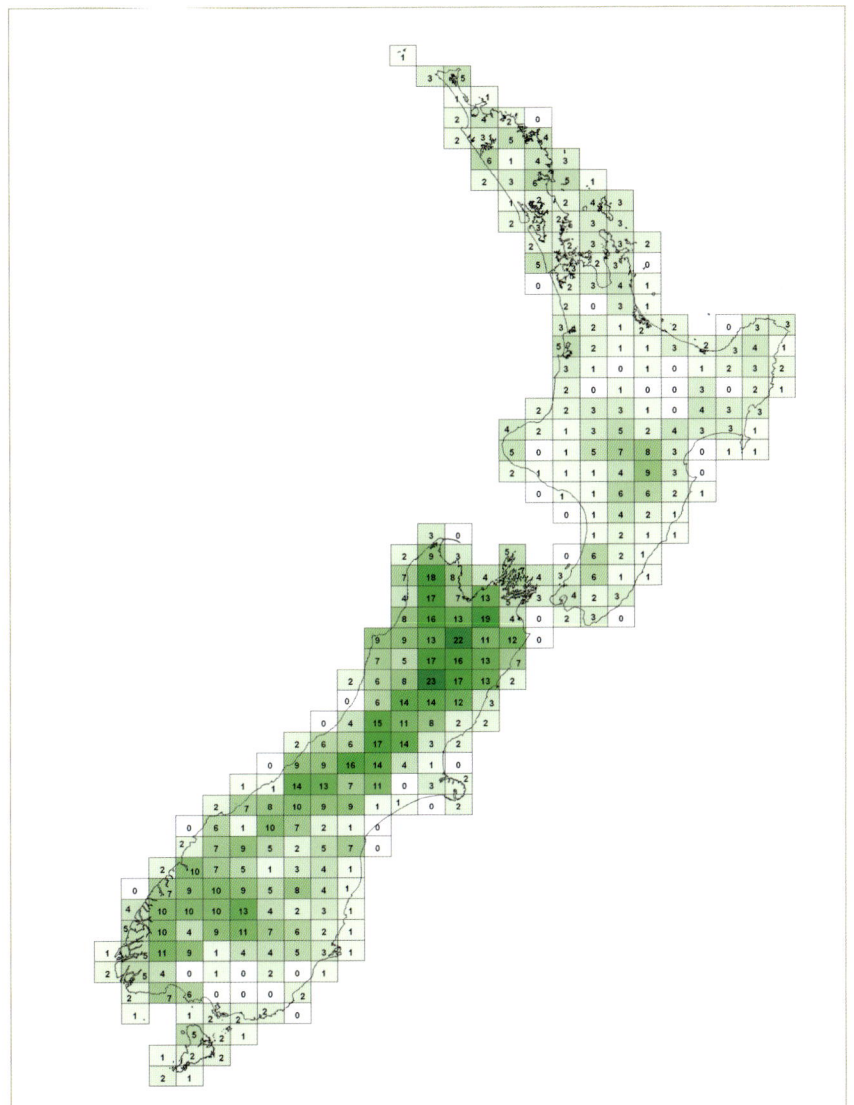

FIG. 10 Map showing the number of *Hebe* and *Leonohebe* species occurring in each of a series of 30 × 40-km grid cells across New Zealand. Each of these grid cells is equivalent to a 1:50 000 NZMS 260 map sheet.

PATTERNS OF SPECIES RICHNESS

Species richness in *Hebe* and *Leonohebe* is generally higher in mountain areas – such as along the spine of South Island, and on the central Volcanic Plateau and surrounding ranges of North Island – than in lowlands (**Fig. 10**). It is greatest in the mountains of northern South Island, near the shared borders of Nelson, Marlborough and Canterbury, particularly in the areas covered by the NZMS 260 map sheets M31 (Lewis), N29 (St Arnaud) and O28 (Wairau). The distributions of many taxa overlap, or are only narrowly separated, in this area. This mix includes both some narrowly distributed and some very widespread taxa. Few taxa have distributions wholly centred there, and a number have their eastern, western, northern or southern distributional limits in this general area. Such high species richness is probably a product of both historical factors that have given rise to high levels of endemism in Nelson/Marlborough (discussed below), and current environmental conditions. There are, for instance, very strong rainfall gradients in this area (generally becoming wetter from east to west), and a range of contrasting habitats, characteristically occupied by taxa variously tolerant of either wetter or drier conditions.

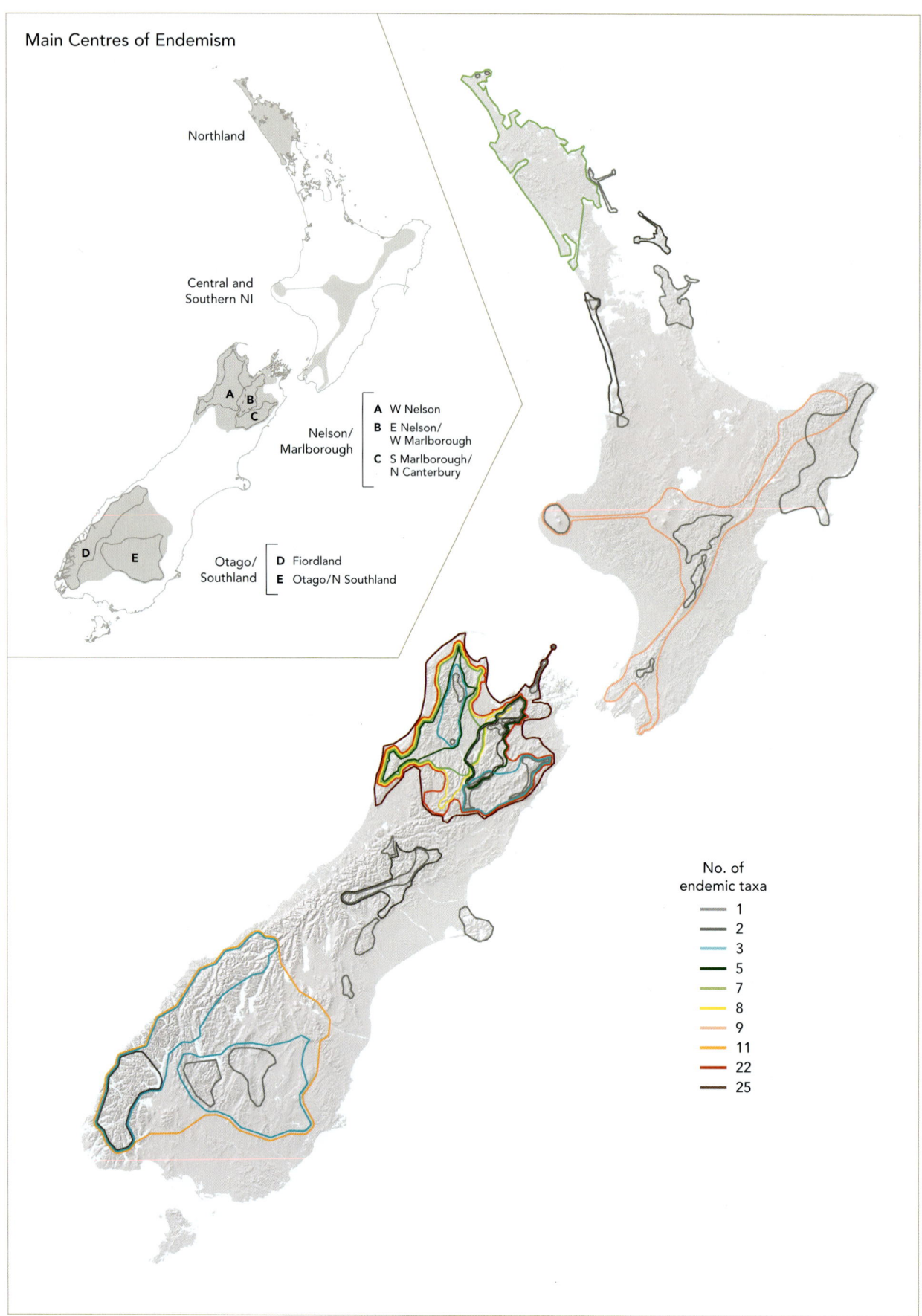

22　AN ILLUSTRATED GUIDE TO NEW ZEALAND HEBES

PATTERNS OF ENDEMISM

Some species of *Hebe* are relatively widespread, but most, and those of *Leonohebe*, have more restricted distributions (as can be seen from the distribution maps provided in the taxonomic section of this book). Within New Zealand, six species (*H. barkeri*, *H. benthamii*, *H. breviracemosa*, *H. chathamica*, *H. dieffenbachii* and *H. insularis*) occur only on offshore islands, twenty-one only on North Island, fifty-eight only on South or Stewart islands, and seven on both North and South islands.

Patterns of endemism in *Hebe* and *Leonohebe* on the main islands of New Zealand are similar to those for the vascular flora as a whole (Wardle 1963, 1988; Burrows 1965; McGlone 1985; McGlone et al. 2001), except for pteridophytes (Brownsey 2001). This is particularly true on South Island, where the species and infraspecific taxa most restricted geographically occur chiefly in two main centres

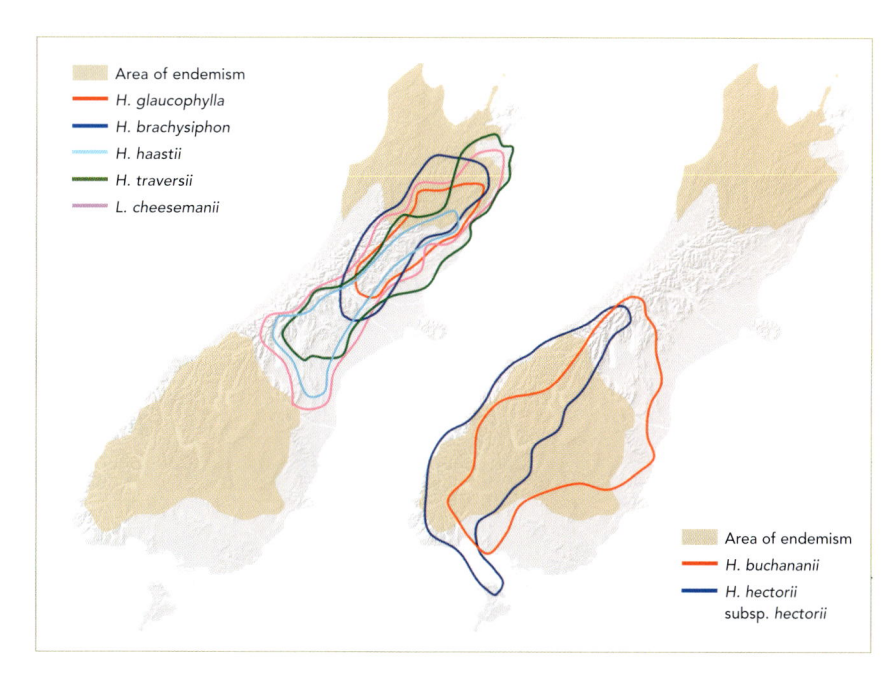

opposite
FIG. 11 Maps showing patterns of regional endemism in *Hebe* and *Leonohebe*. Coloured lines indicate the number of species and infraspecific taxa endemic to an area. Many smaller areas of endemism (e.g. of one or a few taxa) are nested within larger areas. Defining areas of endemism is not straightforward, and it can be done at different scales for different purposes – for example, the whole of South Island, or the whole of New Zealand, could also be treated as areas of endemism. This map has been prepared to highlight some clear patterns of regional endemism and focuses on taxa with more restricted distributions. Some of the most fine-scale patterns, and some large-scale patterns, are not represented.

FIG. 12 Distributions of species that occur in Canterbury and/or Westland, as well as in the Nelson/Marlborough centre of endemism (left) or the Otago/Southland centre of endemism.

below
FIG. 13 South Island distributions of widespread species that occur in both the Nelson/Marlborough and Otago/Southland centres of endemism.

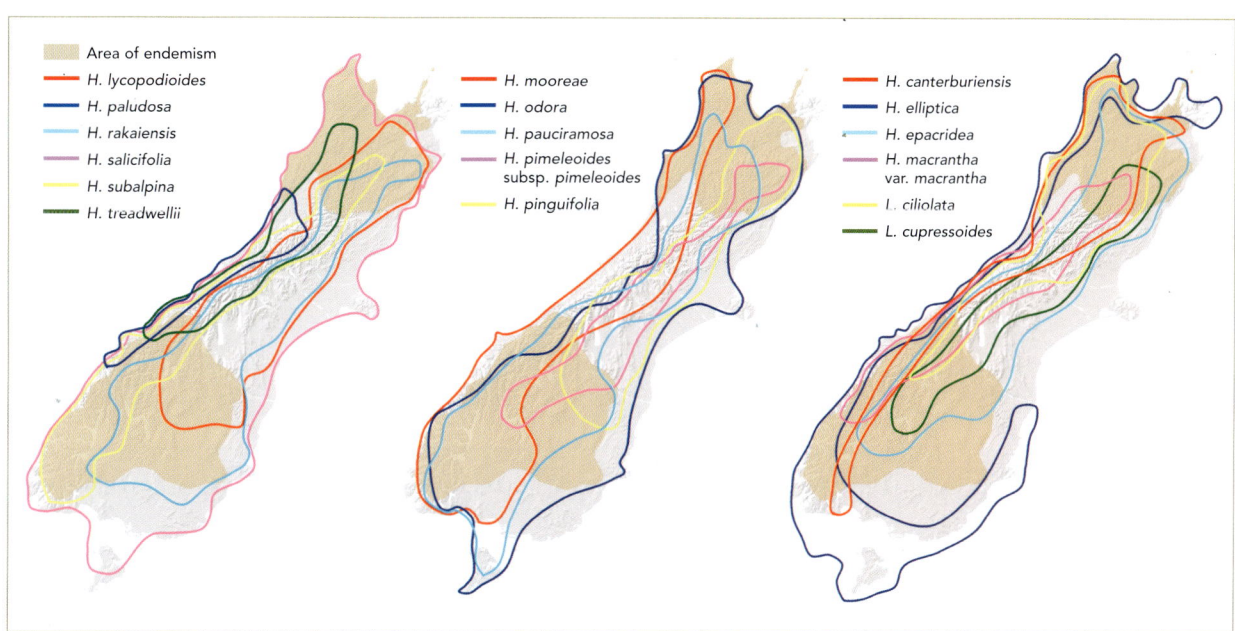

of endemism (**Fig. 11**), one in Nelson/Marlborough (twenty-five taxa), and one in Otago/Southland (eleven taxa). Only a few taxa are endemic to the intervening areas (the "Southern Gap" of Wardle 1988) of Canterbury and Westland, and most that occur in these areas are widespread, also occurring in one (**Fig. 12**) or both (**Fig. 13**) of the main centres of endemism. Within the South Island centres of endemism many restricted taxa have overlapping distributions, but some finer-scale patterns of endemism are, nonetheless, apparent. In the Nelson/Marlborough centre the greatest concentration of restricted taxa is in western Nelson, while other endemics are concentrated in eastern Nelson/western Marlborough or southern Marlborough/north Canterbury. Likewise, in the Otago/Southland centre, most taxa occur in an area centred in Otago and northern Southland, but others are largely endemic to Fiordland.

On North Island, major centres of endemism (**Fig. 11**) are in the mountains of central and southern areas (nine taxa), and in Northland (seven taxa). Other areas with endemic taxa are the East Cape region, Coromandel Peninsula, islands of the outer Hauraki Gulf, and the west coast between about Auckland and Kawhia.

BIOGEOGRAPHIC HISTORY OF *HEBE* AND *LEONOHEBE*

The present distributions of *Hebe* and *Leonohebe*, and patterns of endemism within them, have an historical basis, for which several scenarios have been proposed. With regard to the distribution of *Hebe* and its relatives in the southern hemisphere, two widely differing opinions have been presented. Some authors (e.g. Brown 1935; Melville 1966; Skipworth 1973; Hong 1984; Heads 1989, 1993, 1994*a*, 1994*b*, 1994*c*) suggest that the group was present in the southern hemisphere in the Cretaceous (**Fig. 14** gives a geological timescale) and, therefore, that its current distribution is largely a product of the fragmentation of the former supercontinent of Gondwana. Others (e.g. Raven 1973; Garnock-Jones 1993*a*; Wagstaff & Garnock-Jones 1998; Wagstaff et al. 2002) suggest that the group arrived more recently in the southern hemisphere, dispersed to New Zealand via Australia or Asia, and secondarily dispersed to other southern lands (e.g. *Hebe* to South America and Rapa).

Dispersal of the progenitor(s) of *Hebe* and its allies to New Zealand, subsequent to its separation from Gondwana (which occurred *c.* 85 Mya), is consistent with evidence from the fossil record, the geological and climatic history of New Zealand, and the distributions, ecological tolerances and relationships of extant species. In particular, the following points are of note:

1. For *Hebe*, the earliest appearance in the fossil record is in the Pliocene (*c.* 5.5–1.5 Mya; Mildenhall 1980). For Scrophulariaceae *s. lat.* (i.e. including the *Hebe* complex) it is in the mid-Miocene (*c.* 15 Mya; Tiffney 1985), and for the whole of the order Lamiales (*sensu* APG 1998; which includes Scrophulariaceae *s. lat.*) it is in the mid-Eocene (*c.* 45 Mya; Muller 1981). Although there is always the possibility that older fossils will be found, it would be inconsistent with this record to assume that the *Hebe* complex differentiated prior to the separation of New Zealand from Gondwana.

2. Members of the earliest diverged lineages in the *Hebe* complex in New Zealand are all alpine or montane plants (Wagstaff et al. 2002), most of them occurring in areas above the natural tree-line. If the present ecological requirements of these groups are indicative of those of their past (i.e. assuming that each lineage has not independently and recently adapted to alpine habitats, or seen selective extinction of lowland members), it could be inferred that early

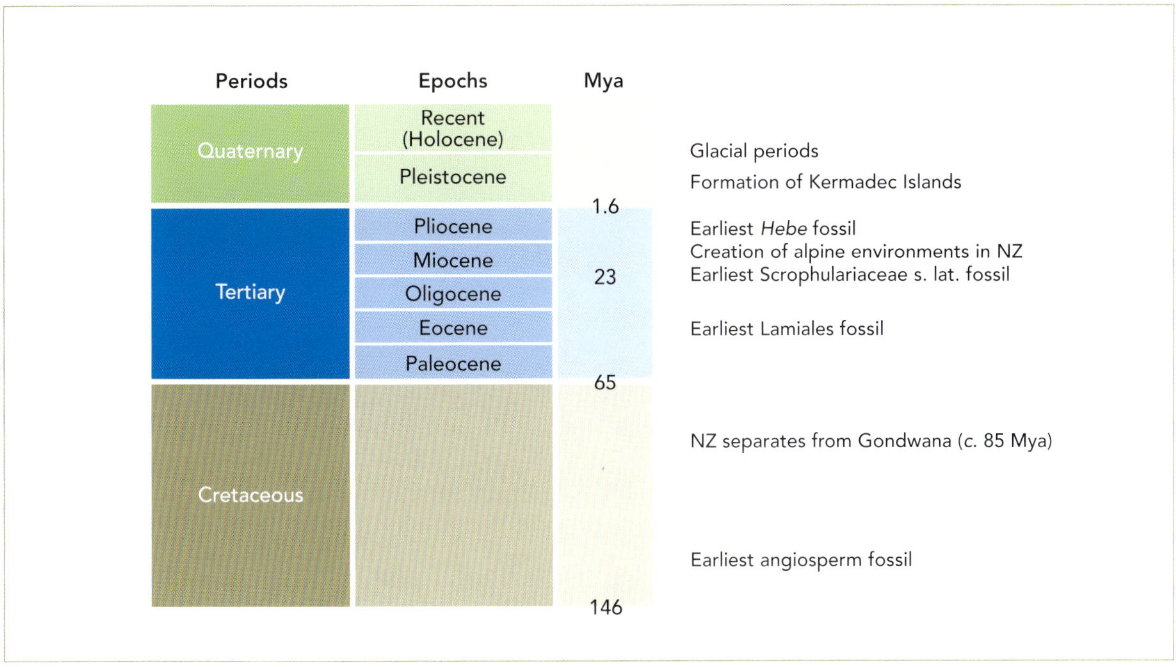

Periods	Epochs	Mya	
Quaternary	Recent (Holocene)		Glacial periods
	Pleistocene		Formation of Kermadec Islands
		1.6	
Tertiary	Pliocene		Earliest *Hebe* fossil
	Miocene		Creation of alpine environments in NZ
	Oligocene	23	Earliest Scrophulariaceae s. lat. fossil
	Eocene		Earliest Lamiales fossil
	Paleocene		
		65	
Cretaceous			NZ separates from Gondwana (c. 85 Mya)
			Earliest angiosperm fossil
		146	

FIG. 14 Geological timescale.

differentiation in the *Hebe* complex in New Zealand occurred in alpine environments, with colonisation of the lowlands being a secondary event. Alpine environments have existed in New Zealand only since the Pliocene or very late Miocene, subsequent to the onset of mountain building, in what was previously relatively low-lying land (Fleming 1979; Ollier 1986).

3. Hebes show clear evidence of a capacity for long-distance dispersal – for example, through their occurrence on the Pleistocene-age Kermadec Islands, 976 km northeast of North Island (Sykes 1977). The occurrence of *H. elliptica* and *H. salicifolia* in both New Zealand and South America is probably also a product of long-distance dispersal (from west to east), given that populations separated by oceans show little or no differentiation in morphology or DNA sequences (Wagstaff & Garnock-Jones 1998; Wardle et al. 2001; Wagstaff et al. 2002).

4. Low levels of DNA sequence divergence within the *Hebe* complex are consistent with recent differentiation (Wagstaff et al. 2002).

With regard to patterns of endemism in *Hebe*, and the vascular flora in general, particularly on South Island, there are also competing theories of biogeographic history. Some authors have suggested that separation of the two main centres of endemism on South Island (**Fig. 11**) and disjunctions in the ranges of species (e.g. *H. mooreae*) are the products of tectonic activity on the Alpine Fault (Heads 1989, 1998; Heads & Craw 2004) in combination with contemporaneous rapid uplift of the Southern Alps (McGlone 1985; **Figs 15A** and **B**). Others (e.g. Cockayne 1926; Wardle 1963; Burrows 1965; McGlone et al. 2001) suggest that these patterns result from the climatic changes of recent glacial periods (**Fig. 15C**). The two theories are similar in that they both propose the formation of a barrier between the previously connected Nelson/Marlborough and Otago/Southland areas. In one case the barrier is the imposition and uplift of newly intervening land through tectonic activity; in the other case it is the formation of extensive glacial ice and a generally harsher (colder) environment. The two theories, however, differ both in the implied ages of the barrier and the speed with which it formed. The tectonic

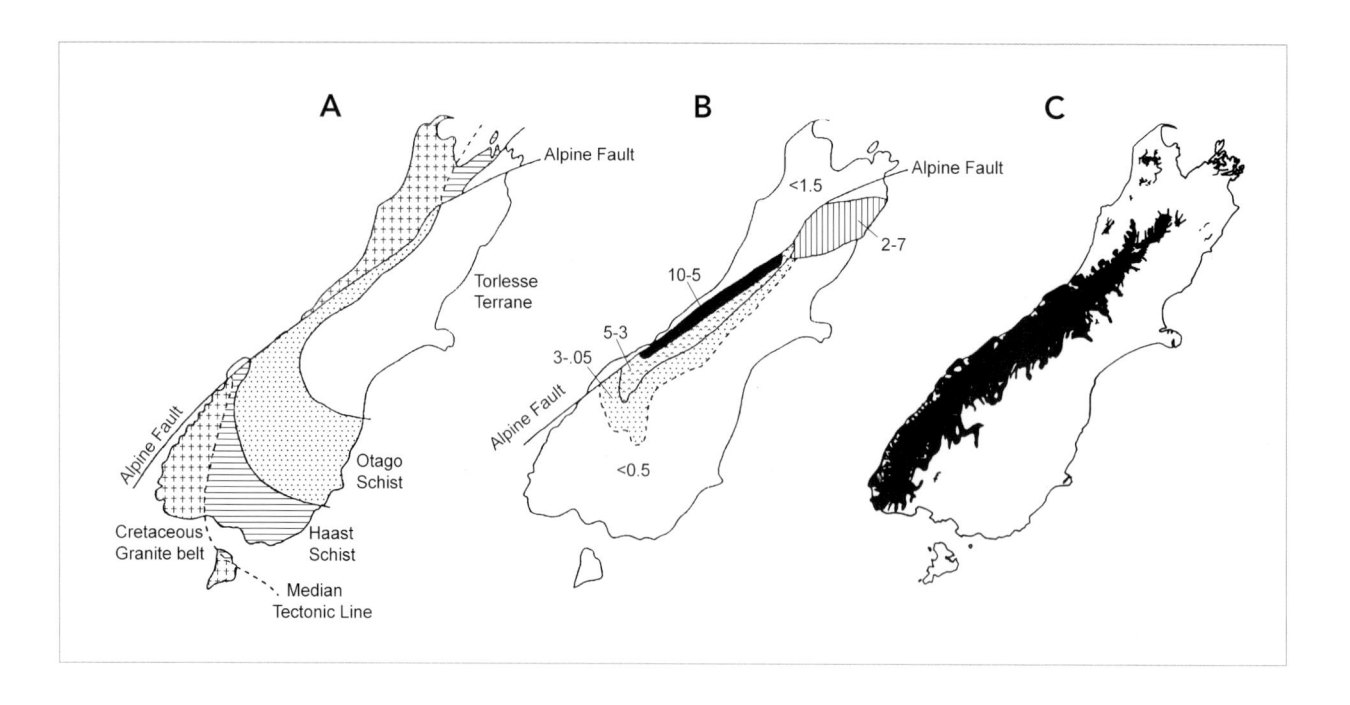

FIG. 15 Maps showing tectonic and glacial features of South Island. **A** broad distribution of rock types relative to the Alpine Fault (after Adams 1979; McGlone et al. 2001). Movement on the fault has resulted in *c.* 470 km of displacement among rocks in the Cretaceous granite belt. **B** current uplift rates (mm yr^{-1}) for South Island (after Wellman 1979). **C** extent of ice cover at the last glacial maximum, *c.* 20 000 years ago, mapped onto the present coastline. Some authors have attributed disjunct species distributions and separation of the South Island centres of endemism (Fig. 11) to the combination of movement on the Alpine Fault and rapid uplift of the Southern Alps (A, B). Others suggest that recent glaciation on South Island (C) has been more important in creating these patterns.

hypothesis implies changes over millions of years: Alpine Fault movement has been greatest since 25 Mya (rates of lateral displacement in the order of 25–30 mm yr^{-1}, resulting in an offset of about 100 km since the Pliocene, and 440–80 km in total), and uplift of the Southern Alps has been most rapid in the last 3–5 Myr (with current rates commonly estimated between 5 and 10 mm yr^{-1} through parts of the Southern Alps; Wellman 1979; **Fig. 15B**). The glacial hypothesis, on the other hand, implies changes over tens to hundreds of thousands of years: glacial maxima through the Quaternary have occurred at intervals of about 100 000 years, with the last maximum reached about 20 000 years ago. Because glacial periods have occurred relatively recently, their effects will be superimposed on earlier distribution patterns that may have been produced by slower moving tectonic processes.

Extensive glaciation (e.g. **Fig. 15C**) and associated climatic changes must have had substantial impacts on distributions of South Island plants. The influence of tectonic events, in comparison, is harder to substantiate.

DNA-based studies will increasingly provide data for testing biogeographic hypotheses, and identifying factors that have been important in shaping the distributions of species. For instance, divergence in DNA sequences within species or species pairs disjunct between northern and southern South Island could be used to estimate the time of separation of these areas. Likewise, patterns of genetic variation in species widespread in Canterbury and Westland (**Figs 12** and **13**) could indicate whether they show evidence of range expansion from northern and/or southern areas, and could possibly be used to estimate the timing of this expansion. Such studies of many species in *Hebe*, and other groups within the flora, could help to build a general picture of the biogeographic history of these areas. The now readily available and expanding range of DNA-based methods for addressing such questions makes this an exciting time for those interested in the evolutionary and geographic history of the flora.

Morphology

P. J. Garnock-Jones

Hebe and *Leonohebe* are parts of a larger generic complex that has probably evolved from a single ancestor in New Zealand (Wagstaff et al. 2002), so the extent of diversity in their vegetative and reproductive morphology is of interest in ecology and evolutionary biology. This chapter describes the basic structure of vegetative and reproductive parts of *Hebe* and *Leonohebe*, and then describes significant deviations. At times, the conditions found in related genera of the *Hebe* complex are also described.

VEGETATIVE MORPHOLOGY

Hairs

Hairs in *Hebe* and *Leonohebe* are of two basic types (Garnock-Jones 1993*a*). Multicellular eglandular hairs are common in all species. These are uniseriate, with usually two to five cells and varying amounts of granular ornamentation on their outer walls; they are usually colourless (appearing white), but may be brownish or yellowish in some species (e.g. *H. cockayneana*). Shorter versions of these hairs are found along leaf margins (e.g. *H. elliptica*) and stems of some species. In some species of *Leonohebe* (e.g. *L. tetrasticha*), eglandular leaf margin hairs are little more than denticles. Irregularly branched eglandular hairs are characteristic of the leaf and calyx margins of *H. pauciflora* (**Fig. 16**).

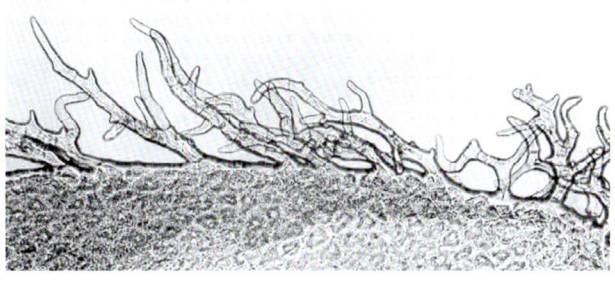

FIG. 16 Irregularly branching eglandular hairs of *Hebe pauciflora*.

The common type of glandular hair in *Hebe* is very small, with a short uniseriate stalk and two-celled head. This type was illustrated and termed testiculate by Garnock-Jones (1993*a*). It is also common elsewhere in Lamiales. In *Hebe*, testiculate glandular hairs are often interspersed with short uniseriate eglandular hairs along bract and calyx margins (**Fig. 17**). Sometimes, very small testiculate hairs occur subappressed on leaf surfaces; these can be seen only with a scanning electron microscope and are not included in the descriptions here. Long glandular hairs with a globular head and uniseriate stalk are found in *Parahebe* and *Heliohebe*, but these are mostly absent from *Hebe* (they are occasionally seen on calyx margins of *H. stenophylla*; Bayly et al. 2000).

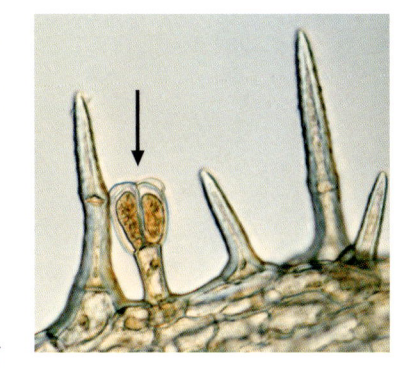

FIG. 17 Margin of calyx lobe with eglandular and (arrowed) twin-headed (testiculate) glandular hairs.

Roots

There appear to be no detailed studies of *Hebe* roots. Hebes produce adventitious roots readily on cuttings (Cockayne 1924; Metcalf 1972) and from the nodes of prostrate stems in contact with the ground (Heads 1994*c*). Their root systems extend a long way both laterally and in depth, and many small fibrous roots are produced. This might be related to observations that hebes need fertile soils; grown in pots, they rapidly deplete soil resources and are quite short-lived.

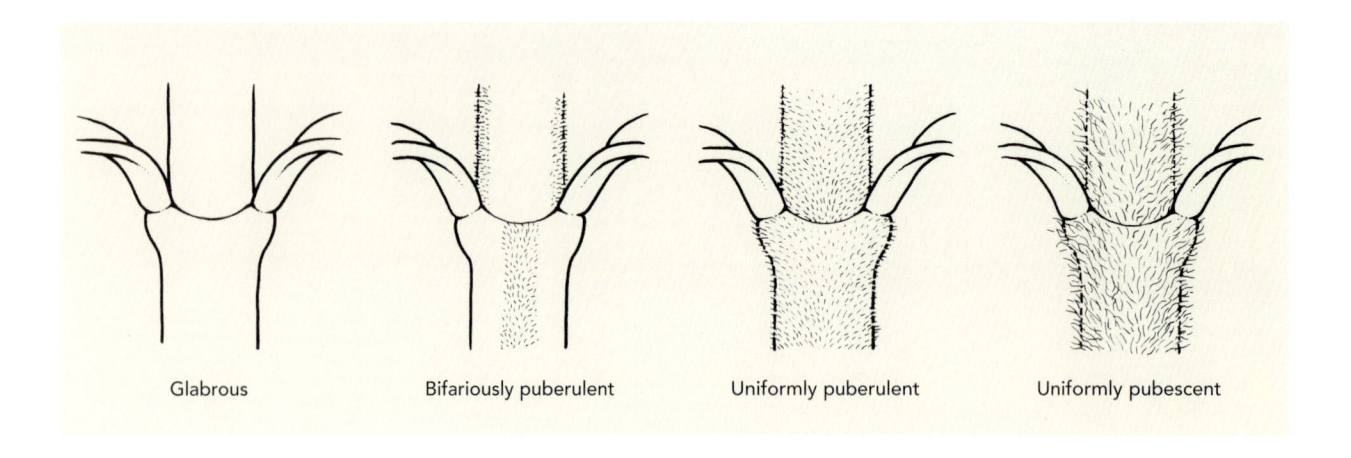

Glabrous   Bifariously puberulent   Uniformly puberulent   Uniformly pubescent

FIG. 18 Distribution and length of hairs on stems.

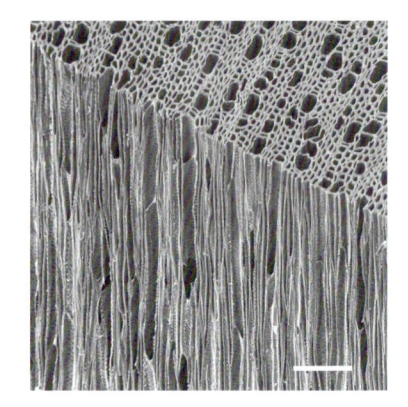

FIG. 19 Scanning electron micrograph of wood of *Hebe salicifolia*, showing a transverse face (above) and a tangential longitudinal face (below). Scale bar = 100 µm.

Roots of *Chionohebe*, *Hebe*, *Parahebe* and *Veronica* have been said to be without root hairs (Hong 1984), which would suggest that mycorrhizae play an important role in root function. At least in *Hebe*, however, root hairs are common (P. J. de Lange pers. comm. 2005); arbuscular endotrophic mycorrhizae are present in *H. stricta*, if not other species too.

Stems

Hebes range in stature from low prostrate shrubs (e.g. *H. buchananii*) to small trees (*H. barkeri*, *H. flavida*, *H. parviflora*) up to 13 m tall. Many hebes are multistemmed shrubs, often with a neat rounded appearance when growing in the open. Older plants, particularly those of the mat-forming habit (e.g. *H. pinguifolia*), tend to spread laterally by layering and may die back in the centre to form a ring-shaped low shrub. Such death of the older parts results from stem predation by a long-horned beetle, *Mesolamia* sp. (Dugdale 1975).

Hebe stems often have pronounced nodes, with prominent leaf-base scars evident, even after many years of growth in thickness. On mature stems, the bark is usually pale greyish, although it is sometimes dark (e.g. *H. buchananii*). In *H. scopulorum* the bark is corky and irregular (Bayly et al. 2002). The internodes (**Fig. 18**) may be glabrous (e.g. *H. pareora*), bifariously hairy (e.g. *H. cockayneana*) or uniformly hairy all around (e.g. *H. leiophylla*). Stem hairs may be very short (e.g. *H. parviflora*) to quite long (e.g. *H. cockayneana*).

Meylan & Butterfield (1978) describe *H. salicifolia* wood (**Fig. 19**) as lacking rays. Rayless wood has been observed in a number of other *Hebe* species (B. G. Butterfield pers. comm. 2004; W. M. Malcolm pers. comm. 2004) and in woody species of *Plantago* (Carlquist 1988); it is thought to be characteristic of woodiness that has evolved among island members of otherwise herbaceous groups and may increase mechanical strength (Carlquist 1988).

In the "Occlusae" and some small-leaved "Apertae", fine, close branching produces a regular rounded shrub. This is often achieved by dieback of the erect main leader and its replacement by a pair of lateral branchlets arising from a subordinate node; the result is a pseudodichotomous branching pattern. In other groups, such as many small-leaved "Apertae" and some "Flagriformes", stems are often erecto-patent rather than erect, and then each main stem and its laterals form a branch that is V-shaped in section and flat-topped by the distally evenly decreasing length of lateral branchlets.

A much greater diversity of stature and habit is found among related New Zealand genera of the *Hebe* complex. Most *Chionohebe* are small alpine cushion

FIG. 20 Apical views of vegetative shoots, showing leaf buds that are: **A** distinct (the most common condition in *Hebe*); **B–D** tightly surrounded by recently diverged leaves.

plants with sessile solitary flowers; *Heliohebe* are subshrubs often with decumbent branches; and most *Parahebe* are subshrubs, although *P. canescens* is a minute, almost herbaceous, creeping plant of lake and tarn margins.

Leaves

Hebe is usually instantly recognisable by its decussate leaf arrangement, although some species (e.g. *H. vernicosa*; **Fig. 119B**) have leaves that tend towards a falsely distichous arrangement (Heads 1994c) by curving of the petioles, especially on prostrate shoots. Shoots usually terminate in a large vegetative bud, with overlapping pairs of developing leaves remaining appressed together at the margins until they are full-sized or nearly so (**Fig. 20A**). In some species the leaves are too small to develop in this way (e.g. "Flagriformes" and *Leonohebe*; **Fig. 20C**) and in others (e.g. *H. macrantha*; **Fig. 20D**) the leaves separate and diverge early in development, as they do in *Heliohebe* and *Parahebe*. Vegetative buds are sometimes grossly enlarged as "big-bud" galls by the action of fly larvae (family Cecidomyidae) (Dugdale 1975); the enclosed bud environment also may provide security, shelter and food for insect larvae.

Moore (1967, in Allan 1961), was the first to describe the leaf bud sinus and to recognise its importance as a taxonomic character. It can be observed by folding down one of the uppermost pair of leaves and examining the margins of the next pair of leaves that are still joined in the bud. The sinus, if present, is a small gap near the base of the closed vegetative bud, formed when the leaf margins are not united for the full length of the bud (**Fig. 21**). A more or less obvious sinus occurs when leaves are petiolate, especially if they have a pronounced shoulder at the junction of the petiole and margin. The size and shape of the sinus varies according to the length and width of the petiole and abruptness of the shoulder. The sinus is an extremely useful character because it varies widely among species and is usually constant within a species. It is thus one of the first attributes that should be looked for in making any *Hebe* identification. The sinus, however, is not particularly informative in a phylogenetic sense, because evolutionary gains and losses of sinuses appear to have been frequent in *Hebe* evolution. In particular, there are pairs of species that appear on other grounds to be closely related, but

Absent	Narrow and acute	Broad and acute	Square to oblong
Small and rounded	Broad and shield shaped	Small and acute	

FIG. 21 Leaf bud sinuses.

which differ in this character (e.g. *H. corriganii* and *H. macrocarpa*), and other species where a sinus is present or absent in different parts of a species' range (e.g. *H. pinguifolia* and *H. pimeleoides*).

Hebe leaves are always simple and usually entire. They vary greatly in size (1–2 mm long in some "Flagriformes" to *c.* 165 mm long in *H. macrocarpa*), shape (e.g. linear in *H. stenophylla*; lanceolate in *H. barkeri*; elliptic in *H. urvilleana*; obovate in *H. speciosa*), colour (glaucous in *H. albicans*; dark green in *H. brevifolia*; pale green in *H. traversii*; often red on the margins in *H. pinguifolia*; yellow at the petiole in *H. flavida*), indumentum (glabrous in most species except for margins and midribs; hairy in *H. amplexicaulis* f. *hirta* and some forms of *H. pubescens*), and thickness (from thin in *H. acutiflora* to coriaceous in *H. speciosa* and rigid in *H. epacridea*). Distinctly toothed leaves are found in some species, especially *H. macrantha*. Incised margins occur in *H. diosmifolia* and sometimes in *H. macrocalyx* and several others, and tiny distant teeth (little more than prominent hydathodes) are a feature of some large-leaved species such as *H. salicifolia*. Hairs on the leaf margins can be distinctive (dense and short, but absent at the apex in *H. elliptica*; long and fringing in *H. gibbsii*; irregularly branched in *H. pauciflora*). Leaf apices and teeth when present sometimes terminate in a hydathode, which consists of a cluster of parenchyma cells at the end of a vein, a small or large sub-epidermal cavity, and a stoma (Adamson 1912).

Leaf margins can be rounded in transverse section or bevelled. In either case, some species (and some populations) have crenulate or erose thickened margins (e.g. *H. mooreae*); these are easily detected by running a fingernail along the margin. Some "Flagriformes", "Buxifoliatae" and *H. pauciflora* (**Fig. 22**) have thick, glossy, yellowish and transparent cuticles.

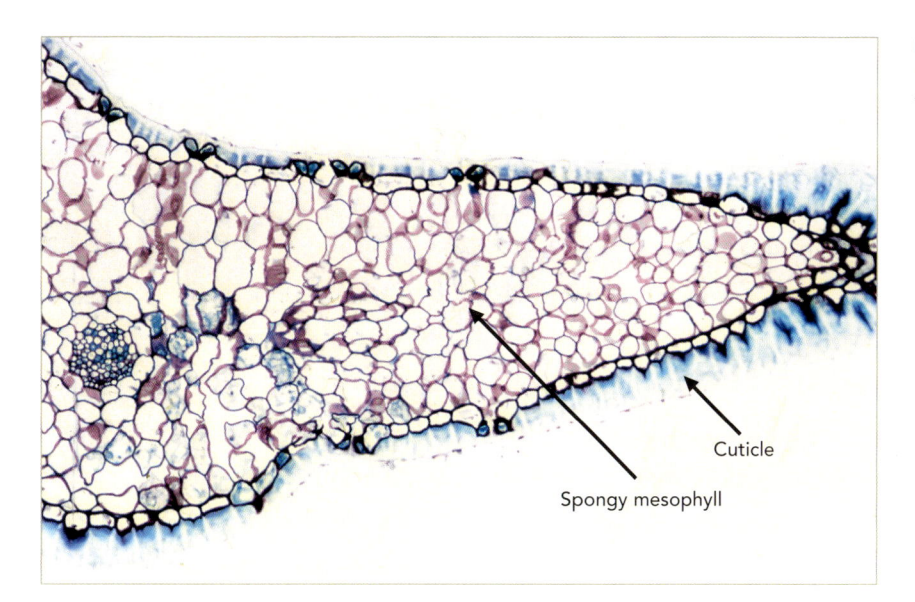

FIG. 22 Transverse section of *Hebe pauciflora* leaf showing thick cuticle.

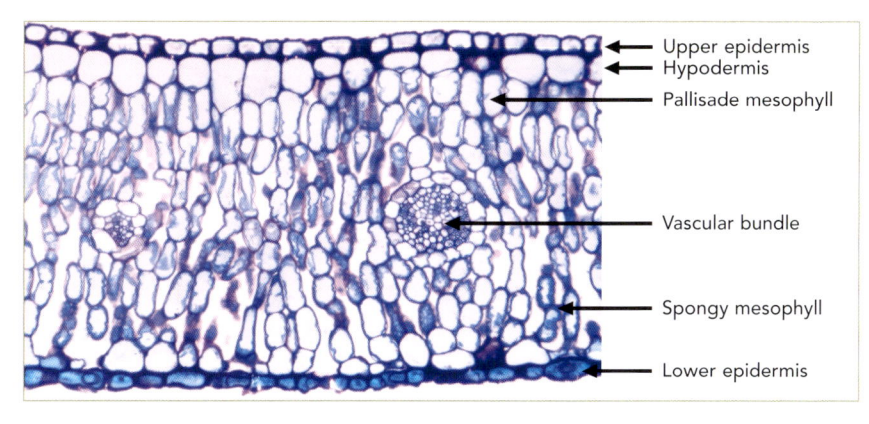

FIG. 23 Transverse section of *Hebe topiaria* leaf showing a distinct hypodermis layer below the epidermis.

According to Adamson (1912), the leaf epidermis in *Hebe* and *Parahebe* consists of cells with straight or slightly curved anticlinal walls, which contrasts with usually sinuate walls in herbaceous *Veronica*. In some species – for example, *H. topiaria* (**Fig. 23**), *H. elliptica* (Adamson 1912), *H. bollonsii* and some forms of *H. pubescens* subsp. *sejuncta* (Bayly et al. 2003), a distinct hypodermis that lacks chloroplasts is present. Palisade mesophyll is one to two cell layers thick and may be weakly or sharply delimited from the spongy mesophyll. In *H. pauciflora* (**Fig. 22**), the mesophyll appears completely spongy and loosely aggregated, with rounded undifferentiated mesophyll cells loosely adhering among air spaces within a bounding epidermis and cuticle.

Although most species of *Hebe* have leafy stems and exposed internodes, "Flagriformes" and *Leonohebe* exhibit the so-called "whipcord" habit because their tiny scale-like appressed leaves make the branches resemble plaited rawhide whips. In "Flagriformes", the leaf anatomy resembles that of the leafy hebes, except that the palisade mesophyll is abaxial to the spongy mesophyll (Adamson 1912). Breitwieser (1993) and Breitwieser & Ward (1998) reported a similar inversion of the mesophyll layers in whipcord species of *Helichrysum* and *Raoulia* (Asteraceae). In some "Flagriformes" there is no clear demarcation between the stem and the scale leaf (e.g. **Fig. 24**), whereas in others a clear nodal joint is present, although often obscured by overlapping leaves (Ashwin, in Allan 1961).

FIG. 24 Whipcord hebes with (left) and without a conspicuous nodal joint.

One species, *H. townsonii*, has intramarginal domatia (Moore, in Allan 1961; Sampson & McLean 1965; **Fig. 149E**). Unlike domatia in most other genera, these are not associated with vein branches. Heads (1993) proposed a novel interpretation for these intramarginal domatia; he considered them to be "the last trace of the epiphyllous inflorescences of the early angiosperms, left over after the hemming-in of growth responsible for the modern angiosperm leaf".

REPRODUCTIVE MORPHOLOGY

Inflorescences

Hebe inflorescences are variable, but they have much in common. The most frequent type is a simple lateral raceme of crowded, spiralled, bracteate, shortly pedicellate flowers (e.g. *H. stricta*). In many species the flowers are loosely aggregated into whorls; that is, there are regions on the inflorescence axis where internodes are short and other regions where they are relatively long. Heads (1994c) referred to these in *H. calcicola* as pseudowhorls but didn't record them for other species. Significant variations in inflorescence structure include compound racemes, apparently terminal inflorescences, opposite flowers, sessile flowers, reductions of the inflorescence to small axillary clusters and even solitary flowers (Moore 1967; Kellow et al. 2003b), and changes in spiral inflorescence phyllotaxis (Garnock-Jones 1993a). Inflorescence bracts in *Hebe* are typically small and entire, although in some plants (e.g. *H. benthamii*) the lowest ones may be larger and almost leaf-like.

Compound lateral racemes are found in a few species (e.g. *H. divaricata*, *H. diosmifolia* and *H. venustula*). These usually have bracteate opposite-decussate branches, and the lower flowers may be opposite as well. Overall, the inflorescence has a compact, almost globular, appearance, with densely clustered flowers.

In a number of species with lateral simple racemes (e.g. *H. arganthera*, *H. cryptomorpha* and *H. pimeleoides*) the flowers are opposite, at least in the case of the lower ones. In some species, there are terminal racemes of opposite flowers subtended by leaf-like bracts (e.g. *H. odora* and *H. hectorii*). Garnock-Jones (1993a) interpreted these as vegetative shoots with solitary lateral flowers representing reduced, leaf-subtended racemes. The terminal inflorescences of the "Connatae" (**Fig. 60**) show stages in this kind of reduction, with some having numerous opposite axillary clusters of flowers and others being simple racemes.

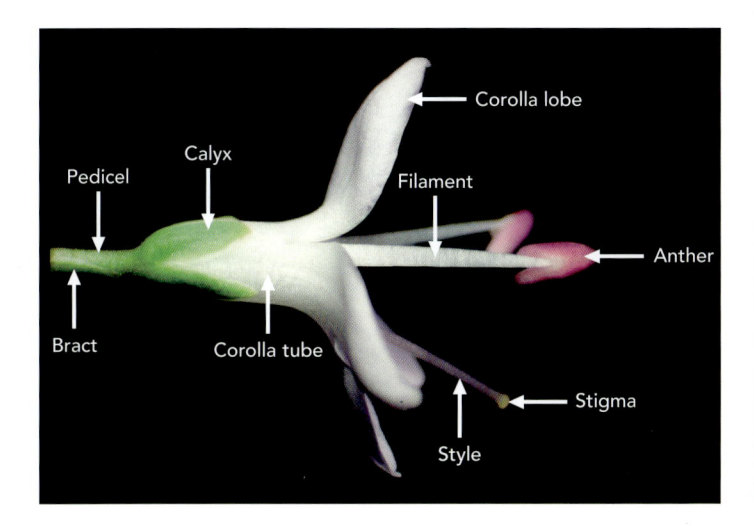

left
FIG. 25 Lateral view of flower, showing some floral parts.

right
FIG. 26 Frontal view of flower, showing terms used for corolla lobes.

Finally, Garnock-Jones (1993*a*) reported variation in phyllotaxis in simple racemes, with some species (e.g. some *Hebe* ser. *Hebe*, sensu Heads 1993) having 3/5 and others (e.g. *H. speciosa*) 5/8 phyllotaxis.

In related genera, *Heliohebe* have terminal compound racemes with opposite branches, at least in the lower parts, and spiralled flowers, whereas most *Parahebe* have lateral simple racemes with spiralled flowers and elongated internodes and pedicels (Moore 1967; Garnock-Jones & Lloyd 2004). Inflorescences of these types are not found in *Hebe*, although *Parahebe* racemes differ from the common *Hebe* type only in the elongation of the axis and the pedicels.

Calyx

In most *Hebe* and *Leonohebe* species there are four calyx lobes (occasionally a rudimentary fifth, posterior, lobe is also present; **Fig. 27**) that are usually only shortly connate. *H. benthamii* has four to six calyx lobes. The anterior pair of calyx lobes may be longer than the posterior pair. The calyx varies in size (1–10.2 mm long) and indumentum. Calyx lobes may be linear, lanceolate, deltoid, ovate, oblong or oblanceolate, and their apices may be obtuse or acute, sometimes varying in the same flower. In a few species the anterior lobes may be fused for some of their length (e.g. *H. dilatata*), and in a few they are almost completely united (e.g. *L. cupressoides* and *H. armstrongii*). In *L. cupressoides* both the anterior and posterior pairs are united, producing a distinctly two-lipped calyx. Saunders (1934) noted the absence of marginal veins in calyx lobes of *L. ciliolata* (as *Veronica gilliesiana*).

Calyx lobe margins are sometimes pale or whitish, or may be tinged pink. They are rarely glabrous (some specimens of *H. colensoi*, *H. pareora* and *H. macrocalyx* only), or eglandular-hairy (e.g. *H. stenophylla*), or, more commonly, have mixed, often alternating, eglandular and glandular hairs (**Fig. 17**).

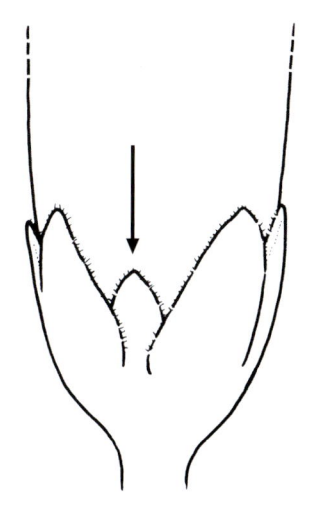

Corolla

The corolla has a short or long tube (**Fig. 25**) and usually four lobes (**Fig. 26**). In many species, the posterior lobe is largest and often forms an upper lip opposite to a lower lip formed by the lateral lobes and the narrow anterior lobe. In *Leonohebe* and a few species of *Hebe* (e.g. *H. macrantha* and *H. pauciflora*) and related genera (*Heliohebe*, *Chionohebe*), the corolla is not obviously divided into two lips. In *Parahebe*, it is divided into a three-lobed upper and one-lobed lower lip. In

FIG. 27 Calyx with a fifth posterior lobe (indicated by arrow). This feature occurs occasionally in a number of taxa (recorded here for twenty-two species).

Hebe, corolla veins branch near the base of the tube, showing the tube to be formed by coalescence of the lobes; this is in contrast to *Chionohebe*, where the corolla vein branches are distal, showing the tube is formed by elongation of the basal part of the corolla (Fig. 6 in Garnock-Jones 1993*a*).

The corolla varies greatly in size, shape and colour, but nectar guides, as seen in *Veronica* (e.g. **Figs 1A** and **B**) and *Parahebe*, are absent. The most unusual species is *Hebe benthamii*, which has four (to six), subequal, blue corolla lobes. Occasional individuals of *H. odora* (especially the cultivar 'Anomala') and some whipcord species have three-lobed corollas, by loss of the anterior lobe (Saunders 1934). Saunders (1934) interpreted the four-lobed corolla common in *Hebe* as derived from an ancestral five-lobed corolla by fusion of the two posterior lobes, citing as evidence the duplex venation in most flowers of most species (*H. odora* [as *V. buxifolia*] and *L. ciliolata* [as *V. gilliesiana*] were cited as exceptions). However, venation of the posterior corolla lobe is variable, with both simplex and duplex conditions found within many species and sometimes within an individual plant.

The largest corollas are found in *H. macrantha* and *H. elliptica*, whereas the smallest are probably females of *L. tetrasticha*. Corolla tubes vary in length from 0.4 mm (*H. rakaiensis*) to 7 mm (*H. petriei*) and in width from 0.7 mm (*H. angustissima*) to 4.5 mm (*H. macrantha*).

Small, rounded corolla lobes are found in many species of "Occlusae"; these are usually widely spreading (the posterior one is sometimes reflexed) and somewhat cupped. Narrow, obtuse and erect or suberect lobes are found in *H. stricta* and related species. Many northern species (e.g. *H. adamsii* and *H. ligustrifolia*), along with the South Island endemic *H. townsonii*, have broad acute lobes, whereas in *H. colensoi* the lobes can be acute and narrow. Corolla tubes and throats may be glabrous or hairy inside, and lobe margins may be ciliate or ciliolate in about a dozen species, including *H. speciosa* (Heads 1993) and *H. rakaiensis* (**Fig. 85E**). In most species the corolla is white, pale mauve or pale pink, but in a few species it is strongly coloured blue (*H. benthamii*), violet (*H. macrocarpa* var. *latisepala*) or magenta (*H. speciosa*, *H. brevifolia*). Rare individuals in otherwise white-flowered populations have pink, blue or mauve flowers, suggesting colour might be governed by rare recessive alleles.

Androecium

There are two stamens (**Fig. 25**), epipetalous at the corolla throat and inserted between the posterior and lateral corolla lobes; their anthers are dorsifixed and introrse. The stamen filaments narrow and are very flexible at the anther attachment. Anther size varies little. Filament length is very variable; the shortest filaments of male-fertile flowers are *c.* 1 mm long in some "Connatae" (e.g. *H. petriei* and *H. benthamii*) and the longest up to 13 mm (in *H. speciosa*). Some species of *Hebe*, particularly among "Occlusae", have stamen filaments that are so elongated in the bud that they are inwardly folded, ready to straighten quickly at anthesis and present pollen far outside the floral tube (Garnock-Jones 1993*a*; **Fig. 43**), but in most species the filaments are straight in the bud and elongate after anthesis. *Hebe* anthers are often dark magenta, sometimes pink or mauve, and rarely pale yellow or white (e.g. *H. arganthera*). Stamen filaments are usually white, but are strongly coloured in some species (e.g. *H. speciosa*). Staminodes in dioecious and gynodioecious species have empty pollen sacs that are smaller (e.g. **Fig. 33**) and often paler (sometimes pale brown or white) than the functional anthers.

Pollen

Pollen grains in the *Hebe* complex have been described by Hong (1984), who compared them with *Veronica* and other related genera, and by Moar (1993). Among the pollen types recognised by Hong (1984), both the *Stenocarpon* type and the *Veronica* type may be found in *Chionohebe*, *Hebe*, *Parahebe* and *Veronica*. Grains were described as tricolpate, usually spheroidal or less often oblate-spheroidal, 21–42 × 16–42 µm, colpus wide, with rounded (*Veronica* type) to pointed (both types) ends, colpus membrane with granular (*Stenocarpon* type) or irregularly shaped (both types) processes, exine *c.* 1 µm thick. The major differences between the types relate to the striations of the tectum. The *Stenocarpon* type has a partial, microreticulate or minutely striate-reticulate tectum, whereas the *Veronica* type has a partial striate-reticulate tectum with straight lirae interconnected by short bridges, or slightly curved lirae (Hong 1984). *Detzneria* was classified in a group on its own, due to its smooth colpus membrane and macroreticulate tectum with large rounded lumina. These features seem not to be related to phylogenetic position (*sensu* Wagstaff et al. 2002) or to geography; they possibly relate instead to functional roles. Moar (1993) did not recognise particular groupings within *Hebe*, noting that the grains are very similar but can vary within species.

Gynoecium

The ovary (**Fig. 28**) is most often included in the corolla tube and surrounded by a usually glabrous nectarial disc (ciliate in *H. barkeri* and sometimes in *H. dieffenbachii*, *H. chathamica*, *H. rapensis*, *H. elliptica*, *L. cheesemanii* and *L. cupressoides*). Heads (1994*c*) discussed variation in the colour, shape and pubescence of nectarial discs in the *Hebe* complex. The ovary and style may be glabrous or hairy. The ovary consists of two locules (two to three in *H. benthamii*) and varies from small and globose (e.g. *H. pauciflora*) to ovoid (most species) or conical (e.g. *H. petriei*). The style is generally about the same length as the stamen filaments and carries a small, weakly bilobed stigma. When the style is long, the stigma may be far-exserted, but in some species (e.g. *H. benthamii*, *H. haastii* and *H. pauciflora*) it is quite short and holds the stigma at the corolla throat. Stigma size varies, so that in some species the stigma is similar in width to the style, whereas in others (e.g. females of *H. subalpina*) it is distinctly capitate. The stigma is usually white, greenish or yellowish, but can be pink, red, magenta or mauve. Ovule numbers range from as low as two ovules per locule in *H. annulata* to as many as sixty-one in *H. elliptica*. Where the ovule number is low, they are all arranged in a single layer at the margins of a flattened placenta, but in species with more ovules they may be in two (to three) layers.

Other descriptions of the variation in form of *Hebe* flower parts can be found in Moore (in Allan 1961), Hutchins (1997) and Garnock-Jones (1993*a*).

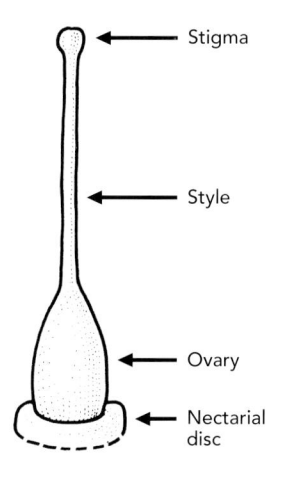

FIG. 28 Nectarial disc and gynoecium (in loculicidal view). The disc (always) and ovary (usually) are surrounded by the corolla tube of a flower (which can be removed to allow closer examination).

Fruits

The fruits of all species of *Hebe* are small, bilocular capsules, except *H. benthamii*, which is sometimes trilocular. The capsule walls vary in thickness from quite woody (e.g. *H. pareora*) to submembranous (e.g. *H. stricta*). In most species the capsule is flattened parallel to the orientation of the septum (**Fig. 29**) so that the septum spans the broad diameter (latiseptate). Latiseptate ovaries and capsules typically have peltate placentae in which the ovules and seeds are confined to one (to two) rows on the margins of a flattened placenta (Jussieu 1789; Moore & Irwin 1978; Garnock-Jones 1993*b*). In *H. macrantha*, *H. pauciflora* and *Leonohebe*, the

FIG. 29 Immature capsules, showing the different types of flattening seen in *Hebe* and *Leonohebe* (after Hutchins 1997). Latiseptate capsules are by far the most common.

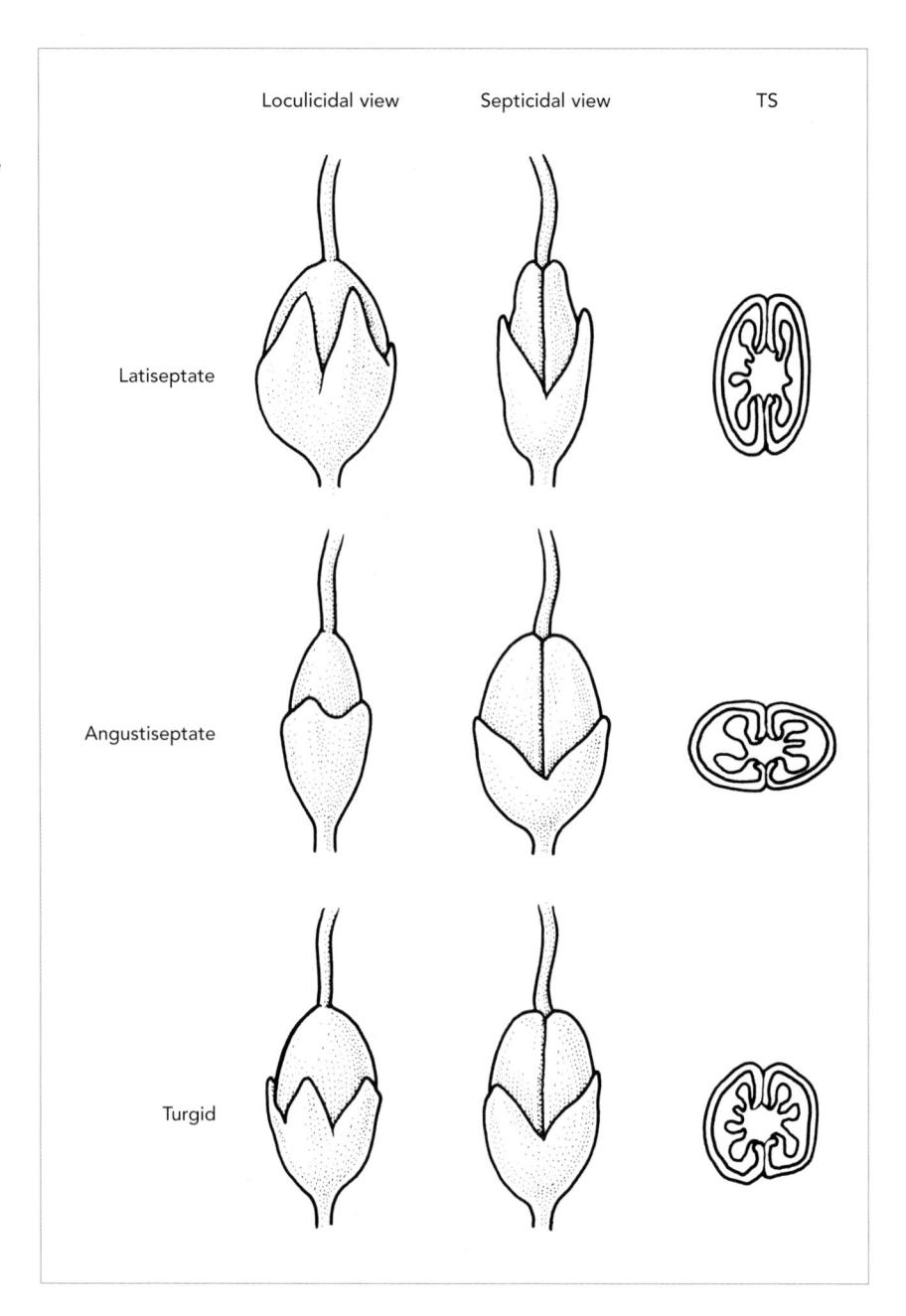

capsule is flattened in the opposite direction, so the septum is narrow (angustiseptate); in these species the ovules are borne in two rows. *H. pareora* (Garnock-Jones & Molloy 1983*a*) sometimes has turgid to angustiseptate capsules. In most species the capsule is acute or subacute at the apex, but in some species with angustiseptate (e.g. *H. pauciflora*) or turgid capsules (e.g. *H. pareora*) it is emarginate or truncate. Fruit function is described in the chapter "Reproductive Biology".

Seeds

The seeds of all species of *Hebe* are small circular or elliptical discs, often lens-shaped and tapering to a narrow indistinct wing (**Fig. 47**). Webb & Simpson (2001) described *Hebe* seeds in detail, and reported a length range of 0.6 mm (*H. arganthera*) to 3.2 mm (*H. macrocarpa*). Simpson (1976) reported a range of mean weights from 0.05 mg (*H. salicifolia*) to 0.19 mg (*H. pinguifolia*) per seed.

The colour varies with the maturity of the seed and the age of the collection, and was described as ranging from pale orange-yellow to dark brown. The testa was described as mammillate or reticulate-mammillate (Webb & Simpson 2001), but generally appears smooth under a dissecting microscope or hand lens. Thieret (1955) compared the seeds of *Veronica* and *Hebe*, and considered the loss of surface reticulation to be a derived characteristic within the group. Seeds of most species of *Veronica* are very similar to those of *Hebe* and *Parahebe*, but seeds of *Heliohebe* and *Chionohebe* have slight differences (Garnock-Jones 1993*a*; Webb & Simpson 2001) that each appear to be derived.

DEVELOPMENTAL MORPHOLOGY

Seedlings and Juvenile Forms (Heteroblasty)

Seedlings and reversion shoots of *Hebe* and *Leonohebe* often produce leaves that are thinner and hairier than the adult leaves, and serrate at the margins (Kirk 1878; Cockayne 1898). These sometimes closely resemble the adult leaves of *Parahebe*, *Heliohebe*, *Hebe macrantha*, and *Veronica*. Entire leaves are formed later in development, before reproductive maturity is attained. In "Flagriformes" and *Leonohebe*, the juvenile and reversion leaves are pinnatifid (Kirk 1878; Cockayne 1898). Kirk (1878) and Cockayne (1898) reported that such pinnatifid leaves can be induced in whipcord hebes under mild temperatures, shade and high humidity.

Heads (1987, and in more detail in 1994*c*) believed the leaves in *Leonohebe* (of which his circumscription included "Flagriformes", "Connatae" and other *Hebe* groups) were modified from fertile or sterile bracts, whereas the "true leaves" were those manifest in juvenile plants and reversal shoots. While heteroblasty in the *Hebe* complex appears to be an example of ontogeny recapitulating phylogeny, there is no evidence that Heads' interpretation is correct. Entire leaves of *Hebe* and *Leonohebe* (*sensu* Heads) often subtend inflorescences rather than the solitary flowers that might be predicted from Heads' interpretation. Further, there are no anatomical differences by which the claimed bract-derived leaves of *Leonohebe* (*sensu* Heads) can be distinguished from true leaves in *Hebe*. Cockayne (1898) argued that rather than representing an example of juveniles passing through an ancestral stage, heteroblasty might be an adaptation for coping with different or changeable conditions.

Flower Development

In spite of similar structure, there are marked differences in the shapes of flowers between *Hebe* and the related genera *Heliohebe*, *Parahebe* and *Chionohebe*, all of which have evolved within *Veronica* (Wagstaff et al. 2002; Albach & Chase 2001). In attempting to understand the evolution of these differences, it would be helpful to know the developmental bases for them (Kampny et al. 1993) and whether they are related to changes in pollinators (Kampny 1995). Garnock-Jones (1993*a*) inferred from corolla vasculature that long corolla tubes in *Hebe* have different modes of development from those in *Chionohebe*, suggesting convergent evolution.

Flavonoid Biochemistry

K. R. Markham

The taxonomic work leading to this book has relied heavily both on the application of traditional characters of plant morphology and on extensive studies of variation in flavonoid biochemistry. Aspects of flavonoid biochemistry have been important contributors to many taxonomic decisions, and are frequently mentioned here in the notes provided under species descriptions. This chapter gives some background on what flavonoids are, the types of flavonoids that occur in *Hebe* and *Leonohebe*, how flavonoids are screened for, and their taxonomic application.

INTRODUCTION TO THE FLAVONOIDS

Flavonoids are a group of plant pigments that are virtually ubiquitous in land-based green plants. For this reason they are thought to have very primitive origins and it has been hypothesised that in order to inhabit the land, plants required components such as flavonoids to protect their proteins and nucleic acids from the damaging effects of the sun's UV-A and UV-B radiation. Studies with extant plants have provided evidence that flavonoids do indeed act as UV filters and/or free-radical scavengers, and that they protect plants from UV damage (Lois & Buchanan 1994; Ryan et al. 1998).

Flavonoids also act as pollinator attractants (as pigments in flowers), pollen-tube growth initiators, stimulators of nodulation (in roots), modulators of enzyme action, phytoprotectants (from viruses, bacteria and fungi) and feeding deterrents (e.g. as tannins, reviewed by Bohm 1998). They range in colour from colourless (e.g. isoflavones, dihydroflavones and dihydroflavonols), through yellow and orange (e.g. flavones, flavonols, chalcones and aurones) to red, purple and blue (anthocyanins).

Many of these flavonoid types are restricted, either to selected plant taxa or to specialised organs (e.g. anthocyanins occur mainly in flowers and fleshy fruits). In the liverworts, mosses, ferns, fern allies, gymnosperms and angiosperms, the predominant flavonoids are the flavones and the flavonols (Markham 1988). Flavonoids are usually accumulated in plants as glycosides (i.e. with sugars attached), and in this water-soluble form are predominantly located in the aqueous environment of the cell vacuole. The less common non-glycosylated flavonoids (aglycones) are relatively non-polar and are normally found associated with the surface waxes of plants.

FLAVONOIDS IN *HEBE* AND *LEONOHEBE*

Flavonoids in leaves of *Hebe* and *Leonohebe* include only flavones, which are ubiquitous, and flavonols, which are absent in *Leonohebe* and found in only some *Hebe* species (Markham et al. 2005). Both of these flavonoid groups originate from the same precursor type, dihydroflavone (**Fig. 30**), but while flavones are

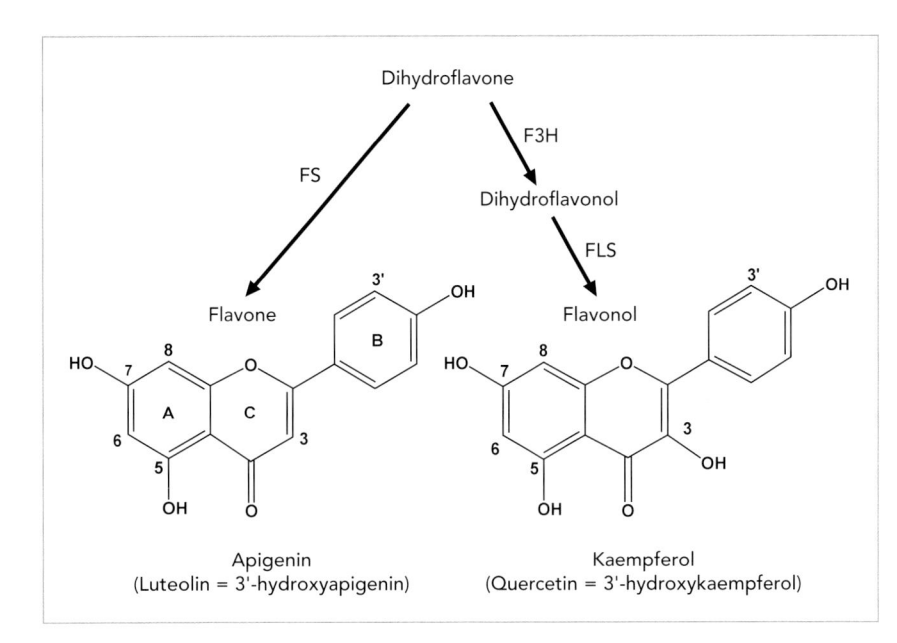

FIG. 30 Branching in the flavonoid biosynthetic pathway that leads to production of the flavone and flavonol series of flavonoids. This scheme relates to the apigenin/kaempferol pathway. An analogous scheme for the biosynthesis of luteolin/quercetin would differ only in that an additional hydroxyl would be present at the 3'-position of each structure. The enzymes involved in each step are as follows: FS = flavone synthase; F3H = flavanone-3-hydroxylase; FLS = flavonol synthase.

derived in one step, the biosynthetic production of flavonols involves the additional introduction of a hydroxyl group at the 3-position on the dihydroflavone nucleus (Heller & Forkmann 1994).

Biosynthetic elaboration of these two structural types, via a series of essentially one-enzyme steps, produces a wide range of flavonoid characters that can be used for taxonomic purposes in *Hebe* and *Leonohebe*. In this process the simplest flavonoid structures are transformed into the more complex in a stepwise fashion, referred to as a biosynthetic sequence. Structural modifications to the basic flavone (apigenin) and flavonol (kaempferol) aglycone types found in *Hebe* and *Leonohebe* are detailed below.

Flavone Series Modifications
- Addition of a hydroxyl group at the 3'-position of **apigenin** (at an earlier stage in the biosynthesis) produces **luteolin** and thereby a new series of flavones.
- Both apigenin and luteolin are found further modified with an extra hydroxyl group at the 6'-position (i.e. **6-hydroxyapigenin** or **6-hydroxyluteolin**), or at the 8-position (i.e. **8-hydroxyapigenin** or **8-hydroxyluteolin**).
- Methylation of the hydroxyl groups at the 4'-position of the apigenin and luteolin aglycones gives rise to **4'-O-methyl-apigenin** (not found in *Leonohebe*) and **4'-O-methyl-luteolin** respectively. Methylation of the 6-hydroxyl group in 6-hydroxyluteolin produces **6-methoxyluteolin**.

The above flavones occur in *Hebe* and *Leonohebe* as glycosides, although the aglycones themselves are found in the whipcords and in *H. odora*, probably residing in the leaf waxes. Glycosides are produced in the final stages of biosynthesis through the enzymatic transfer of the sugars glucose, glucuronic acid, xylose and rhamnose (in *Hebe* and *Leonohebe*) to any of the hydroxyl groups on the aglycone other than the inaccessible 5-hydroxyl. One or more sugars may be attached to any of the hydroxyl groups, but in *Hebe*, as in many other plants, the maximum number appears to be three (see Markham et al. 2005). Thus, mono-, di- and tri-glycosides are encountered. The linkages between the sugars in di- and tri-glycosides may also vary. For example, both xylo(1-2)glucosides and xylo(1-6)glucosides are found in *Hebe* (Mitchell et al. 1999, 2001).

Flavonol Series Modifications

Hebe species display a much more restricted range of flavonol variants. Only one hydroxyl additional to those in **kaempferol** is seen, and this is at the 3'-position, giving rise to the **quercetin** series of flavonols. No methylation of hydroxyl groups is evident in either flavonol series, and sugars (always present) are found attached only to the 3- and 7-hydroxyl groups.

A vast number of characters of value in taxonomic studies of *Hebe* and *Leonohebe* are provided by the variation of flavonoid content and structure. Variable features include the aglycone structural types (e.g. flavone, flavonol, number and position of hydroxyl groups, methylation, etc.), the nature of the sugars, the number of sugars attached and the positions on the aglycone to which they are attached, and finally, the inter-sugar linkages. Even with the limited range of variables found in *Hebe* and *Leonohebe* (at least fifteen aglycone types and forty-three glycoside types), the number of possible flavonoid-sugar combinations provides a large pool of potentially useful taxonomic characters. Although not all possible combinations have been found, the total number of flavonoids from *Hebe* and *Leonohebe* useful for taxonomic purposes is well in excess of eighty (**Table 4** lists all the significant flavonoids), and many are, to date, unique to these taxa. The distribution of flavonoids between taxa is variable; one flavonoid, luteolin-7-*O*-glucoside (compound 4; **Table 4**), is found in the majority of species, while others occur in only one or two species. The combinations of flavonoids in different species provide the variation in plant flavonoid profiles used in our taxonomic studies.

SCREENING FOR FLAVONOIDS

Screening for the presence/absence of flavonoids is most easily done by two-dimensional paper chromatography (2D-PC). This technique involves the extraction of oven-dried leaf material with methanol:water (80:20) and the application of the extract from 25 mg of dry leaf to a 460 × 570 mm Whatman 3MM paper sheet as described by Markham (1982). The extract spot on the paper is then separated into its constituent flavonoids by 2D-PC using TBA (*t*-butanol:acetic acid:water, 3:1:1) in the first direction and 15 per cent acetic acid in the second. The two-dimensional profile of flavonoid spots is then seen by viewing the 2D-PC over a 366 nm UV lamp through UV-absorbing glasses (see details in Markham 1982).

The flavonoids of *Hebe* and *Leonohebe* all appear as dark UV-absorbing spots, whereas the accompanying hydroxy-cinnamic acid derivatives appear as light blue fluorescent spots. The positions of flavonoid spots on 2D-PCs, and their colour reaction with ammonia vapour, can be used to distinguish most of the commonly encountered flavonoids of *Hebe* and *Leonohebe* from one another. A diagram of a composite 2D-PC containing the spots of all the significant flavonoids identified in *Hebe* is presented in **Fig. 31**, and the compound structures, numbers and their colour changes in ammonia vapour are detailed in **Table 4**.

It will be noted from **Fig. 31** that some flavonoids run in close proximity to one another on a 2D-PC. In cases where the colour reaction does not distinguish alternative structures, distinction may be achieved by reference to the data available on flavonoids previously identified in other taxa. Where there is still doubt over the identification, it is necessary to excise the spot, extract the flavonoid(s) and analyse them by high-pressure liquid chromatography (HPLC), which will provide retention time (RT) and UV absorption spectroscopic data for each

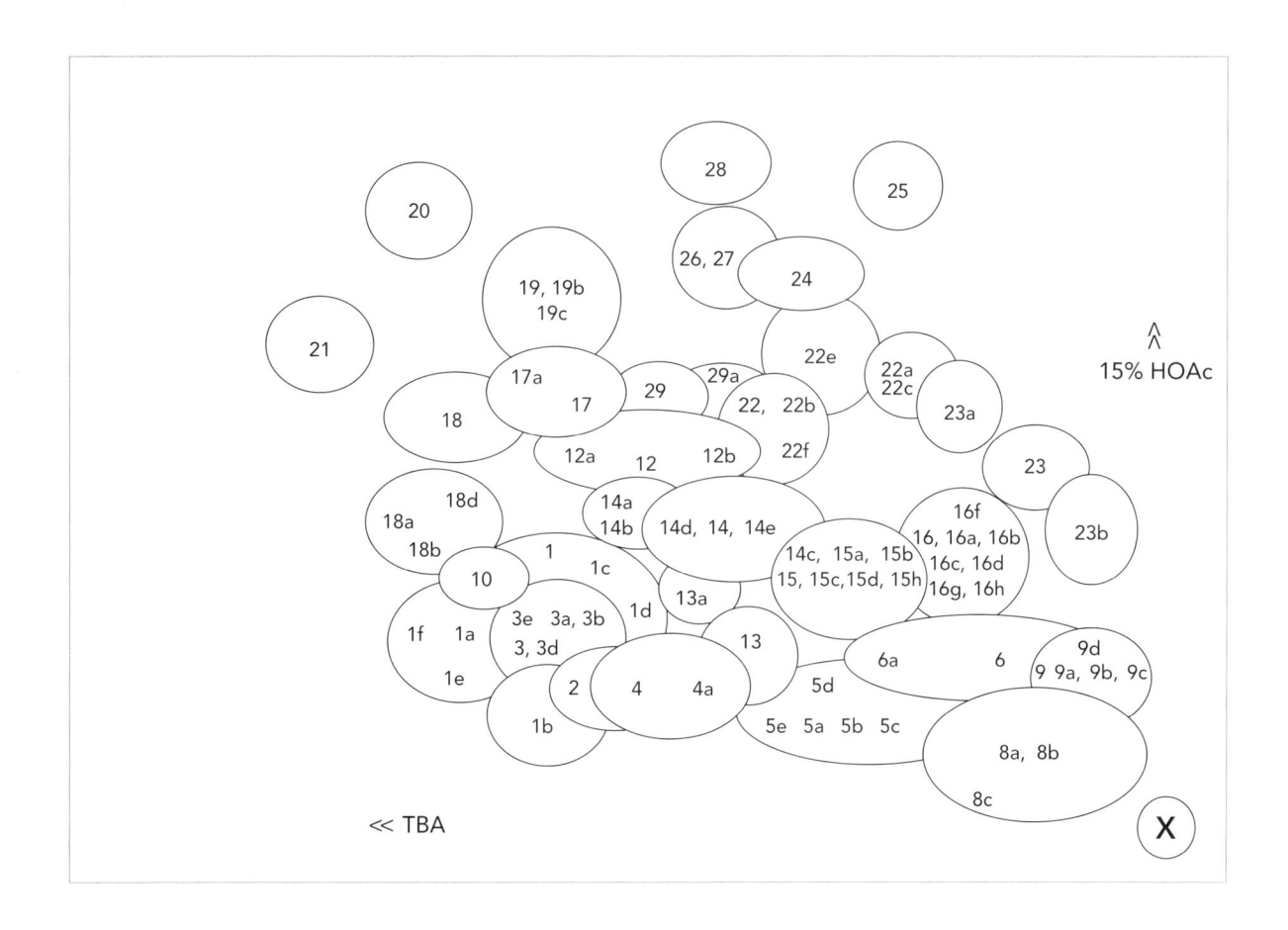

FIG. 31 A composite, diagrammatic representation of a theoretical 2D-PC showing all significant flavonoids found in *Hebe* and *Leonohebe*.

flavonoid. Comparison of these data with available data for the authentic compounds will generally lead to an identification. This approach was used routinely in studies of *Hebe* and *Leonohebe* to confirm the identity of all the flavonoids present in each species, and occasionally led to the discovery of more than one flavonoid in a 2D-PC spot. The HPLC techniques are described in detail in previous publications on *Hebe* flavonoids (e.g. Mitchell et al. 2001; Bayly et al. 2003), and a compilation of RT data is presented in **Table 4**.

TAXONOMIC APPLICATION OF FLAVONOID CHARACTERS

The application of flavonoids as taxonomic characters dates back to the early twentieth century (Swain 1963; Harborne 1967), but their widespread acceptance as legitimate characters for plant taxonomy did not occur until the 1960s. This acceptance was triggered by the work of a research group at the University of Texas in Austin, which demonstrated by 2D-PC that flavonoid patterns were species-specific (Alston & Turner 1963) and that a range of environmental variables had no qualitative effect on the flavonoid profile (McClure & Alston 1964). A number of examples of more recent successful applications of flavonoid characters to plant taxonomy and phylogeny have been summarised by Bohm (1998).

In our studies of *Hebe* and *Leonohebe*, flavonoid pattern variation in species did not seem to result from differences in sampling time, analytical technique used or sample age. Thus, HPLC and 2D-PC analyses of herbarium specimens as old as twenty-four years, of samples collected and analysed early in this study, and of plants re-collected from sites previously sampled, all produced reliable and consistent flavonoid patterns.

TABLE 4 Significant flavonoids in *Hebe* and *Leonohebe*: their structures, distribution and some physico-chemical properties. A, apigenin; K, kaempferol; L, luteolin; Q, quercetin; T, tricetin (5'OH luteolin); 4'Me, 4'-methyl (i.e. 4'-hydroxyl methylated); 6OH, 6-hydroxy; 6OMe, 6-methoxy; 8OH, 8-hydroxy; 8OMe, 8-methoxy; ?, structure tentatively identified; Fl, fluorescent; sh, shoulder.

Flavonoid number[1]	Flavonoid structure	2D-PC spot colour (ammonia)	HPLC retention time	Absorption maxima (nm)	No. of occurrences (in species)
1	L-4'-O-[6-O-rhamnoglucoside]?	Dark	32.2	242/247, 267, 288sh, 337	4
1a	L-4'-O-glucoside	Dark	33.8	247, 267, 288sh, 337	58
1b	L-3'-O-glucoside	Olive	35	240, 267, 288sh, 341	23
1c	A-7-O-glucuronide	Green	34.3	267, 337	29
1d[2]	6OMeL-7-O-glucoside-acetate?	Indefinite	≈33	255, 271, 350	2
1e[2]	8OHA-8-O-glucoside	Olive	35.3	270, 297, 331	33
1f	L-7-O-glucoside-acetate	Yellow	36.6	254, 267, 348	2
2	8OHL-8-O-glucoside	Olive	35.1	257, 270, 292sh, 352	44
3	6OMeL-7-O-glucoside	Olive	31.5	255, 271, 347	57
3a	6OMeL-7-O-glucuronide	Olive	34.5	253, 272, 343	6
3b	8OHA-7-O-glucoside	Olive	32.2	277, 305, 327	7
3d	4'MeL-7-O-glucoside?	Dark	34.1	252, 267, 347	13
3e	6OHA-like-O-glycoside[3]	Indefinite	32.3	233, 255sh, 285, 337	4
4	L-7-O-glucoside	Yellow	30.8	255, 266, 348	87
4a	L-3'-O-glucuronide	Olive	≈35	273, 340	2
5a	8OHL-7-O-glucoside	Olive	30.1	255sh, 275, 296sh, 342	69
5b[2]	6OHA-7-O-[2-O-xyloxyloside]	Olive	35.4	230sh, 282, 334	8
5c	T-7-O-glucuronide?	Yellow	34.5	248, 267, 352	1
5d	6OHA-7-O-glucoside	Olive	29.9[4]	230sh, 282, 334	4
5e	6OHA-like-O-glycoside[3]	Indefinite	34.4	233, 255sh, 285, 337	4
6[2]	6OHA-7-O-[2-O-xyloglucoside]	Olive	31.8	230sh, 282, 334	25
6a[2]	8OHL-7-O-[6-O-xyloglucoside]?	Olive	29	255sh, 275, 296sh, 342	9
8a[2]	6OHL-7-O-[2-O-xyloxyloside]	Olive	32.2	230sh, 255, 282, 345	26
8b[2]	6OHL-7-O-[2-O-xyloglucoside]	Olive	29.7	230sh, 255, 282, 345	62
8c	6OHL-7-O-glucoside	Olive	28.5	230sh, 255, 280, 343	43
9	6OHL-7-O-[2-O-glucoglucoside]	Olive	27.2	230sh, 255, 282, 345	58
9a[2]	6OHL-7-O-[6-O-xyloglucoside]	Olive	≈28	230sh, 255, 280, 343	8
9b[2]	4'Me6OHL-7-O-[2-O-xyloglucoside]	Dark	RT?	230sh, 255, 280, 345	0[5]
9c[2]	6OHL-7-O-[2-O-glucoglucuronide]	Olive	28.8[4]	230sh, 255, 282, 345	6
9d	6OHL-7-O-triglycoside?	Olive	26.5	230sh, 255, 282, 345	9
10	A-7-O-glucoside	Green	33.7	267, 337	62
12	A-7-O-[6-O-xyloglucoside]	Green	32.6	267, 337	24
12a	L-7-O-[2-O-rhamnoglucoside]	Yellow	30.9	255, 273, 349	28
12b[2]	8OHL-7-O-[2-O-rhamnoglucoside]	Olive	39.5	255sh, 275, 296sh, 342	5
13	L-7-O-glucuronide	Yellow	31.3	254, 267, 348	43
13a[2]	6OMeL-7-O-[6-O-xyloglucoside]?	Olive	31.6	255, 271, 347	23
14	L-7-O-[6-O-rhamnoglucoside]	Yellow	30.2	255, 267, 348	26
14a	4'MeL-7-O-[6-O-rhamnoglucoside]	Dark	≈33	252, 267, 347	8
14b[2]	8OHA-7-O-[6-O-rhamnoglucoside]?	Dark	31.3	277, 305, 327	4
14c	6OMeL-7-O-[6-O-rhamnoglucoside]	Olive	30.4	255, 271, 347	4
14d	L-4'-O-diglycoside	Dark	32.3	250, 268, 288sh, 337	3
14e[2]	L-4'-O-diglucoside	Olive	30.8	248, 268, 288sh, 337	2
15	L-7-O-[6-O-xyloglucoside]	Yellow	30.1	255, 267, 348	40
15a[2]	4'MeL-7-O-[6-O-xyloglucoside]	Dark	33.4	252, 267, 347	8
15b	8OHL-7-O-[6-O-rhamnoglucoside]	Olive	29.1	255sh, 275, 296sh, 342	4

Flavonoid number[1]	Flavonoid structure	2D-PC spot colour (ammonia)	HPLC retention time	Absorption maxima (nm)	No. of occurrences (in species)
15c	6OHL-7-O-[2-O-rhamnoglucoside]?	Olive	27.5	230sh, 255, 282, 345	4
15d	6OHA-7-O-[2-O-glucoglucoside]	Olive	29	230sh, 282, 334	14
15h	4'Me6OHL-7-O-[rhamnodixyloside]?	Indefinite	30.1	230sh, 255, 280, 345	3
16	L-7,3'-di-O-glucoside	Fl green	28.7	240/246, 268, 341	28
16a	L-7-O-[2-O-glucoglucuronide]	Yellow	28.1	255, 267, 348	34
16b	L-7-O-[2-O-xyloxyloside]-3'-O-xyloside?	Fl green	30	240, 267, 342	8
16c	L-7,3'-di-O-glucuronide	Fl green	29.4[4]	238, 268, 341	1
16d	L-7-O-[2-O-glucuronoglucuronide]	Yellow	26.5[4]	255, 267, 348	3
16f	4'MeL-7-O-[2-O-glucuronoglucuronide]?	Dark	30.7	252, 267, 347	3
16g	L-7,3'-di-[glucoside-glucuronide][6]	Fl green	28.1[4]	238/246, 267, 342	1
16h	L-7,3'-di-[glucoside-glucuronide][6]	Fl green	28.8[4]	240, 267, 341	1
17	A-7-O-[6-O-rhamnoglucoside]	Green	32.4	267, 337	15
17a	4'MeA-7-O-[6-O-rhamnoglucoside]	Dark	37.9	267, 334	2
18	Q-3-O-glucoside	Dull yellow-green	30.6	255/265, 354	28
18a[2]	6OMeA-7-O-glucoside	Olive	33.5	273, 334	16
18b	A-4'-O-glucoside?	Dark	33.5	268, 325	7
18d	Q-3-O-xyloside	Dull yellow-green	31.2	255/265, 354	2
19	Q-3-O-[6-O-rhamnoglucoside]	Dull yellow-green	29.9	257/265, 354	18
19b	6OHA-7-O-[2-O-rhamnoglucoside]?	Olive	31.7	230sh, 282, 334	1
19c[2]	6OMeL-7-O-[2-O-rhamnoglucoside]	Olive	30.6	255, 271, 347	10
20	K-3-O-[6-O-rhamnoglucoside]	Dull yellow-green	31.7	234sh, 265, 297sh, 342	3
21	K-3-O-glucoside	Dull yellow-green	32.8	234sh, 265, 297sh, 342	2
22	A-7-O-[2-O-glucuronoglucuronide]	Olive-green	≈29	267, 337	20
22a	L-7,4'-di-O-glucoside	Dark	27.1	250, 268, 338	15
22b[2]	A-7-O-[2-O-glucoglucuronide]	Green	28.7[4]	267, 337	1
22c[2]	L-7-O-[2-O-gluco-6-O-rhamnoglucoside]	Yellow	29.2	255, 267, 350	1
22e	A-6-C-xyloside-8-C-glucoside	Olive	26.7	271, 336	5
22f[2]	6OHL-7-O-[2-O-glucuronoglucuronide] Me ester?	Olive	29.5	230sh, 255, 282, 345	5
23[2]	L-7-O-[2-O-gluco-6-O-xyloglucoside]	Yellow	27.6	255, 267, 348	16
23a	L-6,8-di-C-glucoside	Olive/yellow	22.5	258, 270, 350	7
23b[2]	8OHL-7-O-[2-O-gluco-6-O-rhamnoglucoside]?	Olive	28.3	255sh, 275, 296sh, 342	2
24	A-6,8-di-C-glucoside	Olive	≈25	271, 336	23
25[2]	Q-7-O-glucuronide-3-O-glucoside	Yellow-green	≈21	255/265, 353	7
26[2]	L-7-O-[2,6-di-O-rhamnoglucoside]	Yellow	29.5	255, 267, 350	7
27[2]	L-7-O-[2-O-rhamno-6-O-xyloglucoside]	Yellow	29.2	255, 267, 350	1
28	6OMeL-7-O-[2,6-di-O-rhamnoglucoside]?	Olive	29.5	255, 271, 347	2
29[2]	4'MeA-7-O-[2-O-glucoglucuronide]	Dark	35.3	267, 333	1
29a	4'MeA-7-O-[2-O-glucuronoglucuronide]	Dark	35.2	267, 333	1

1 Compound numbers are used for convenience only; the same numeral indicates close proximity on 2D-PCs.
2 Not previously reported in other plant genera, as detailed in Williams & Harborne (1994) and Williams (2006).
3 Absorption spectra similar to, but not identical with, those of 6OHA-7-O-glycosides.
4 Retention time obtained using the same HPLC system, but a different column than for the others.
  The retention times are approximately 1 minute shorter than would be expected using the original column.
5 This flavonoid is only known from the hybrids H. calcicola x H. salicifolia and H. calcicola x H. albicans.
6 These structures may be interchangeable.
RT? indicates that the compound was identified using a different HPLC system, and so relative RT is unknown.

Species/Subspecies Delineation in Hebe *and* Leonohebe

Flavonoid datasets (profiles) were determined through 2D-PC (**Fig. 31** and **Table 4**) and commonly HPLC for *c.* 700 samples of *Hebe* and *Leonohebe*. Profile variation within a species was generally found to be less than between species, and species-specific patterns could, on occasions, be clearly defined. Although the flavonoid profiles were widely used in this study to provide general support for species delineation based on traditional taxonomic characters, there are a number of examples where flavonoid characters played a more prominent role in the taxonomy. These include: taxonomic revision of the *H. parviflora* complex (Bayly et al. 2000; Mitchell et al. 2001); recognition of *H. calcicola* (Bayly et al. 2001); delineation of new taxa in "Subdistichae" (small-leaved "Apertae" in this treatment; Bayly et al. 2002); definition of the geographic ranges of *H. bollonsii* and *H. pubescens*, and recognition of three subspecies of *H. pubescens* (Bayly et al. 2003); recognition of two subspecies in *H. pimeleoides* (Kellow et al. 2003*a*); and, in "Connatae", redefinition of the geographic range of *H. epacridea* and reinstatement of *H. macrocalyx*, including two varieties (Kellow et al. 2003*b*).

Hybrid Identification

Studies of the flavonoid constituents have also provided valuable confirmatory evidence for the definition of hybrids and their parentage. The flavonoid profile of a hybrid normally represents an approximate superimposition of the flavonoid profiles of the two parent species (Alston & Turner 1962; Bohm 1998), due to the inheritance of biosynthetic capability from both parents. In some cases this has also led to the production in the hybrid of distinctive "hybrid" compounds that are not found in either parent (Alston et al. 1965; Markham et al. 1992). Pioneering investigations by Alston & Turner (1962, 1963) revealed the true value of flavonoid characters in the determination of the parentage of hybrids when, with a complex population of four species of *Baptisia* (Leguminosae), they were able to identify all six possible two-way hybrids from 2D-PC studies.

In the present study, a number of suspected hybrids have been found and their flavonoid profiles assessed. The most thoroughly examined of these profiles relate to *H. calcicola* (Bayly et al. 2001), which was demonstrated to hybridise with *H. albicans* and *H. salicifolia*. The identified hybrids were shown not only to possess distinctive flavonoid profiles as a result of inheriting biosynthetic capability from both parents, but also their own unique hybrid marker flavonoid (9b).

Other suspected hybrids found in the present study were not investigated as fully, but inspection of the flavonoid distribution revealed that, although hybrid compounds are rare, the superimposition of much of the two parental flavonoid patterns is frequently a good indicator of hybridisation. The putative *H. corriganii* × *H. parviflora*, *H. odora* × *H. tetragona* subsp. *subsimilis* and *H. salicifolia* × *H. albicans* hybrids all possess flavonoid profiles that can be rationalised from a combination of the relevant parental profiles.

Chromosomes

Chromosome numbers of *Hebe* and *Leonohebe* are well known. The chromosome numbers of all species of *Leonohebe* have been counted at least once. The same is true for the accepted species, subspecies and varieties of *Hebe*, except for *H. rapensis* and *H. carnosula*; a count previously published for *H. carnosula* (Hair 1967) is referable to *H. pinguifolia*.

Details of published chromosome counts for *Hebe* and *Leonohebe* are presented in **Table 5**. This table updates the most recent compendium of chromosome numbers (Dawson 2000) by using the currently accepted taxonomy and incorporating recently published counts (by Bayly et al. 2002; de Lange & Murray 2002; Kellow et al. 2003*a*; de Lange et al. 2004*c*). It also presents, where possible, information on the provenance of plants counted, and their voucher specimens, information omitted from Dawson's large compendium of chromosome numbers for New Zealand spematophytes. We have had the opportunity to examine many of the voucher specimens for chromosome counts, and in some instances have reidentified them, thereby creating further differences between our list and that of Dawson.

Species of *Leonohebe* have $n = 21$ chromosomes. Those of *Hebe* have numbers based on $n = 20$ (sixty-eight species, including most "Apertae", most "Occlusae", all "Subcarnosae", some "Flagriformes" and one "Connatae") or $n = 21$ (fourteen species, including "Grandiflorae", "Pauciflorae", most "Buxifoliatae", two small-leaved "Apertae", some "Flagriformes" and most "Connatae"), or numbers presumably derived through aneuploidy, sometimes with polyploidy, from either $n = 20$ or 21 – these are $n = 59$ (*H. masoniae* and *H. brevifolia*), $n = 61$ (*H. topiaria*) and $n = 62$ (*H. ochracea*). Among species with chromosome numbers based on $n = 20$, there are thirty-nine diploids, thirteen tetraploids, possibly seven both diploid and tetraploid (*H. albicans*, *H. barkeri*, *H. buchananii*, *H. diosmifolia*, *H. pimeleoides*, *H. pinguifolia* and *H. stricta*), six hexaploids, and one both tetraploid and hexaploid (*H. macrocarpa*, in which ploidy is possibly correlated with segregate varieties). Among those based on $n = 21$, there are thirteen diploids, one tetraploid, one both diploid and tetraploid (*H. odora*), and one hexaploid.

Phylogenetic studies (Garnock-Jones 1993*a*; Wagstaff & Garnock-Jones 1998, 2000; Wagstaff & Wardle 1999; Wagstaff et al. 2002) suggest that $n = 21$ is the plesiomorphic chromosome number in the New Zealand *Hebe* complex. They also suggest that some groups (e.g. "Buxifoliatae" and some "Connatae") may have secondarily derived $n = 21$ chromosomes (from $n = 20$). For other species – for example, *H. vernicosa* and *H. societatis* (both small-leaved "Apertae"), and *H. annulata*, *H. armstrongii* and *H. salicornioides* ("Flagriformes") – it is not clear whether possession of $n = 21$ chromosomes represents retention of the ancestral state or a secondary derivation/reversal. Better resolved and supported phylogenetic hypotheses would allow for greater understanding of chromosome evolution in *Hebe*, as could studies of chromosome morphology and composition (e.g. using methods of *in situ* hybridisation).

It is not clear how often polyploidy has arisen in *Hebe*. However, given that it is found in taxa with both $x = 20$ and 21 chromosomes, morphologically divergent taxa with the same base chromosome number, and within at least some species, it seems likely that independent origins of polyploidy have been relatively common. Species with more than one ploidy level are potentially worthy of further taxonomic study; that is, they might, in some cases, be further taxonomically divided. They may also prove useful in studies of evolutionary processes within species.

TABLE 5 Chromosome counts for *Hebe* and *Leonohebe*

Species	n	2n	Locality	Voucher (and duplicates)[1]	Original reference
Hebe					
H. acutiflora	20	—	Kerikeri Falls, N Auckland	CHR 103042! (WELT 81945!; CHR 198177!, 198178!)	Hair (1967: 344)
	20	—	Kerikeri Falls, N Auckland	CHR 103158!	Hair (1967: 344)
	—	40	Trounson Kauri Park, Dargaville, Northland	AK 235668	Murray & de Lange (1999: 513, 518)
H. adamsii	40	80	Unuwhao Bush, The Pinnacle, North Auckland	AK 250799	de Lange & Murray (2002: 5)
H. albicans	20	—	Mt Peel, NW Nelson	CHR 103017 (WELT 82211!, 82206!)	Hair (1967: 347)
	20	—	Aorere R., NW Nelson	CHR 103003! (WELT 82239!)	Hair (1967: 347); as *H. recurva*
	20	—	Cult. Otari Gardens, Wellington	CHR 103093! (WELT 82232!, 82231!)	Hair (1967: 347); as *H. recurva*
	20	—	Cult., ex. Gouland Downs	CHR 103085 (AK 125959, 125960)	—[2]
	—	80	Gorge Stm, NW Nelson	CHR 462371!	Dawson & Beuzenberg (2000: 6, 17); as *H.* aff. *glaucophylla*
	—	80	Hoary Head, Arthur Ra., Nelson	CHR 438366!	Dawson & Beuzenberg (2000: 6, 17); as *H.* aff. *glaucophylla*
	—	80	Wakamarama Ra., Mt Burnett, Nelson	AK 252966	de Lange & Murray (2002: 5); as *H.* aff. *albicans*
H. amplexicaulis	—	40	—	—	Frankel in Darlington & Wylie (1955: 311)
H. amplexicaulis f. *hirta*	20	—	Mt Peel, S Canterbury	CHR 103044!	Hair (1967: 347); as *H. allanii*
	—	40	Geraldine, Mt Peel, Canterbury	AK 250719	de Lange & Murray (2002: 5)
H. angustissima	—	40	Otaki Gorge, Wellington	CHR 103272	Hair (1967: 345); as *H. stricta* var.?
	—	40	Otaki Gorge, Wellington	AK 233637	Murray & de Lange (1999: 513, 518); as *Veronica salicifolia* var. *angustissima*

1 ! indicates the specimen was sighted and its identity verified by the authors. Specimens in italics are from the same clone, but collected at a different time to the original chromosome vouchers (i.e. they are not strictly duplicates).

2 Not published? Vouchers not seen, record based on database at AK.

Species	n	2n	Locality	Voucher (and duplicates)[1]	Original reference
H. angustissima continued	—	40	Motu R., Gisborne	CHR 462369!	Dawson & Beuzenberg (2000: 6, 15); as *H.* aff. *stricta*
H. annulata	—	42	Takitimu Mts, Southland	CHR 103038!	Hair (1967: 350)
H. arganthera	—	40	Fiordland NP, Murchison Mts, Takahe Va., Southland	WELT 81845!	de Lange & Murray (2002: 5)
H. armstrongii	—	84	Cult. Botanic Gardens, Christchurch	CHR 103259!	Hair (1967: 350)
H. armstrongii?[3]	—	124	—	—	Frankel in Darlington & Wylie (1955: 312); as *H. armstrongii*
H. barkeri	—	40	Rangaika, Chatham Id	CHR 517724[4]	Dawson & Beuzenberg (2000: 6, 15)
	40	—	Cult. Christchurch Botanic Gardens, ex. Chatham Ids	CHR 7446!	—[5]
H. barkeri?	—	80	—	—	Frankel in Darlington & Wylie (1955: 312); as *H. gigantea*
H. benthamii	20	40	Adams Id, Auckland Ids	CHR 103263[6]	Hair (1967: 351)
H. biggarii	20	—	Eyre Mts, Southland	CHR 103033!	Hair (1967: 352)
H. bishopiana	—	40	Waitakere Ra., Auckland	CHR 462362	Beuzenberg in de Lange (1996: 189); Dawson & Beuzenberg (2000: 6, 15)
H. bollonsii	20	—	Poor Knights Ids	CHR 103011!	Hair (1967: 345)
	20	—	Poor Knights Ids	CHR 103112!	Hair (1967: 345)
	20	—	Poor Knights Ids	CHR 103115!	Hair (1967: 345)
H. brachysiphon	60	—	Cult. Botanic Gardens, Christchurch	CHR 74420![7] (= HV 166)	Frankel & Hair (1937: 672); as *H. montana*
	60	—	Cass	CHR 74448! (= HV 45)	Frankel & Hair (1937: 672); as *H. traversii*
	—	120	Ghost Ck, Castle Hill	CHR 74426! (= HV 225)	Frankel (1940: 172); as *H. traversii*
	60	—	Waimakariri R., Canterbury	CHR 103197!	Hair (1967: 342–3)
	—	120	Big Hill, Molesworth, Marlborough	CHR 103215	Hair (1967: 342–3)
	—	120	Cass, Canterbury	CHR 103214	Hair (1967: 342–3)
	—	120	Mt Oxford, N Canterbury	CHR 103242[8]	Hair (1967: 342–3)
	—	120	Cult., ex. Red Hills Ridge	WELT 82560!	de Lange et al. (2004c: 882)
	—	120	Hodder Va.	WELT 81685!	de Lange et al. (2004c: 882)
	—	120	Cult., ex. Jacks Pass	WELT 82896![9]	de Lange et al. (2004c: 882)
H. brevifolia	59	—	North Cape	CHR 103080!	Hair (1967: 345); as *H. macrocarpa* var. *brevifolia*
	—	118	Surville Cliffs, North Cape	AK 227425!	Murray in de Lange (1997: 3); Murray & de Lange (1999: 513, 519)
H. aff. *brevifolia*[10]	—	120	Surville Cliffs, North Cape	AK 235669	Murray & de Lange (1999: 513, 519)
H. breviracemosa	20	40	Kermadec Ids, Raoul Id, Hutchinson Bluff	AK 246814	de Lange & Murray (2002: 5)

3 No voucher cited, and the identity of the plant used is unknown. Given the unusual chromosome number, the count may be for *H. ochracea*, not *H. armstrongii*.
4 Grown from cuttings of MJB 699 (= WELT 80541!).
5 It is not clear whether this count relates to any that have been published.
6 Could not locate the cited voucher, but CHR 101153!, also from Adams Id, is labelled "voucher specimen" and "n = 20".
7 Voucher is probably *H. brachysiphon*, but it is difficult to be certain without mature flowers or details of locality. Published as *H. subalpina?* by Dawson (2000).
8 Could not locate the cited voucher, but CHR 74469! (= HV 296), also from Mt Oxford, is labelled n = 59–59½.
9 Voucher incorrectly cited by de Lange et al. (2004c) as WELT 82816.
10 Identification uncertain (see comments of Murray & de Lange 1999).

Species	n	2n	Locality	Voucher (and duplicates)[1]	Original reference
H. buchananii	—	80	—	—	Frankel in Darlington & Wylie (1955: 312)
	20	—	Rock and Pillar Ra., Central Otago	CHR 103052!	Hair (1967: 348); as *H.* aff. *buchananii*
	40	—	Dunstan Mts, Central Otago	CHR 103053!	Hair (1967: 348)
	40	—	Mts above Lindis Pass, Otago	CHR 103032!	Hair (1967: 348)
	40	—	Darran Mts, Southland	CHR 103131!	Hair (1967: 348)
	40	—	Canyon Ck, Otago	CHR 103082! (WELT 82152a & b!)	Hair (1967: 348); as *H.* aff. *buchananii*
H. aff. *buchananii*[11]	40	—	Mt Maude, Central Otago	CHR 103089	Hair (1967: 348); as *H.* aff. *buchananii*
H. calcicola	—	80	Mt Arthur, Nelson	CHR 462364[12]	Dawson & Beuzenberg (2000: 8); as *H.* (indet.; *H.* sp. (o) in Druce 1980 and Eagle 1982; *H.* "marble" in Druce 1993)
H. canterburiensis	20	—	Arthur's Pass	HV 32	Frankel & Hair (1937: 672); as *H. vernicosa* var. *canterburiensis*
	20	—	Mt Holdsworth, Tararua Ra.	CHR 103076!	Hair (1967: 343)
	20	—	Spenser Ra.	CHR 103190!	Hair (1967: 343)
	20	—	Jacks Hut, Arthur's Pass, Canterbury	CHR 103118!	Hair (1967: 343)
	20	—	Margarets Tarn, Arthur's Pass, Canterbury	CHR 103119!	Hair (1967: 343)
	20	—	Near Twin Cks, Arthur's Pass, Canterbury	CHR 103121!	Hair (1967: 343)
	—	40	Iron Hill, Cobb Va., Nelson	CHR 103227!	Hair (1967: 343)
	—	40	Red Hills, Wairau Va., Marlborough	CHR 103229	Hair (1967: 343
H. chathamica	20	—	Chatham Ids	CHR 103244 (*CHR 198197!*, *198198!*)	Hair (1967: 346)
	—	40	Chatham Ids	CHR 103267!	Hair (1967: 346)
H. cockayneana	—	120	—	—	Frankel in Darlington & Wylie (1955: 312)
	60	—	Upper Clinton R., McKinnon Pass, Fiordland	CHR 103051!	Hair (1967: 343)
H. colensoi	20	—	Rangitikei R., Kaimanawa Mts	CHR 103005!	Hair (1967: 342); as var. *colensoi*
	—	40	Rangitikei R., Kaimanawa Mts	CHR 103220	Hair (1967: 342); as var. *colensoi*
	—	40	Ngaruroro R., near Kuripapango	CHR 103266	Hair (1967: 342); as var. *hillii*
H. corriganii	40	—	Lower Pouakai Ra., Taranaki	CHR 103091!	Hair (1967: 344)
H. crenulata	—	80	Lk. Peel, Peel Ra.	WELT 81743!	Bayly et al. (2002: 580); de Lange et al. (2004c: 882)
	—	80	Spenser Mts, near Ada Pass	WELT 81695!	Bayly et al. (2002: 580); de Lange et al. (2004c: 882)
H. cryptomorpha	—	40	Cult., original locality unknown	CHR 462365[13]	Dawson & Beuzenberg (2000:; 7) as *H.* aff. *rigidula* (*H.* sp. (q))
	—	40	Richmond Ra., near Richmond Saddle Hut	WELT 81664!	Bayly et al. (2002: 580); de Lange et al. (2004c: 882)
H. decumbens	20	—	Mt Fyffe, Seaward Kaikouras	CHR 103015!	Hair (1967: 347)
	20	—	Mt Terako, Seaward Kaikouras	CHR 103022!	Hair (1967: 347)
	—	40	Langridge, Awatere Va., Marlborough	CHR 171715!	Hair (1967: 347)

11 Identification uncertain.
12 Voucher is probably *H. calcicola*, but leaves are not ciliate (and do not look mature), and there are no flowers or fruit, so it cannot be identified with certainty.
13 Voucher identity probably correct, but cannot be certain.

Species	n	2n	Locality	Voucher (and duplicates)[1]	Original reference
H. dieffenbachii	20	—	Chatham Ids	CHR 103012! (WELT 81941!; CHR 198196!; AK 126027!)	Hair (1967: 346); as H. barkeri
	20	—	Chatham Ids	CHR 103192! (WELT 81943!; CHR 198195!)	Hair (1967: 345)
	—	40	Near Waitangi, Chatham Ids	CHR 103270!	Hair (1967: 346); as H. barkeri
	—	40	Chatham (Rekohu) Id, Ocean Mail, Lk. Wharemanu	AK 250803!	de Lange & Murray (2002: 5)
H. dilatata	60	—	Spence Burn, Takitimu Mts, Southland	AK 250792	de Lange & Murray (2002: 5); as H. crawii
	—	120	Gem Lk., Umbrella Mts, Otago	AK 250804	de Lange & Murray (2002: 5)
	—	120	Lk. Laura, Garvie Ra., Otago	AK 252990	de Lange & Murray (2002: 5)
H. diosmifolia	—	40	—	—	Frankel (1941: 118)
	20	—	Wairua Falls, Titoki, N Auckland	CHR 103031!	Hair (1967: 342)
	20	—	Hatea R., Whangarei	CHR 103167!	Hair (1967: 342)
	40	—	Doubtless Bay, N Auckland	CHR 103019!	Hair (1967: 342)
	20	40	Pukenui, Northland	AKU 21374a–d	Murray et al. (1989: 589)
	—	40	Woodhill, Auckland	AKU 21373a–h	Murray et al. (1989: 589)
	—	40	Waipoua Forest, Northland	AKU 21378	Murray et al. (1989: 589)
	—	40	Wairua Falls, Northland	AKU 21371a–e	Murray et al. (1989: 589)
	—	40	Ngaitonga [Ngaiotonga] State Forest, Northland	AKU 21372	Murray et al. (1989: 589)
	40	—	Rarawa, Northland	AKU 21376a	Murray et al. (1989: 589)
	—	80	Karikari Peninsula, Northland	AKU 21377	Murray et al. (1989: 589)
	—	80	Te Paki, Northland	AKU 15954	Murray et al. (1989: 589)
	—	80	Te Paki, Northland	AKU 21379	Murray et al. (1989: 589)
	—	80	Te Paki, Northland	AKU 21375a–c	Murray et al. (1989: 589)
H. diosmifolia?[14]	—	24?	—	—	Huber (1927: 372)
H. divaricata	40	—	Pelorus Va., Marlborough Sounds	CHR 103025! (WELT 82144!)	Hair (1967: 342)
	40	80	Nr Attempt Hill, D'Urville Id,	WELT 81680!	de Lange et al. (2004c: 882)
	—	80	West slopes of Mt Richmond, Richmond Ra.	WELT 81658!	de Lange et al. (2004c: 882)
	—	80	SW of Richmond Saddle, Richmond Ra.	WELT 81668!	de Lange et al. (2004c: 882)
	—	80	Cobb Va.	WELT 80690!	de Lange et al. (2004c: 882)
H. elliptica	20	—	—	—	Frankel & Hair (1937: 683)
	20	—	Punakaiki, Westland	CHR 103009! (WELT 82204!, 82205!)	Hair (1967: 343); as H. elliptica var. elliptica
	20	—	Pahia Coast, Invercargill	CHR 103104! (WELT 82201!)	Hair (1967: 343); as H. elliptica var. elliptica
	20	—	Campbell Id	CHR 103066!	Hair (1967: 343); as H. elliptica var. elliptica
	20	—	Titahi Bay, Wellington	CHR 103029! (WELT 81939!)	Hair (1967: 343); as H. elliptica var. crassifolia
	—	40	Whaririki [Wharariki] Beach, NW Nelson	CHR 103226!	Hair (1967: 343); as H. elliptica var. elliptica

continued

14 Count or identity uncertain; see Frankel (1941: 118) for comments.

Species	n	2n	Locality	Voucher (and duplicates)[1]	Original reference
H. elliptica continued	—	40	Kahurangi Point, NW Nelson	CHR 103224!	Hair (1967: 343); as *H. elliptica* var. *elliptica*
	—	40	Jackson Bay, S Westland	CHR 103223!	Hair (1967: 343); as *H. elliptica* var. *elliptica*
	—	40	Mussel Bay, Southland	CHR 103218!	Hair (1967: 343); as *H. elliptica* var. *elliptica*
	—	40	The Snares	CHR 103185!	Hair (1967: 343); as *H. elliptica* var. *elliptica*
	20	—	Westland	CHR 103086! (WELT 82202!, 82203!)	—[15]
H. epacridea	21	—	Mt Torlesse, Canterbury	CHR 103048!	Hair (1967: 350)
	—	42	Lk. Tekapo, Mackenzie County	CHR 103221	Hair (1967: 350)
H. evenosa	—	120	—	—	Frankel in Darlington & Wylie (1955: 312)
	60	—	Tararua Mts	CHR 103049	Hair (1967: 346)
H. flavida	—	40	Hauturu Trig, Hokianga, Northland	AK 235951	Murray & de Lange (1999: 513, 518); as *H*. aff. *stricta* (A)
H. gibbsii	20	—	Ben Nevis, Nelson	CHR 103045!	Hair (1967: 347)
H. glaucophylla	40	—	Castle Hill, Canterbury	CHR 103043!	Hair (1967: 347)
H. haastii	—	42	Head of Craigieburn R., Canterbury	CHR 103312!	Beuzenberg & Hair (1983: 16); as *H. haastii* var. *haastii*
H. hectorii	—	40	Huxley Ra., Otago	CHR 103253!	Hair (1967: 349); as *H. hectorii* var. *hectorii*
H. hectorii subsp. *hectorii*	—	40	Green Lk., Fiordland	CHR 103254!	Hair (1967: 349); as *H. hectorii* var. *hectorii*
	—	40	Stewart Id	CHR 103060	Hair (1967: 349); as *H. laingii*
	—	40	Takahe Va., Murchison Mts, Fiordland	CHR 103255![16]	Hair (1967: 349); as *H. laingii*
H. hectorii subsp. *coarctata*	—	40	Lk. Sylvester, Cobb Va., Nelson	CHR 103235	Hair (1967: 349) as *H. coarctata*
H. hectorii subsp. *demissa*	20	—	Rock and Pillar Ra., Central Otago	CHR 103057!	Hair (1967: 349); as *H. hectorii* var. *demissa*
	—	40	Garvie Mts, Southland	CHR 103269!	Hair (1967: 352); as *H. subulata*
H. imbricata	—	40	Green Lk., Fiordland	CHR 103239[17]	Hair (1967: 350)
	—	40	Old Man Ra., Otago	—	Dawson & Beuzenberg (2000: 6); as *imbricata* subsp. *poppelwellii*
H. insularis	20+f	—	Great Id, Three Kings Ids	CHR 103200!	Hair (1967: 342)
H. leiophylla	40	—	Boulder Lk., NW Nelson	CHR 103243![18]	Hair (1967: 344); as *H. gracillima*
	—	80	Red Hills, Wairau Va., Marlborough	CHR 103230![19]	Hair (1967: 347); as *H. glaucophylla*
H. leiophylla?[20]	—	40, 80	—	—	Frankel in Darlington & Wylie (1955: 311)
H. ligustrifolia	20	—	—	—	Frankel & Hair (1937: 683)[21]
	20	—	Spirits Bay, N Auckland	CHR 103161! (WELT 82217!)	Hair (1967: 344)

15 Not published? Record based on notes on voucher specimen.
16 Resembles specimens from Stewart Id.
17 Voucher has reversion leaves only, so identification is uncertain, although probably correct.
18 Voucher has larger leaves than those usually seen in the species.
19 Voucher has small leaves, but its densely hairy branchlets and small leaf bud sinus support its identification as *H. leiophylla*. *H. leiophylla*, but not *H. glaucophylla*, is otherwise known from the Red Hills Ridge.
20 These counts are probably from two different species.
21 Although there is mention of *H. ligustrifolia* being n = 20 in Frankel & Hair (1937), no voucher details are given and it is not clear whether Frankel & Hair made the count themselves. They included it in their hybridisation study.

Species	n	2n	Locality	Voucher (and duplicates)[1]	Original reference
H. ligustrifolia continued	20	—	Kiripaki [Kiripaka], Whangarei	CHR 103166![22]	Hair (1967: 345); as *H. stricta* var. *stricta*
	—	40	Motu Arohia [Motuarohia] Id, Bay of Ids	CHR 103208!	Hair (1967: 344)
	—	40	Whangarei, N Auckland	CHR 462363!	Dawson & Beuzenberg (2000: 7); as *H.* (indet.; *H.* sp. (m) in Druce 1980 and Eagle 1982; *H.* "Whangarei" in Druce 1993; *H.* "Manaia"
	—	40	N Auckland, North Cape, Surville Cliffs	AK 251839	de Lange & Murray (2002: 5); as *H.* aff. *ligustrifolia*
H. lycopodioides	20	40	Cult. Botanic Gardens, Christchurch	—	Frankel & Hair (1937: 672)
	20	—	Spenser Mts	CHR 103037	Hair (1967: 350); as *H. lycopodioides* var. *patula*
	—	40	Mt Terako, Seaward Kaikouras	CHR 103237	Hair (1967: 350); as *H. lycopodioides* var. *lycopodioides*
	—	40	Mt Torlesse, Canterbury	CHR 103256!	Hair (1967: 350); as *H. lycopodioides* var. *lycopodioides*
	—	40	Waterfall Ck, Craigieburn Mts, Canterbury	CHR 103257!	Hair (1967: 350); as *H. lycopodioides* var. *lycopodioides*
	—	40[23]	Ruahine Ra.[24]	CHR 103034!	Hair (1967: 349); as *H. tetragona*
	—	40	Trovatore Basin, Spenser Mts	CHR 103258!	Hair (1967: 350); as *H. lycopodioides* var. *patula*
	—	40	Mt Technical, Lewis Pass	CHR 103134!	Hair (1967: 350); as *H. lycopodioides* var. *patula*
H. macrantha var. *macrantha*	21	—	Wilberforce R., Canterbury	CHR 103069! (WELT 82209!, 82210!, 82208!)	Hair (1967: 351)
	21	—	Head of Temple R., Lk. Ohau	CHR 103078! (WELT 82218!)	Hair (1967: 351)
H. macrantha var. *brachyphylla*	21	—	Mt Misery, Nelson Lakes District	CHR 103103! (WELT 82219!)	Hair (1967: 351)
H. macrocalyx var. *macrocalyx*	21	—	Temple Basin, Arthur's Pass, Canterbury	CHR 103262!	Hair (1967: 351); as *H. haastii* var. *macrocalyx*
H. macrocalyx var. *humilis*	21	—	Mt Hoaryhead [Hoary Head], NW Nelson	CHR 103073!	Hair (1967: 351); as *H. haastii* var. *humilis*
H. macrocarpa var. *macrocarpa*	40	—	Mt Tamahunga, Rodney County, N Auckland	CHR 103156!	Hair (1967: 345); as *H. macrocarpa* var. *macrocarpa*?
	40	—	Cuttygrass Rd, Waitakere Ra., Auckland	CHR 103154! (WELT 82186!)	Hair (1967: 345); as *H. macrocarpa* var. *macrocarpa*?
	40	—	Cult., Auckland, orig. unknown	CHR 103079! (WELT 82187!)	Hair 1967: 345); as *H. macrocarpa* var. *macrocarpa*?
	40	80	St Paul, Great Barrier Id	AK 251702	de Lange & Murray (2002: 5; as *H. macrocarpa* var. *macrocarpa*
H. macrocarpa var. *latisepala*	60	—	Cult. Auckland	CHR 103006! (CHR 198189!, 198190!)	Hair (1967: 345); as *H. macrocarpa* var. *latisepala*
	—	120	Mt Hobson, Windy Canyon Track, Great Barrier Id	AK 250790!	de Lange & Murray (2002: 5); as *H. macrocarpa* var. *latisepala*
H. macrocarpa var. *latisepala*?	—	120	Mt Kohukohunui, Hunua Ra., Auckland	CHR 103222	Hair (1967: 345); as *H. macrocarpa* var. *latisepala*?

22 CHR 198646! and 198179! appear to be further vouchers (not cited by Hair 1967) of *H. ligustrifolia* (originally identified as *H. stricta*) from Kiripaka with $n = 20$ chromosomes.
23 Published as "$2n = 40$", annotated on the voucher specimen as $n = 20$.
24 Stated locality is probably not correct.

Species	n	2n	Locality	Voucher (and duplicates)[1]	Original reference
H. macrocarpa var.?	—	80	New Zealand	—	Frankel in Darlington & Wylie (1955: 312)
	—	80	Great Barrier Id, Auckland	CHR 462366	Dawson & Beuzenberg (2000: 8); as H. (indet.; H. sp. (w) in Druce 1980; H. "Great Barrier" in Druce 1993)
	40	—	Great Barrier Id, Northern Bush	AK 250801! (WELT 81557, 82580)	de Lange & Murray (2002: 5); as H. aff. macrocarpa var. latisepala
	—	80	Cult., ex. Great Barrier Id, Tryphena, The Needles?[25]	AK 250802!	de Lange & Murray (2002: 5); as H. aff. macrocarpa var. latisepala
	—	120	Te Moehau Ra., S Auckland	AK 251852!	de Lange & Murray (2002: 5); as H. macrocarpa var. latisepala
H. masoniae	59	—	Mt Peel, NW Nelson	CHR 103035!	Hair (1967: 349); as H. pauciramosa var. masoniae
H. matthewsii?[26]	20	—	—	—	Simonet (1934: 1154); as Veronica matthewsii
H. mooreae	63	—	—	—	Hair (1966: 571); as part of the H. odora complex[27]
	63	—	Gouland Downs, NW Nelson	CHR 103040!	Hair (1967: 349); as H. odora sens. lat.
	—	126	Lk. Monk, Fiordland	CHR 174766!	Hair (1967: 349); as H. odora sens. lat.
H. murrellii	—	42	Takahe Va., Murchison Mts, Fiordland	CHR 103260!	Hair (1967: 350); as H. petriei var. petriei
H. obtusata	20	—	Piha, Auckland	CHR 103127!	Hair (1967: 345)
	20	—	Cult. Auckland	CHR 103047!	Hair (1967: 345)
	20	—	Cult. Auckland	CHR 103101	Hair (1967: 345)
H. ochracea	—	124	Mt Peel, NW Nelson	CHR 103039	Hair (1967: 350)
H. odora	21	42	Arthur's Pass	HV 216	Frankel & Hair (1937: 672); as H. buxifolia var. pauciramosa
	—	42	New Zealand	—	Frankel in Darlington & Wylie (1955: 312); as H. buxifolia
	—	84	New Zealand	—	Frankel in Darlington & Wylie (1955: 312); as H. buxifolia var. odora
	—	84	New Zealand	—	Frankel in Darlington & Wylie (1955: 312); as H. buxifolia var. prostrata
	21	—	—	—	Hair (1966: 571); as part of the H. odora complex[27]
	42	—	—	—	Hair (1966: 571); as part of the H. odora complex[27]
	21	—	Ruahine Mts	CHR 103063!	Hair (1967: 348–9); as H. odora
	21	—	Tararua Mts	CHR 103137!	Hair (1967: 348–9); as H. odora sens. lat.
	21	—	Gouland Downs, NW Nelson	CHR 103041!	Hair (1967: 348–9); as H. odora sens. lat.
	21	—	Ranui Cove, Auckland Ids	CHR 103184!	Hair (1967: 348–9); as H. odora sens. lat.
	—	42	Cobb Reserve, NW Nelson	CHR 103186!	Hair (1967: 348–9); as H. odora sens. lat.
	—	42	Jacks Pass, nr Hanmer	CHR 103195!	Hair (1967: 348–9); as H. odora sens. lat.
	—	42	Near Ranui Cove, Auckland Ids	CHR 103198!	Hair (1967: 348–9); as H. odora sens. lat.
	42	—	Twin Cks, Arthur's Pass, Canterbury	CHR 103122!	Hair (1967: 348–9); as H. odora sens. lat.
	42	—	Twin Cks, Arthur's Pass, Canterbury	CHR 103123!	Hair (1967: 348–9); as H. odora sens. lat.

25 Locality uncertain; flowers dark blue.
26 A name placed incertae sedis here. Without a voucher it is not known whether the count is from a plant matching the type of that name, or from another species.
27 The count is mentioned in the text of this paper, but no collection or voucher details are given. It could refer to the count subsequently published by Hair (1967).

Species	n	2n	Locality	Voucher (and duplicates)[1]	Original reference
H. odora continued	42	—	Porters Pass, Canterbury	CHR 103100!	Hair (1967: 348–9); as *H. odora* sens. lat.
	42	—	Head of Clinton R., Fiordland	CHR 103072!	Hair (1967: 348–9); as *H. odora* sens. lat.
	—	84	Mt Anglem, Stewart Id	CHR 103179!	Hair (1967: 349); as *H. odora* var. *prostrata*
	—	84	Stewart Id	CHR 103181!	Hair (1967: 349); as *H. odora* var. *prostrata*
	—	42	Pouakai Ra., Taranki NP, Taranaki	CHR 421294!	Dawson & Beuzenberg (2000: 7, 16); as *H. odora* "diploid"
	—	42	Dew Lakes, Maungatapu, Nelson	CHR 483588	Dawson & Beuzenberg (2000: 7, 16); as *H. odora* "diploid"
	—	42	Upper Boulder Stm, Wairau R. Va., Marlborough	CHR 401679	Dawson & Beuzenberg (2000: 7, 16); as *H. odora* "diploid"
	—	42	Headwaters of Omaka R., Black Birch Ra., Marlborough	CHR 470181!	Dawson & Beuzenberg 2000: 7, 16; as *H. odora* "diploid"
	—	42	Headwaters of Omaka R., Black Birch Ra., Marlborough	CHR 470182!	Dawson & Beuzenberg (2000: 7,16); as *H. odora* "diploid"
	42	—	Arthur's Pass	—[28]	Frankel & Hair (1937: 672); as *H. buxifolia* var. *odora*
	42	—	Maungatua	CHR 74471! (= HV 63)	Not explicitly published? Attributed by Dawson & Beuzenberg (2000: 15) to Frankel & Hair (1937)
	42	—	Mt Oxford	CHR 74476! (= HV 316)	Not explicitly published? Attributed by Dawson & Beuzenberg (2000: 15) to Frankel & Hair (1937)
	42	—	Tims Ck, Castle Hill	CHR 74474! (= HV 183)	Not explicitly published? Attributed by Dawson & Beuzenberg (2000: 15) to Frankel & Hair (1937)
	43	—	Mt Oxford	CHR 74475! (= HV 295)	Not explicitly published? Attributed by Dawson & Beuzenberg (2000: 15) to Frankel & Hair (1937)
	—	c. 84	Island Pass, Nelson	—	Dawson & Beuzenberg (2000: 7, 16); as *H. odora* "tetraploid"
	—	84	Mt Herbert, Banks Peninsula, Canterbury	CHR 483585	Dawson & Beuzenberg (2000: 7, 16); as *H. odora* "tetraploid"
	—	84	Glaisnock Wilderness Area, Fiordland NP	CHR 483586	Dawson & Beuzenberg (2000: 7, 16); as *H. odora* "tetraploid"
	—	84	Mt Burns, Southland	CHR 483584	Dawson & Beuzenberg (2000: 7, 16); as *H. odora* "tetraploid"
	—	c. 84	Blue Mts, Otago	CHR 465612!	Dawson & Beuzenberg (2000: 7, 16); as *H. odora* "tetraploid"
	—	84	Tower Peak, Takitimu Ra., Southland	CHR 483582	Dawson & Beuzenberg (2000: 7, 16); as *H. odora* "tetraploid"
	—	c. 84	S Princess Mts, Lk. Poteriteri, Southland	CHR 483583	Dawson & Beuzenberg (2000: 7, 16); as *H. odora* "tetraploid"
	—	84	Nr summit, Mt Anglem, Stewart Id	—	Dawson & Beuzenberg (2000: 7, 16); as *H. odora* "tetraploid"
	—	84	Nr summit, Table Hill, Stewart Id	CHR 483587	Dawson & Beuzenberg (2000: 7, 16); as *H. odora* "tetraploid"
	—	84	Toitoi Flat, Stewart Id	—	Dawson & Beuzenberg (2000: 7, 16); as *H. odora* "tetraploid"; Beuzenberg in Druce (1980: 41)
H. odora?[29]	40	—	—	—	Simonet (1934: 1145); as *Veronica buxifolia*

28 No voucher is cited, but this could be CHR 74473! (= HV 137), Twin Creek, Arthur's Pass, labelled "n = 42" (an interpretation also accepted by Dawson & Beuzenberg 2000).
29 Either the count or identification should be treated as uncertain.

Species	n	2n	Locality	Voucher (and duplicates)[1]	Original reference
H. paludosa	—	80	Mt Hercules, Westland	AK 231325	Murray & de Lange (1999: 513, 518); Murray in Norton & de Lange (1998: 533, 537)
H. pareora	20	—	Pareora Gorge, S Canterbury	CHR 103210![30]	Hair (1967: 347); as H. amplexicaulis var. amplexicaulis
H. parviflora	—	80	Cult. Christchurch Botanic Gardens	—	Frankel (1940: 172)
	40	—	Kaimanawa Mts	CHR 103004!	Hair (1967: 346); as H. parviflora var. ?
	40	—	Woodside Gorge, Marlborough	CHR 103126!	Hair (1967: 346); as H. parviflora var. ?
	—	80	Karori Hills, Wellington	CHR 103203! (AK 125953!, 125954!)	Hair (1967: 346); as H. parviflora var. arborea
H. parviflora?[31]	20	—	—	—	Simonet (1934: 1154); as Veronica parviflora
H. pauciflora	—	42	Lk. Mike, Dusky Sound, Fiordland	CHR 103247!	Hair (1967: 349)
	—	42	Above Thompson Sound, Fiordland	CHR 103248!	Hair (1967: 349)
H. pauciramosa	21	—	Arthur's Pass, Canterbury	CHR 103050	Hair (1967: 349); as H. pauciramosa var. pauciramosa
	21	—	Canyon Ck, Otago	CHR 103083!	Hair (1967: 349); as H. pauciramosa var. pauciramosa
	—	42	Upper Von R., Eyre Mts, Southland	CHR 483593	Dawson & Beuzenberg (2000: 7, 15)
H. perbella	—	40	Ahipara gumlands, Tanutanu Stm, Northland	AK 231962!	Murray in de Lange (1998: 401); Murray & de Lange (1999: 513, 518)
	—	40	Waima Ra., near Hokianga, Northland	AK 230113!	Murray in de Lange (1998: 401); Murray & de Lange (1999: 513, 518)
	—	40	Kaipara, Maungaraho Rock, Northland	AK 230137!	Murray in de Lange (1998: 401); Murray & de Lange (1999: 513, 518)
	—	40	West of Herekino, N Auckland	CHR 462367!	Dawson & Beuzenberg (2000: 7, 15)
H. petriei	—	42	Tower Peak, Takitimu Ra., Southland	CHR 465672a & b!	Dawson & Beuzenberg (2000: 7, 15)
	—	42	Upper Eyre Ck, Eyre Mts, Southland	CHR 512498![32]	Dawson & Beuzenberg (2000: 7, 15)
H. pimeleoides subsp. pimeleoides	20	40	Cult. Botanic Gardens, Christchurch	HV 244	Frankel & Hair (1937: 672); as H. pimeleoides var. minor
	20	—	Bill Hill, Molesworth, Marlborough	CHR 103116!	Hair (1967: 348)
	20	—	Castle Hill, Canterbury	CHR 103061!	Hair (1967: 348)
	20	—	Lk. Heron, Canterbury	CHR 103136!	Hair (1967: 348)
	20	—	Lk. Heron, Canterbury	CHR 103191!	Hair (1967: 348)
	20	—	Potts R., Canterbury	CHR 103096!	Hair (1967: 348)
	—	40	Cult., orig. unknown	CHR 462377!	Dawson & Beuzenberg (2000: 7, 16); as H. pimeleoides var. glauco-caerulea
	—	40	Lk. Heron, Canterbury	CHR 462378!	Dawson & Beuzenberg (2000: 7, 15–16); as H. pimeleoides var. minor
	—	40	Spider Lks, Canterbury	CHR 462379!	Dawson & Beuzenberg (2000: 7, 15–16); as H. pimeleoides var. minor
	—	80	Spider Lks, upper Ashburton R., Canterbury	CHR 462376!	Dawson & Beuzenberg (2000: 7, 16); as H. aff. pimeleoides

30 Flowers on the voucher are shortly pedicellate (typical for the species), but the peduncle is minutely puberulent (not typical).
31 No voucher is cited, meaning the identity of the plant used cannot be confirmed, but it was probably not H. parviflora given the chromosome number (H. stenophylla is one species likely to be misidentified as that).
32 The cited voucher is probably sterile (although some young inflorescences with many sterile bracts might be present) and thus difficult to identify with certainty. However, the original wild collection from which this material was propagated (CHR 465651; P. J. Garnock-Jones 1905) is certainly H. petriei.

Species	n	2n	Locality	Voucher (and duplicates)[1]	Original reference
H. pimeleoides subsp. *pimeleoides* continued	—	80	Mt John Flats, Mackenzie Basin, S Canterbury	CHR 462375!	Dawson & Beuzenberg (2000: 7, 16); as *H.* aff. *pimeleoides*
	—	40	Boltons Gully, Canterbury	WELT 82449!	de Lange in Kellow et al. (2003a: 253); de Lange et al. (2004c: 882)
	—	40	Spider Lks, Canterbury	WELT 82471!	de Lange in Kellow et al. (2003a: 253); de Lange et al. (2004c: 882)
H. pimeleoides subsp. *faucicola*	20	—	Alexandra, Central Otago	CHR 103036!	Hair (1967: 348)
	40	—	Near Roxburgh, Otago	CHR 103059!	Hair (1967: 348); as *H. pimeleoides* aff. var. *rupestris*
H. pimeleoides var. *glauca-caerulea*[33]	20	—	—	—	Simonet (1934: 1154); as *Veronica pimeleoides* var. *glauca-caerulea*
H. pinguifolia	40	—	Castle Hill	HV 75	Frankel & Hair (1937: 672)
	40	—	Mt Binser, Canterbury	CHR 103070! (WELT 82191!, 82192!)	Hair (1967: 347)
	40	—	Kakanui Mts, Otago	CHR 103075	Hair (1967: 347)
	—	80	Mt Robert, Nelson	CHR 103172![34]	Hair (1967: 352); as *H. carnosula*
	—	40	Mt Hutt, Canterbury	—	Dawson & Beuzenberg (2000: 7); as *H. pinguifolia* "diploid"
H. cf. *pinguifolia*[35]	—	40	Mt Peel, S Canterbury	CHR 102506!	Dawson & Beuzenberg (2000: 7); as *H. pinguifolia* "diploid"; Beuzenberg in Garnock-Jones & Molloy (1983a: 392)
	—	80	Near Chapmans Ck, Mt Somers, Canterbury	CHR 461353	Dawson & Beuzenberg (2000: 7); as *H. pinguifolia* "tetraploid"
	—	80	Summit of Mt Winterslow, Canterbury	CHR 461354	Dawson & Beuzenberg (2000: 7); as *H.* aff. *pimeleoides*
	—	80	Mt Winterslow, Canterbury	CHR 461355	Dawson & Beuzenberg (2000: 7); as *H.* aff. *pimeleoides*
H. propinqua	20	—	Garvie Mts, Southland	CHR 103054!	Hair (1967: 350)
H. pubescens subsp. *pubescens*	20	—	Coromandel Peninsula	CHR 103027!	Hair (1967: 344)
H. pubescens subsp. *rehuarum*	—	40	Fitzroy/Harataonga Rd, Great Barrier Id	AK 250788!	de Lange & Murray (2002: 5); as *H.* aff. *pubescens*
H. pubescens subsp. *sejuncta*	—	40	Motukino (Fanal) Id, Mokohinau Ids	AK 234626!	Murray & de Lange (1999: 513); as *H.* aff. *bollonsii* (A)
	—	40	Tirikakawa Stm, Little Barrier Id	AK 234625!	Murray & de Lange (1999: 513); as *H.* aff. *bollonsii* (B)
	—	40	Mokohinau Ids, N Auckland	CHR 462368!	Dawson & Beuzenberg (2000: 8); as *H.* (indet.; *H.* sp. (v) in Druce 1980 and Eagle 1982; *H.* "Mokohinau" in Druce 1993 and Cameron et al. 1995; *H.* aff. *bollonsii* in de Lange et al. 1999; *H.* aff. *bollonsii* (A) in Murray & de Lange 1999)
H. rakaiensis	40	—	Strathtaeri [Strath Taieri], E Otago	CHR 103030 (*CHR 199113, 199115*)	Hair (1967: 347)
	40	—	Head of Shag R., Kakanui Mts, E Otago	CHR 103088!	Hair (1967: 347)
H. ramosissima	—	42	E Hapuka [Hapuku], Kaikoura Ra., Marlborough	CHR 103261!	Hair (1967: 350)

33 The application of this name is uncertain.
34 Variant with a small sinus.
35 Identity uncertain. These populations require further investigation.

Species	n	2n	Locality	Voucher (and duplicates)[1]	Original reference
H. rigidula var. *rigidula*	20	—	Pelorus R. Bridge, Marlborough	CHR 103065!	Hair (1967: 342)
	—	40	Ben Nevis, Nelson	CHR 103231!	Hair (1967: 342)
H. rigidula var. *sulcata*	—	40	D'Urville Id, Attempt Hill	AK 252335!	de Lange & Murray (2002: 5); as *H.* aff. *rigidula* (b)
	—	40	Rangitoto ki te Tonga (D'Urville Id), Nelson	CHR 462372!	Dawson & Beuzenberg (2000: 7, 17); as *H.* aff. *rigidula*
H. rupicola	20	—	Mt Terako, Seaward Kaikouras	CHR 103008!	Hair (1967: 342)
H. salicifolia	20	—	—	—	Simonet (1934: 1154); as *Veronica salicifolia*
	20	—	Arthur's Pass	HV 52[36]	Frankel & Hair (1937: 672); as *H. salicifolia* var. *communis*
	20	—	Spenser Ra.	CHR 103189! (WELT 82207!)	Hair (1967: 344)
	20	—	Bealey, Canterbury	CHR 103124! (WELT 82156!)	Hair (1967: 344)
	20	—	Banks Peninsula, Canterbury	CHR 103105!	Hair (1967: 344)
	20	—	Lk. Hawea, Central Otago	CHR 103090!	Hair (1967: 344)
	—	40	Cass, Canterbury	CHR 103213	Hair (1967: 344)
	—	40	Mt Hercules, Westland	AK 236066	Murray & de Lange (1999: 513, 518)
	20	—	Mistake Ck, Eglinton Va.	CHR 74428! (= HV 350)	—[15]
	20	—	Nr Dunedin	CHR 74447! (= HV 153)	—[15]
	20	—	Okuku, Fox Ck	CHR 74467! (= HV 118)	—[15]
	20	—	Arthur's Pass	CHR 74427! (= HV 54)	—[15]
H. salicornioides	—	42	New Zealand	—	Frankel in Darlington & Wylie (1955: 312)
	21	—	Jacks Pass, nr Hanmer	CHR 103236!	Hair (1967: 350)
H. scopulorum	—	40	Kawhia district, S Auckland	CHR 177355!	Hair (1967: 342); as *H. rigidula*
	—	40	Awaroa Va., Rock Peak	AK 251134!	de Lange & Murray (2002: 5); as *H.* aff. *rigidula* (a)
H. societatis	—	42	Braeburn Ra., Mt Murchison	WELT 82621!	Bayly et al. (2002: 580); de Lange et al. (2004c: 882)
H. speciosa	20	—	—	—	Simonet (1934: 1154); as *Veronica speciosa*
	20	—	Unknown (cult. Lincoln)	CHR 103098!	Hair (1967: 343)
H. stenophylla var. *stenophylla*	20	—	Buller Gorge, Nelson	CHR 103010![37] (AK 125955!, 125956!)	Hair (1967: 346); as *H. parviflora* var. *angustifolia*
	—	40	D'Urville Id	CHR 103204	Hair (1967: 346); as *H. parviflora* var. *angustifolia*
	—	40	Cult., orig. unknown	CHR 462370![38]	Dawson & Beuzenberg (2000: 7, 16); as *H.* aff. *parviflora* var. *angustifolia*
H. stenophylla var. *hesperia*	—	40	Paturau R. mouth, Nelson	AK 252992[39] (WELT 82639!)	de Lange & Murray (2002: 5)

36 Voucher not seen, but CHR 74427! (= HV 54; see below) is a voucher specimen from the same locality.
37 A specimen of *H. diosmifolia* (CHR 103019) was, prior to Oct. 2003, incorrectly databased with this number.
38 Leaf margins on the voucher specimen are hairy near the apex (not the common condition in this var.), but the identification is supported by the long corolla tubes, which are glabrous inside, and by the long, narrow leaves.
39 Erroneously published by de Lange & Murray (2002) as AK 252966.

Species	n	2n	Locality	Voucher (and duplicates)[1]	Original reference
H. stenophylla var. *oliveri*	—	40	Stephens Id	WELT 82518!	de Lange & Murray (2002: 5)
H. stricta var. *stricta*	20	—	Totara North, N shore, Whangarei Harbour	CHR 103159!	Hair (1967: 345)
	20	—	Head of Kerikeri Inlet, N Auckland	CHR 103164![40]	Hair (1967: 345)
	20	—	Kohukohu, Hokianga Harbour	CHR 103162!	Hair (1967: 345)
	20	—	Waipoua Swamp, N Auckland	CHR 103163!	Hair (1967: 345)
	20	—	Mangawai Gorge, N Auckland	CHR 103165!	Hair (1967: 345)
	20	—	Piha, Auckland	CHR 103128! (WELT 82260!, 82261!)	Hair (1967: 345)
	—	40	Pirongia, S Auckland	CHR 103225![41]	Hair (1967: 345
	20	—	W of Paihia, Bay of Islands	CHR 103160a![42]	Hair (1967: 345); as *H. stricta* group
	40	—	W of Paihia, Bay of Islands	CHR 103160b![42] (WELT 82173!, 82172!)	Hair (1967: 345); as *H. stricta* group
	—	80	William Hewitt Reserve, Pipiwai, Northland	AK 232946 (WELT 82007!)	Murray & de Lange (1999: 513, 518); as *H.* aff. *stricta* (C)
	—	80	Kaniwhaniwha, Mt Pirongia, S Auckland	AK 236442	Murray & de Lange (1999: 513, 518); as *H.* aff. *stricta* (C)
	—	40	Tirikakawa Stm, Little Barrier Id	AK 235237	Murray & de Lange (1999: 513, 518)
	20	—	West coast, Auckland	CHR 74430! (= HV 138)	—[15]
H. stricta?[43]	20	—	Punakaiki	CHR 74432! (= HV 15)	—[15]
H. stricta var. *atkinsonii*	—	40[44]	Pelorus Sound	CHR 74429! (= HV 370)	Frankel (1940: 172); as *H. salicifolia* var. *communis*[45]
	20	—	Picton, Marlborough	CHR 103081!	Hair (1967: 345)
	—	40	D'Urville Id	CHR 103205!	Hair (1967: 345)
	—	40	D'Urville Id	CHR 103206	Hair (1967: 345)
	20	—	Wellington	CHR 74431! (= HV 84)	—[15]
	20	—	Ruamahanga R., Wellington	CHR 74434! (= HV 11)	—[15]
	20	—	Ruamahanga R., Wellington	CHR 74437! (= HV 10)[46]	—[15]
	—	40	Ruamahanga R., Wellington	CHR 74436! (= HV 7)[46]	—[15]
H. stricta var. *egmontiana*	—	80	New Zealand	—	Frankel in Darlington and Wylie (1955: 312); as *H. salicifolia* var. *egmontiana*
	40	—	Mt Taranaki	CHR 103002! (*CHR 198183!*)	Hair (1967: 345)
H. stricta var. *lata*	40	—	Mt Whanakawa, East Cape District[47]	CHR 103182! (*CHR 198184, 198185*)	Hair (1967: 345)

40 Corolla lobes on the voucher are quite long, but it is probably *H. stricta*.
41 Corolla tubes on voucher are short (≈ calyx length)
42 Voucher has hairs on the outside of calyx lobes and long corolla lobes.
43 Voucher looks like *H. stricta*, but the stated locality could not be correct for that species.
44 Count published as "2n = 40", but annotated on voucher specimen as n = 20.
45 Also listed by Dawson (2000) as *H. salicifolia*.
46 Original determination by Frankel was "*H. parviflora* ( ) *salicifolia*".
47 This locality is presumably "Whanokao" on NZMS 260 maps and "Honokawa" on NZMS 1 and NZMS 18 maps.

Species	n	2n	Locality	Voucher (and duplicates)[1]	Original reference
H. stricta var. *macroura*	20	—	Mahia Peninsula	CHR 103007 (AK 126007!; CHR 198181!, 199110!)	Hair (1967: 345)
	20	—	Mahia Peninsula	CHR 103024! (WELT 82174!, 82181!)	Hair (1967: 345)
	—	40	Coast S of Kawhia, S Auckland	CHR 103193	Hair (1967: 345)
	20	40	East Cape, Hicks Bay, Gisborne	AK 246173	de Lange & Murray (2002: 5)
	20	—	Tolaga Bay	CHR 74441! (= HV 17)	—[15]
H. aff. *stricta*	—	80	Hikurangi Swamp, Northland	AK 232218	Murray & de Lange (1999: 513, 518); as *H.* aff. *stricta* (B)
H. strictissima	40	—	Little River Rd, Canterbury	CHR 103099! (WELT 82157!, 82158!; CHR 199112, 199114)	Hair (1967: 347)
	40	—	Church Bay, Banks Peninsula	CHR 74407! (= HV 148)	—[48]
	40	—	Mt Herbert	CHR 74406! (= HV 55)	—[15]
H. subalpina	40	—	Arthur's Pass	HV 49	Frankel & Hair (1937: 672)
	—	80	Arthur's Pass	CHR 74403! (= HV 46)	Frankel (1940: 172)
	—	80	Arthur's Pass	HV 211	Frankel (1940: 172)
	40	—	Arthur's Pass, Canterbury	CHR 103021! (WELT 82178!, 82179!)	Hair (1967: 346)
	40	—	Homer Tunnel, Southland	CHR 103067!	Hair (1967: 346)
	40	—	Hunter R., Central Otago	CHR 103102! (WELT 82169!, 82180!)	Hair (1967: 346)
	40	—	Mistake Ck, Eglinton Va., Southland	CHR 103106!	Hair (1967: 346)
	40	—	Lk. Wanaka	CHR 103001!	Hair (1967: 346); as *H. fruticeti*
	—	80	Garvie Mts, Southland[49]	CHR 103130!	Hair (1967: 346)
H. subalpina?	—	80, 120[50]	New Zealand	—	Frankel in Darlington & Wylie (1955: 312); as *H. subalpina*
H. tairawhiti	40	—	Ahimanu, East Cape District	CHR 103219!	Hair (1967: 345); as *H. stricta* var?
	—	80	Mangaeone [Mangaone] Va., NE of Wairoa	CHR 103271	Hair (1967: 345); as *H. stricta* var?
H. tetragona subsp. *tetragona*	—	40	Desert Rd, Volcanic Plateau	CHR 103249!	Hair (1967: 349)
H. tetragona subsp. *subsimilis*	—	40	Whana Huia [Whanahuia], Ruahine Ra.	CHR 103250!	Hair (1967: 349); as *H. subsimilis* var. *subsimilis*
	—	40	Mt Holdsworth, Tararua Ra.	CHR 103251!	Hair (1967: 349); as *H. subsimilis* var. *astonii*
	—	40	Tararua Ra.	CHR 103252!	Hair (1967: 349); as *H. subsimilis* var. *astonii*

48 Not published? Record based on notes on voucher specimen, which was originally identified as *H. leiophylla*.
49 The voucher specimen has no flowers or fruits, but the leaf shape and size, and the glabrous leaf margins, match *H. subalpina*, a species not otherwise recorded from the Garvie Mts. The voucher is from a plant cultivated at Lincoln from material previously cultivated at Otari. Perhaps there was some confusion as to the wild provenance of this material?
50 The count of 2n = 120 is probably not for *H. subalpina* (most likely *H. brachysiphon*?), but this cannot be ascertained without a voucher specimen.

Species	n	2n	Locality	Voucher (and duplicates)[1]	Original reference
H. topiaria	61	—	Amuri Pass, Nelson	CHR 103013! (WELT 82197!, 82198!)	Hair (1967: 347)
H. townsonii	20	—	Hills N of Westport	CHR 74445! (HV 206)	Not published? Or part of the collection cited by Frankel (1940: 172)?
	—	40	Hills near Westport	HV 157	Frankel (1940: 172)
	20	—	Mt Messenger, Taranaki[51]	CHR 103064! (WELT 82176!, 82177!)	Hair (1967: 343)
	20	—	Hills near Westport	CHR 103062! (WELT 82175!; CHR 198166!)	Hair (1967: 343)
H. traversii	20	—	Lees Va., Canterbury	CHR 103125!	Hair (1967: 346)
	20	—	Lees Va., Canterbury	CHR 103169	Hair (1967: 347)
	20	—	Avoca, Canterbury	CHR 103016!	Hair (1967: 347)
	20	—	Ashley Gorge, Canterbury	CHR 103132!	Hair (1967: 347)
	—	40	Limestone Ck, Awatere Va., Marlborough	CHR 103212[52]	Hair (1967: 346)
	—	40	W of Clarence Reserve, Marlborough	CHR 103157!	Hair (1967: 346)
	20	—	Fox Ck, Okuku, Canterbury	CHR 74416! (= HV 113)	—[15]
	20	—	Lk. Rotoiti, Nelson	CHR 74417! (= HV 374)	—[15]
	20	—	Lk. Rotoiti, Nelson	CHR 74411! (= HV 107)	—[15]
	20	—	Ashley Gorge, Canterbury	CHR 74412! (= HV 105)	—[15]
	20	—	Castle Hill, Canterbury	CHR 74408! (= HV 209)	—[15]
H. traversii?[53]	60	—	—	—	Simonet (1934: 1154); as *Veronica traversii*
H. traversii?[54]	—	40, 80, 120	New Zealand	—	Frankel in Darlington & Wylie (1955: 312); as *H. traversii*
H. treadwellii	—	40	Bald Knob Ridge, NW Nelson	—	Dawson & Beuzenberg (2000: 7); as *H.* aff. *treadwellii* (*H.* "Bald Knob Ridge" in Druce 1993 and Cameron et al. 1995)
	20	—	Mts above Amuri Pass, Nelson	CHR 103028! (WELT 82196!, CHR 199162!)	Hair (1967: 352); as *H. brockiei*
H. truncatula	40	—	Ruahine Mts	CHR 103155!	Hair (1967: 346)
H. urvilleana	—	120	Collins Va., Nelson	CHR 103211	Hair (1967: 346)
H. venustula	—	120	New Zealand	—[55]	Frankel in Darlington & Wylie (1955: 312); as *H. laevis*
	60	—	Desert Road, Volcanic Plateau	CHR 103201!	Hair (1967: 342)
	60	120	Mt Ruapehu	CHR 103026!	Hair (1967: 342)
	60	—	Tararua Ra.[56]	CHR 103049!	—[15]

51 Locality unlikely. See notes under this species in Taxonomic Treatment.
52 Voucher not found, but CHR 171716!, from the same locality, is labelled "= G 6624, Diploid".
53 This count is probably not from *H. traversii* (most likely *H. brachysiphon*?), but this cannot be ascertained without a voucher specimen.
54 Counts of 2n = 80 and 120 are probably not from *H. traversii*, but this cannot be ascertained without voucher specimens.
55 No vouchers are cited, but these could be CHR 74443! (= HV 377) and 74444!, both of which are labelled "n = 60".
56 Locality incorrect? *H. venustula* is not otherwise known from the Tararua Ra.

Species	n	2n	Locality	Voucher (and duplicates)[1]	Original reference
H. vernicosa	21	—	Nelson	HV 13	Frankel & Hair (1937: 672); as H. vernicosa var. gracilis
	21	—	Mt Hoaryhead [Hoary Head], NW Nelson	CHR 103014!	Hair (1967: 343)
	21	—	Mt Piripiri, Marlborough Sounds	CHR 103092!	Hair (1967: 343)
	—	42	Travers Ra., Nelson	CHR 103144	Hair (1967: 343)
Leonohebe					
L. cheesemanii	—	42	Headwaters Omaka R., Mt Altimarlock [Altimarloch], Marlborough	CHR 174933	Hair (1967: 352); as Hebe cheesemanii
L. ciliolata	21	42	Arthur's Pass	HV 212–13	Frankel & Hair (1937: 672); as Hebe ciliolata
	—	42	Victoria Ra., W Nelson	CHR 103241!	Hair (1967: 351); as Hebe ciliolata
L. cupressoides	21	42	Cult. Christchurch Botanic Gardens	—	Frankel & Hair (1937: 672); as Hebe cupressoides
	21	—	Grandview, Grandview Ra., upper Clutha	CHR 103187!	Hair (1967: 350); as Hebe cupressoides
L. tetrasticha	—	42	Castle Hill	HV 82	Frankel & Hair (1937: 672); as Hebe tetrasticha
	—	42	Cult. Botanic Gardens, Christchurch	CHR 103263	Hair (1967: 351); as Hebe tetrasticha
	—	42	Craigieburn Mts, Canterbury	CHR 103264!	Hair (1967: 351); as Hebe tetrasticha
L. tumida	—	42	Ben Nevis, Nelson	CHR 103232	Hair (1967: 352); as Hebe tumida

Reproductive Biology
P. J. Garnock-Jones

There have been several detailed studies of reproductive biology in *Hebe* and *Leonohebe*. In particular, Lynda Delph (1988, 1990*a*, 1990*b*, 1990*c*; Delph & Lively 1992; Delph & Lloyd 1991) used *Hebe* as a model to study the evolution and maintenance of gynodioecy, listed breeding systems for many species, and provided considerable detail for *H. stricta*, *H. subalpina* and *H. strictissima*. Seed morphology (Webb & Simpson 2001) and germination (Simpson 1976) are quite well known. However, we still know very little about the breeding systems of many species, their pollinators, the development of their flowers, and the timing of anther dehiscence and stigma receptivity, and there are no detailed comparative studies on fruit structure and function.

FLOWER BIOLOGY

The flowers of all species of *Hebe* are based on the same overall structure and they function in similar ways, but there is nevertheless considerable variation within the genus, both in the form of the flowers and in the details of their functioning.

Kampny (1995) and Kampny & Dengler (1997) reviewed pollination and flower diversity in Scrophulariaceae *s. lat.*, which includes many genera now treated in Plantaginaceae, such as *Hebe*. From their descriptions, *Hebe* and *Leonohebe* lack many of the specialised features found in other genera, including closed corollas, feeding anthers, nectar guides (present in *Parahebe*; Garnock-Jones 1976*b*; Garnock-Jones & Lloyd 2004) and nectar spurs. *Hebe* and its relatives, it seems, thus retain generalised pollination systems, but among them there is plenty of variation.

Corolla

Most species of *Hebe* and *Leonohebe* have pale or white flowers, compared with the frequent blue colour in northern hemisphere *Veronica* (Lloyd 1985; Wardle 1991). Exceptions include *H. benthamii*, endemic to the Auckland and Campbell islands, which has blue flowers, and a few mainland species that have strongly coloured corollas (e.g. magenta in *H. speciosa*, magenta to deep violet in *H. brevifolia*, and deep violet in some populations of *H. macrocarpa*).

Scent and Nectar

Most *Hebe* flowers are not noticeably scented (Garnock-Jones 1992, 1993*a*), although some species with clustered small flowers, such as *H. bishopiana* (de Lange 1996), *H. adamsii*, *H. stricta* var. *stricta*, *H. stenophylla* (P. J. de Lange pers. comm. 2005), *H. petriei*, and *H. murrellii* (Delph 1988; Garnock-Jones & Clarkson 1994) may have a sweet scent (see also Thomson 1881). *Leonohebe* sect. *Leonohebe* and most "Connatae" have a stale scent (Delph 1988; Garnock-Jones 1992) resembling that of ivy (*Hedera*) flowers. Some plants of *H. speciosa* have a fruity scent, reminiscent of peaches (Garnock-Jones 1992). Hooker (1844) described a "delicious fragrance" in *H. odora* flowers, but I have not noticed this.

Hebe flowers generally produce small amounts of visible nectar from a small nectarial disc at the base of the ovary. There have been no detailed studies of nectar, but some plants of *H. speciosa* produce copious quantities (M. Bayly pers. comm. 2005), and Thomson (1881) reported nectar production in plants he referred to as *V. traversii*, *V. buxifolia*, *V. salicifolia* and especially in *V. elliptica*.

Flower Phenology

Many northern, lowland species seem to flower in winter to early spring (e.g. *H. macrocarpa*, *H. pubescens* and *H. flavida*), whereas southern and alpine species are spring- and summer-flowering (e.g. *H. canterburiensis*, *H. hectorii* subsp. *coarctata*, *H. stenophylla* and *L. ciliolata*; Simpson 1976). In *H. odora*, flower buds are initiated later than March (Mark 1970), and flowering in many cultivated hebes is initiated by winter cooling (Noack et al. 1996). In *H. strictissima* (studied intensively by Delph 1990*b*), individual plants flower for two to five weeks between December and February, with most inflorescences on a plant opening at the same time. Capsules then develop quickly and dehisce in autumn, although a few capsules develop more slowly. Primack (1983) reported that flowers of *H. epacridea* lasted on average 3.5 days, and *H. pinguifolia* 3.2 days. There are no comparable data for other species. Records from cultivated plants and herbarium specimens might give a false impression of the timing and duration of flowering, respectively (since plants may flower out of season in cultivation, and herbarium collectors may actively seek out flowering specimens in populations where most plants are not flowering).

Dichogamy and Herkogamy

All species of *Hebe* exhibit mechanisms that separate the production and receipt of pollen in time (dichogamy) or space (herkogamy) or both (Lloyd & Webb 1986; Webb & Lloyd 1986; Delph 1990*a*). Most species of *Hebe* are protandrous (anther development and dehiscence preceding stigma receptivity, but with some overlap) (Thomson 1881; Garnock-Jones & Molloy 1983*b*; Delph 1988, 1990*a*) whereas protogyny (stigma receptivity preceding anther dehiscence) is more common in related plants (Kampny 1995). In *Hebe* "Connatae", however, stigmas are presented beyond the anthers throughout the life of the flower (Delph 1990*a*), which might indicate protogyny. In several species of *Hebe*, the stigma is tucked within a folded or rolled anterior corolla lobe at anthesis (e.g. **Fig. 32**), and does not pull free and move upwards until later in the flower's life (Delph 1988). In some species, the anthers move away from the centre of the flower as the style elongates (e.g. **Figs 79G** and **85E**) to present the stigma centrally (Delph 1988, 1990*a*), or anthers are abscised (Lloyd & Webb 1986), thus avoiding interference between the two functions of pollen dissemination and receipt.

Self-incompatibility and Inbreeding Depression

Delph (1988, 1990*a*) reported self-compatibility in *Hebe*, but some studies are suggestive of self-incompatibility (Frankel & Hair 1937; de Lange & Cameron 1992) and inbreeding depression (Garnock-Jones & Molloy 1983*b*, Delph & Lloyd 1996). In natural populations, outbreeding varies, from 9 per cent in *Hebe subalpina* to 50 per cent in *H. traversii* (Delph 1988).

FIG. 32 Flowers of *Hebe divaricata*. The stigma is tucked into the folded anterior calyx lobe (arrow) on young flowers.

FLOWER VISITORS AND POLLINATORS

Hebe flowers are visited by a wide range of unspecialised and promiscuous insects (Thomson 1927; Heine 1937; Dugdale 1975; Primack 1978, 1983; Delph 1988, 1990*b*; de Lange & Cameron 1992). Beetles and flies, especially syrphids, collect mainly pollen, whereas bees visit the flowers for both pollen (female bees) and nectar, and tachinid flies and Lepidoptera (moths and butterflies) primarily collect nectar (Delph 1988). Flowers of *H. speciosa* sometimes attract birds, but are thought to be insect-pollinated (de Lange & Cameron 1992). Overall, small native solitary bees (*Lasioglossum*, *Leioproctus* and *Hylaeus*) and flies (especially Tachinidae and Syrphidae) are the most numerous flower visitors (Primack 1983; Thomson 1927), particularly at low altitudes, but both tachinid flies and beetles are more common visitors at subalpine and alpine altitudes (Delph 1988). In the study by Delph (1988), isozyme data showed that bee pollination in *H. traversii* achieved higher levels of outcrossing than fly and beetle pollination in *H. subalpina*. It seems possible that the lack of bees at high altitudes may have led to the higher frequency of gender dimorphism in *Hebe* and *Leonohebe* there (Delph 1990*a*), through the greater fitness of the outcrossed progeny that result.

Some morphological differences between species, especially short (e.g. *H. subalpina*) and long (e.g. *H. decumbens* and *H. albicans*) corolla tubes, might be related to different pollinators (see Kampny 1995 for a general discussion of flower shapes in relation to pollinators for related genera).

REPRODUCTIVE BIOLOGY

GENDER

Hebe and *Leonohebe* populations can be cosexual or gender dimorphic. In cosexual populations, both sexes are combined in every individual (in *Hebe* all flowers of cosexual populations are hermaphrodite). In dimorphic populations, individuals specialise as either males or females. Complete gender specialisation (where plants reproduce only as either males or females) is termed dioecy, whereas populations with both female plants and where some or all male-fertile plants produce fruit are termed gynodioecious (see Webb et al. 1999 for detailed definitions). Male-fertile plants of gynodioecious species, although technically hermaphrodite, often produce reduced amounts of fruit when compared with females of the same species. In addition, they are the pollen parents of all offspring in the populations, and thus their reproductive fitness is primarily derived from their male sex function (Delph & Lloyd 1991; Delph & Lively 1992).

The existence of separate sexes and of sexual dimorphism in *Hebe* was first reported by Hooker (1864). Frankel & Hair (1937) and Frankel (1940) subsequently reported male sterility in some species, specifically citing (between them) plants identified as *H. subalpina*, *H. traversii*, *H. parviflora*, *H. salicifolia* var. *communis* and *H. townsonii*, as well as a hybrid. Later, the gender conditions of many species were reported for the first time by Delph (1988, 1990*a*). The following sections briefly summarise her findings, and other recent studies.

Taxonomic Distribution of Gender Dimorphism

No gender dimorphism has been found in "Flagriformes", "Grandiflorae" or *Leonohebe* sect. *Aromaticae*. All species of "Connatae" (gynodioecious) and *Leonohebe* sect. *Leonohebe* (dioecious) are dimorphic in gender. Groups where some species are dimorphic and others cosexual are "Apertae" (both small- and large-leaved), "Occlusae", "Buxifoliatae" and "Subcarnosae". The last two groups were recorded as cosexual by Delph (1990*a*), but recent observations suggest that at least some populations of *H. odora* ("Buxifoliatae") (E. M. L. Low pers. comm. 2005) and *H. pinguifolia* ("Subcarnosae") (e.g. near Lake Lyndon, WELT 81677; **Figs 71E–H**) include male-sterile plants. There has been no detailed study of gender in "Pauciflorae".

Differences between Males and Females in Dimorphic Populations

In most dimorphic species of *Hebe*, and in most flowering plants in general, male-functioning plants produce flowers with larger corollas (e.g. **Fig. 33**) and anthers than females do. In gynodioecious species, flowers on male-fertile plants have functional ovaries and set fruit, although they usually set fewer fruit when compared with flowers on female plants. Flowers on female plants produce fruit, but not pollen, and females invest more resources than male-functioning plants in gynoecium and fruit production. Overall, female allocation to reproduction can be nearly twice that of male-functioning plants (Delph 1990*c*; Delph & Lively 1992). For example, even though flowers on hermaphrodite plants of *H. subalpina* are larger than those on female plants (**Fig. 33**), and seed produced from self-pollination is equal in size to the seed produced from outcrossing, females, on average, produce in excess of three times more fruit than hermaphrodites, over seven times more seed, and their seed fitness is nine times greater (Delph & Lloyd 1991).

FIG. 33 Hermaphrodite (left) and female inflorescences of *Hebe subalpina*.

Sex Ratios

Delph (1988) recorded sex ratios for fourteen species of *Hebe* and *Leonohebe*. In these species, sex ratios close to 1:1 were found in dioecious species and in those gynodioecious species that had very low fruit set on hermaphrodite plants. In gynodioecious species where hermaphrodites set substantial amounts of fruit, the sex ratio had a significant excess of hermaphrodites.

In gynodioecious *H. strictissima* and *H. subalpina*, fruit set on hermaphrodite plants, but not on female plants, is highest in vigorous individuals (Delph 1990*b*, 1990*c*). This leads to high proportions of hermaphrodites in favourable sites where females cannot out-compete hermaphrodites in seed production.

Alpine species of *Hebe* are more likely than lowland ones to have gender dimorphism (Delph 1990*a*). In addition, lowland dimorphic species more often have an excess of hermaphrodites, and some populations have lost females altogether, whereas alpine species have higher numbers of females. Similarly, in many species in *Hebe* "Occlusae" that have gynodioecious populations, there is a range of seed set on hermaphrodites and a sex ratio that has an excess of hermaphrodite individuals, a situation that differs only slightly from cosexuality (Delph & Lloyd 1991). In *H. subalpina*, however, the sex ratio is 1:1 and hermaphrodites often set few seeds, a situation that approaches dioecy.

Evolution of Gender Dimorphisms

Delph (1988, 1990*a*) suggested that gender is inherited partly cytoplasmically in *H. brachysiphon* and *H. stricta*, and that nuclear genes restore male fertility in some offspring of female plants in *H. stricta*. She suggested that evolution of dioecy in the *Hebe* complex followed the commonly proposed gynodioecy pathway (cosexual–gynodioecious–subdioecious–dioecious), although a clear transition along this path is not evident in phylogenetic treatments (Wagstaff & Garnock-Jones 1998; Wagstaff et al. 2002), and that gynodioecy has been retained for a long time in some lineages. She also hypothesised that separate sexes (either dioecy or gynodioecy) evolved a minimum of four times within the *Hebe* complex from the ancestral condition of cosexuality. Phylogenetic analyses (Wagstaff & Garnock-Jones 1998; Wagstaff et al. 2002) also support multiple origins of gender dimorphism, although, given limits of taxon sampling and poor resolution of some relationships, the exact number is not clear.

FRUIT AND SEED BIOLOGY

Fruits

All species of *Hebe* have xerochastic (dry dehiscent) capsule-opening (Garnock-Jones 1993a) that is only weakly reversible by wetting. In some species, the capsules are borne on recurved pedicels so they hang downwards and disperse seeds by gravity (e.g. *H. salicifolia*; Simpson 1976); in most, the capsules are erect, and shaking by wind is necessary to remove the seeds (e.g. *H. diosmifolia* and *H. pinguifolia*). In many species, seeds disperse in a few days after ripening (Simpson 1976), although capsules stay on the plants for up to a year, becoming pale and eroded with time.

Most *Hebe* capsules open at the apex by septicidal splits that divide the septum along the junction of the two carpels, and by loculicidal splits that split the carpels, at least partly, down their midribs (**Fig. 45**). The carpel margins may also become detached from each other apically, which splits the septum. The septum contracts as the capsule dries, pulling apart the opening of each carpel into an apical spout, through which the seeds are shaken.

Seeds

The small size and flattened shape of the seeds (**Fig. 47**) suggests effective wind dispersal, although Simpson (1976) questioned its effectiveness in sheltered habitats. In the case of two species with transoceanic distributions, dispersal in mud adhering to the feet, legs, or plumage of migratory birds was suggested by Godley (1967).

Simpson (1976) reported seed ripening times of three-and-a-half months in *H. salicifolia* and three months in *H. elliptica*, and these times are probably similar in most species. She reported that ripe seed was available in the wild during the months of March and April for many species. Seed set was reported as good, with no damage by fungal pathogens and little by insect predators.

Hebe seeds exhibit epigeal germination, and germination requirements for a large number of species of *Hebe*, *Leonohebe*, *Heliohebe* and *Parahebe* were studied by Simpson (1976). A group of lowland and some alpine species germinated freely at 25°C in light. Another group, of mostly alpine species, responded to cooler temperatures (10°C) or to cold pre-treatment. *Hebe salicifolia* and *H. elliptica* seeds lost viability rapidly approaching two years of age; *H. stricta* lost viability more gradually between one and two years of age. Other species exhibited gradual declines in viability over time, or seemed to germinate better at certain times of the year.

Conservation

P. J. de Lange

CONSERVATION STATUS

The conservation status of New Zealand native plants is regularly assessed by teams of invited specialists (e.g. de Lange et al. 1999, 2004*a*, 2004*b*) using the criteria of the New Zealand threat classification system (de Lange & Norton 1998; Molloy et al. 2002; **Fig. 34**). This system was developed to identify clearly New Zealand conservation priorities. It is distinct from the system of the International Union for the Conservation of Nature (IUCN 1994, 2000), which obscures conservation priorities by not adequately differentiating between taxa under immediate human-induced threat and taxa that are naturally geographically restricted or uncommon, but not necessarily threatened (de Lange & Norton 1998). Taxa in this last category are numerous in the geographically isolated and highly endemic New Zealand flora.

Despite the fact that *Hebe* is the largest indigenous genus of New Zealand plants, few taxa within it qualify as seriously threatened (de Lange et al. 2004*a*, 2004*b*). Just nine species (10 per cent) of *Hebe* and one of *Leonohebe* are listed as "Acutely Threatened" (de Lange et al. 2004*a*, 2004*b*), and only *Hebe pimeleoides* subsp. *faucicola* is rated as "Chronically Threatened (Gradual Decline)" (**Table 6**). Outside these higher categories of threat, a further thirty-four *Hebe* species, subspecies, varieties and forms, and one *Leonohebe* species, *L. tumida*, are considered "At Risk" (**Table 6**). Taxa in this last category are not directly threatened, but their listing is an indication that they require frequent monitoring to ensure

FIG. 34 New Zealand threat classification system (after Molloy et al. 2002). Boxes denote categories in the classification.

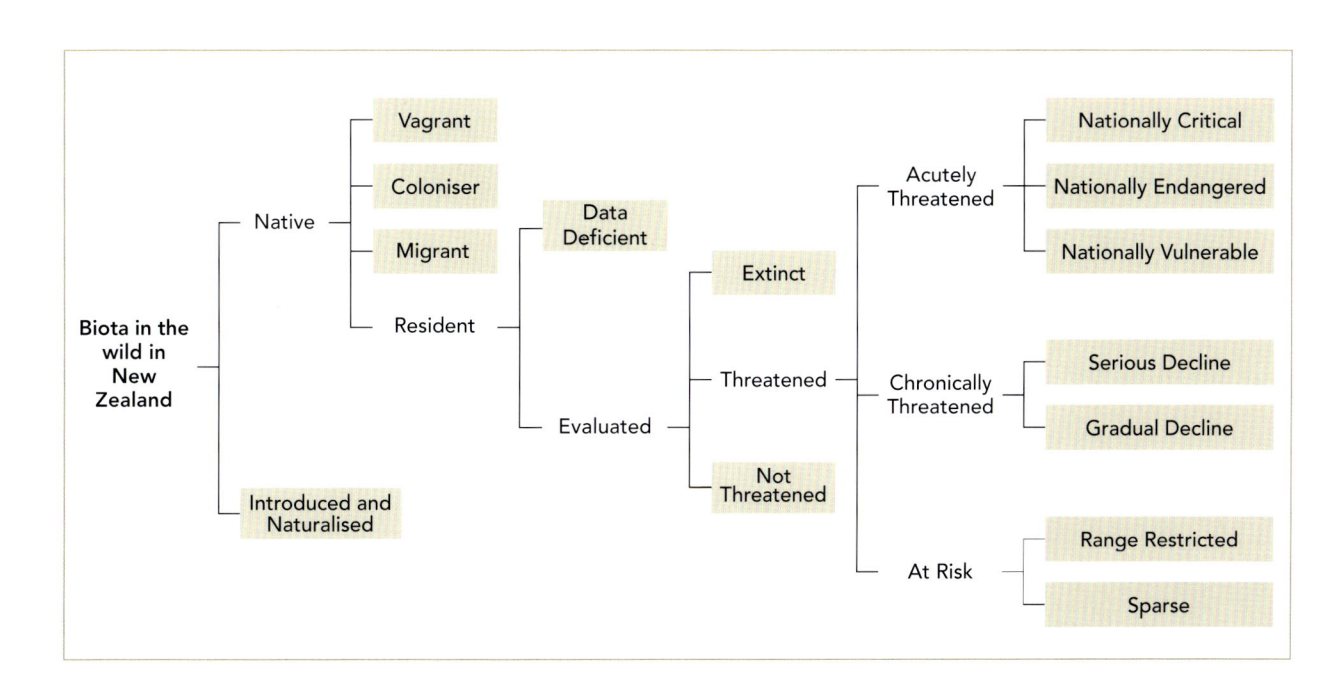

TABLE 6 Threatened and uncommon *Hebe* and *Leonohebe* (after de Lange et al. 2004a, 2004b)

Acutely Threatened

Nationally Critical
- *H. breviracemosa*
- *H. societatis*

Nationally Endangered
- *H. armstrongii*
- *H. salicornioides*
- *H. speciosa*

Nationally Vulnerable
- *H. barkeri*
- *H. bishopiana*
- *H. perbella*
- *H. scopulorum*
- *L. cupressoides*

Chronically Threatened

Gradual Decline
- *H. pimeleoides* subsp. *faucicola*

At Risk

Sparse
- *H. annulata*
- *H. dilatata*
- *H. tairawhiti*

Range Restricted

H. acutiflora	*H. dieffenbachii*	*H. pubescens* subsp. *rehuarum*
H. adamsii	*H. elliptica* var. *crassifolia*	*H. pubescens* subsp. *sejuncta*
H. amplexicaulis f. *amplexicaulis*	*H. evenosa*	*H. ramosissima*
H. amplexicaulis f. *hirta*	*H. gibbsii*	*H. rigidula* var. *rigidula*
H. arganthera	*H. imbricata*	*H. rigidula* var. *sulcata*
H. benthamii	*H. insularis*	*H. stenophylla* var. *hesperia*
H. biggarii	*H. macrocalyx* var. *macrocalyx*	*H. stenophylla* var. *oliveri*
H. brevifolia	*H. obtusata*	*H. townsonii*
H. calcicola	*H. ochracea*	*H. urvilleana*
H. carnosula	*H. pareora*	*L. tumida*
H. chathamica	*H. pauciflora*	

Taxonomically Indeterminate

Nationally Critical
- *H.* aff. *bishopiana* (AK 202263; Hikurangi Swamp) (= *H.* aff. *stricta*)

Nationally Endangered
- *H.* aff. *albicans* (AK 252966; Mt Burnett) (= *H. albicans*)

Sparse
- *H.* aff. *diosmifolia* (AK 215221; tetraploid) (= *H. diosmifolia*)

Range Restricted
- *H.* aff. *ligustrifolia* (AK 207101; Surville Cliffs) (= *H. ligustrifolia*)
- *H.* aff. *pinguifolia* (CHR 461354; "high flyer") (*incertae sedis*)

Data Deficient
- *H. matthewsii* (*incertae sedis*)
- *H.* aff. *brevifolia* (AK 235669; Surville Cliffs) (*incertae sedis*)
- *H.* aff. *treadwellii* (CHR 394533; Bald Knob Ridge) (= *H. treadwellii*)

that their populations remain stable. Obviously, should circumstances change for the worse, they are strong candidates for a higher threat rating.

Acutely Threatened

In terms of immediate conservation action, only those species rated as "Acutely Threatened" are considered management priorities. Of these, six species occupy lowland or offshore island habitats (indeed two, *Hebe breviracemosa* and *H. barkeri*, are island endemics), and three are confined to montane habitats, where they grow in damp sites on forest margins or amongst tall tussocks (*H. salicornioides*), in grey scrub (*Leonohebe cupressoides*) or amongst bog pine on poorly drained intermontane floodplains (*H. armstrongii*). Only one species, the recently described and apparently very localised *H. societatis* (Bayly et al. 2002), occurs in low penalpine grassland.

Among lowland species of the larger New Zealand islands, *H. bishopiana*, *H. perbella* and *H. scopulorum*, all rated as "Nationally Vulnerable", are confined to rock tor, cliff and bluff habitats in lowland forest, while the spectacular magenta-flowered *H. speciosa* is confined to coastal sites, where it grows only on steep slopes and cliff faces. The first two of these lowland "cliff dwellers" are placed at high risk because they occur in widely scattered small populations that are actively threatened by weed invasion and browsing animals. For the Waitakere endemic *H. bishopiana*, weeds posed another unexpected risk; until very recently, routine roadside weed control operations themselves were threatening *H. bishopiana* populations at the type locality. The third species, *H. scopulorum*, is probably the least common of the three, and also the most seriously threatened. It is known from ten limestone outcrops (at only six of which it is reasonably abundant) south of Kawhia Harbour, where it is at risk from the spread of Mexican daisy (*Erigeron karvinskianus*) and from browsing by goats. Goats, pigs and possums are also threatening the species through their combined browsing pressure on the adjoining lowland forest, which is now on the verge of collapse. The loss of the forest has greatly altered the cool bluff climate *H. scopulorum* is thought to require, and has also facilitated the spread of light-demanding weeds such as Mexican daisy onto the limestone bluffs, where they compete with young seedlings of *H. scopulorum*. Only two populations of *H. scopulorum* occur within reserves.

H. speciosa, though well known in cultivation, particularly as an important progenitor of many highly coloured cultivars, is even more seriously threatened. Formerly recorded from fourteen sites spanning the west coast of New Zealand from Scott Point, Northland to the Marlborough Sounds, it now survives at six locations, two of which are in serious decline despite active management. Recent genetic fingerprinting of the remaining wild populations suggests that only three of them, one near Muriwai, one at Maunganui Bluff and one at South Head, Hokianga Harbour, are truly natural, and that the rest of the extant populations, and very probably all the recently extinct ones, stem from deliberate Māori plantings of this unusual magenta-flowered species (Armstrong & de Lange 2005). It is now believed that these plantings lack the genetic diversity necessary to persist without direct human intervention. The situation is made worse by the fact that the remaining three "natural" populations are not secure; two are very small, and one has declined markedly through the effects of coastal erosion. All known populations are subjected to competition from invasive weeds and browsing animals.

The Chatham Island endemic *H. barkeri* is unusual within New Zealand hebes as one of the few truly arborescent species. The early writings on the vegetation of the Chatham Islands by pioneering New Zealand plant ecologist Leonard Cockayne indicate that *H. barkeri* was once widespread across the island (Cockayne 1902). It was an important component of the canopy of interior swamp forests, which developed away from drying coastal winds on thin lenses of peat associated with exposures of basalt, limestone and schist rocks. In these forests it often grew in association with *Brachyglottis huntii*, within canopy gaps, clearings and around lake margins. Although still highly threatened by browsing animals, and virtually extinct over the northern two-thirds of the main island, it remains a conspicuous tree of some forest remnants along the southern tablelands, notably the Tuku Nature Reserve, and at Rangaika Reserve. There, despite the strong gales, sizeable populations have been discovered as canopy emergents within dense *Dracophyllum arboreum/Myrsine chathamica/Olearia chathamica* forest clinging to the cliff sides of that reserve. Although it is now actively managed, *H. barkeri* is still at risk on the two main Chatham islands from cattle, sheep and possum browsing, inadequate reservation, and recruitment failure within isolated plants and/or populations outside protected areas.

The other island endemic, *H. breviracemosa*, from Raoul Island, is the most seriously threatened of the species discussed here. Even by the time it was discovered and described as a new species of *Veronica* by Oliver (1910), it was already scarce due to goat browsing. Subsequent field surveys during the 1960s and 1970s failed to find any plants, so it was listed as extinct (Given 1981). The situation changed when in 1983 a single seedling was discovered by a goat hunter (de Lange & Stanley 1999). Further plants were discovered in 1997 on a nearby cliff face, increasing the known population to around fifty wild plants. Despite these new discoveries and the observation that seed was easily germinated, the species remained seriously threatened through recruitment failure. The possibility that kiore (Polynesian rat; *Rattus exulans*) might be responsible for this failure by eating seedlings, although never resolved, seems to be the answer. With the eradication of kiore in 2002, seedlings began to appear in the vicinity of wild and planted specimens. In 2004, the world population of this species stood at 160 individuals.

In contrast to the coastal, lowland or offshore island endemics, the recently described *H. societatis* (Bayly et al. 2002) is a penalpine species. At present it is considered highly threatened because it is known only from one very small population. This population, while not directly threatened, is placed at risk by feral pigs, which root up the carpet grass (*Chionochloa australis*) through which *H. societatis* grows.

The remaining "Acutely Threatened" species, the two whipcords *H. armstrongii* and *H. salicornioides*, together with *L. cupressoides*, are placed at risk through a combination of habitat loss, animal browsing, weed competition, recruitment failure and uncontrolled fire. Of the three, *H. armstrongii* is probably the most seriously threatened. This species was long believed extinct in the wild, and was known only from garden material purportedly gathered from the headwaters of the Rangitata River, until a natural population was discovered during the 1970s within a bog pine (*Halocarpus bidwillii*) remnant near Castle Hill village, Arthur's Pass. That population, despite intensive conservation management, has never thrived, probably as a consequence of water loss from the bog pine remnant. This has allowed the spread of aggressive introduced grasses, such as brown-top (*Agrostis capillaris*), which prevent the establishment of *Hebe armstrongii* seedlings. Fortunately, another more extensive population has been discovered in the Nigger

Valley, in and around some alpine tarns. These plants, also associated with bog pine, are in better condition, and some recruitment has occurred. However, even here the population remains at risk from the spread of brown-top, potential changes in the surrounding catchment hydrology, cattle browsing, feral pig rooting and fire.

Fire has probably caused the major historical declines of *L. cupressoides* and *H. salicornioides*. In particular, *L. cupressoides*, whose pungent blue-grey foliage smells of terpenes, is highly flammable wet or dry. Once recorded from the upper Wairau River to Queenstown (Widyatmoko & Norton 1997), the species is now known only from fragmented remnants around the Boyle and Henry rivers in north Canterbury, Castle Hill Basin, at Lake Lyndon in mid-Canterbury, and near lakes Tekapo and Pukaki in south Canterbury. Further south, there are several large populations in the upper catchment of the Shotover River, and some smaller stands occur in the Remarkables near Lake Wakatipu. Recruitment failure remains a major problem at virtually all sites. In contrast, *H. salicornioides*, while a more local endemic, is probably less threatened, though recruitment failure remains a problem throughout its range. It prefers wet ground, flushes and bogs, and is usually found amongst red tussock (*Chionochloa rubra* subsp. *occulta*) with *H. pauciramosa*. The species is now virtually confined to the upper Clarence and Wairau rivers, and smaller streams and bogs near Lake Tennyson.

Chronically Threatened

The only hebe in this category is *H. pimeleoides* subsp. *faucicola*. This endemic of the Manuherikia, Clutha and Kawarau river gorges in central Otago is locally common. However, plants are often seriously damaged by browsing animals and, while there have been no detailed studies, field observations suggest that recruitment failure might be a serious problem in some populations.

At Risk

Thirty-four hebes and one *Leonohebe* are listed as "At Risk". From a conservation perspective, while none of these taxa is actively threatened, they are treated as "At Risk" because they occupy very small portions of the country, or they are never particularly common, making them especially vulnerable to any change to their environment. At present, two subcategories of "At Risk" are distinguished: "Sparse" and "Range Restricted". Three *Hebe* species are recognised as sparse, and thirty-one as range restricted. A good example of a sparse species is *H. dilatata*, which is restricted to a few southerly mountain ranges where it is never very common (Bayly et al. 2002). There are a number of situations that qualify a species as range restricted. For example, *H. insularis*, although very common, is endemic to the Three Kings Islands, northwest of Cape Reinga, while *H. carnosula* is restricted by its geological tolerances to ultramafic rocks of the Red Hills, Marlborough and Dun Mountain, Nelson. A further example, *H. adamsii*, whilst not constrained by geology, is confined to cliff faces in lowland forest at Te Paki, Northland. In this case the species is uncommon simply because the number of suitable relatively open cliff-face habitats is restricted by surrounding forest vegetation.

Aside from these main conservation listings there are a further eight hebes (**Table 6**) treated as "Taxonomically Indeterminate" by de Lange et al. (2004*a*, 2004*b*). The threatened plant panel has taken a different view of the taxonomic status of these entities to that adopted by the authors of this book, and they are not discussed further here.

CONSERVATION MANAGEMENT

The most intensively managed species are *Hebe armstrongii*, *H. breviracemosa*, *H. bishopiana* and *Leonohebe cupressoides*. At present, two of these, *H. bishopiana* and *L. cupressoides*, are actively managed by recovery plans (de Lange 1999; Norton 2000) administered by the Waitakere City Council and DoC, respectively. The Raoul Island endemic *H. breviracemosa* and Chatham Island endemic *H. barkeri* are actively managed as part of those islands' overall vegetation and threatened species restoration objectives. One population of *H. armstrongii* has been the subject of more than twenty years of intensive research and management initiated by the former DSIR Botany Division and continued by DoC. Management for these species includes weed control, habitat restoration, replanting of existing populations and the translocation of plants to secure sites, as well as ongoing research into the species' autecology, threat mitigation and habitat requirements. Of the remaining threatened species, only *H. perbella* is not specifically part of some management programme.

Though in its infancy, study of gene-flow within and between populations of *Hebe* species has started to help determine minimum population sizes for future translocations, and to provide a guide toward assessing the long-term genetic stability of what are often highly disjunct, fragmented populations. At this stage, studies of the reproductive biology and capabilities of the more highly threatened *Hebe* species are urgently needed before a truly holistic conservation management programme can be initiated. Those species whose range has been reduced to very small, widely scattered populations or individuals (e.g. *H. barkeri* and *H. breviracemosa*) remain a priority for this type of research. However, to help set a standard, studies of those species with apparently naturally small populations (e.g. *H. adamsii* and *H. scopulorum*) are also needed.

Cultivation

Apart from their diversity in form, hebes are also popular horticultural subjects because most species are generally easy to grow. As a group they are tolerant of a wide range of soil, light and temperature conditions, although they usually don't do well in full shade, and different species have different requirements. Most species grow readily from cuttings, the most widely used method of propagation.

Aspects of cultivation are largely outside the scope of this book, but detailed information can be found in the works of Metcalf (1987), Chalk (1988), Hutchins (1997) and Wheeler & Wheeler (2002). Additional information can also be obtained from the Hebe Society, based in the United Kingdom. The society has produced a booklet on the cultivation of hebes and also publishes a quarterly magazine, *Hebe News*. Contact details can be obtained from the organisation's website (www.hebesoc.vispa.com) and from Hutchins (1997: 298).

FIG. 35 Potted hebes in a Wellington garden centre.

PART B

Identification and Description of Species

Materials and Methods

SCOPE OF SURVEY

This book treats all species, subspecies and varieties recognised in *Hebe* and *Leonohebe*. The descriptions and notes for these are based on the study of herbarium specimens, field observations and observations of cultivated plants. Herbarium specimens at WELT, CHR and AK were used most extensively, but some specimens from AKU, BAB, BM, BISH, C, CANU, CHBG, K, MEL, NZFRI, OTA, P, SGO, WAIK and WELTU were also examined (herbarium abbreviations follow Holmgren et al. 1990). A list of representative specimens for each taxon is available in hard copy at AK, CHR, OTA and WELT, and in electronic form from the authors on request. A large number of the specimens examined have also been annotated with determinavit or confirmavit slips. Field observations and collections were made throughout most of New Zealand, including Stewart and Chatham islands; **Fig. 36** provides an indication of the geographic coverage of collections made on the three main islands. Material from many of the field collections, and some obtained from other botanists, was propagated and grown in the authors' gardens for further study. Cultivated plants, mostly of known provenance, were also studied in gardens at Otari-Wilton's Bush (Wellington), Percy's Reserve (Lower Hutt) and on the grounds of Landcare Research (Lincoln).

CONCEPTS OF SPECIES AND INFRASPECIFIC TAXA

Different botanists have different opinions on how the various ranks of species, subspecies, variety and forma should be used. Much has been written on the matter, and many different "species concepts" are discussed in biological literature. At the taxonomic coalface, theory and practice are often mismatched. Botanists often need to make taxonomic decisions without detailed knowledge of the evolutionary relationships of taxa, and usually apply criteria based on morphological resemblance, which is most readily observed.

Here, taxa are mostly defined such that they are morphologically recognisable. An underlying assumption (except for taxa at the rank of form, discussed below) is that the differences in morphology used for taxonomic recognition generally reflect underlying reproductive or evolutionary processes; in other words, that most taxa are distinct biological entities or lineages. In some cases, further support for this assumption has been gained from studies of leaf flavonoids or chromosome number; morphological variation correlated with these independent traits allows greater confidence that differences reflect evolutionary change, rather than random variation.

Infraspecific ranks are used only where it seems likely that taxa "belong together" and are closely related. The ranks of subspecies and variety are often more or less interchangeable as used here. In general, we would argue, given the order of the taxonomic hierarchy, that varieties should be less distinct or more likely to intergrade than subspecies, but such things are difficult to quantify. Because of both the uncertainty about some patterns of geographic variation, and the differing views on how infraspecific ranks should be used, we have avoided changing the ranks of infraspecific taxa in this book; we have generally opted to apply existing infraspecific names, thus minimising the number of new names or combinations published here. The rank of forma is used only in *H. amplexicaulis*, where forma names already existed (Garnock-Jones & Molloy 1983*a*) to describe variation in indumentum that potentially represents variation in a single gene, and does not reflect the divergence of evolutionary lineages.

FIG. 36 Map showing the geographic spread of field collections of *Hebe* and *Leonohebe* made by the authors on North Island, South Island and Stewart Island, between 1996 and 2002 (generated from data on herbarium specimens at WELT).

This treatment includes several exceptions to the general guidelines presented here on taxon recognition. Notable among these are the circumscriptions of *H. crenulata* and *H. cryptomorpha*, and of *H. stricta* var. *macroura*, which are discussed in detail in the notes for these species.

MORPHOLOGICAL DESCRIPTIONS

Data for species descriptions were stored in an electronic database using the DELTA system (Dallwitz et al. 1993). This database holds details of morphological character states and the species in which they were found. Written descriptions were produced automatically from the database, but with some secondary editing, automated using a series of custom-written macros in Microsoft Word, to remove awkward or redundant phrases. The characters on which the descriptions are based are listed in **Table 7**. Descriptions of genera and infraspecific taxa were written manually, but compiled using data in the DELTA dataset. Descriptions were generally compiled from subsets of the specimens examined, chosen to encompass most of the variation in each species.

ARRANGEMENT OF SPECIES

Species are arranged into informal infrageneric groups, as discussed in the chapter "Classification and Evolution". The order of these groups follows that of the synopsis (see below), which was constructed with identification (and not phylogenetic relationships) in mind. Within groups, species are arranged in systematic order, with similar species placed close together. Obviously, a linear sequence of species won't reflect all patterns of similarity accurately – for example, some species are about equally similar to several others (which can't all be placed next to one another), while some differ markedly from all others (but still need to be placed somewhere in the sequence). The order used here is adopted primarily to assist comparison and identification. It reflects a general impression of which species are similar, and might be closely related, but has not been the subject of detailed analysis.

TREATMENT OF HYBRIDS

A large number of hybrid combinations have been recorded both in wild (e.g. Cockayne & Allan 1934) and cultivated hebes (Hutchins 1997; Metcalf 2001). Cultivated forms are mostly outside the scope of this book, and wild hybrids are not comprehensively dealt with. Although many previous records of wild hybrids are soundly based, at least some reflect imprecise concepts of species, and subsequent revisionary work (e.g. Moore, in Allan 1961) has shown them to be in error. Identifying hybrids (and their parents), particularly from herbarium specimens, is not straightforward, and a comprehensive list of verified hybrids has not been attempted here. Some hybrid combinations, mostly those about which we have first-hand knowledge, are mentioned in the notes under parent species. Likewise, botanical names that are known or suspected to be based on hybrids are indicated in the nomenclatural part of this book.

DISTRIBUTION MAPS

Distribution maps are based on verified herbarium specimens (*c*. 9000 in total), the details of which, including spatial coordinates, are available in the list of representative specimens (see above). When possible, locality data on herbarium labels were estimated or approximated to the nearest minute of latitude and longitude (even when more accurate coordinates were available on specimens). Approximate latitudes and longitudes were converted to other formats – for example, New Zealand Map Grid references (NZMS 260) – as required, using the New Zealand Map Grid Reference Conversion Utility for Windows (Pickard 1996*a*). These data were initially mapped and tested using MapInfo Professional Version 5.5 or Distribution Plotter for Windows (Pickard 1996*b*). The final maps for this book, based on the same data, were prepared by GeographX (NZ) Ltd (Wellington).

In general, no distinction is made between historical locality records and those supported by recently collected specimens. This is primarily because it was not practical to verify the current occurrence of all species at all recorded localities. There are a few historical collections that are considered anomalous or potentially erroneous in their locality information. In most cases these records are omitted from the distribution maps, and this is mentioned in the notes provided under the species descriptions. Where current (as opposed to historical) species' distributions have been the subject of intensive study, as with *L. cupressoides* (Widyatmoko & Norton 1997) and *H. speciosa* (de Lange & Cameron 1992), references are provided to the relevant works, but these are usually not discussed in detail, and their information is not reflected in the distribution maps.

FLOWERING TIMES

Data on flowering times are derived from verified herbarium specimens, supplemented by some photographic records (when the identity of the plant and time of the photo are known with certainty). Unless indicated otherwise (by dates in square brackets), these data relate only to plants in wild populations, not to those in cultivation.

PHOTOGRAPHY AND IMAGING

The images used to illustrate species were obtained using a range of techniques and equipment. Habit photos of plants were mostly taken with an Olympus OM-2n camera, fitted with 50 or 28 mm lenses, using Fujichrome Sensia film (of various speeds). Images of plant sprigs (those generally in the top right-hand corner of each species plate), and some of inflorescences and infructescences, were obtained using a Hewlett Packard ScanJet 5100C flatbed scanner. For these images, plant material was placed directly on the scanner platen and surrounded by a box painted black inside, basically as described by Malcolm & Garnock-Jones (2000). Macro photographs were taken mostly indoors through an Olympus zoom dissection microscope. Specimens were posed against a matt-black background and illuminated with a remote flash unit. All but a very few of these images were recorded on Fujichrome film rated at ASA 100, with the remainder recorded digitally.

Credits for photographs and images are given on page 380. Details of voucher specimens are listed in Appendix 5.

Scale bars are included on a few images in the plate for each species, which should be sufficient to give an indication of the size of the leaves, flowers and fruits. Scale bars on digital scans and sprigs are accurate, but those on photographs taken by Bill Malcolm are estimates. Scale bars for his photos were calculated from measurements of leaves and flowers (in particular calyces, which do not shrink much on drying) on his voucher specimens. Average measurements were used where there was substantial variation on a specimen.

TABLE 7 Characters used in morphological descriptions of species. States shown in *italics* are scored as implicit in the DELTA dataset; these are generally common states and are omitted from the descriptions of many species.

Habit and form

1. Habit: 1. subshrub; 2. spreading low shrub; 3. bushy shrub; 4. small tree (see Glossary for details).
2. Up to (height): __ m tall.
3. General form: 1. of whipcord form; 2. of semiwhipcord form. Scored only for whipcords and semiwhipcords.

Branches

4. Branches (habit): 1. prostrate; 2. spreading; 3. decumbent; 4. ascending; 5. erect; 6. pendent (Fig. 165).
5. Old stems (colour): 1. brown; 2. red-brown; 3. black; 4. grey. Scored only for non-whipcords.
6. Branchlets (colour): 1. brown; 2. red-brown; 3. black; 4. green; 5. grey; 6. yellowish; 7. purplish; 8. orange; 9. red. Scored only for non-whipcords; for youngest portion of branchlets only.
7. Branchlets (hairiness): 1. puberulent; 2. pubescent; 3. *glabrous* (Fig. 18).
8. Branchlet hairs (distribution): 1. bifarious; 2. uniform; 3. *absent* (Fig. 18).
9. Internodes (length): __ mm long. Measured between any pair of successive leaves or leaf scars. For whipcords and semiwhipcords, this is estimated by halving the distance between the apices of two adjacent leaves lying in the same plane (Fig. 37).
10. Leaf decurrencies (prominence): 1. obscure; 2. evident; 3. swollen; 4. evident and extended for length of internode. Scored for non-whipcords only.
11. Branchlets, including leaves, (diameter): __ mm wide. Scored only for whipcords and semiwhipcords, from non-woody branchlets, excluding the apical portions of very youngest branchlets. Measured across the widest point of any leaf pair (Fig. 38).
12. Branchlets (profile in TS): 1. square in TS (e.g. Figs 156B and D); 2. cruciform in TS (e.g. Figs 157B and D). Scored for semiwhipcords only.
13. Connate leaf bases: 1. hairy; 2. glabrous. Scored only for whipcords and semiwhipcords (instead of character for branchlet hairs).
14. Nodal joint (distinct or obscure): 1. distinct; 2. obscure (Fig. 24). Scored only for whipcords. It records whether the nodal joint is distinctly marked (usually by a line or change of colour), or whether the leaves are more or less confluent with the stem. In some species the nodal joint, though distinct, may be hidden by the apices of the leaf pair below (see next character).
15. Nodal joint (hidden or exposed): 1. exposed (e.g. Fig. 53C); 2. hidden (e.g. Fig. 55C). Scored only for whipcords. It records whether the nodal joint (which can be distinct or obscure – see previous character) is hidden by the apices of the leaf pair below (common where leaves are tightly overlapping) or exposed (seen when leaves are more widely spaced on stem).
16. Leaves (abscission): 1. *abscising at nodes*; 2. abscising above nodes and lower part of petioles remaining attached to stem; 3. not readily abscising and persistent along the stem for some distance.

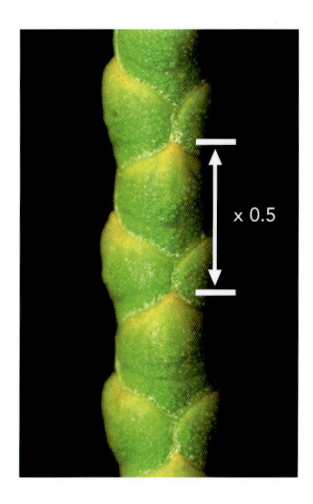

left
FIG. 37 Distance measured (and then halved) to determine internode length in whipcords and semiwhipcords.

right
FIG. 38 Distance measured as branchlet width in whipcords and semiwhipcords.

Leaf bud and sinus

17 Leaf bud (distinctiveness): 1. distinct (e.g. Fig. 20A); 2. tightly surrounded by recently diverged leaves (e.g. Fig. 20B–D).
18 Sinus (shape; Fig. 21): 1. absent; 2. narrow and acute; 3. broad and acute; 4. square to oblong; 5. small and rounded; 6. broad and shield-shaped; 7. small and acute.

Leaves

19 Leaves (arrangement): 1. *decussate* (e.g. Figs 106B and 118B); 2. subdistichous (e.g. Fig. 119B).
20 Leaves (fusion of bases): 1. connate (e.g. Figs 53C, 62E and F); 2. *free at base*.
21 Leaves (posture; Fig. 168): 1. appressed; 2. erect; 3. erecto-patent; 4. patent; 5. recurved; 6. reflexed.
22 Lamina (shape; Fig. 166): 1. linear; 2. lanceolate; 3. oblanceolate; 4. ovate; 5. obovate; 6. oblong; 7. elliptic; 8. circular; 9. deltoid; 10. spathulate; 11. rhomboid; 12. semicircular; 13. sub-circular.
23 Lamina (thickness): 1. thin; 2. rigid; 3. subcoriaceous; 4. coriaceous; 5. fleshy.
24 Lamina (folding): 1. flat; 2. concave; 3. m-shaped in TS.
25 Lamina (length): __ mm long. Measured on mature leaves.
26 Lamina (width): __ mm wide. Measured at the widest point on a leaf, using only mature leaves.
27 Lamina (apical thickening; Fig. 48): 1. thickened near the apex; 2. not thickened near the apex. Scored only for whipcords.
28 Apex (of lamina, shape; Fig. 167): 1. acute; 2. subacute; 3. obtuse; 4. rounded; 5. mucronate; 6. subapiculate; 7. apiculate; 8. plicate; 9. retuse; 10. truncate; 11. acuminate.
29 Base (of lamina, shape): 1. *cuneate*; 2. truncate; 3. subcordate; 4. cordate; 5. amplexicaul.
30 Venation (of leaves): 1. *evident in fresh leaves*; 2. not evident in fresh leaves.
31 Evident venation in fresh leaves including: 1. *midrib only*; 2. two secondary laterals arising from base (as well as midrib); 3. brochidodromous secondary veins (as well as midrib). Only scored for species with evident leaf veins.
32 Midrib (thickening): 1. *depressed to grooved above and thickened below*; 2. thickened above; 3. thickened below; 4. thickened above and below; 5. not thickened; 6. depressed to grooved above.
33 Margin (of leaves, cartilaginous): 1. cartilaginous; 2. not cartilaginous.
34 Margin (of leaves, thickened): 1. not thickened; 2. thickened. Scored only for "Connatae", to distinguish the thickened margins of *H. epacridea* (Figs 61A and 62E) from those of other taxa.
35 Margin (of leaves, indumentum): 1. glabrous; 2. minutely papillate; 3. puberulent; 4. ciliolate; 5. ciliate; 6. glandular-ciliate; 7. pubescent.
36 Margin (of leaves, colour): 1. tinged red; 2. *not tinged red*.
37 Margin (of leaves, dissection): 1. *entire*; 2. shallowly toothed; 3. deeply toothed; 4. distantly denticulate; 5. minutely crenulate (Fig. 39).
38 Upper surface (of leaves, colour): 1. green; 2. glaucescent; 3. glaucous; 4. bronze-green; 5. light green; 6. yellowish-green; 7. dark green.
39 Upper surface (of leaves, gloss): 1. glossy; 2. dull; 3. vernicose (thick glossy cuticle, e.g. *H. pauciflora*).
40 Upper surface (of leaves, stomata): 1. without evident stomata; 2. with few stomata; 3. with many stomata.
41 Upper surface (of leaves, pitted): 1. *not pitted*; 2. sparsely pitted (often only along margins) with small depressions that each contain a twin-headed glandular hair; 3. conspicuously pitted with small depressions that each contain a twin-headed glandular hair (e.g. Fig. 77B). See notes (below) for similar character for lower leaf surface.
42 Upper surface (of leaves hairs): 1. glabrous; 2. hairy along midrib; 3. hairy toward base; 4. covered with minute glandular hairs; 5. uniformly eglandular pubescent; 6. covered with a mixture of eglandular and glandular hairs.
43 Lower surface (of leaves, domatia): 1. with a regular series of short oblique domatia just within margin (e.g. Figs 149D and E); 2. *without domatia*.
44 Lower surface (of leaves, colour): 1. green; 2. glaucescent; 3. glaucous; 4. bronze-green; 5. light green; 6. yellowish-green; 7. dark green; 8. pinkish.
45 Lower surface (of leaves): 1. with prominent shallow veins that give a ribbed or striped appearance; 2. veins not visible. Scored only for whipcords.
46 Lower surface (of leaves, gloss): 1. glossy; 2. *dull*.
47 Lower surface (of leaves, stomata): 1. with few stomata; 2. *with many stomata*.
48 Lower surface (of leaves, pitted): 1. *not pitted*; 2. faintly pitted with small depressions that each contain a twin-headed glandular hair; 3. conspicuously pitted with small depressions that each contain a twin-headed glandular hair (e.g. Fig. 77D). A hand lens (at least) or dissecting microscope is required to see pits (and the hairs are us. not readily visible). In fresh leaves pits are most evident under oblique light, and are often more visible toward leaf margins. They are more obvious on dried leaves.
49 Lower surface (of leaves): 1. *glabrous*; 2. hairy along midrib; 3. hairy toward base; 4. covered with minute glandular hairs; 5. uniformly eglandular pubescent; 6. covered with a mixture of eglandular and glandular hairs.

FIG. 39 Leaf margins. **A** entire; **B** shallowly toothed; **C** deeply toothed; **D** distantly denticulate; **E** minutely crenulate.

50   Petiole (length): __ mm long. Only scored for taxa with conspicuous petioles.
51   Petiole (hairs): 1. glabrous; 2. hairy along margins; 3. hairy above; 4. hairy below; 5. hairy above and below.

Juvenile and reversion leaves

52   Juvenile leaves (shape): 1. pinnatifid; 2. incised; 3. serrate; 4. crenate; 5. denticulate; 6. entire.
53   Juvenile leaves (indumentum): 1. glabrous; 2. ciliate; 3. ciliolate; 4. pubescent; 5. puberulent.
54   Reversion leaves (shape): 1. pinnatifid; 2. incised; 3. serrate; 4. crenate; 5. denticulate; 6. entire.
55   Reversion leaves (indumentum): 1. glabrous; 2. ciliate; 3. ciliolate; 4. pubescent; 5. puberulent.

Inflorescences

56   Inflorescences with (flower number): __ flowers. Scored on intact inflorescences showing no sign of damage. This can also be scored from infructescences, counting all scars left by detached flowers or fruits.
57   Inflorescences (position; Fig. 40): 1. terminal; 2. lateral; 3. terminal and lateral.
58   Inflorescences (branching): 1. unbranched; 2. tripartite (e.g. Fig. 126G, left); 3. with more than three branches (e.g. Fig. 137H).
59   Inflorescences (length): __ cm long. Measured on inflorescences with at least some open flowers (and including those with mature fruit); younger inflorescences can be considerably shorter. For species with short, dense inflorescences (e.g. semiwhipcords and "Connatae"), this length includes the floral parts of apical flowers.
60   Inflorescences (length cf. leaves): 1. shorter than subtending leaves; 2. about equal to subtending leaves; 3. *longer than subtending leaves.*

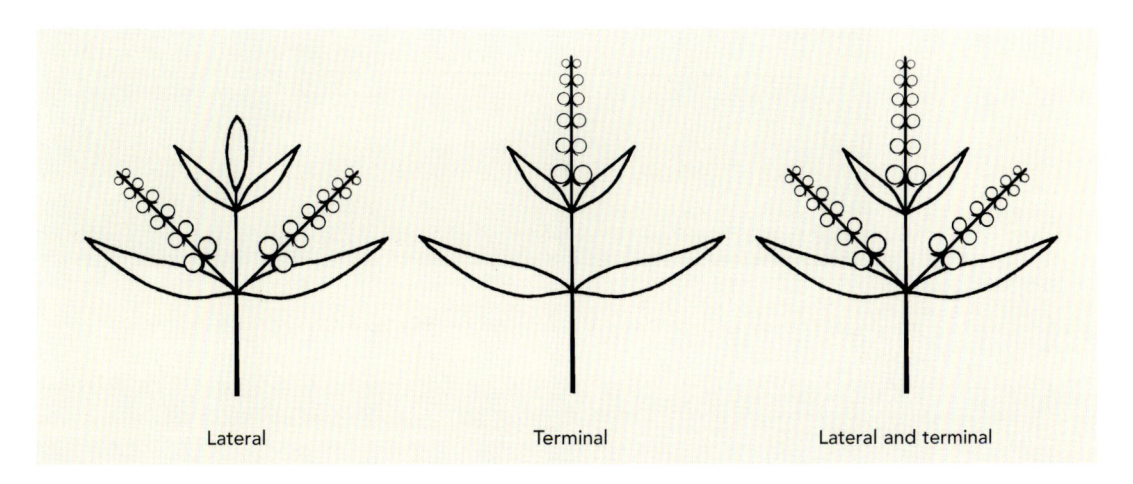

FIG. 40 Inflorescence positions. (Note that inflorescences are all shown as spikes, but they are often racemes, and may be simple or branched.)

61 Inflorescences (apex): 1. *with all flowers (including those near the apex) generally developing to maturity*; 2. sometimes with a conspicuous number of unopened flowers toward the apex (leaving a protruding "rat's tail").
62 Peduncle (length): __ cm long. Measured on inflorescences with at least some open flowers (and including those with mature fruit); younger inflorescences can be considerably shorter.
63 Peduncle (indumentum): 1. *hairy*; 2. glabrous.
64 Rachis (length): __ cm long. Measured on inflorescences with at least some open flowers (and including those with mature fruit); younger inflorescences can be considerably shorter.
65 Rachis (indumentum): 1. *hairy*; 2. glabrous.
66 Bracts (arrangement): 1. opposite and decussate; 2. alternate (spiralled); 3. opposite and decussate below, becoming alternate above; 4. alternate except for a basal whorl of three; 5. subopposite to alternate; 6. lowermost pair opposite, then subopposite or alternate above.
67 Bracts (separation at base): 1. connate; 2. *free*. Scored for species with opposite bracts; otherwise inapplicable.
68 Bracts (shape): 1. linear; 2. lanceolate; 3. elliptic; 4. ovate; 5. deltoid; 6. narrowly deltoid; 7. oblong; 8. obovate; 9. sub-circular; 10. semicircular; 11. oblanceolate.
69 Bracts (apex): 1. obtuse; 2. subacute; 3. acute; 4. acuminate; 5. apiculate; 6. subapiculate.
70 Bracts (indumentum of margins): 1. margins glabrous; 2. *margins hairy*.
71 Bracts (indumentum of outer surface): 1. *glabrous outside*; 2. hairy outside. This refers only to the outer surface, not to the margins.

Flowers

72 Flowers (when produced): __. Text character indicating flowering months. See notes on page 80.
73 Flowers (sexuality): 1. male or female (on different plants); 2. hermaphrodite or female (on different plants); 3. hermaphrodite.

Pedicels

74 Pedicels (length cf. bracts): 1. *varying from shorter than to longer than bracts (either on different plants or on the same plant)*; 2. always shorter than bracts; 3. shorter than or equal to bracts; 4. longer than or equal to bracts; 5. always longer than bracts; 6. absent. Scored only from open flowers or fruit; pedicels can be shorter on unopened buds.
75 Pedicels (length): __ mm long. Scored only from open flowers or fruit; pedicels can be shorter on unopened buds.
76 Pedicels (indumentum): 1. glabrous; 2. *hairy*.
77 Pedicels (posture at fruiting): 1. *not recurved in fruit*; 2. recurved in fruit.

FIG. 41 Lateral view of flower showing distances measured for calyx length and anther length.

Calyx

78 Calyx (length): __ mm long. Measured on mature flowers or fruits (not unopened buds); it represents total length (Fig. 41), from apex of lobes to the apex of pedicel (i.e. including the fused portion at base of the calyx).
79 Calyx (number of lobes): 1. 2-lobed; 2. 3-lobed; 3. *4-lobed*; 4. 5-lobed; 5. 6-lobed.
80 Calyx (fusion of anterior lobes): 1. *with anterior lobes free for most of their length*; 2. with anterior lobes united to one-third the way to apex; 3. with anterior lobes united to two-thirds the way to apex; 4. with anterior lobes united to apex.

81 Calyx (fusion of posterior lobes): 1. *with posterior lobes free for most of their length*; 2. with posterior lobes united.
82 Calyx lobes (shape): 1. linear; 2. lanceolate; 3. elliptic; 4. ovate; 5. deltoid; 6. oblanceolate; 7. obovoid; 8. oblong; 9. obovate.
83 Calyx lobes (apex): 1. acuminate; 2. acute; 3. subacute; 4. obtuse; 5. emarginate.
84 Calyx lobes (indumentum of margins): 1. glandular ciliate; 2. glandular ciliolate; 3. eglandular ciliate; 4. eglandular ciliolate; 5. margins glabrous; 6. *with mixed glandular and eglandular cilia*. Requires detailed examination. Glandular hairs can usually be seen with a dissecting microscope, but confirmation that they are lacking usually requires examination with a compound microscope.
85 Calyx lobes (indumentum of outer surface): 1. hairy outside; 2. *glabrous outside*. This refers only to the outside of calyx lobes, not to the basal part of the calyx or the lobe margins.

Corolla

86 Corolla tube (indumentum): 1. hairy outside; 2. hairy inside; 3. glabrous. Hairs can generally be seen in fresh flowers with a hand lens or dissecting microscope. Confirmation that hairs are lacking (particularly on herbarium specimens) usually requires examination with a compound microscope.
87 Corolla tube (length):[1] __ mm long. Measured on open flowers, mostly from fresh material (Fig. 42A). It is best measured on detached corollas, so the base of the tube is not obscured by the calyx. Tubes are often slightly longer on the anterior side (most pronounced in taxa with short tubes); values include this range of variation.
88 Corolla tube (width):[1]; __ mm wide. Measured on open flowers, mostly from fresh material (Fig. 42B).
89 Corolla tube (shape):[1] 1. cylindric; 2. funnelform; 3. contracted at base; 4. contracted at mouth; 5. expanded in lower half.
90 Corolla tube (length cf. calyx):[1] 1. shorter than calyx; 2. equalling calyx; 3. longer than calyx.
91 Corolla lobes (number): 1. *four*; 2. five; 3. six.
92 Corolla lobes (colour at anthesis): 1. white at anthesis; 2. pale blue at anthesis; 3. blue at anthesis; 4. magenta at anthesis; 5. pink at anthesis; 6. violet at anthesis; 7. mauve at anthesis.
93 Corolla lobes (colour with age): 1. *white with age*; 2. pale blue with age; 3. blue with age; 4. magenta with age; 5. pink with age; 6. violet with age; 7. mauve with age; 8. pale pink with age.
94 Corolla lobes (shape): 1. circular; 2. elliptic; 3. oblong; 4. lanceolate; 5. linear; 6. rhomboid; 7. ovate; 8. obovate; 9. deltoid; 10. oblanceolate.
95 Corolla lobes (apex): 1. acute; 2. subacute; 3. obtuse.
96 Corolla lobes (posture): 1. erect; 2. suberect; 3. patent; 4. recurved.
97 Corolla lobes (length cf. tube): 1. longer than corolla tube; 2. equalling corolla tube; 3. shorter than corolla tube.
98 Corolla lobes (indumentum): 1. *glabrous*; 2. ciliate; 3. ciliolate; 4. hairy outside; 5. hairy inside; 6. papillate outside; 7. papillate inside; 8. with a few hairs toward base on inner surface. Examination for papillae used a compound microscope (e.g. Fig. 86G).
99 Corolla throat (colour): 1. *white*; 2. yellow; 3. pink; 4. magenta; 5. violet; 6. blue; 7. mauve.

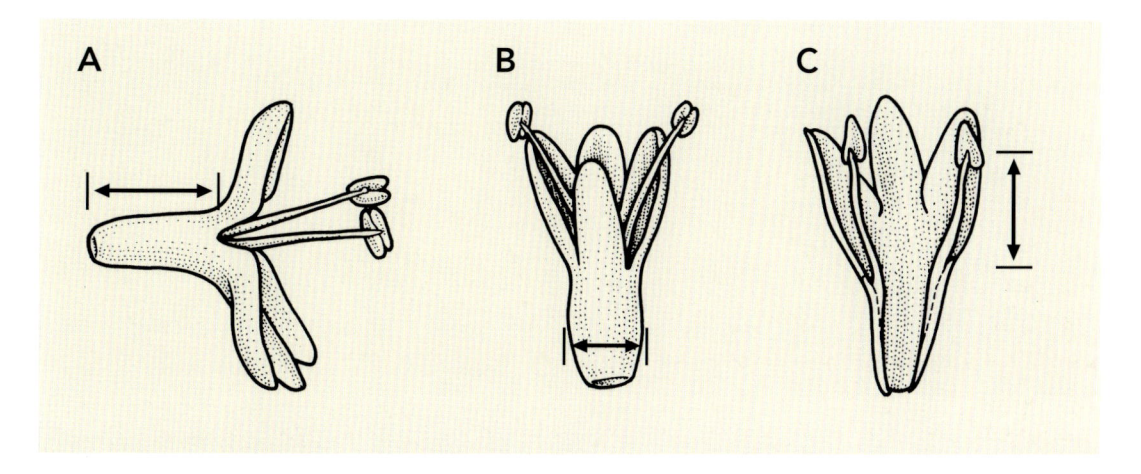

FIG. 42 Distances measured as: **A** corolla tube length; **B** corolla tube width; **C** filament length (corolla tube shown with posterior portion removed).

---

1 Where possible these characters were scored separately for hermaphrodite, male and female flowers.

Stamens

100 Stamen filaments (colour): 1. *white*; 2. coloured.
101 Stamen filaments (curving in bud): 1. incurved at apex in bud (e.g. Fig. 43) [becoming straight after anthesis]; 2. *straight at apex in bud*.
102 Stamen filaments (divergence): 1. remaining erect (e.g. Fig. 155E); 2. *diverging with age* (e.g. Figs 79G and 85E).
103 Stamen filaments (length): __ mm long. Measured on open flowers, mostly from fresh material. This measurement is for the free portion of the filament only – in other words, the distance from the apex to the point of insertion on the corolla (Fig. 42C). It is very short in some taxa (e.g. some "Connatae"), in which filaments are adnate to the corolla for most of their length.
104 Anthers (colour at anthesis): 1. white; 2. yellow; 3. pink; 4. magenta; 5. blue; 6. mauve; 7. violet; 8. purple; 9. pale brown; 10. buff; 11. orange.
105 Anthers (length): __ mm long. Measured on open flowers, mostly on fresh material (Fig. 41). Anthers are often slightly longer just prior to dehiscence than subsequently; measurements should cover this range (but not shrivelled anthers on herbarium specimens).
106 Sterile anthers of female flowers (colour): 1. white; 2. yellow; 3. pink; 4. magenta; 5. blue; 6. mauve; 7. violet; 8. purple; 9. light brown; 10. buff.
107 Sterile anthers (length): __ mm long. Measured on open flowers, mostly on fresh material.

Nectarial disc, ovary and style

108 Nectarial disc (indumentum): 1. ciliate; 2. ciliolate; 3. *glabrous*; 4. glandular-ciliate.
109 Ovary (shape): 1. *ovoid*; 2. globose; 3. ellipsoid; 4. conical.
110 Ovary (indumentum): 1. *glabrous*; 2. hairy.
111 Ovary (length): __ mm long. Measured on open flowers, but before the corolla is shed.
112 Ovary apex (in septum view): 1. acute; 2. subacute; 3. truncate; 4. emarginate; 5. obtuse; 6. didymous; 7. *not emarginate or didymous* (Fig. 44). States 1–5 were only scored for polymorphic taxa that are sometimes didymous or emarginate.
113 Ovary (locule number): 1. *bilocular*; 2. trilocular.
114 Ovules (number): __ per locule. Requires dissection of ovary and examination under a dissecting microscope. It can also be scored from immature fruit, counting all shrivelled and undeveloped ovules.
115 Ovules (arrangement on placenta): 1. *marginal on a flattened placenta*; 2. scattered on a hemispherical placenta; 3. in two vertical rows on placenta. Scored from open flowers, before fruit development.

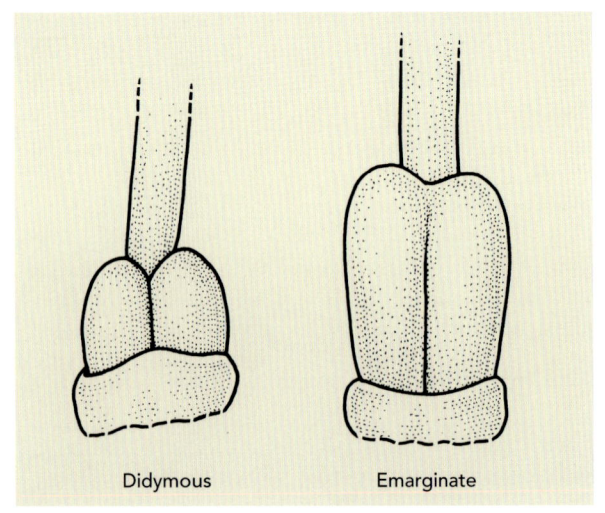

left
FIG. 43 Scanning electron micrograph of a newly opened flower with strongly incurved stamen filaments (these straighten as the flower matures).

right
FIG. 44 Septum (lateral) view of ovaries that are didymous or emarginate at the apex (both infrequent conditions in *Hebe* and *Leonohebe*).

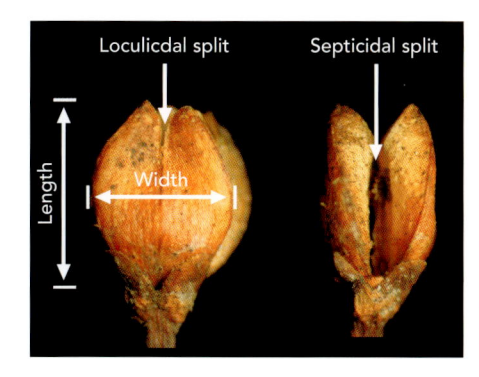

from left to right

FIG. 45 Latiseptate capsule (the most common type in *Hebe*) shown in loculicidal (left) and septicidal view.

FIG. 46 Angustiseptate capsule (cf. Fig. 45) shown in loculicidal (left) and septicidal view. Distances measured as width and thickness are indicated (measurements of the latter are included only in descriptions for species with angustiseptate or turgid capsules).

FIG. 47 Measurements used in seed descriptions.

116 Ovules (number of layers, on margins of flattened placenta): 1. *in one layer*; 2. in two layers; 3. in three layers. Scored from open flowers, before fruit development.

117 Style (length): __ mm long. Measured on mature flowers (or fruits) after the corolla is shed; the style can elongate dramatically in the early stages after flower opening.

118 Style (indumentum): 1. *glabrous*; 2. hairy.

Capsules

119 Capsules (when present): __. Text character indicating months in which capsules have been found on specimens. See notes on page 80.

120 Capsules (flattening): 1. *latiseptate*; 2. angustiseptate; 3. turgid (Fig. 29).

121 Capsules (apex): 1. acute; 2. subacute; 3. truncate; 4. emarginate; 5. didymous; 6. obtuse.

122 Capsules (length): __ mm long. Measured on mature (dehisced) capsules (Fig. 45).

123 Capsules (width, parallel to septum): __ mm wide. Measured on mature (dehisced) capsules (Fig. 45).

124 Capsules (thickness, at 90 degrees to septum): __ mm thick. Measured on mature (dehisced) capsules (Fig. 46), and only for species with angustiseptate or turgid capsules (in other taxa, variation is more dependent on capsule age and the extent of dehiscence, and is not as meaningful).

125 Capsules (indumentum): 1. *glabrous*; 2. hairy.

126 Septicidal split of capsule extending (distance to base): 1. one-quarter the way to base; 2. one-third the way to base; 3. halfway to base; 4. three-quarters the way to base; 5. *to base*. Measured on mature (dehisced) capsules (Fig. 45).

127 Loculicidal split of capsule extending (distance to base): 1. one-quarter the way to base; 2. one-third the way to base; 3. halfway to base; 4. two-thirds the way to base; 5. three-quarters the way to base; 6. to base. Measured on mature (dehisced) capsules (Fig. 45).

Seeds

128 Seed characters: 1. *recorded*; 2. not recorded.

129 Seeds (flattening): 1. weakly flattened; 2. flattened; 3. strongly flattened.

130 Seeds (shape): 1. ellipsoid; 2. ovoid; 3. obovoid; 4. irregular; 5. discoid; 6. oblong.

131 Seeds (winging): 1. winged; 2. *not winged (may be flattened)*; 3. weakly winged (± complete narrow hyaline border).

132 Seeds (surface): 1. *smooth*; 2. finely papillate.

133 Seeds (colour): 1. straw-yellow; 2. pale brown; 3. brown; 4. dark brown.

134 Seeds (length): __ mm long (Fig. 47). Measured only on well-developed seed.

135 Seeds (width): __ mm wide (Fig. 47). Measured only on well-developed seed.

136 MR (micropylar rim): __ mm (Fig. 47). Measured only on well-developed seed. This is a conspicuous, ± circular area surrounding the micropyle. It is equivalent to the "endosperm plateau" of Thieret (1955).

Taxonomic Treatment

SYNOPSIS

The following synopsis (opposite) divides *Hebe* and *Leonohebe* into eleven groups for the purpose of identification. To identify a specimen, first decide what group it belongs to, then proceed to the separate key for that group. The couplet about capsule compression divides the groups into two primary subdivisions. Within each of these subdivisions, the synopsis should be read from top to bottom, until a match to a specimen is found.

IMPLICIT CHARACTER STATES

To avoid extensive repetition in species descriptions, the most common states of some characters have been omitted. Where a species description provides no contradictory information, the following character states can be assumed to apply:
- **Leaves** abscising at nodes, decussate, free at base; *base* cuneate; venation evident in fresh leaves, including midrib only; *midrib* depressed to grooved above and thickened below; *margin* not tinged red, entire; *upper surface* not pitted; *lower surface* without domatia, dull, with many stomata, not pitted, glabrous.
- **Inflorescences** longer than subtending leaves, with all flowers (including those near the apex) generally developing to maturity; peduncle hairy; rachis hairy.
- **Bracts** free, margins hairy, glabrous outside.
- **Pedicels** varying from shorter to longer than bracts, hairy, not recurved in fruit.
- **Calyx** four-lobed, with anterior lobes free for most of their length, and with posterior lobes free for most of their length; *lobes* with mixed gl. and egl. cilia, glabrous outside.
- **Corolla lobes** four, white with age, glabrous; corolla throat white.
- **Stamen filaments** white, straight at apex in bud, diverging with age.
- **Nectarial disc** glabrous.
- **Ovary** ovoid, glabrous, apex (in septum view) not emarginate or didymous, bilocular; ovules marginal on a flattened placenta, in one layer; style glabrous.
- **Capsules** latiseptate, glabrous, septicidal split extending to base.
- **Seeds** not winged, smooth.

OTHER NOTES ON DESCRIPTIONS

Measurements connected by a multiplication sign indicate length by breadth. A single measurement for an organ indicates its length (unless stated otherwise). Numbers enclosed in parentheses are extreme values that are outside the usual range. The DELTA software automatically removes superfluous decimal units from measurements (e.g. presenting values of 6.3–7.0 mm as 6.3–7 mm); the degree of precision for any range of measurements is indicated by the most precise value given in that range. Where corolla tube length is compared to calyx length, this is not an absolute measurement, but rather refers to whether the corolla tube extends beyond the tips of the calyx lobes. Square brackets are used to indicate unusual flowering times seen only on cultivated specimens, or those recorded by Moore (in Allan 1961) that are outside the range of our observations (the latter are indicated by the phrase "in Allan"). Dates given for capsules are those times when specimens have been found with fruit on them (some of which may long since have dehisced).

Hebe

Capsule latiseptate (Fig. 29), except sts for *H. pareora*

Plants of whipcord form with small, scale-like, usually overlapping leaves (e.g. Figs 54 and 59); inflorescences terminal. — **"Flagriformes"** (the Whipcords) — PAGE 91

Leaf bases connate (e.g. Figs 62E, F and 65H), although sometimes only scarcely (e.g. Figs 67G and 64E); leaf buds often indistinct (e.g. Figs 62L and 64C); inflorescences terminal, sometimes also lateral; usually low-growing, decumbent or prostrate shrubs (us. < 40 cm tall) of alpine rocks. — **"Connatae"** — PAGE 113

Leaf bud sinus usually absent (e.g. Figs 73C and 74C), although sometimes present in *H. pimeleoides* (e.g. Fig. 75E) and *H. pinguifolia*; leaves dull, usually waxy and glaucous, often ± fleshy; inflorescences lateral; plants mostly low-growing, often decumbent or sprawling, only sometimes erect (but then not forming a dense, rounded bush), usually <1 m tall. — **"Subcarnosae"** — PAGE 129

Leaf bud sinus absent; leaves glossy or dull but not glaucous (except in *H. albicans*, *H. topiaria* and *H. glaucophylla*). — **"Occlusae"** — PAGE 144

Leaf bud sinus ± shield-shaped (e.g. Figs 115C and 118C); flowers sessile; bracts opposite and ± leaf-like; leaves usually abscising just above base, leaving a fragment of petiole attached to stem (e.g. insets of Figs 115B and 118B); inflorescences terminal, lateral or both. — **"Buxifoliatae"** — PAGE 215

Leaf bud sinus conspicuous, often narrow and acute; leaves usually < 4 cm long. — **Small-leaved "Apertae"** — PAGE 224

Leaf bud sinus conspicuous, rounded to acute; leaves usually > 4 cm long. — **Large-leaved "Apertae"** — PAGE 266

Capsule angustiseptate (Fig. 29)

Leaves conspicuously toothed (e.g. Fig. 151D). — **"Grandiflorae"** (*Hebe macrantha*) — PAGE 286

Leaves entire, abruptly narrowed to an obvious petiole (e.g. Fig. 152D). — **"Pauciflorae"** (*Hebe pauciflora*) — PAGE 288

Leonohebe

Low-growing subshrubs, to *c.* 30 cm tall (e.g. Fig. 155A); leaves overlapping on stem, inflorescences lateral (e.g. Figs 157G and H). — **Sect. *Leonohebe*** (the Semiwhipcords) — PAGE 291

Bushy shrub (e.g. Fig. 158A); branchlets ± glaucous; leaves widely spaced; inflorescences terminal. — **Sect. *Aromaticae*** (*Leonohebe cupressoides*) — PAGE 300

Hebe Juss.

DESCRIPTION Subshrubs to small trees, 0.05–13 m tall. **Branches** prostrate to erect; branchlets glabrous to pubescent, hairs bifarious or uniform; internodes 0.4–58 mm; leaf decurrencies obscure, or evident, or swollen. **Leaf bud** distinct (us.), or tightly surrounded by recently diverged leaves; sinus absent, or present. **Leaves** decussate or subdistichous, connate or free at base, appressed (and scale-like) or erect to recurved, linear or lanceolate or oblanceolate or ovate or obovate or oblong or elliptic or circular or deltoid or spathulate or rhomboid or sub-circular, thin to rigid or coriaceous or fleshy, 1.2–163 × 0.7–51 mm; *lamina surfaces* glossy to glaucous, glabrous or hairy; *apex* acuminate to retuse; *base* cuneate or truncate or sub-cordate or amplexicaul; *margin* glabrous or papillate to pubescent, entire to deeply toothed; venation evident (us.) or indistinct in fresh leaves; *petiole* 0–8.5 mm. **Juvenile leaves and reversion leaves** pinnatifid to entire, glabrous to pubescent. **Inflorescences** with 2–300 flowers, lateral or terminal or both, unbranched (us.) or branched, to 24.5 cm, shorter to longer than subtending leaves; peduncle to 6 cm; rachis to 20.5 cm; bracts opposite and decussate to alternate, free or connate. **Flowers** either all ☿, or ☿ or ♀ (on different plants). **Pedicels** shorter to longer than bracts, 0–14.7 mm. **Calyx** 1–10.2 mm, (3–)4(–6)-lobed, anterior lobes free for most of their length or partly or wholly united, posterior lobes free for most of their length. **Corolla** tube sts hairy inside and sts hairy outside, 0.4–7 × 0.7–4.5 mm, cylindric or funnel-form, shorter to longer than calyx; *lobes* white (us.) or coloured, acute to obtuse, erect to recurved, shorter to longer than corolla tube, sts papillate (inside) and/or hairy; corolla throat white (us.) or coloured. **Stamen filaments** 2, epipetalous and inserted on either side of posterior corolla lobe, white (us.) or coloured, straight or incurved at apex in bud, remaining erect or diverging with age, 0.1–13 mm; anthers white or yellow or pink or magenta or blue or mauve or violet or purple or buff or orange, 0.7–3.5 mm; sterile anthers of ♀ flowers white or yellow or pink or magenta or mauve or purple or light brown or buff, 0.6–1.8 mm. **Nectarial disc** glabrous (us.) or ciliate. **Ovary** ovoid or globose or ellipsoid or conical, sts hairy, 0.5–2.5 mm, apex (in septicidal view) acute to didymous, bi- or trilocular; ovules 2–61 per locule, marginal on a flattened placenta or scattered on a hemispherical placenta, in 1–3 layers; style 1–15 mm, sts hairy; stigma capitate to no wider than style. **Capsules** us. latiseptate or rarely angustiseptate or turgid, acute to truncate or emarginate or didymous, 1.5–10 × 0.8–6.6 mm, glabrous or hairy, septicidal split extending from ¾ to all the way (us.) to base, loculicidal split extending ¼ to all the way to base. **Seeds** weakly to strongly flattened, ellipsoid or ovoid or obovoid or discoid or oblong or irregular, winged or not winged, smooth or finely papillate, straw-yellow to dark brown, 0.6–3.2 × 0.5–2.2 mm, MR 0.1–1 mm. $2n$ = 40, 42, 80, 84, 118, 120, 122, 124 or 126.

NUMBER OF SPECIES 88

DISTRIBUTION (Fig. 8) Throughout New Zealand (Kermadec Ids, Three Kings Ids, NI, SI, Stewart Id, Snares Ids, the Auckland Ids, Campbell Id and Chatham Ids), with *H. elliptica* and *H. salicifolia* extending to southern South America, and *H. rapensis* endemic on Rapa (Austral Ids, French Polynesia). Some species or hybrids are also naturalised in Europe (Webb 1972), Tasmania (Rozefelds et al. 1999), Hawai'i, and on the Monterey Peninsula, California, USA (de Lange & Cameron 1992).

ETYMOLOGY Taken from Hebe of Gk mythology, the goddess of youth, daughter of Zeus and Hera, wife of Heracles and, for a time, cup bearer to the gods on Olympus.

"Flagriformes" (the Whipcords)

CRITICAL FEATURES Shrubs, decumbent or erect, to *c.* 1 m tall; leaf buds indistinct and overlapped by recently diverged leaves; leaf bases connate; **leaves appressed, scale-like**, entire, **not glaucous**, not readily abscising, persisting on the stem for some time; **inflorescences terminal**; **capsule latiseptate**.

NUMBER OF SPECIES 9

DISTRIBUTION NI (south from Raukumara Ra.), SI and Stewart Id

KEY TO SPECIES

1 Leaves with prominent, shallow veins, giving them a ribbed or striped appearance (e.g. Figs 54 and 55) 2
  Leaves sometimes wrinkled when dry but not regularly ribbed or striped (although veins sometimes evident on bracts) 3

2 Leaf apices mucronate (e.g. Fig. 54C1), apiculate (e.g. Fig. 54C2) or acute .... 4. *Hebe lycopodioides*
  Leaf apices rounded (e.g. Fig. 55C) to subacute .... 5. *Hebe imbricata*

3 Plant with *both* nodal joint well marked (e.g. Figs 49D and 53D [arrow]), though often hidden by overlap of leaves, *and* anterior calyx lobes only fused basally for *c.* ⅓ of their length (e.g. Figs 49E and 53E) 4
  Plant with *either* nodal joint obscure (e.g. Figs 56D and 59D), though often exposed, the leaf appearing continuous with internode below, *or* anterior calyx lobes almost or completely fused (e.g. Figs 56E and 59E) 6

4 Leaves often well separated on stem, with much of internode and nodal joint exposed (e.g. Fig. 53C) .... 3. *Hebe propinqua*
  Leaves usually partly overlapping, internodes and nodal joint (although present) either not, or only slightly, exposed (e.g. Figs 49C and 52C) 5

5 Leaf lamina thickened at the apex (e.g. Fig. 48B) [NI] .... 1. *Hebe tetragona*
  Leaf lamina not thickened at the apex (e.g. Fig. 48A) [SI] .... 2. *Hebe hectorii*

6 Overall colour of live plants orange/yellowish-brown (e.g. Fig. 56A) [western Nelson] .... 6. *Hebe ochracea*
  Overall colour of live plants green [eastern Nelson, Marlborough, Canterbury, Otago, Southland] 7

7 Leaves usually slightly mucronate (e.g. Figs 58C and D) .... 8. *Hebe armstrongii*
  Leaves rounded to subacute at the apex 8

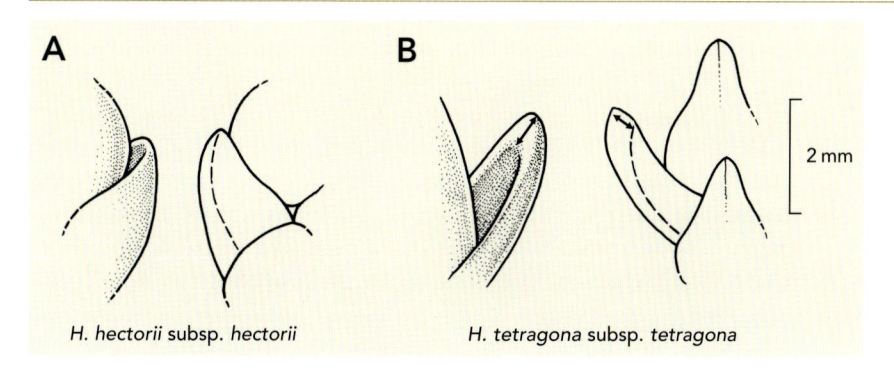

8 Leaves usually scarcely overlapping (e.g. Figs 57B, C and F) and not strongly protruding from stem .... 7. *Hebe salicornioides*
  Leaves strongly (e.g. Fig. 59E and F) to slightly (e.g. Fig. 59C) overlapping, and often somewhat protruding from stem .... 9. *Hebe annulata*

FIG. 48 Examples of whipcord hebes without (**A**) and with (**B**) the leaf lamina thickened at the apex (indicated by arrows).

1. *Hebe tetragona* (Hook.) Andersen

DESCRIPTION Spreading low or bushy shrub to 0.6 m tall, of whipcord form. **Branches** erect or ascending; internodes 0.5–1.6 mm; branchlets, including leaves, 1.8–3.5(–4.5) mm wide; connate leaf bases hairy; nodal joint distinct, often hidden and/or exposed (can vary on one branch); leaves not readily abscising, persistent along the stem for some distance. Leaves connate, appressed or erect; *lamina* 1.4–3.2(–4) mm, thickened near the apex; *apex* subacute (us.) or acute (± keeled at apex); *margin* ciliate; *lower surface* yellowish-green or dark green, veins not visible, glossy. Reversion leaves incised or entire, glabrous. **Inflorescences** with 2–12 flowers, terminal, unbranched, 0.35–1.2 cm. **Bracts** opposite and decussate, connate, ovate or deltoid or oblong, obtuse to subacute (sts ± attenuate toward apex), sts hairy outside (near basal, connate portion). **Flowers** Dec–Feb(–Apr), ♀. **Calyx** 2–3.2 mm, 4–5-lobed (5th lobe small, posterior); *lobes* ovate or elliptic. **Corolla** tube hairy inside, 1.5–2.1 × 1.7–2 mm, funnelform, shorter than (us.) or equalling calyx; *lobes* white at anthesis, ovate or elliptic (sts broadly), obtuse, erect to recurved, longer than corolla tube. **Stamen filaments** 3.3–3.7 mm; anthers magenta or purple, 1.4–1.7 mm. **Ovary** 0.6–0.8 mm, apex (in septum view) obtuse or slightly emarginate; ovules 10–12 per locule, in 1–2 layers; style *c.* 3.4–6 mm. **Capsules** (Jan–)Feb–May(–Nov), obtuse or truncate, 1.5–3 × 1.7–2.4 mm, loculicidal split extending ¼–½-way to base. **Seeds** flattened, ellipsoid (sts broadly) or irregular, ± smooth, pale brown, (0.9–)1.1–1.5 × (0.7–)0.8–1.1 mm, MR 0.2–0.4 mm. $2n = 40$.

Subsp. *tetragona*
Branchlets strongly tetragonous to cruciform in cross section; maximum width of ultimate branchlets (2.4–)2.8–3.5(–4.5) mm; internodes (0.5–)0.6–1.2(–1.6) mm long; **lamina (1.8–)2–3.2(–4) mm long, strongly keeled and narrowed toward elongated tip** (Fig. 49C1); apex acute to subacute.

Subsp. *subsimilis* (Colenso) Bayly et Kellow
Branchlets weakly to strongly tetragonous in cross section; maximum width of ultimate branchlets 1.8–3 mm; internodes 0.5–1.5 mm long; **lamina 1.4–2.3(–2.5) mm long, not keeled or keeled at apex; apex "boat-shaped" in "side view"** (Fig. 49C2), subacute.

DISTRIBUTION AND HABITAT Subsp. *tetragona* occurs on mountains of NI, including the Raukumara Ra., volcanoes of the central NI, Kaimanawa Ra., Kaweka Ra. and northern Ruahine Ra. (Otupae Ra.). Subsp. *subsimilis* occurs on the Ruahine Ra. (south from the Mokai Patea Ra.), Tararua Ra., and on the Pouakai Ra. near Mt Taranaki. Both subspecies grow in subalpine shrubland/penalpine grassland.

NOTES Key features of the species include: anterior calyx lobes free for most of their length; leaves not obviously ribbed, with conspicuous nodal joints; internodes mostly hidden. It is most similar to *H. hectorii*, from which it is distinguished by having leaves that are thickened at their apices (to varying extents), and a distinctive flavonoid profile (Markham et al. 2005). It is the only whipcord species of NI. It sometimes hybridises with *H. odora* (Mitchell et al. in prep.; Fig. 50).

A range of views have been presented on the classification of plants included under this species and *H. hectorii*. Two species (*H. tetragona* with two subspecies, and *H. hectorii* with three subspecies) are accepted here. However, Ashwin (in Allan 1961) accepted five species (two of which had two varieties), Druce (1980, 1993) accepted one species (with six varieties/subspecies), Heads (1994a, as *Leonohebe*) accepted five species (but noted that they are "perhaps better treated as five or six subspecies of a single species"), and Wagstaff & Wardle (1999) accepted two species (one with six subspecies). Defining two species on the basis of leaf apex thickening, flavonoid profile and geographic distribution seems useful for the present, but may not be a long-term solution. Further data are needed to assess whether *H. tetragona* and *H. hectorii* are most closely related to each other, and whether each is monophyletic (i.e. *H. hectorii* might be paraphyletic with respect to *H. tetragona* or vice versa). It may be that, when more is known about the relationships of these taxa, the most appropriate scheme will include all under one species, as suggested by Druce (1980, 1993) and Heads (1994a). This would require publication of several new subspecific combinations under *H. tetragona* (the name with priority).

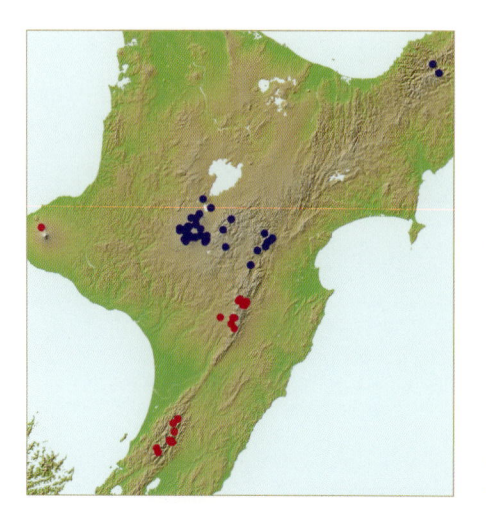

- Subsp. *tetragona*
- Subsp. *subsimilis*

Hebe tetragona   FLAGRIFORMES

FIG. 49   A subsp. *tetragona* at Littles Clearing, Kaweka Ra. B sprig of subsp. *tetragona*. C branchlets (at same scale as those shown for other whipcord spp.): C1, subsp. *tetragona* from Kaweka Ra.; C2, subsp. *subsimilis* from Whanahuia Ra. D close-up of leaves with evident nodal joints: D1, subsp. *tetragona* from Kaweka Ra.; D2, subsp. *subsimilis* from Whanahuia Ra. E terminal inflorescence, subsp. *tetragona*; inset shows anterior calyx lobes free for most of their length. F infructescence, subsp. *subsimilis*.

FIG. 50 *Hebe odora* (**A**), *H. tetragona* subsp. *subsimilis* (**C**), and the probable hybrid between them (**B**). All of these plants are from the Whanahuia Ra., Ruahine Forest Park.

Analyses of ITS sequences published by Wagstaff & Wardle (1999) suggested that *H. tetragona s. str.* is more closely related to some larger-leaved members of *Hebe* than to the other whipcords (including subsp. *subsimilis*). Although supported in the context of their data, it seems unlikely the result is a correct assessment of relationships. ITS sequence variation within the whipcords is low and some characters are homoplasious. Trees showing the whipcords as monophyletic (obtained by reanalysis of the ITS data) are only one step longer than Wagstaff and Wardle's shortest trees; if other relevant data (e.g. on leaf form and inflorescence structure) were also considered in such analyses the results would be different.

ETYMOLOGY *Tetragona* means four-angled (Gk *tetragonum* = quadrangle, tetragon) and refers to the leafy branchlets; *subsimilis* means "somewhat similar" (L. *sub* = almost, approaching; *similis* = like, resembling, similar), a reference to its similarity to *H. tetragona s. str.*

2. *Hebe hectorii* (Hook.f.) Cockayne et Allan

DESCRIPTION Spreading low or bushy shrub to 1 m tall, of whipcord form. **Branches** erect or ascending or spreading; internodes 0.6–1.9(–2.7) mm; branchlets, including leaves, 1.3–4(–4.6) mm wide; connate leaf bases hairy (at least when young; but sts connate portion deeply furrowed and hairs not visible); nodal joint distinct, either hidden or exposed; leaves not readily abscising, persistent along the stem for some distance. **Leaves** connate, appressed; *lamina* 1.2–2.7(–3.1) mm, not thickened near the apex; *apex* obtuse or subacute or apiculate or mucronate; *margin* ciliate or ciliolate (at least when young, but hairs often deciduous with age); *lower surface* dark green or bronze- or yellowish-green, veins not visible, glossy. **Reversion leaves** incised or entire, glabrous. **Inflorescences** with 4–16 flowers, terminal, unbranched, 0.35–1.5 cm; rachis densely hairy (with long, white, tangled hairs). **Bracts** opposite and decussate, connate, ovate or deltoid, obtuse or apiculate or subacute, sts hairy outside (near basal, connate portion). **Flowers** (Nov–)Dec–Mar(–Apr), ♀. **Calyx** 1.8–3.5(–4.2) mm, 4–5-lobed (5th lobe small, posterior), with anterior lobes free for most of their length or united to ⅓–⅔-way to apex; *lobes* elliptic, obtuse or subacute, with mixed gl. and egl. cilia (gl. hairs us. obscured by long egl. hairs). **Corolla** tube hairy inside, 1.5–3.3 × 1.3–2.6 mm, cylindric or funnelform, slightly shorter to slightly longer than calyx; *lobes* white at anthesis, ovate or elliptic, obtuse, suberect to recurved, longer than corolla tube. **Stamen filaments** 3.5–4.2 mm; anthers magenta or purple or pink, 1.1–1.7 mm. **Ovary** ovoid or somewhat globose, 0.6–1 mm, apex (in septum view) obtuse or slightly emarginate or didymous; ovules *c.* 18–34 per locule, in 1–3 layers; style 3.3–6.5 mm. **Capsules** (Jan–)Feb–Jun(–Dec), obtuse or subacute, 1.8–3.2 × 1.8–2.5 mm, loculicidal split extending ¼–½-way to base. **Seeds** flattened, ellipsoid to oblong, ± smooth, straw-yellow or brown, 0.9–1.4 × 0.5–0.8 mm, MR 0.2–0.5 mm. $2n = 40$.

Hebe hectorii subsp. *hectorii* FLAGRIFORMES

FIG. 51 *Hebe hectorii* subsp. *hectorii*. **A** plant near Brewster Hut, Westland. **B** flowering sprig. **C** branchlets (at same scale as those shown for other whipcord spp.): C1, plant from Humboldt Mts; C2, plant from Mt Anglem, Stewart Id (type locality of *H. laingii*). **D** close-up of leaves with evident nodal joints: D1, plant from Humboldt Mts; D2, plant from Mt Anglem. **E** terminal inflorescence; inset shows anterior calyx lobes free for most of their length. **F** infructescence.

TAXONOMIC TREATMENT

KEY TO SUBSPECIES

1   Apices of at least some leaves with an apiculus or mucro > 0.05 mm long (e.g. Figs 52C2 and 52C3) .... **subsp. *demissa***

      Leaf apices subacute to obtuse or smoothly rounded but not, or barely, apiculate/mucronate (e.g. Figs 51C and 52C1) 2

2   Maximum width of ultimate branchlets (e.g. Fig. 38) (1.6–)2–4(–4.6) mm at widest point, 1.2–2.7(–3.1) mm at narrowest point; internodes 0.7–1.9(–2.7) mm long; leaf apex subacute to rounded, leaves (1.2–)1.7–2.7(–3.1) mm long (Canterbury, Westland, Otago, Southland, Stewart Id) .... **subsp. *hectorii***

      Maximum width of ultimate branchlets 1.3–2.7(–3.6) mm at widest point, 0.9–1.6(–1.9) mm at narrowest point; internodes 0.6–1.8 mm long; leaf apex subacute to obtuse, leaves 1.2–2(–2.7) mm long (Nelson) .... **subsp. *coarctata***

DISTRIBUTION AND HABITAT Subsp. *hectorii* occurs on southwest SI, from the Aoraki/Mt Cook area southwards, and on Mt Anglem, Stewart Id. Subsp. *coarctata* occurs on northwest SI, from near Boulder Lk. south to *c.* 42° 3' S (Paparoa Ra. in the southwest, southern St Arnaud Ra. in the southeast). Subsp. *demissa* occurs in Otago and Southland, from the Waitaki Va. (an historical record) and Rock and Pillar Ra. in the east, to the Forbes Mts in the west. All subspecies grow in penalpine grassland and subalpine shrubland.

NOTES Key features of the species include: anterior calyx lobes free for most of their length; leaves not obviously ribbed; conspicuous nodal joints; and internodes mostly hidden. It is most similar to *H. tetragona*; notes under *H. tetragona* discuss the accepted limits of the two species, and the differences between them.

The two southernmost subspecies (*demissa* and *hectorii*) probably intergrade. The shape of the leaf apices varies almost continuously, from rounded to just perceptibly apiculate, to very prominently mucronate. Mucronate-leaved plants (subsp. *demissa*) generally occur on drier mountains in the east, and obtuse-leaved plants (subsp. *hectorii*) occur on wetter mountains in the west, with some overlap (e.g. in the Forbes Mts). Clearly demarcating the two subspecies is not straightforward and different circumscriptions (or no division at all) could be argued for. The type of *H. hectorii*, in particular, is among a group of specimens that are most difficult to place – that is, those with barely perceptible apicula/mucros. The circumscriptions adopted here preserve the traditional uses of the names *hectorii* and *demissa*.

Two additional subspecies of *H. hectorii* (described as distinct species by Cockayne 1909; Simpson 1952) were recognised in the recent treatment of Wagstaff & Wardle (1999). One, subsp. *laingii*, was distinguished on the basis of branchlet width. The other, subsp. *subulata*, was distinguished on the basis of mucro length. Variation in these characters, including substantial variation within single populations and small geographic areas (Appendix 2), is such that no clear grounds have been found for the recognition of these subspecies. Subsp. *laingii* is included here under subsp. *hectorii*, and subsp. *subulata* is included under subsp. *demissa*.

• Subsp. *hectorii*

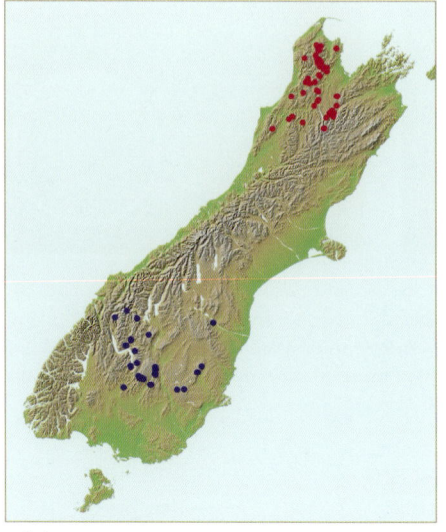

• Subsp. *coarctata*
• Subsp. *demissa*

Hebe hectorii subsp. *coarctata* and subsp. *demissa*   FLAGRIFORMES

FIG. 52 *Hebe hectorii* subsp. *coarctata* and subsp. *demissa*. **A** habit: A1, subsp. *coarctata* on Mt Arthur; A2, subsp. *demissa* in the Garvie Mts. **B** sprigs of subsp. *coarctata* upper) and subsp. *demissa*. **C** branchlets (at same scale as those shown for other whipcord spp.): C1, subsp. *coarctata* from Peel Ra., Nelson; C2, subsp. *demissa* from Old Man Ra., Otago; C3, subsp. *demissa*, cultivated plant originally from Old Man Ra. **D** close-up of leaves, with evident nodal joints: D1, subsp. *coarctata*; D2, subsp. *demissa*. **E** terminal inflorescence, subsp. *coarctata*; inset shows anterior calyx lobes free for most of their length. **F** infructescence, subsp. *demissa*.

Some specimens of subsp. *hectorii* from Fiordland with narrow branchlets (i.e. matching subsp. *laingii*) are very similar to specimens of subsp. *coarctata*. Differences between the two are worthy of further investigation. They are retained here as distinct taxa primarily because of their geographic separation (some similarities are possibly independently derived in the two subspecies).

ETYMOLOGY *Hectorii* honours Sir James Hector (1834–1907), first director of the Colonial Museum and collector of the type specimen; *coarctata* (L. *coarctatus* = crowded) presumably refers to the leaves; *demissa* (L. *demissus* = low-lying, hanging) refers to the habit.

3. *Hebe propinqua*   (Cheeseman) Cockayne et Allan

DESCRIPTION Spreading low or bushy shrub to 1 m tall, of whipcord form. **Branches** erect or ascending; internodes (0.8–)1–2.8 mm; branchlets, including leaves, (1.3–)1.6–2.4 mm wide; connate leaf bases hairy (us.), or glabrous; nodal joint distinct, exposed; leaves not readily abscising, persistent along the stem for some distance. **Leaves** connate, appressed; *lamina* not thickened near the apex; *apex* obtuse; *midrib* not thickened; *margin* ciliolate to ciliate; *lower surface* green, veins not visible, dull or sts ± glossy. **Reversion leaves** incised or crenate or entire, glabrous. **Inflorescences** terminal, unbranched, 0.3–0.9 cm; rachis densely hairy (with long, white, tangled hairs). **Bracts** opposite and decussate, connate (at least lowermost), broadly deltoid or ovate, obtuse. **Flowers** (Oct–)Jan–Feb, ⚥. **Calyx** 1.5–2.5 mm, with anterior lobes free for most of their length or united to ⅓-way to apex; *lobes* elliptic, obtuse. **Corolla** tube hairy inside, 1.5–2.1 × 1.5–2.1 mm, funnelform, equalling calyx; *lobes* white at anthesis, elliptic (often broadly) to almost circular, obtuse, suberect to recurved, longer than or equalling corolla tube. **Stamen filaments** 3.5–4.5 mm; anthers magenta, 0.9–1.6 mm. **Ovary** 0.6–1.1 mm, apex (in septum view) didymous; ovules 11–24 per locule, marginal on a flattened placenta (possibly scattered when many ovules present), in 1–3 layers; style 2.7–5.8 mm. **Capsules** (Dec–)Feb–Mar, obtuse, (1.5–)2.5–2.9 × (1.5–)2.1–2.2 mm, loculicidal split extending ¼–⅓-way to base. **Seeds** flattened, ellipsoid (sts broadly), ± finely papillate, pale brown, 0.6–1 × 0.5–0.7 mm, MR 0.1–0.2 mm. $2n = 40$.

DISTRIBUTION AND HABITAT Mountains of Otago and Southland, SI, in an area roughly bounded by Mt Ida in the northeast, Mt Maungatua in the southeast and the Mararoa Va. in the west. It grows in penalpine grassland and subalpine shrubland.

NOTES A distinctive species, distinguished from other whipcords by the combination of: dark green, unribbed, closely appressed, rounded leaves; prominent nodal joints; internodes that are usually prominently exposed; and anterior calyx lobes that are free for most of their length. Plants differ greatly in stature, depending on growing conditions, from low open shrubs *c.* 15 cm tall (e.g. places in the Rock and Pillar Ra.; Fig. 53B) to dense rounded bushes *c.* 1 m high and 1 m wide (Fig. 53A). It sometimes occurs with, or close to, *H. annulata*, *H. hectorii* and *H. imbricata*.

ETYMOLOGY L., meaning related to or like another (*propinquus* = near, neighbouring); it is not clear why Cheeseman (1906) gave the species this name.

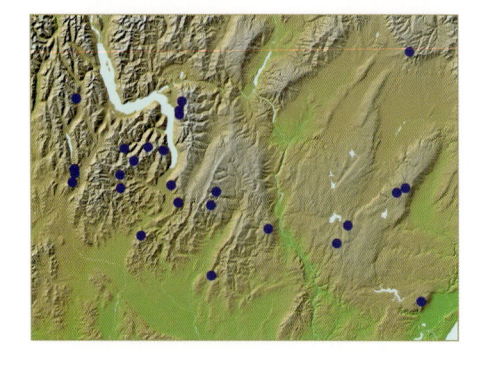

Hebe propinqua   **FLAGRIFORMES**

FIG. 53   **A** plant at Wye Ck, The Remarkables. **B** low-growing plant on the Rock and Pillar Ra. **C** Sprig. **D** branchlet (at same scale as those shown for other whipcord spp.), with prominent nodal joints (e.g. arrow). **E** terminal inflorescence; inset shows anterior calyx lobes free for most of their length. **F** infructescence.

4. *Hebe lycopodioides* (Hook.f.) Andersen

DESCRIPTION Spreading low or bushy shrub to 1 m tall, of whipcord form. **Branches** ascending or decumbent or erect; internodes (0.55–)0.8–1.3(–1.55) mm; branchlets, including leaves, (1.8–)2.3–3.3(–4.2) mm wide; connate leaf bases hairy; nodal joint distinct, us. hidden (but sts barely) or exposed; leaves not readily abscising, persistent along the stem for some distance. **Leaves** connate, appressed; *lamina* not thickened near the apex; *apex* mucronate (us.) to subacute; *margin* ciliate or ciliolate; *lower surface* dark green to yellowish-green, with prominent shallow veins that give a ribbed or striped appearance (at least faintly), dull to slightly glossy. **Juvenile leaves** crenate to pinnatifid, ciliate (near base and on lower surface). **Reversion leaves** entire or incised to pinnatifid, glabrous. **Inflorescences** with (4–)6–16(–20) flowers, terminal, unbranched, (0.35–)0.5–1.6(–1.9) cm; rachis hairy (with long, white, tangled hairs). **Bracts** opposite and decussate, connate, broadly deltoid, acuminate to subacute. **Flowers** [Nov–]Dec–Feb(–Apr), ♀. **Calyx** 2.8–3.5 mm, 4–5-lobed (5th lobe small, posterior); *lobes* lanceolate or elliptic or oblong, obtuse to acute, with mixed gl. and egl. cilia (gl. hairs us. obscured by long egl. hairs). **Corolla** tube hairy inside, 2.5–3.2 × 1.1–1.3 mm, cylindric, longer than or *c.* equalling calyx; *lobes* white at anthesis, elliptic or ovate, obtuse (posterior sts emarginate), suberect to patent, shorter to longer than corolla tube. **Stamen filaments** 2.5–3.6 mm; anthers magenta, *c.* 1–1.3 mm. **Ovary** 0.7–0.8 mm, apex (in septum view) didymous; ovules *c.* 13–16 per locule, marginal on a flattened placenta (but sts recurved and appearing scattered), in 1–2 layers; style 2.5–7 mm. **Capsules** Jan–Apr(–Dec), obtuse, (1.7–)2.2–3.4 × (1.3–)1.8–2.4 mm, loculicidal split extending ¼–⅓-way to base. **Seeds** flattened, ellipsoid, ± finely papillate, pale brown, 0.9–1.5 × *c.* 0.7 mm, MR *c.* 0.2 mm. $2n = 40$.

DISTRIBUTION AND HABITAT Mountains of SI, chiefly on or east of the Main Divide, from the Bryant Ra. in the north to the Kakanui Mts in the south. It grows in penalpine grassland and subalpine shrubland.

NOTES Similar to *H. imbricata*, from which it is distinguished by its strongly mucronate, acute or apiculate leaf apices. Plants from near Lewis Pass (e.g. Fig. 54C2) were included in var. *patula* (Simpson & Thomson 1943; Ashwin, in Allan 1961) or subsp. *patula* (Wagstaff & Wardle 1999), on the basis of their less mucronate leaves, often slender branchlets and usually low-growing habit. Despite obvious geographic trends in these characters (e.g. Fig. 162), specimens cannot be separated into clear-cut morphological groups (Fig. 164), and no infraspecific taxa are recognised here.

Historical specimens of H. J. Matthews (WELT 17415, 17420; AK 8215, 8216) suggest the species may also occur in the Greenstone Va. and Humboldt Mts, western side of Lk. Wakatipu, but these localities have not been substantiated by recent collections (and at least some of Matthews' specimens are based on cultivated plants, and there might have been confusion regarding original provenance).

ETYMOLOGY Resembling clubmoss, of the genus *Lycopodium*, family Lycopodiaceae (the Gk suffix -*oides* = resembling, like).

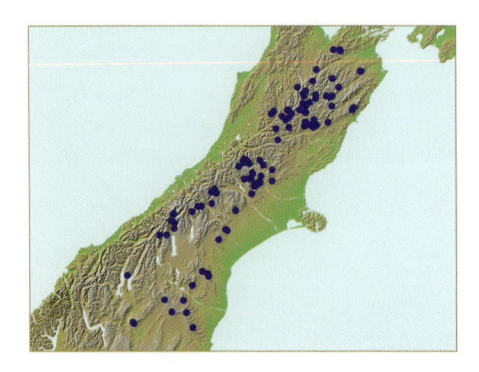

Hebe lycopodioides   FLAGRIFORMES

FIG. 54   **A** plant on Mt Nimrod, Hunters Hills. **B** sprig. **C** branchlets (at same scale as those shown for other whipcord spp.): C1, plant from Four Peaks Ra.; C2, cultivated plant originally from near Lewis Pass. **D** close-up of leaves, with prominent nodal joints: D1, cultivated plant originally from the Craigieburn Ra.; D2, cultivated plant originally from near Lewis Pass. **E** terminal inflorescence; inset shows anterior calyx lobes free for most of their length. **F** infructescence.

5. *Hebe imbricata*   Cockayne et Allan

DESCRIPTION Spreading low or bushy shrub to 0.8 m tall, of whipcord form. **Branches** erect or ascending; internodes 0.5–1.25(–1.4) mm; branchlets, including leaves, 2–3.2(–3.5) mm wide; connate leaf bases hairy; nodal joint distinct, almost always hidden; leaves not readily abscising, persistent along the stem for some distance. **Leaves** connate, appressed; *lamina* not thickened near the apex; *apex* obtuse (sts strongly keeled) or subapiculate; *margin* ciliate to minutely ciliolate; *lower surface* yellowish-green, with prominent shallow veins that give a ribbed or striped appearance, dull or slightly glossy. **Reversion leaves** incised or entire, glabrous. **Inflorescences** with 6–12(–18) flowers, terminal, unbranched, 0.3–1.3 cm. **Bracts** opposite and decussate, connate, ovate (sts broadly), obtuse or almost acuminate. **Flowers** (Dec–)Jan–Mar, ⚥. **Pedicels** absent. **Calyx** 2.2–3 mm, 4–5-lobed (5th lobe small, posterior); *lobes* broadly oblong or elliptic, obtuse or subacute, with mixed gl. and egl. cilia (egl. hairs long, flattened and tangled). **Corolla** tube hairy inside, 1.7–2.5 × 1–1.6 mm, shortly cylindric or funnelform, shorter than or equalling calyx; *lobes* white at anthesis, elliptic (sts broadly) or circular, obtuse (posterior sts emarginate), patent to recurved, longer than or equalling corolla tube. **Stamen filaments** 2.5–4.5 mm; anthers magenta, 1–1.6 mm. **Ovary** globose or ovoid, 0.5–1.3 mm, apex (in septum view) didymous; ovules 11–24 per locule, in 1–2 layers; style 3.6–5 mm. **Capsules** Feb[–Jul](–Dec), obtuse, 2–2.5(–3) × 1–2 mm, loculicidal split extending ¼–½-way to base. **Seed** characters not recorded. $2n = 40$.

DISTRIBUTION AND HABITAT Mountains of Otago and Southland, from the Rock and Pillar Ra. in the northeast to near Green Lk. in the southwest, including the Lammermoor Ra., Lammerlaw Ra., Mt Benger, Old Man Ra., Garvie Mts and Eyre Mts. It grows in penalpine grassland and low shrubland.

NOTES Similar to *H. lycopodioides*, from which it is distinguished by its rounded to subacute leaf apices. The circumscription adopted here includes plants recognised as *H. poppelwellii* by Ashwin (in Allan 1961). In the strict sense, *H. imbricata* includes only plants from between Lk. Monowai and Mt Burns, the southwesternmost points on the distribution map. Plants from this area tend to have greener, longer, less strongly keeled leaves and (despite appearances in Fig. 55C) stouter branchlets than plants from localities further east and north, traditionally included in *H. poppelwellii*. The two taxa should possibly be treated as distinct at an infraspecific rank, as they were by Wagstaff & Wardle (1999), but, despite some obvious geographic trends, consistent differences are difficult to identify.

ETYMOLOGY L. (*imbricatus* = overlapping like roofing tiles), refers to the leaves.

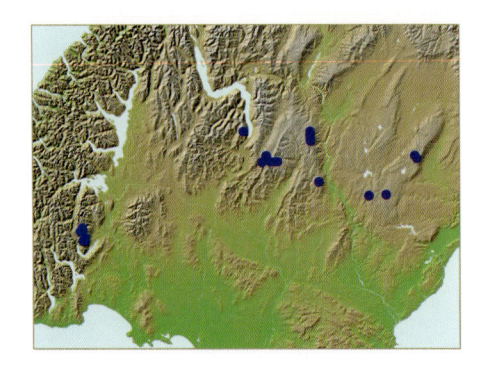

Hebe imbricata FLAGRIFORMES

FIG. 55 **A** plant on Mt Cuthbert. **B** sprig. **C** branchlets (at same scale as those shown for other whipcord spp.): C1, plant from Mt Burns; C2, plant from the Old Man Ra. **D** close-up of leaves, showing prominent nodal joint. **E** terminal inflorescence; inset shows anterior calyx lobes free for most of their length. **F** infructescence.

TAXONOMIC TREATMENT 103

6. *Hebe ochracea* Ashwin

DESCRIPTION Spreading low or bushy shrub to 0.4 m tall, of whipcord form. **Branches** ascending or spreading (with numerous short and erect secondary branches arising from upper surface); internodes (0.4–)1–2.7(–3) mm; branchlets, including leaves, 1.4–3.4 mm wide; connate leaf bases us. hairy or sts glabrous; nodal joint us. obscure (but sts apparent in older leaves), exposed; leaves not readily abscising and fragments persistent along the stem for some distance. **Leaves** connate, appressed; *lamina* not thickened near the apex; *margin* densely ciliate; *lower surface* dark green (and ochre-coloured at tips), veins not visible, glossy. **Inflorescences** with 4–8 flowers, terminal, unbranched, (0.2–)0.3–0.85 cm. **Bracts** opposite and decussate, connate, broadly ovate, obtuse or subacute. **Flowers** [Nov–]Jan–Feb, ⚥. **Calyx** 2–2.4 mm, with anterior lobes united to apex; *lobes* ovate (fused anterior lobe very broadly oblong-ovate), subacute (posterior) or obtuse (or slightly emarginate, anterior). **Corolla** tube hairy inside, 1.2–1.4 × 1.5–1.6 mm, funnelform, shorter than or equalling calyx; *lobes* white at anthesis, obovate or elliptic, obtuse or subacute (posterior sts emarginate), suberect to recurved (with age), longer than corolla tube. **Stamen filaments** straight or possibly slightly incurved at apex in bud, 2.8–3.5 mm; anthers pink to orange, 1.3–1.4 mm. **Ovary** globose, sts hairy, 0.6–0.7 mm, apex (in septum view) didymous; ovules 5–9 per locule; style 3–4.5 mm. **Capsules** Feb–Apr, obtuse or truncate or didymous, 1.7–2.6 × 1.7–2.5 mm, sts hairy, loculicidal split extending ⅓–¾-way to base (mostly c. ⅓). **Seed** characters not recorded. $2n = 124$.

DISTRIBUTION AND HABITAT Mountains of western Nelson, SI, from the Anatoki Ra. to Mt Owen, with a disjunct southern occurrence in the Paparoa Ra. (based on a single historical record, CHR 331850, Mt Buckland [Buckland Peaks], W. Thomson). It grows in grassland or shrubland, usually over limestone or marble rocks.

NOTES Distinguished from other whipcord species by the combination of: usually fused anterior calyx lobes; lack of a conspicuous nodal joint, except sometimes on older leaves; leaves that are not obviously ribbed; and the overall ochre colour of fresh plants, a product of the colour of the leaf tips. It is probably most similar to the group of related species comprising *H. salicornioides*, *H. armstrongii* and *H. annulata*, with which it usually shares the first three of these features. It is geographically distinct from those species, and differs in overall coloration, the relative size, shape and arrangement of leaves, as well as in chromosome number (Table 5), and ITS sequences (Wagstaff & Wardle 1999).

ETYMOLOGY L., refers to the overall colour of fresh plants (*ochraceus* = ochre-yellow, yellowish-brown).

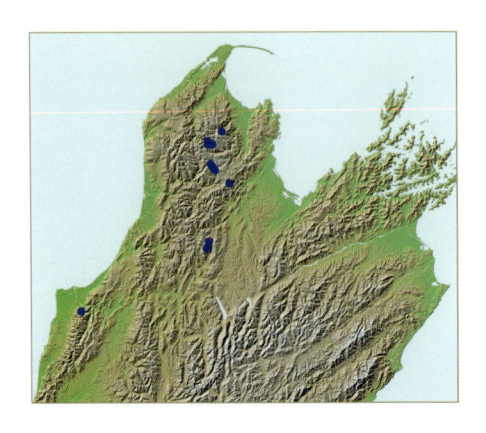

Hebe ochracea   FLAGRIFORMES

FIG. 56   **A** plant on Mt Arthur. **B** sprig. **C** branchlet (at same scale as those shown for other whipcord spp.). **D** close-up of leaf with obscure nodal joint. **E** terminal inflorescence; inset shows anterior calyx lobes fused throughout their length. **F** infructescence.

TAXONOMIC TREATMENT   105

7. *Hebe salicornioides* (Hook.f.) Cockayne et Allan

DESCRIPTION Spreading low or bushy shrub to 1 m tall, of whipcord form. **Branches** ascending (us.) or erect; internodes 1.2–3.8(–6) mm; connate leaf bases hairy (but often not near apex of connature); nodal joint obscure, exposed; leaves not readily abscising, persistent along the stem for some distance. **Leaves** connate, appressed (when fresh, but us. spreading when dry); *lamina* not thickened near the apex; *apex* obtuse and rounded; *margin* ciliate; *lower surface* light to dark green, veins not visible, dull or slightly glossy. **Reversion leaves** incised, glabrous. **Inflorescences** with (2–)6–8(–10) flowers, terminal, unbranched, 0.4–1.6 cm. **Bracts** opposite and decussate, connate, semi-circular or ovate, obtuse. **Flowers** [Nov–]Jan, ⚥. **Pedicels** absent. **Calyx** 1.9–3 mm, 4–5-lobed (5th lobe small, posterior), with anterior lobes free for most of their length or united between ⅓ and all the way to apex; *lobes* broadly oblong, obtuse, apparently egl. ciliate or with mixed gl. and egl. cilia, sts hairy outside (on posterior lobes). **Corolla** tube hairy inside, 1.3–1.7 × 1.2–1.5 mm, funnelform, slightly shorter to slightly longer than calyx; *lobes* white at anthesis, elliptic, obtuse, erect to recurved, longer than corolla tube. **Stamen filaments** 2.1–3.8 mm; anthers white or magenta or purple, 0.8–1.1 mm. **Ovary** 0.9–1.2 mm, apex (in septum view) obtuse or very slightly emarginate; ovules 14–19 per locule, in 1–2 layers; style 3.2–5 mm. **Capsules** [Jan–]Feb, obtuse, 2.5–4.3 × 1.5–3 mm, loculicidal split extending ¼–½-way to base. **Seed** characters not recorded. $2n = 42$.

DISTRIBUTION AND HABITAT Mountains of west Marlborough, SE Nelson and north Canterbury, SI, where it is known from Mt Severn, Mt Tarndale, Lk. Tennyson, Mt St Patrick, Jacks Pass and Mt Charon. It grows in red tussock grassland, often in flushes or slightly boggy sites.

NOTES A distinctive species, distinguished from all others by the combination of generally obtuse, strongly appressed, widely spaced, and not visibly ribbed leaves, and the lack of a conspicuous nodal joint. The anterior calyx lobes vary from being almost free to completely fused. The species is probably closely related to *H. annulata* and *H. armstrongii*, which also have chromosome numbers based on $n = 21$ (Table 5), similar ITS sequences (Wagstaff & Wardle 1999), tend to have fused anterior calyx lobes (probably more consistently so than *H. salicornioides*), and lack a nodal joint (the last two features are also shared with *H. ochracea*). *H. salicornioides* is geographically separated from these species, and differs in leaf size and arrangement, as described above.

Specimens labelled "above Lk. Harris, *T. Kirk*" (e.g. WELT 17479) and "Humboldt Mts" (WELT 17477, Herb. L. Cockayne) are probably *H. salicornioides*. These records are not shown on the distribution map because it is unlikely that the species occurs in Otago.

ETYMOLOGY Resembling the genus *Salicornia*, coastal glassworts (the Gk suffix *-oides* = resembling, like).

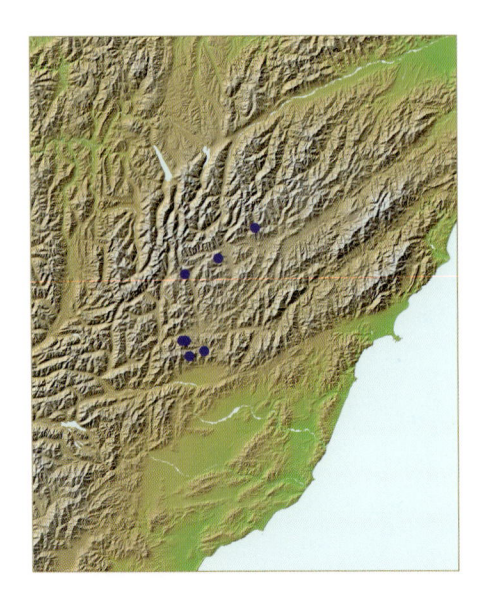

Hebe salicornioides   **FLAGRIFORMES**

FIG. 57   **A** plant on Mt St Patrick, St James Ra. **B** sprig. **C** branchlet (at same scale as those shown for other whipcord spp.). **D** close-up of leaf with obscure nodal joint. **E** terminal inflorescence; inset shows anterior calyx lobes fused throughout their length. **F** infructescence.

TAXONOMIC TREATMENT   107

8. *Hebe armstrongii* (J.B.Armstr.) Cockayne et Allan

DESCRIPTION Bushy shrub to 1 m tall, of whipcord form. **Branches** erect or ascending; internodes (0.65–)0.9–2.3 mm; branchlets, including leaves, 1.5–2(–2.7) mm wide; connate leaf bases hairy; nodal joint distinct (mostly weakly) or obscure, us. exposed (apart from very youngest portions of branchlets); leaves not readily abscising, persistent along the stem for some distance. **Leaves** connate, appressed (when fresh, but more spreading and ± encircling branchlet when dry); *lamina* not thickened near the apex; *apex* obtuse and apiculate or subapiculate; *margin* ciliate; *lower surface* yellowish-green, veins not visible. **Reversion leaves** incised or entire, glabrous. **Inflorescences** with (2–)6–8(–10) flowers, terminal, unbranched, (0.25–)0.4–0.7(–0.85) cm. **Bracts** opposite and decussate, connate, semi-circular or ovate, obtuse or apiculate or subapiculate. **Flowers** [Oct–]Jan, ⚥. **Pedicels** absent. **Calyx** 1.5–2.1 mm, 3-lobed, i.e. with anterior lobes united to apex (or very nearly so, sts split secondarily); *lobes* ovate or oblong (fused anterior lobe very broadly), obtuse or emarginate. **Corolla** tube hairy inside, *c.* 1–1.7 × *c.* 1.3–1.6 mm, funnelform and contracted at base, equalling or shorter than calyx; *lobes* white or mauve at anthesis, white or mauve with age, ovate or elliptic to broadly oblong, obtuse (posterior sts emarginate), suberect to patent, longer than corolla tube; corolla throat white or mauve. **Stamen filaments** 2–3 mm; anthers yellow or tinged pink, 1.4–1.6 mm. **Ovary** globose, 0.8–1 mm, apex (in septum view) didymous; ovules 5–7 per locule; style 2–4 mm. **Capsules** [Dec–]Jan–May(–Nov), obtuse, 2–2.5(–3) × 1.4–2.2 mm, loculicidal split extending ¼–¾-way to base. **Seeds** weakly flattened, ± ellipsoid-oblong or discoid, ± smooth, straw-yellow to pale brown, 0.9–1.3 × 0.5–0.8 mm, MR 0.2–0.4 mm. $2n = 84$.

DISTRIBUTION AND HABITAT Mountains of east Canterbury; known from near Poulter Hill, Esk River, Castle Hill and, historically, from the Jollie Ra. It grows on river terraces and in bogs, often with *Halocarpus bidwillii* (bog pine).

NOTES Key features of this species include: fused anterior calyx lobes; leaves that are not conspicuously ribbed; and a nodal joint that is often not, but sometimes faintly, evident. It is probably closely related to *H. annulata* and *H. salicornioides*, but also shares features with *H. ochracea* (see notes under the last two species). Morphological differences from *H. annulata* are only slight (see notes under that species), but in comparison *H. armstrongii* tends to have branchlets that are generally more slender, and leaves that are slightly mucronate, often less imbricate and not so tightly appressed (cf. Figs 58C and 59C). A specimen (not mapped) labelled "Kurow Mts" is discussed under *H. annulata*. *H. armstrongii* has been widely cultivated in New Zealand.

ETYMOLOGY Presumably honours Joseph F. Armstrong (1820–1902), who collected the type specimen.

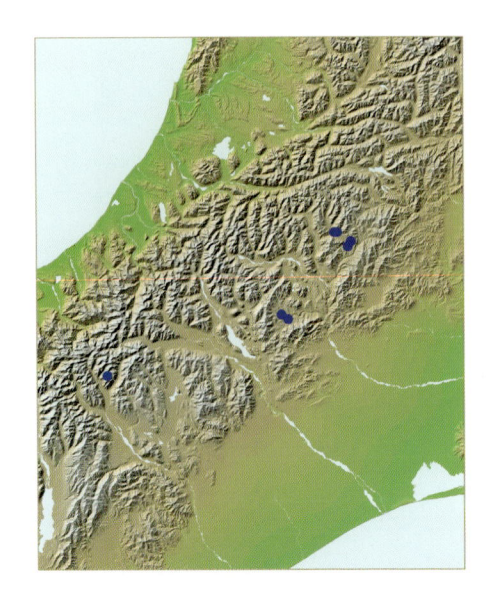

Hebe armstrongii   **FLAGRIFORMES**

FIG. 58   **A** plant in protective cage at Enys Reserve, near Castle Hill. **B** sprig. **C** branchlet (at same scale as those shown for other whipcord spp.). **D** close-up of leaves with nodal joint either obvious (**D1**) or obscure (**D2**). **E** terminal inflorescence; inset shows anterior calyx lobes fused throughout their length. **F** infructescence.

TAXONOMIC TREATMENT   109

9. *Hebe annulata* (Petrie) Andersen

DESCRIPTION Spreading low shrub to 0.5 m tall, of whipcord form. **Branches** erect or ascending; internodes (0.5–)1.3–1.8(–2.2) mm; branchlets, including leaves, 1.8–2.9 mm wide; connate leaf bases hairy; nodal joint obscure, exposed (us.) or hidden; leaves not readily abscising, persistent along the stem for some distance. **Leaves** connate, appressed; *lamina* not thickened near the apex; *apex* obtuse; *margin* conspicuously ciliate; *lower surface* green or yellowish-green, veins not visible. **Inflorescences** with 4–10 flowers, terminal, unbranched, 0.25–0.7 cm. **Bracts** opposite and decussate, connate, semi-circular, obtuse. **Flowers** [Oct–]Dec–Jan, ⚥. **Pedicels** absent. **Calyx** 1.5–2.2 mm, 3-lobed, i.e. with anterior lobes united to apex (forming one large lobe, sts split secondarily); *lobes* broadly oblong (but fused anterior lobe much larger than posterior), obtuse or slightly emarginate (e.g. anterior lobe when secondarily split). **Corolla** tube hairy inside, 1.5–1.7 × *c.* 1.2 mm, funnelform, shorter than (us.) or equalling calyx; *lobes* white at anthesis, oblong (broadly) or elliptic or obovate, obtuse, erect to recurved, longer than corolla tube. **Stamen filaments** 2.6–3.5 mm; anthers pink to purple, 1–1.2 mm. **Ovary** globose, 0.6–0.8 mm, apex (in septum view) didymous; ovules 2–4 per locule; style 2.5–4.2 mm. **Capsules** Dec–Mar, obtuse, 1.8–2.7 × 1.5–1.9 mm, loculicidal split extending ¼–½-way to base. **Seeds** weakly flattened, obovoid or narrow and irregular, ± smooth, pale brown, 0.9–1.3 × 0.5–0.8 mm, MR *c.* 0.2 mm. $2n = 42$.

DISTRIBUTION AND HABITAT Dry mountains of Otago and Southland, where it is known from the Criffel Ra., The Remarkables, Hector Mts and Takitimu Mts. It possibly also occurred historically near Kurow in the Waitaki Va. (the northernmost point on the distribution map). It grows in sparse shrubland on rocky sites.

NOTES Key features of this species include: fused anterior calyx lobes; lack of a nodal joint; and leaves that are not visibly ribbed. It is probably closely related to *H. armstrongii* and *H. salicornioides*, and is also similar to *H. ochracea* (see notes under those species). Of these it is most similar to *H. armstrongii*, from which it can be difficult to distinguish. In *H. annulata* the branchlets are generally stouter, and the leaves are generally more imbricate and strongly appressed, and have apices that are not so frequently mucronate (cf. Figs 58C and 59C). The two are retained here as separate species on the basis of these minor morphological characters, together with the reported differences in ploidy (diploid in *H. annulata*, tetraploid in *H. armstrongii* – based on single counts of each), geographic separation and possible ecological differences (in contrast to *H. armstrongii*, *H. annulata* generally occurs on sites that are rocky and well-drained).

A specimen (AK 256600) labelled "Kurow Mts, Otago 3000 ft, *D. Petrie*" was previously mounted with a piece of *H. armstrongii* on AK 8251. It seems unlikely that the two species co-occurred at this locality, and P. J. de Lange has suggested (annotations on sheet) that the *H. armstrongii* specimen may have been from "…a cultivated plant put on the sheet for comparison with the other two pieces [of *H. annulata*]", since "Petrie had garden plants of *H. armstrongii*".

ETYMOLOGY L., refers to the arrangement of the leaves, which often appear as tight, overlapping rings on the branchlets (*annulatus* = marked with rings, surrounded by raised rings or bands).

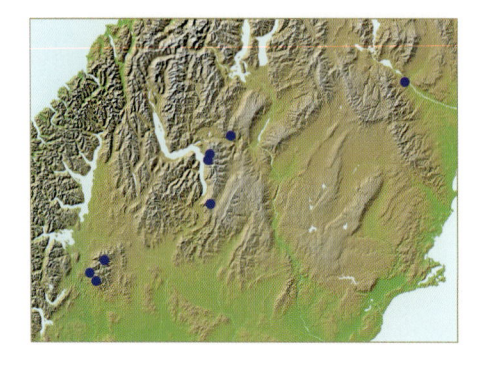

Hebe annulata **FLAGRIFORMES**

FIG. 59 **A** plant at Wye Ck, The Remarkables. **B** sprig. **C** branchlet (at same scale as those shown for other whipcord spp.). **D** close-up of leaf with obscure nodal joint. **E** terminal inflorescence; inset shows anterior calyx lobes fused throughout their length. **F** infructescence.

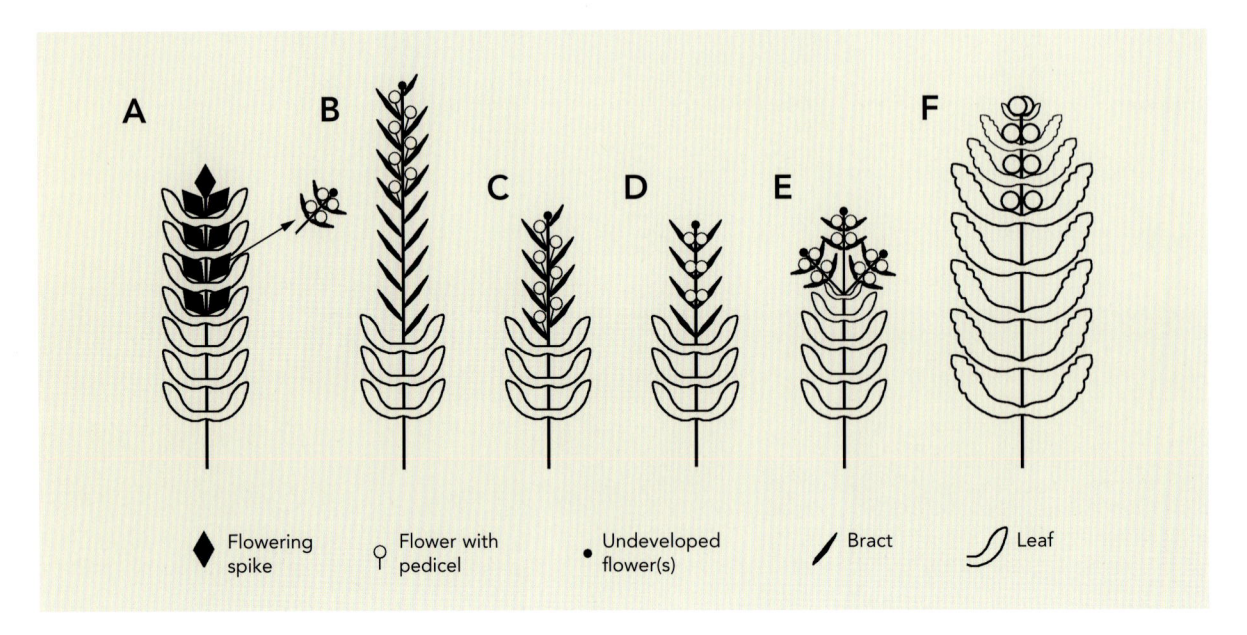

FIG. 60 Inflorescence types in "Connatae". **A** "flowering head" with small lateral spikes in the axils of little-altered leaves, surmounted by a terminal spike (e.g. *Hebe haastii*, *H. macrocalyx*, *H. epacridea*). **B** and **C** simple terminal inflorescence with alternate bracts and pedicellate flowers; sometimes with sterile bracts towards base (e.g. *H. petriei*, *H. murrellii*). **D** terminal inflorescence with usually opposite bracts and sessile flowers, sometimes with sterile bracts at base (e.g. *H. ramosissima*). **E** as for D but terminal inflorescence pedunculate and subtended by two lateral inflorescences (e.g. *H. ramosissima*). **F** terminal inflorescence with flowers in the axils of leaf-like bracts (e.g. *H. benthamii*).

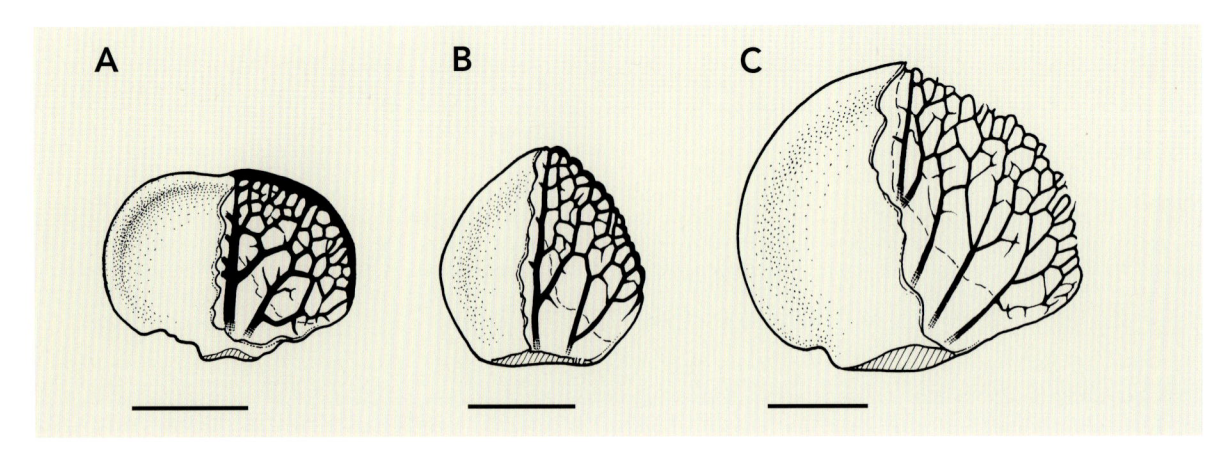

FIG. 61 Leaf venation in *Hebe epacridea*. **A** with a highly thickened and smooth marginal vein. **B** with a less thickened and scalloped marginal vein. **C** without a continuous marginal vein. Scale bar = 2 mm.

"Connatae"

CRITICAL FEATURES Subshrubs or shrubs, low-growing, decumbent or prostrate (us. < 40 cm tall); leaf buds often indistinct; **leaf bases connate, although sometimes only scarcely**; **leaves often small but not appressed or scale-like**, entire or toothed, dull or glossy, but not glaucous, not readily abscising (us.); **inflorescences terminal, and sometimes also lateral**; **capsule latiseptate** (sometimes turgid in *H. benthamii*).

NUMBER OF SPECIES 7

DISTRIBUTION Mountains of SI

KEY TO SPECIES

1	Inflorescences arranged into dense terminal "heads" of lateral and terminal sessile spikes in the axils of slightly altered leaves (e.g. Fig. 60A)	2
	Inflorescences terminal (sometimes to a short axillary shoot) racemes or spikes; if lateral inflorescences also present, then inflorescences pedunculate (e.g. Figs 60B–F)	4
2	Leaves bright green, fleshy, usually narrowed into a petiole; calyx usually linear, glabrous or minutely ciliolate, 4.5–9 mm long	11. *Hebe macrocalyx*
	Leaves dark green, coriaceous or rigid, not, or only slightly, narrowed into a petiole; calyx elliptic to lanceolate, minutely ciliolate or ciliate, 3.4–5(–6) mm long	3
3	Leaves coriaceous, often shallowly toothed, (6.6–)8.2–13 mm long, (4.2–)5.5–9.3(–11.6) mm wide, not strongly keeled (e.g. Fig. 65D) or recurved, margins not thickened, virtually glabrous; leaf bases glabrous or minutely ciliolate; calyx lobes minutely ciliolate (e.g. Fig. 65I)	13. *Hebe haastii*
	Leaves rigid, entire, (2.5–)4–8(–9) mm long, 2.5–5.5(–7) mm wide, often strongly keeled and recurved with thickened margins (usually Nelson, Marlborough, Canterbury; e.g. Fig. 62E), but sometimes only slightly keeled and erecto-patent, with margins not thickened (usually Aoraki/Mt Cook NP, Otago, Southland; e.g. Fig. 62F); leaf bases ciliate; calyx lobes long-ciliate (e.g. Fig. 62M)	10. *Hebe epacridea*
4	Flowers blue, often with 5 or 6 corolla and calyx lobes; bracts leaf-like, becoming less so towards the inflorescence apex (e.g. Fig. 60F); leaves usually toothed, fringed with dense hairs (Campbell Id and the Auckland Ids)	16. *Hebe benthamii*
	Flowers white with 4 corolla and calyx lobes; bracts not leaf-like; leaves entire, margins glabrous or minutely ciliate (SI)	5
5	Flowers pedicellate; corolla tube funnelform, ≤ 2 mm long; fertile anthers dark purple, well exserted from corolla tube; inflorescences terminal racemes (Figs 60B and C)	15. *Hebe murrellii*
	Flowers pedicellate or sessile; corolla tube cylindrical, > 2 mm long; fertile anthers magenta, held at throat of corolla tube; inflorescences terminal racemes or spikes (Figs 60B–D) or terminal spikes subtended by 2(–4) lateral spikes (Fig. 60E)	6
6	Corolla tube (4.2–)5–5.5(–7) mm long, calyx 4–5(–6.5) mm long, bracts alternate, flowers pedicellate; inflorescence terminal only (e.g. Figs 60B and C); leaves minutely puberulent on upper surface near base (Otago, Southland)	14. *Hebe petriei*
	Corolla tube 2.8–3.5 mm long, calyx 2.5–4 mm long, bracts usually opposite (at least lowermost pair), flowers sessile; inflorescences on each flowering shoot usually consisting of a pedunculate terminal spike subtended by 2(–4) lateral spikes (e.g. Fig. 60E), or sometimes terminal only, often with a few sterile bracts at base (e.g. Fig. 60D); leaves glabrous on upper surface (Marlborough, Nelson, Canterbury)	12. *Hebe ramosissima*

10. *Hebe epacridea* (Hook.f.) Andersen

DESCRIPTION Spreading low shrub (sts ± mat-like) to 0.4 m tall. **Branches** decumbent or ascending, old stems brown; branchlets green or purplish, puberulent to pubescent or glabrous (rarely), hairs bifarious (us.) or uniform; internodes 1–3(–4.5) mm; leaves not readily abscising, persistent along the stem for some distance. **Leaf bud** tightly surrounded by recently diverged leaves. **Leaves** decussate, connate, us. patent to recurved or erect to erecto-patent; *lamina* broadly oblong or ovate or elliptic, rigid, somewhat concave or flat (plants from Otago or Aoraki/Mount Cook National Park lack thickened margins), (2.5–)4–8(–9) × 2.5–5.5(–7) mm; *apex* obtuse or subacute; *midrib* thickened and evident below (us. forming a prominent keel, except on plants without a thickened leaf margin); *margin* not cartilaginous, conspicuously thickened (the outward manifestation of a rigid intramarginal vein) or not thickened (on plants that lack a marginal vein), commonly ciliate (toward base and, on one plant from Roys Peak, along entire margin) or minutely papillate or glabrous, sts tinged red, entire (us.) or minutely crenulate (rarely) or shallowly toothed (seen on one plant from Otago only); *upper surface* dark to light green, dull, with many stomata, glabrous; *lower surface* green, hairy toward base (along connate portion). **Inflorescences** with 2–8 flowers (per spike), terminal and lateral (arranged as spikes in the axils of little-altered leaves, forming a compact terminal flowering head), unbranched, (0.5–)0.8–2.6 cm (whole flowering head). **Bracts** opposite and decussate or lowermost pair opposite, then subopposite or alternate above, connate, ovate or deltoid, obtuse or subacute or acuminate, sts hairy outside. **Flowers** Dec–Feb(–Apr), ☿ or ♀ (on different plants). **Pedicels** absent. **Calyx** 3.4–5.8 mm; *lobes* oblong or ovate or elliptic or lanceolate, obtuse or subacute or acuminate. **Corolla** tube glabrous; tube of ☿ flowers 3.8–4.8(–5.4) × 1.6–2.2 mm, cylindric and contracted at base, equalling or longer than calyx; tube of ♀ flowers 2.4–4 × 1.3–1.9 mm, cylindric or funnelform, shorter than (only slightly) or equalling calyx; *lobes* white at anthesis, elliptic or ovate or obovate (narrowly), obtuse or subacute, suberect to recurved, shorter than corolla tube. **Stamen filaments** remaining erect, 0.1–1.2 mm (c. 0.8–1.2 mm for stamens of ☿ flowers, 0.1–0.4 mm for staminodes of ♀ flowers); anthers yellow or pink to purple, 1.2–2.1 mm; sterile anthers of ♀ flowers pink, 0.8–1.1 mm. **Ovary** sts hairy, 0.8–1.4 mm, apex (in septum view) obtuse or slightly emarginate; ovules 8–18 per locule, in 1–2 layers; style 2.5–6(–7) mm (generally longer in ☿ flowers than in ♀ flowers), rarely hairy (esp. toward base); stigma more prominent in ♀ flowers. **Capsules** Dec–Apr(–Sep), subacute, 2.7–4.5 × 1.5–2.6 mm, sts hairy, septicidal split extending ¾-way to base or completely to base, loculicidal split extending ¼- (mostly) to ⅓-way to base. **Seeds** weakly flattened, ellipsoid or ovoid or obovoid, straw-yellow, 0.8–1(–1.1) × 0.5–0.7 mm, MR 0.2–0.3 mm. $2n = 42$.

DISTRIBUTION AND HABITAT SI mountains, chiefly on or east of the Main Divide, from the Devil Ra., NW Nelson, to the Eyre and Livingstone Mts, Southland. It grows in open alpine areas on rock debris or scree. Together with *Parahebe birleyi* and *Ranunculus grahamii*, it grows at the highest altitudes known for any vascular plant in New Zealand (c. 2900 m asl in the Malte Brun Ra., Aoraki/Mt Cook NP).

NOTES A widespread and somewhat variable species, distinguished from other members of "Connatae" by: its small rigid leaves, which do not narrow towards the base; retained dead leaves along the length of the stem; and bracts and calyx lobes fringed by long hairs. Leaves are usually strongly keeled with a thickened margin (Figs 61A and 62E), both characters the result of very thick, woody leaf veins. Plants from Aoraki/Mt Cook NP and Otago (e.g. Figs 61B, C and 62F) often lack these leaf characters, causing frequent confusion in their identification. Variation is discussed in detail by Kellow et al. (2003b).

ETYMOLOGY Suggests resemblance to the southern heath family, Epacridaceae (now generally included in Ericaceae), many members of which have small, rigid leaves.

Hebe epacridea — CONNATAE

FIG. 62 **A** plant on Mt St Patrick, St James Ra. **B** sprig. **C** lower leaf surface. **D** upper leaf surface. **E** connate leaf bases (typical form with characteristically thickened leaf margins and midvein). **F** connate leaf bases (form lacking prominently thickened leaf margins and midvein). **G** frontal view of hermaphrodite flower. **H** frontal view of female flower. **I** lateral view of hermaphrodite flower. **J** lateral view of female flower. **K** inflorescence. **L** top view of shoot apex. **M** calyx, with lobe margins fringed with long hairs.

TAXONOMIC TREATMENT 115

11. *Hebe macrocalyx* (J.B.Armstr.) G.Simpson

DESCRIPTION Spreading low shrub to 0.2 m tall. **Branches** prostrate to decumbent, old stems brown; branchlets green to purplish, glabrous or puberulent, hairs bifarious; internodes 1–5.5(–10.9) mm; leaves not readily abscising, persisting on stem, or decaying leaving basal parts attached. **Leaf bud** tightly surrounded by recently diverged leaves. **Leaves** decussate to slightly subdistichous, connate, erecto-patent to patent; *lamina* obovate to spathulate (var. *macrocalyx* us. spathulate) or elliptic to ovate or rhomboid, fleshy, concave, (4–)6–11 × (2–)3–8(–9) mm; *apex* obtuse to rounded or slightly retuse; *midrib* slightly thickened below and slightly depressed to grooved above; *margin* cartilaginous (var. *macrocalyx*) or not cartilaginous (var. *humilis*), not thickened, glabrous or gl.-ciliate and sts minutely papillate (var. *humilis* only), sts tinged red, entire or rarely shallowly toothed; *upper surface* green, dull to glossy, with many stomata, glabrous; *lower surface* green, dull to glossy; *petiole* (1–)2–5 mm, hairy along margins. **Inflorescences** with 2–12 flowers per spike, 3–12 spikes per flowering head, terminal and lateral (arranged, often laxly, in a flowering head), unbranched, (0.5–)1–3(–4.3) cm (total length of flowering head), spikes about equal to subtending leaves (flowering head us. longer than subtending leaves); peduncle *c.* 0.1 cm, hairy or glabrous; rachis 0.3–0.4(–0.6) cm (longest when growing in shade). **Bracts** lowermost pair opposite, then subopposite or alternate above, connate or rarely free, lanceolate to linear or sts deltoid, subacute (us.) or obtuse, margins glabrous or hairy. **Flowers** [Sep–]Nov–Mar, ⚥ or ♀ (on different plants). **Pedicels** absent. **Calyx** 4.5–7(–9) mm, 4–5-lobed (5th lobe small, posterior); *lobes* linear-lanceolate to linear (us.) or oblong or deltoid (sts, in Nelson populations only), subacute to obtuse or occasionally acute, with minute mixed gl. and egl. cilia (us.) or glabrous. **Corolla** tube glabrous; tube of ⚥ flowers (2.5–)4–6 × 1.5–2.5 mm, cylindric, shorter to longer than calyx (sts); tube of ♀ flowers 3–4 × 1.5–2.2 mm, funnelform, shorter than calyx; *lobes* white at anthesis, ovate (us.) to elliptic, subacute to obtuse, patent to recurved (with age), shorter than (us.) to longer than (rarely) corolla tube. **Stamen filaments** remaining erect, 0.5–1.3 mm; anthers magenta, 1–1.9 mm; sterile anthers of ♀ flowers white, 0.8–1.1 mm. **Ovary** narrowly ovoid to conical, 2–2.5 mm; ovules 18–28 per locule, in 1–3 layers; style 3–8 mm on ⚥ flowers, 5–6 mm on ♀ flowers; stigma larger in ♀ flowers. **Capsules** [Nov–]Dec–Apr(–Oct), acute to subacute, 3.5–4.5(–5.5) × 2–3.5 mm, loculicidal split extending ¼–½-way to base. **Seeds** flattened, ellipsoid or ± discoid, straw-yellow, 0.7–1 × 0.5–0.9 mm, MR 0.1–0.3 mm. $2n = 42$.

Var. *macrocalyx*
Mat-forming subshrub to 20 cm tall, 1 m diameter. **Leaves** not distinctly keeled, usually spathulate, 5–16 × (2.5–)3.5–9 mm; apex rounded to slightly retuse; **margins** green or colourless, smoothly cartilaginous (Fig. 63C), entire. **Calyx lobes** green, linear, 5–9 mm long.

Var. *humilis* G.Simpson
Subshrub to 20 cm high. **Leaves** often slightly keeled, elliptic or rhomboid to spathulate, 5.5–11(–13.5) × (2–)3–7.5(–8.5) mm; apex obtuse; **margins** usually red near shoot apex, slightly papillate, or rounded and/or minutely erose (Fig. 63E), occasionally obscurely toothed. **Calyx lobes** red at apex, oblong to linear, 4.5–7 mm long.

DISTRIBUTION AND HABITAT Var. *macrocalyx* occurs near Arthur's Pass, Canterbury, SI, between Mt Alexander in the north and upper Bealey Va. in the south, on alpine rock debris and scree. Var. *humilis* occurs on mountains of Nelson and Marlborough, SI, south from the Anatoki Ra. to the Spenser Mts, and as far east as Mt Richmond, in rocky alpine herbfields, on rock debris or scree. On the basis of cultivated specimens, var. *humilis* was also recorded by Simpson (1952) from Mt French, west Otago (CHR 243479), and Mt Elliot, Southland (CHR 97169); neither of these localities is represented on the distribution map.

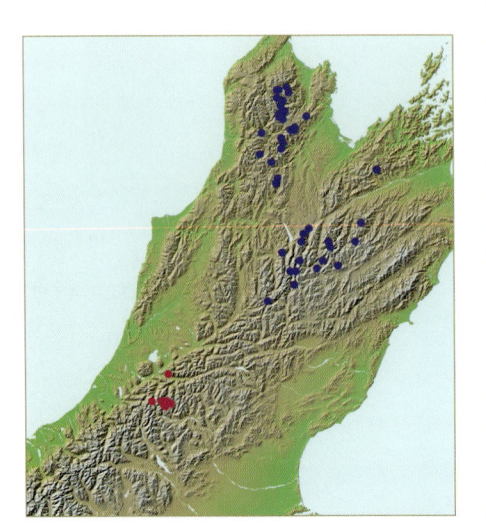

NOTES Distinguished from *H. haastii*, in which it was included by Cheeseman (1906) and Moore (in Allan 1961), by its: bright green, fleshy, petiolate and often subdistichous leaves (Figs 63B, C and E); sometimes lax flowering head; usually linear calyx lobes (Figs 63H and I); narrow acute capsule (Figs 63J and K). Var. *macrocalyx* is relatively uniform in morphology, all specimens having bright green, spathulate leaves with cartilaginous leaf margins and green, linear calyx lobes. Var. *humilis* is more variable in terms of leaf shape (elliptic or rhomboidal to spathulate; Appendix 4), leaf margins (smooth to papillose or erose), and calyx shape and length (lanceolate to

• Var. *humilis*
• Var. *macrocalyx*

Hebe macrocalyx   **CONNATAE**

FIG. 63   **A** var. *macrocalyx*, upper Bealey Va. **B** sprig, var. *macrocalyx*. **C** lower and upper leaf surfaces, var. *macrocalyx*; magnified inset shows cartilaginous leaf margin. **D** connate leaf bases. **E** lower and upper leaf surfaces of var. *humilis*. **F** frontal view of hermaphrodite flower. **G** frontal view of female flower. **H** lateral view of hermaphrodite flower. **I** lateral view of female flower. **J** septicidal view of capsule. **K** loculicidal view of capsule.

TAXONOMIC TREATMENT   117

linear). In northwest Nelson, var. *humilis* sometimes grows in shaded rock crevices and sinkholes, and these "shade form" plants have an etiolated, more sprawling habit with much longer internodes, and darker, spathulate and often toothed leaves (Appendix 4, WELT 80745). Some specimens from Mt Rolleston (e.g. CHR 497122, WELT 17583), although clearly related to *H. macrocalyx* and geographically close to specimens of var. *macrocalyx*, do not fit comfortably in either of the varieties as circumscribed here. The species commonly grows near *H. epacridea*.

ETYMOLOGY *Macrocalyx* means "large calyx" (Gk *macros* = large, long, tall; *calyx* = calyx); *humilis* (L.) means "low-growing".

12. *Hebe ramosissima*   G.Simpson et J.S.Thomson

DESCRIPTION Subshrub or spreading low shrub to 0.15 m tall. **Branches** decumbent, old stems brown; branchlets red-brown or purplish or green, puberulent, hairs bifarious; internodes 0.9–5(–7.5) mm; leaves not readily abscising, persistent along the stem for some distance. **Leaf bud** tightly surrounded by recently diverged leaves. **Leaves** decussate to subdistichous, connate (sts barely), erecto-patent to recurved; *lamina* elliptic to obovate (often narrowly), slightly fleshy, ± flat or slightly concave, 3.3–7.5(–9.5) × (1.5–)2–5.5 mm; *apex* subacute to rounded (often dimpled at apex); *midrib* sts evident in fresh leaves, slightly thickened below; *margin* not cartilaginous, not thickened, glabrous or gl.-ciliate (sts minutely), sts tinged red, entire or shallowly toothed; *upper surface* green to dark green, dull or slightly glossy, with many stomata, glabrous; *lower surface* green to dark green, dull or slightly glossy; *petiole* hairy along margins. **Inflorescences** us. terminal and lateral (i.e. a pedunculate terminal spike subtended by 2(–4) lateral spikes; more than 2 laterals are present only on Mt Lyford and Mt Terako specimens) or sts only terminal (and often with a few sterile bracts at base), unbranched, 0.8–1.7(–2) cm; peduncle 0.15–0.4 cm; rachis 0.35–0.5(–1.1) cm. **Bracts** opposite and decussate or lowermost pair opposite, then subopposite or alternate above, connate or free, elliptic to ovate or oblong, subacute to obtuse. **Flowers** Dec–Feb, ⚥ or ♀ (on different plants). **Pedicels** absent. **Calyx** 2.5–4 mm; *lobes* elliptic to lanceolate or narrowly oblong, subacute (us.) or obtuse. **Corolla** tube glabrous; tube of ⚥ flowers 2.8–3.5 × 1.4–1.5 mm, cylindric, shorter than (us.) to longer than calyx; tube of ♀ flowers *c.* 3 × 1.5 mm, cylindric, equalling to shorter than calyx; *lobes* white at anthesis, ovate, obtuse, patent or becoming recurved, ± equalling corolla tube. **Stamen filaments** remaining erect, 0.5–1 mm; anthers magenta, 1–1.2 mm; sterile anthers *c.* 0.7 mm. **Ovary** ovoid (often very narrowly), *c.* 2 mm; ovules 8–12(–18) per locule; style 2–3.5 mm. **Capsules** (Dec–)Feb–Mar, acute to subacute, 3.7–4 × 1.8–2 mm, loculicidal split extending ¼-way to base. **Seeds** flattened, ellipsoid, pale brown, 0.7–0.8 × 0.5–0.6 mm, MR 0.1–0.3 mm. $2n = 42$.

DISTRIBUTION AND HABITAT SI, mountains of east Marlborough, SE Nelson and north Canterbury, where it occurs on the Inland and Seaward Kaikoura ranges, and near Mt Weld, Mt Terako and Mt Lyford. It grows on alpine rocks and scree, often in moist places.

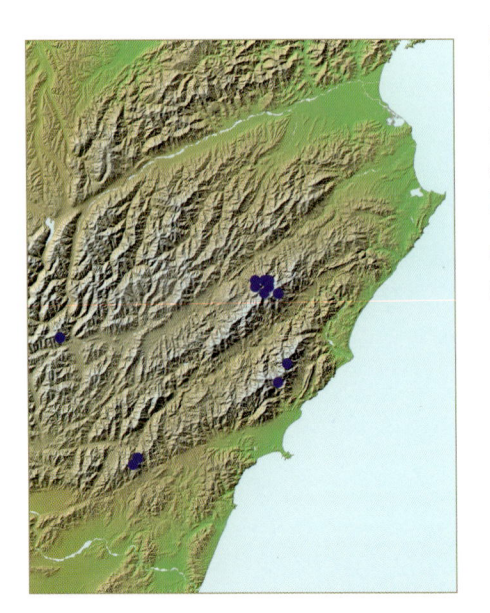

NOTES The inflorescences of *H. ramosissima* (e.g. Figs 60D and E) distinguish it from other "Connatae". The species is probably most closely related to *H. macrocalyx*, which some sterile specimens (e.g. CHR 405231) strongly resemble. It grows near *H. epacridea* at a range of localities, and probably co-occurs with *H. haastii* on Mt Terako (based on herbarium specimens only). Resemblance to *H. petriei* is discussed under that species.

ETYMOLOGY L. (*ramosus* = with many branches; *-issimus* = most so, to the greatest degree), means most branched.

Hebe ramosissima   **CONNATAE**

FIG. 64   **A** plant at Staircase Stm, Inland Kaikoura Ra. **B** sprig. **C** top view of shoot apex showing indistinct leaf bud. **D** lower and upper leaf surfaces. **E** very shortly connate leaf bases. **F** inflorescence showing sessile, opposite flowers. **G** lateral view of flower. **H** frontal view of flowers. **I** flowering shoot showing pedunculate inflorescences.

13. *Hebe haastii* (Hook.f.) Cockayne et Allan

DESCRIPTION Spreading low shrub to 0.2 m tall. **Branches** decumbent or ascending or spreading, old stems brown; branchlets purplish or green or brown, puberulent, hairs bifarious; internodes 2.5–7(–10) mm; leaves not readily abscising, persisting on stem (us.) or decaying leaving basal parts attached. **Leaf bud** tightly surrounded by recently diverged leaves. **Leaves** decussate, connate, erecto-patent to patent; *lamina* elliptic to obovate or ovate or spathulate (rarely, on lowermost leaves), coriaceous to fleshy, flat or concave, (6.6–)8.2–13 × (4.2–)5.5– 9.3(–11.6) mm; *apex* rounded to subacute; *midrib* slightly thickened below, sts evident in fresh leaves (below); *margin* not cartilaginous, not thickened, glabrous, often tinged red, entire or shallowly to deeply toothed; *upper surface* green to dark green (sts tinged dark red), dull or glossy, with many stomata, glabrous; *lower surface* green to dark green (sts tinged dark red), dull or glossy. **Inflorescences** with 4–6 flowers per spike, (8–)12–19(–25) spikes per flowering head, terminal and lateral (arranged in a terminal flowering head), unbranched, (0.85–)1.3– 3.3 cm (total length of flowering head), spikes about equal to subtending leaves (flowering head longer than subtending leaves); peduncle 0–0.1 cm; rachis 0.2–0.3 cm. **Bracts** lowermost pair opposite, then subopposite or alternate above, connate, oblong to deltoid or lanceolate, acute to subacute or rarely obtuse. **Flowers** Dec–Jan(–Feb), ☿ or ♀ (on different plants). **Pedicels** absent. **Calyx** 4–5(–6) mm; *lobes* oblong or elliptic to lanceolate, subacute to obtuse. **Corolla** tube glabrous; tube of ☿ flowers 4–5.5(–6) × 1.8–2 mm, cylindric, shorter than or equalling calyx; *lobes* white at anthesis, elliptic or ovate, sub-acute, suberect to patent, shorter than corolla tube. **Stamen filaments** remaining erect, 0.1–0.4 mm; anthers pink, 1.1–1.2 mm. **Ovary** rarely hairy, 1.5–2 mm; ovules 24–30 per locule, in 1–2 layers; style 2–2.4(–4) mm. **Capsules** Jan–Apr(–Aug), subacute, 5–6 × 2.5–3.7 mm, loculicidal split extending ¼–½-way to base. **Seeds** ± flattened, ± broad ellipsoid, straw-yellow, 0.9–1.3 × 0.6–0.9 mm, MR 0.1–0.2 mm. $2n = 42$.

DISTRIBUTION AND HABITAT Chiefly on mountains of Canterbury, SI, from the Craigieburn Ra. to The Hunters Hills, with a disjunct, northernmost occurrence on Mt Terako, Marlborough. It grows in open, alpine sites on rock debris and scree. One specimen (CHR 510021, *W. Lee*) is incorrectly labelled "Eyre Mts" (W. Lee pers. comm. 2003); its true origin is unknown.

NOTES Distinguished from other "Connatae" by both flavonoid and morphological characters (Kellow et al. 2003b); it is most similar to *H. macrocalyx* and *H. epacridea*. It is distinguished from the former by darker coloured, decussate leaves that are not narrowed into a petiole, shorter calyx lobes, and larger but more compact flowering heads. It is distinguished from the latter by larger, toothed and less rigid leaves, which are never keeled and do not have thickened margins, and by minutely ciliolate (rather than long-ciliate) calyx lobes.

ETYMOLOGY Honours the New Zealand geologist and botanist Sir Julius von Haast (1822–87), who first collected the species and whose specimen is the type.

Hebe haastii   CONNATAE

FIG. 65   **A** plant at Mt Hutt ski field. **B** sprig. **C** inflorescences, comprising "terminal flowering head". **D** lower leaf surface. **E** upper leaf surface. **F** lateral view of flower. **G** frontal view of flowers. **H** connate leaf bases. **I** calyx; magnified inset shows minutely ciliate margins. **J** infructescence. **K** septicidal view of capsule. **L** loculicidal view of capsule.

TAXONOMIC TREATMENT   121

14. *Hebe petriei* (Buchanan) Cockayne et Allan

DESCRIPTION Subshrub or spreading low shrub to 0.3 m tall. **Branches** decumbent, old stems red-brown or brown; branchlets green or purplish, glabrous or minutely puberulent, hairs bifarious; internodes 1.5–6(–9) mm; leaves not readily abscising, persisting on stem, or decaying leaving basal parts attached. **Leaf bud** ± indistinct and tightly surrounded by recently diverged leaves; sinus narrow and acute. **Leaves** decussate to subdistichous, connate, erecto-patent to recurved; *lamina* obovate or oblong or elliptic (often narrowly), slightly fleshy, concave (shallowly), (4–)5–7.5(–12) × (2–)2.5–5(–5.5) mm; *apex* obtuse to rounded; *midrib* slightly thickened below; *margin* not cartilaginous, not thickened, minutely ciliolate, occasionally tinged red; *upper surface* green, dull, without evident or with few stomata, hairy toward base; *lower surface* green; *petiole* (0.5–)1–2.5(–3) mm, hairy along margins and above. **Inflorescences** with 18–62 flowers, terminal, unbranched (sts with numerous sterile bracts towards the base), 0.8–2.5(–6) cm; peduncle 0.1–0.7 cm; rachis 0.5–4.2 cm. **Bracts** alternate, linear to narrowly ovate, sts minutely hairy outside (esp. on lower, sterile bracts). **Flowers** [Oct–]Dec–Feb(–Mar), ⚥ or ♀ (on different plants). **Pedicels** always shorter than bracts, 0.5–2.5 mm (longest towards base of inflorescence). **Calyx** 4–5(–6.5) mm; *lobes* linear (us.) to narrowly oblong, acute to subacute. **Corolla** tube glabrous; tube of ⚥ flowers (4.2–)5–5.5(–7) × 1.5–2 mm, cylindric, longer than (us.) or equalling calyx; *lobes* white at anthesis, elliptic (us. narrowly), subacute to obtuse (posterior sts emarginate), suberect to recurved, equalling or shorter than corolla tube. **Stamen filaments** remaining erect, 1.3–1.7 mm (⚥ flowers); anthers magenta, 0.9–1.5(–2) mm; sterile anthers of ♀ flowers magenta. **Ovary** conical, 2–2.5 mm; ovules 8–15 per locule; style 3–6 mm; stigma more conspicuous in ♀ flowers. **Capsules** (Nov–)Mar, acute, 4–4.5 × 1.7–2.3 mm, loculicidal split extending ⅓-way to base. **Seeds** flattened, ellipsoid to discoid, brown, 0.7–1.1 × 0.6–0.9 mm, MR 0.2–0.3 mm. $2n = 42$.

DISTRIBUTION AND HABITAT Mountains of Otago and Southland, SI, including Mt Repulse, the Pisa Ra., and the Forbes, Humboldt, Livingstone, Hector, Eyre, Garvie and Takitimu mts. It grows on alpine rocks and scree.

NOTES Similar to *H. murrellii* (see notes under that species). It is also similar to *H. ramosissima*, in having shortly connate leaves, sterile bracts sometimes present at the base of inflorescences, and magenta anthers held at the mouth of the cylindrical corolla tube. Differences between the two species are outlined in the key on page 113; the two are geographically disjunct, and analysis of ITS sequences (e.g. Wagstaff et al. 2002) does not suggest that they are closely related.

ETYMOLOGY Honours Donald Petrie (1846–1925), pioneering New Zealand botanist, who discovered the species and whose specimen is the type.

Hebe petriei CONNATAE

FIG. 66   A plant on the Takitimu Mts. B sprig. C leaf bud with narrow acute sinus. D lower and upper leaf surfaces. E frontal view of hermaphrodite flowers. F frontal view of female flowers. G connate leaf bases. H lateral view of hermaphrodite flower. I inflorescence with sterile bracts at base.

15. *Hebe murrellii* G.Simpson et J.S.Thomson

DESCRIPTION Subshrub or spreading low shrub to 0.2 m tall. **Branches** decumbent, old stems brown or grey; branchlets green to brown, puberulent, hairs bifarious; internodes 1–4(–6) mm; leaves abscising above nodes and lower part of petioles remaining attached to stem. **Leaf bud** tightly surrounded by recently diverged leaves; sinus narrow and acute (to acuminate). **Leaves** decussate to subdistichous, very shortly connate, erecto-patent to patent; *lamina* obovate or oblong or elliptic, subcoriaceous or slightly fleshy, flat or slightly concave, 3–9 × 2–5(–6) mm; *apex* rounded or slightly retuse; *midrib* slightly thickened below and depressed to grooved above; *margin* not cartilaginous, not thickened, ciliolate; *upper surface* light green, glossy or dull, with many stomata (not always clearly visible), hairy along midrib; *lower surface* light green, glossy or dull, glabrous or occasionally hairy along midrib. **Inflorescences** with 12–50 flowers, terminal, unbranched, 0.7–3.5 cm; peduncle 0–1 cm; rachis 1–3.2 cm. **Bracts** alternate or lowermost pair opposite, then subopposite or alternate above, linear or lanceolate to narrowly elliptic, obtuse to acute. **Flowers** Dec–Mar, ☿ or ♀ (on different plants). **Pedicels** always shorter than bracts, 0.5–2 mm. **Calyx** 2.5–3.5 mm; *lobes* elliptic to oblong or occasionally deltoid, subacute to obtuse. **Corolla** tube glabrous; tube of ☿ flowers 1.5–2 × 1.5–2 mm, funnelform, shorter than or equalling calyx; tube of ♀ flowers 1–2 × 1.5–2 mm, funnelform, shorter than to equalling calyx; *lobes* white at anthesis, elliptic (sts broadly) or obovate or deltoid (anterior only), obtuse, patent to recurved, longer than corolla tube. **Stamen filaments** 1–3 mm; anthers purple, 2 mm; sterile anthers of ♀ flowers magenta or purple, 0.9–1.2 mm. **Ovary** 1–1.2 mm; ovules 7–16 per locule, in 1–2 layers; style (1–)3–5 mm. **Capsules** Jan–Apr, acute or subacute, 3–5 × 2–3 mm, loculicidal split extending up to ⅓-way to base. **Seeds** flattened, ellipsoid to discoid, straw yellow to pale brown, 0.8–1.1 × 0.5–0.8 mm, MR *c.* 0.2 mm. $2n = 42$.

DISTRIBUTION AND HABITAT SI, mountains of Southland where it is known from the Earl Mts, Murchison Mts, Kepler Mts, Lk. Annie, Takitimu Mts and Mt Burns. It grows on alpine rock outcrops, boulder fields and scree.

NOTES Most similar to *H. petriei*, in which it was included by Moore (in Allan 1961), and with which it co-occurs in the Takitimu Mts. It is most readily distinguished from that species by its flowers, in which corolla tubes are shorter and the dark purple, fertile anthers (and even sterile anthers on female flowers) are held well outside the corolla tube (Figs 67E, F and H). In contrast, *H. petriei* has longer corolla tubes and magenta anthers held at or near the mouth of the corolla tube (Figs 66E, F and H). Additional notes on the two species are provided by Garnock-Jones & Clarkson (1994).

ETYMOLOGY Honours Robert Murrell (?–1943), Fiordland explorer.

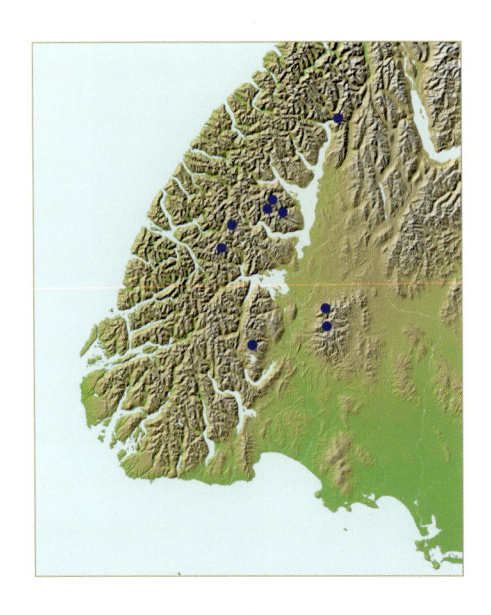

Hebe murrellii   **CONNATAE**

FIG. 67   **A** plant on Mt Burns. **B** sprig. **C** leaf bud with acute sinus. **D** lower and upper leaf surfaces. **E** hermaphrodite flowers with exserted anthers. **F** female flowers. **G** narrowly connate leaf bases. **H** lateral view of hermaphrodite flower. **I** inflorescence with sterile bracts at base. **J** septicidal view of capsule. **K** loculicidal view of capsule.

TAXONOMIC TREATMENT   125

16. *Hebe benthamii* (Hook.f.) Cockayne et Allan

DESCRIPTION Bushy or spreading low shrub to 1 m tall. **Branches** decumbent or ascending; branchlets pubescent (hairs white, ± strap-like and multicellular) or glabrous, hairs bifarious (us.) or uniform; internodes 1–13(–15.6) mm; leaves ± abscising at nodes (but leaving a small amount of connate region – between petioles of a leaf pair – attached to stem). **Leaf bud** tightly surrounded by recently diverged leaves (surrounding leaves us. overtopping bud). **Leaves** connate, erecto-patent to reflexed; *lamina* elliptic or obovate, coriaceous, flat, 10–33 × 3.5–14.5 mm; *apex* obtuse or truncate (and us. with a prominent apical gland (hydathode?)); *midrib* thickened below and depressed to grooved above (but groove us. not extending to leaf apex); *margin* conspicuously pubescent, shallowly to deeply toothed; *upper surface* green, without evident stomata, glabrous or hairy along midrib or hairy toward base. **Inflorescences** with 11–27 flowers, terminal (but sts terminal to short axillary branches, and therefore appearing almost lateral), unbranched or with 3 or more branches (sts up to 4 lateral branches, but never compound branching); peduncle 0.8–1.9 cm; *rachis* 1.6–9.3 cm. **Bracts** opposite and decussate, free (us.) or connate (lowermost, sts), obovate (us.) or elliptic, apex with a prominent gland, obtuse or subacute (or occasionally emarginate, the bract appearing to be a fused structure, with two vascular strands). **Flowers** (Oct–)Nov–Feb(–May). **Pedicels** 1–4 mm, hairy or sts glabrous. **Calyx** 3–8.5 mm, 4–6-lobed; *lobes* oblong or obovate, obtuse or subacute (with a prominent apical gland), egl. ciliate (with long, tangled, white hairs), glabrous outside (but us. hairy inside). **Corolla** tube glabrous, 2–3.2 × 3.5–3.9 mm, cylindric (and somewhat dorso-ventrally compressed), shorter than calyx; *lobes* 4 to 6, sky-blue to violet at anthesis, blue with age, obovate or circular, obtuse (posterior sts emarginate), erect to patent, longer than corolla tube; corolla throat blue or white. **Stamen filaments** coloured, remaining erect, 1–1.5 mm; anthers blue, 1.2–1.6 mm. **Ovary** 1.8–2.3 mm, bi- to trilocular (when ovary is 3-chambered, 1 chamber us. appears bigger than the other 2); ovules 10–13 per locule, marginal on a flattened placenta (or at least appearing this way in largest locules; ovules in smaller locules appear ± scattered), in 1–2 layers; style 2.1–3.2 mm. **Capsules** (Nov–)Dec–Apr(–Oct), latiseptate (2 locules) or turgid (3 locules), subacute, 4.5–6 mm, hairy, septicidal split sts extending only ¾-way to base, loculicidal split extending ¼–¾-way to base (us. less than ½-way). **Seeds** strongly flattened, broad ellipsoid to discoid, winged, straw-yellow or dark brown, 1.2–1.9 × 1.3–1.6 mm, MR 0.3–0.6 mm. $2n = 40$.

DISTRIBUTION AND HABITAT Restricted to Campbell Id and the Auckland Ids in the New Zealand subantarctic. It grows in tall tussock grassland (often with *Polystichum*), or low shrubland and rocky places.

NOTES A highly distinctive species, not readily confused with any other. Key features include: leaves densely fringed by white hairs, often toothed; terminal inflorescence with leaf-like bracts; flowers with blue corollas, 4–6 corolla and calyx lobes; and ovaries and fruit with 2–3 locules. Further illustrations are provided by Hooker (1844), Hombron & Jacquinot (1845) and Eagle (1982).

ETYMOLOGY Honours George Bentham (1800–84), one of the most prolific botanists of the nineteenth century.

Auckland Islands

Campbell Island

Hebe benthamii   **CONNATAE**

FIG. 68   **A** plant on Campbell Id. **B** sprig. **C** leaf bud. **D** lower and upper leaf surfaces. **E** dorsal view of flower. **F** frontal view of flower showing five corolla lobes. **G** top view of shoot. **H** loculicidal view of capsule with three locules. **I** flowering shoots.

FIG. 69 *Hebe buchananii* near Mt Cerberus, Livingstone Mts, Southland.

"Subcarnosae"

CRITICAL FEATURES Shrubs, **often decumbent or sprawling**, but sometimes erect, usually <1 m tall; **leaf buds distinct, or tightly surrounded by recently diverged leaves; sinus usually absent** (e.g. Figs 73C and 74C), only sometimes present in *H. pimeleoides* (e.g. Fig. 75E) and *H. pinguifolia*; leaf bases free; leaves not appressed or scale-like, entire, **usually waxy and glaucous**, abscising at nodes; inflorescences lateral; capsule latiseptate (us.) or turgid to angustiseptate (sts in *H. pareora* only).

NUMBER OF SPECIES 7

DISTRIBUTION Mountains of SI

KEY TO SPECIES

1   Corolla lobes violet, blue or pale mauve (e.g. Fig. 75C) when young,
        sometimes fading to pale pink or almost white with age .... **22.** *Hebe pimeleoides*
    Corolla lobes white or only lightly tinged with colour when young, white when old 2

2   Flowers, at least the lowermost flowers on each inflorescence, borne on obvious pedicels 3
    Flowers usually sessile (short pedicels only rarely present in *H. amplexicaulis* and *H. buchananii*) 7

3   Peduncles glabrous .... **18.** *Hebe pareora*
    Peduncles at least minutely hairy 4

4   Ovaries and capsules minutely hairy [southeast Marlborough to mid-Canterbury]   **32.** *Hebe glaucophylla*
                                                                                    (treated under small-leaved "Occlusae")
    Ovaries and capsules glabrous [NW Nelson to Lewis Pass; or Southland] 5

5   Corolla tube not or scarcely exceeding calyx [Southland] .... **23.** *Hebe biggarii*
    Corolla tube of male-fertile flowers longer than calyx [NW Nelson to Lewis Pass] 6

6   Corolla tubes ≤ 2.5 mm; erect, bushy shrub with a dense, rounded habit
        (the product of extensive branching toward shoot tips) [subalpine shrubland] .... **33.** *Hebe topiaria*
                                                                                    (treated under small-leaved "Occlusae")
    Corolla tubes > 2.5 mm; plant sprawling or erect, usually not dense or rounded
        (usually sparingly branched toward shoot tips) [rocky places, lowland to subalpine] .... **34.** *Hebe albicans*
                                                                                    (treated under small-leaved "Occlusae")

7   Leaf margins fringed with long hairs (e.g. Figs 72B and D); hairs mostly absent
        from leaf lamina (only sometimes present on the upper leaf surface, and then
        concentrated near the base, above the midvein, or near the apex) .... **19.** *Hebe gibbsii*
    Leaf margins glabrous or with short hairs; if longer hairs are present, these also cover
        upper and lower leaf surfaces (e.g. Fig. 70F) 8

8[1] Leaf buds closely surrounded by several imbricate, partly separated leaf pairs
        (e.g. Figs 74B, C and H); corolla tube ≤ 2 mm .... **21.** *Hebe buchananii*
    Leaf buds usually not closely surrounded by several imbricate leaf pairs
        (e.g. Figs 70B and 73B); corolla tube ≥ 2 mm 9

9   Corolla tube 2–3 mm long .... **20.** *Hebe pinguifolia*
    Corolla tube 3.4–4.8 mm long .... **17.** *Hebe amplexicaulis*

---

1 This couplet will not give correct identifications for some specimens of *H. pinguifolia* from Marlborough (see notes under that species).

17. *Hebe amplexicaulis* (J.B.Armstr.) Cockayne et Allan

DESCRIPTION Spreading low shrub (mostly branching near base) to 0.5 m tall. **Branches** decumbent (us.) or erect, old stems brown; branchlets green to black, pubescent or glabrous, hairs bifarious or uniform (or sts with just a few hairs around leaf bases); internodes (1–)3–8(–10) mm; leaf decurrencies evident, obscure, or sts ± swollen. **Leaf bud** distinct; sinus absent. **Leaves** erecto-patent to patent (us.) or reflexed; *lamina* broadly oblong (us.) or ovate, fleshy, flat or slightly concave, (8–)12–21(–25) × (4–)9–12(–17) mm; *apex* obtuse (us.) or subacute; *base* amplexicaul (or ± narrowed to branchlet width in plants corresponding to var. *erecta*); *midrib* weakly thickened below, weakly evident on underside of fresh leaves; *margin* glabrous or minutely papillate or puberulent (if leaf pubescent, but then hairs no more concentrated on margin than on any other area of leaf surface), sts tinged red; *both surfaces* glaucous, glabrous or uniformly egl. pubescent; *upper surface* with many stomata. **Inflorescences** with 10–25 flowers, lateral, unbranched, 2.1–4.6 cm; peduncle 0.6–2.1 cm; rachis 1–3.2 cm. **Bracts** opposite and decussate or opposite and decussate below and becoming alternate above, ovate or deltoid or oblong, obtuse or subacute, sts hairy outside. **Flowers** (Oct–)Nov(–Jan) [Nov–Mar in Allan], ⚥. **Pedicels** almost always absent or if present then always shorter than bracts (obvious pedicels are seen only in plants from the upper Rangitata matching var. *erecta*, and then only on lowermost flowers of an inflorescence), 0–0.5 mm. **Calyx** 3–4.2 mm; *lobes* ovate (narrowly) or oblong, obtuse or subacute, sts hairy outside. **Corolla** tube glabrous or hairy inside (seen only in plants of forma *hirta*), 3.4–4.8 mm, cylindric, equalling or longer than calyx; *lobes* white at anthesis, acute to obtuse, sts hairy inside. **Stamen filaments** *c.* 3.8–4 mm; anthers purple or dark magenta, *c.* 1.3–1.7 mm. **Ovary** globose, sts hairy (sparsely to densely), *c.* 0.5 mm, apex (in septum view) slightly emarginate; ovules *c.* 10–12 per locule, marginal on a flattened placenta; style 4–10 mm, often hairy (us. only near base). **Capsules** Feb–Apr, narrowly latiseptate, obtuse, 3.5–4 × 2–2.5 mm, often hairy, loculicidal split extending ¼-way to base. **Seeds** flattened, ellipsoid to discoid (or oblong or obovoid), not winged to only weakly winged, ± smooth, straw-yellow to brown, 1.1–1.8 × 0.8–1.5 mm, MR 0.3–0.7 mm. $2n = 40$.

DISTRIBUTION AND HABITAT SI, endemic to south Canterbury, where it is known from Mt Somers, Mt Peel, Orari Gorge and the Four Peaks Ra. It grows on rock outcrops, chiefly in subalpine scrub but sometimes in forest.

NOTES Similar to *H. pareora* (differences are discussed under that species). It also resembles *H. pinguifolia*, from which it is distinguished by its longer corolla tubes, usually more amplexicaul leaf bases and frequently broader leaves.

Plants either have glabrous or hairy leaves (Fig. 70). Those with hairy leaves, which usually grow in mixed populations with glabrous-leaved plants, were treated by Cockayne & Allan (1926b) and Moore (in Allan 1961) as a distinct species, *H. allanii*. Garnock-Jones & Molloy (1983a) concluded that differences in hairiness are the product of a genetic polymorphism (with hairy dominant to glabrous), and that the two morphs are conspecific. They provided the name *H. amplexicaulis* forma *hirta* for those wishing to refer, by name, to the hairy morph. This morph may resemble *H. gibbsii*, but differs in that both the upper and lower leaf surfaces are covered with hairs, not just the leaf margins (in *H. gibbsii*, hairs are only sometimes present on the upper leaf surface, and then are concentrated either near the base, or above the midvein, or near the apex).

Evidence of inbreeding depression in a line of glabrous morphs was provided by Garnock-Jones & Molloy (1983b). Varieties of *H. amplexicaulis* described by Cockayne & Allan (1926b), based largely on differences in habit, are not recognised here.

ETYMOLOGY L. (*amplexicaul* = stem clasping), refers to the characteristic leaf bases of the species.

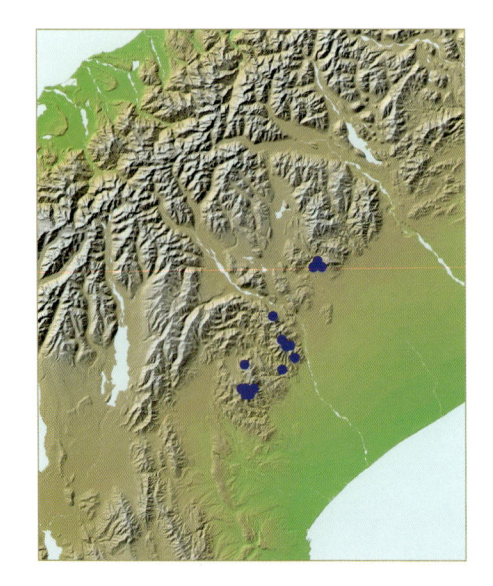

Hebe amplexicaulis **SUBCARNOSAE**

FIG. 70 **A** plant on Four Peaks Ra. **B** sprig. **C** leaf bud with no sinus. **D** lower and upper leaf surfaces. **E** flowering shoot. **F** leaf of f. *hirta*, covered with hairs. **G** lateral view of flower. **H** young (left) and mature infructescences, showing peduncles that are hairy (cf. *H. pareora*, Fig. 71) and comparatively long, and the condensed, opposite-decussate arrangement of sessile fruits/flowers.

18. *Hebe pareora* Garn.-Jones et Molloy

DESCRIPTION Spreading low shrub to 0.5 m tall (or hanging branches to 1(–3) m long). **Branches** pendent or ascending, old stems grey; branchlets green to brown or grey, glabrous; internodes (1.5–)6–10(–18) mm; leaf decurrencies evident. **Leaf bud** distinct; sinus absent. **Leaves** erecto-patent to patent; *lamina* ovate or obovate or oblong or elliptic, coriaceous, concave, (15–)20–30 × (10–)15–18 mm; *apex* rounded (us.) or subacute; *base* amplexicaul; 2 or more lateral *secondary veins* evident at base of fresh leaves; *midrib* not thickened; *margin* glabrous; *upper surface* glaucous, with many stomata, glabrous; *lower surface* glaucous. **Inflorescences** with (12–)20–40(–60) flowers, lateral, unbranched, (3–)4–7 cm, about equal to or longer than subtending leaves; peduncle 1.5–2.5(–3) cm, glabrous; rachis (2–)3–5 cm, glabrous or sparsely hairy (toward apex). **Bracts** alternate except sts for a basal whorl of 3, lanceolate to narrowly deltoid, acute or sts subacute, margins hairy (sparsely) or glabrous. **Flowers** [Nov–]Jan, ⚥. **Pedicels** shorter than or equal to bracts, 1–4.5 mm, glabrous (us.) or sparsely hairy (rarely). **Calyx** (2–)3–4.5 mm; *lobes* linear or oblanceolate, acute, margins glabrous or egl. ciliolate. **Corolla** tube glabrous, 2.5–3(–4) × 0.9–1.1 mm, cylindric to slightly funnelform, slightly longer than calyx; *lobes* white at anthesis, elliptic or oblong or broadly lanceolate, obtuse to acute, suberect to recurved, longer than corolla tube. **Stamen filaments** 4–5.5 mm; anthers magenta, 2–2.3 mm. **Ovary** ovoid to ellipsoid, 1.2–1.4 mm, apex (in septum view) slightly emarginate; ovules c. 15–25 per locule, marginal on a flattened placenta (but us. recurved and appearing somewhat scattered), in 1–2 layers; style 6.7–8 mm. **Capsules** Jan–Feb, latiseptate or angustiseptate to turgid, emarginate to didymous, 3.5–5 mm long, 2.5–3.5 mm wide, 1.5–2.8 mm thick, loculicidal split extending ½-way to base. **Seeds** weakly flattened, ellipsoid to obovoid, straw-yellow to pale brown, 1.3–1.5 × 0.8–0.9 mm. $2n = 40$.

DISTRIBUTION AND HABITAT SI, endemic to south Canterbury, where it is known from Rocky Gully, the upper Pareora River, White Rock River and Nimrod Stm, all of which drain the northeast slopes of The Hunters Hills. It grows on rock outcrops and cliffs in river and stream gorges. It probably also occurs in the gorge of the Opihi R. and near Blue Duck Stm (Garnock-Jones & Molloy 1983*a*), but there are no specimens from these localities.

NOTES Most similar to *H. amplexicaulis*, from which it is distinguished by (among other features) its glabrous peduncles and pedicellate (at least the lowermost) flowers (for both features cf. Figs 71H and 70H), usually larger leaves and often longer stems. Other notes on recognition are provided by Garnock-Jones & Molloy (1983*a*).

ETYMOLOGY refers to the Pareora R., which is the type locality and the centre of the narrow distribution of the species.

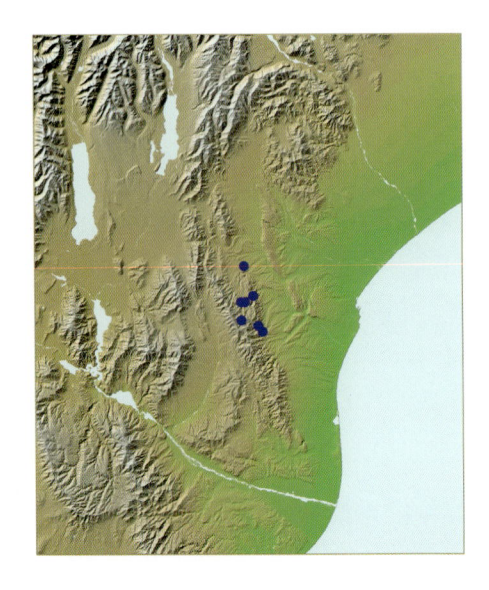

Hebe pareora   **SUBCARNOSAE**

FIG. 71   **A** plant at Mt Nimrod Scenic Reserve. **B** sprig. **C** leaf bud with no sinus. **D** lower and upper leaf surfaces. **E** top view of inflorescence. **F** lateral view of flower. **G** frontal view of flower. **H** young (left) and mature infructescences, showing that the peduncle is glabrous, and that at least the lowermost of the opposite-decussate fruits/flowers have conspicuous pedicels (cf. *H. amplexicaulis*, Fig. 70).

TAXONOMIC TREATMENT   133

19. *Hebe gibbsii* (Kirk) Cockayne et Allan

DESCRIPTION Sparsely branched, spreading low shrub to 0.35 m tall. **Branches** decumbent; branchlets green to orange-brown, pubescent (with long, multi-celled hairs), hairs bifarious or uniform; internodes (1–)2–6(–8) mm; leaf decurrencies evident and extended for length of internode (stem rounded and smooth). **Leaf bud** distinct; sinus absent. **Leaves** erect to patent (sts recurved with age); *lamina* ovate or elliptic (sts broadly), coriaceous or fleshy, ± concave, 9–20 × 4–13 mm; *apex* subacute (mostly) or obtuse or acute; *base* broadly cuneate or slightly amplexicaul; venation evident on underside of fresh leaves, us. not evident above, sts including 2 secondary laterals arising from base; *midrib* often slightly thickened below or not thickened; *margin* long ciliate, sts tinged red; *upper surface* glaucous, with many stomata, glabrous or hairy along midrib; *lower surface* glaucous. **Inflorescences** with (6–)11–25(–30) flowers, lateral, unbranched, 1.5–3.5 cm, longer than or about equal to subtending leaves; peduncle 0.5–2.1 cm; rachis 0.8–2.2 cm. **Bracts** lowermost pair opposite, then subopposite or alternate above, narrowly deltoid, acute (us.) or subacute, sts hairy outside. **Flowers** [Oct–]Dec–Feb(–Mar), probably ⚥. **Pedicels** absent or when present always shorter than bracts, 0–1 mm. **Calyx** *c.* 2.5–3.5 mm, with anterior lobes free for most of their length or united to ⅓–⅔-way to apex; *lobes* mostly narrowly deltoid, acute or subacute, often hairy outside. **Corolla** tube glabrous, 2.5–4 × 1.5–2 mm, cylindric, longer than or sts equalling calyx; *lobes* white at anthesis, elliptic or ovate, obtuse or subacute, suberect to patent, longer than or equalling corolla tube. **Stamen filaments** 6–6.5 mm; anthers magenta or dark purple or cream, 2–2.6 mm. **Ovary** sts hairy, *c.* 1–1.3 mm; ovules 13–19 per locule, in 1–2 layers; style 6.5–9.5 mm, sts hairy. **Capsules** Jan–May(–Nov), acute or subacute, 2.5–4 × 1.6–2 mm, sts hairy, loculicidal split extending ¼–½-way to base. **Seeds** flattened, ovoid-ellipsoid to discoid, ± smooth, brown (sts pale), 0.8–1.7 × 0.6–0.9 mm, MR 0.2–0.5 mm. $2n = 40$.

DISTRIBUTION AND HABITAT Mountains of eastern Nelson and western Marlborough, SI, where it is known with certainty only from Mt Starveall, Ben Nevis, Mt Rintoul and near Mt Patriarch. It grows in open, rocky areas.

NOTES Distinguished from other species by its thick glaucous leaves, with margins fringed with long hairs (see also notes under *H. amplexicaulis*).

Notes on cultivated specimens suggest that *H. gibbsii* may also occur on Dun Mountain (CHR 132094, WELT 17008), a locality also implied by a figure caption (but not the text) provided by Salmon (1992), and on Gordons Knob (CHR 332379), as also suggested by Martin (1932). A further specimen (WELT 67926) that lacked an original label was associated, prior to mounting, with collections made by F. G. Gibbs on Mt Franklin, Spenser Mts, 30 Jan 1896. It remains uncertain whether the specimen came from that locality (*c.* 50 km south of known localities), or was accidentally mixed with the other Gibbs collections.

A specimen from "Mt 'Z'", Wairau Va. (on the ridge running northwest from Mt Patriarch) is unusual. It has the stem and leaf characters of *H. gibbsii*, which is common in the area, but differs from other collections in having longer, sometimes branched, inflorescences and some flowers that are conspicuously pedicellate. The specimen might possibly be a hybrid (*H. divaricata*, which has branched inflorescences and pedicellate flowers, is common in this area), or the product of a developmental abnormality.

ETYMOLOGY Honours Frederick G. Gibbs (1866–1953).

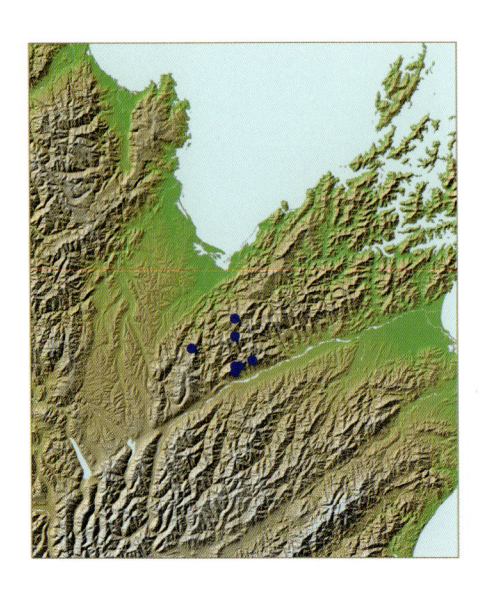

Hebe gibbsii   SUBCARNOSAE

FIG. 72   **A** plant near Mt Patriarch, Marlborough. **B** sprig; inset shows a variant with red leaf margins. **C** leaf bud with no sinus. **D** lower and upper leaf surfaces; magnified inset shows conspicuous hairs on leaf margin. **E** inflorescence showing frontal view of flower. **F** lateral view of flower. **G** capsule in septicidal (left) and loculicidal view. **H** infructescence and inflorescence, showing compact spikes on elongated peduncles.

20. *Hebe pinguifolia* (Hook.f.) Cockayne et Allan

DESCRIPTION Spreading low shrub (openly branched, or compact) to 0.4(–0.8) m tall. **Branches** decumbent or spreading or erect, old stems dark brown or grey; branchlets green (tinged maroon, esp. at nodes) or red-brown, puberulent, hairs bifarious; internodes (0.5–)1–7(–10) mm; leaf decurrencies evident. **Leaf bud** distinct; sinus absent (us.), or small and acute to rounded. **Leaves** erect or erecto-patent; *lamina* lanceolate (often broadly) to ovate or obovate, fleshy, concave, (3–)7–16(–22) × (2–)4–9(–12) mm; *apex* rounded or sts subacute; *midrib* very slightly thickened below; *margin* us. minutely papillate and rarely gl.-ciliate (toward leaf base), often tinged red; *upper surface* glaucous (us.) or glaucescent, with many stomata, glabrous; *lower surface* glaucous (us.) or glaucescent. **Inflorescences** with (4–)12–22 flowers, lateral, unbranched, 1–2.8(–3.4) cm, about equal to or longer than subtending leaves; peduncle 0.3–1.5(–2.1) cm; rachis 0.3–1(–1.5) cm. **Bracts** opposite and decussate (or apparently so) or lowermost pair opposite, then subopposite or alternate above, ovate (often narrowly) or deltoid, subacute. **Flowers** Dec–Feb(–Apr), ☿ or ♀ (on different plants). **Pedicels** absent or if evident then always shorter than bracts, 0–0.8 mm. **Calyx** (2–)2.7–3.2(–4) mm; *lobes* elliptic or oblong or ovate, subacute to obtuse. **Corolla** tube glabrous; tube of ☿ flowers 2–3 × *c.* 1–1.5 mm, cylindric or narrowly funnelform, *c.* equalling calyx; tube of ♀ flowers 2–2.5 × *c.* 1.5–1.8 mm, cylindric or narrowly funnelform, *c.* equalling calyx; *lobes* white at anthesis, ovate or lanceolate or elliptic, obtuse, suberect to recurved, longer than corolla tube. **Stamen filaments** 4.5–5 mm; anthers magenta, 2.1–2.3 mm; sterile anthers of ♀ flowers magenta or buff, 1.4–1.6 mm. **Ovary** ovoid or globose, hairy, 0.5–1.1 mm, apex (in septum view) obtuse or slightly didymous; ovules *c.* 8–13 per locule, in 1 layer (but sts a few ± overlapping); style (4–)5–7.5 mm, hairy. **Capsules** Jan–Apr, obtuse or truncate, 3–4.5 × 2.5–3.2 mm, us. hairy, loculicidal split extending ¼-way to base. **Seeds** flattened, ellipsoid to oblong, ± smooth, brown (sts pale), 0.9–1.7 × 0.6–1.1 mm, MR 0.3–0.6 mm. $2n = 40$ or 80.

DISTRIBUTION AND HABITAT Mountains of SI, east of the Main Divide, from the Bryant Ra. to the Kakanui Mts. It grows in open alpine areas, on rocks and debris slopes, sometimes in grassland.

NOTES A variable species, distinguished from most others by the combination of the shape and size of the glaucous leaves, glabrous leaf margins, sessile flowers, and the length of bracts relative to calyces. The limits of the species are not well defined, and differences from *H. buchananii* (see below) are problematic. It has sometimes been confused with *H. carnosula* (see below), and specimens are sometimes misidentified as *H. decumbens* and vice versa (see notes under that species).

No single character has been found to distinguish *H. pinguifolia* and *H. buchananii* consistently, and they are generally distinguished here on combinations of characters. *H. pinguifolia* plants are often taller (although sprawling, they do not tend to form dense mats) and usually have more distinct leaf buds, these not closely surrounded by recently diverged leaf pairs (except in some Marlborough specimens). They mostly have larger leaves (although shape is variable) that are not keeled when fresh (although they may appear so when dry, as the fleshy lamina shrinks away from the midrib). They may have more slender, less corky stems, and bracts and calyces that are usually shortly ciliolate with glandular hairs (but sometimes long-ciliate with eglandular hairs). In contrast, *H. buchananii* tends to be more mat-forming (except for "var. *exigua*-like" plants) and lower growing, with leaf buds closely surrounded by recently diverged leaves. It often has smaller leaves (although shape is variable) that are more keeled. It also often has thicker, more corky stems, and has calyces and bracts that often have longer cilia. Some specimens of *H. pinguifolia*/ *H. buchananii* have not been identified with certainty, and the distribution maps for both species are based only on specimens about whose identities we are reasonably confident. As defined here, there is some geographic overlap between the two species. Further investigation of their variation, relationships and circumscriptions would be worthwhile.

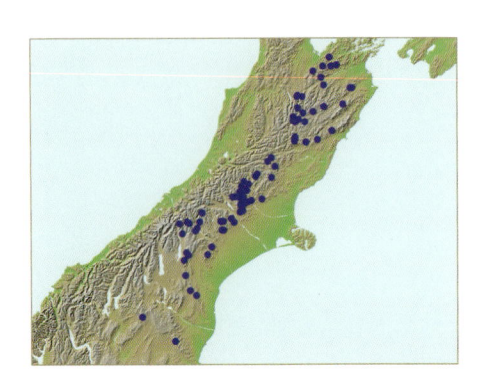

Included here in *H. pinguifolia* are specimens from the north of the species' range (e.g. Mt Starveall, Travers Ra., St Arnaud Ra., Hodder Va., Black Birch Ra.) that sometimes have a small but distinct sinus in the leaf bud (a feature seen only rarely on plants from other areas). In this respect, these specimens resemble *H. carnosula*, a name that has sometimes been applied to them. They can be distinguished from that species by usually red-edged leaves that are paler green (under a glaucous bloom), larger, strictly opposite bracts, and blunt and usually hairy capsules. Some of these northern specimens (particularly those from Black Birch Ra.) are quite

Hebe pinguifolia **SUBCARNOSAE**

FIG. 73 **A** plant on Mt Hutt. **B** sprig. **C** leaf bud with no sinus. **D** lower and upper leaf surfaces. **E** inflorescence showing frontal view of hermaphrodite flowers. **F** inflorescence showing frontal view of female flowers. **G** lateral view of hermaphrodite flower. **H** lateral view of female flower. **I** young infructescence showing opposite arrangement of bracts. **J** septicidal view of capsule. **K** loculicidal view of capsule.

small-leaved and, in this respect, may also resemble *H. buchananii*. However, given their geographic distance from that species, the resemblance is probably coincidental, and a close relationship does not seem likely.

Both diploid and tetraploid chromosome numbers are recorded in *H. pinguifolia* (Table 5), but chromosome variation has not been correlated with variation in morphology. Some vouchers for chromosome counts (diploid from Mt Peel, Canterbury; tetraploid from Mt Somers and Mt Winterslow) have been identified here as *H.* cf. *pinguifolia* (Table 5). These specimens are cultivated and sterile and cannot be identified with certainty. *H. pinguifolia* does, on the basis of other specimens, occur on Mt Peel and Mt Somers but is not otherwise known from Mt Winterslow (which is not included on the distribution map).

ETYMOLOGY L. (*pinguis* = fat, plump; *folium* = leaf), refers to leaf thickness/texture.

21. *Hebe buchananii* (Hook.f.) Cockayne et Allan

DESCRIPTION Spreading low shrub (often ± mat-like, but sts more upright) to 0.3 m tall. **Branches** decumbent (us.) or erect, old stems dark grey or brown or black; branchlets red-brown, puberulent to pubescent, hairs bifarious; internodes (0.5–)1–4(–9) mm; leaf decurrencies swollen. **Leaf bud** tightly surrounded by recently diverged leaves; sinus absent. **Leaves** erect to erecto-patent; *lamina* obovate to broadly elliptic or rarely almost circular, fleshy to rigid, concave, (1.5–)3–6(–8) × (1–)3–5(–6) mm; *apex* obtuse to rounded; *midrib* slightly keeled or thickened below, only sts evident in fresh leaves; *margin* glabrous or ciliate and often minutely papillate, sts tinged red (on young leaves); *upper surface* glaucescent or glaucous, with many stomata, glabrous; *lower surface* glaucescent or glaucous (us. not quite as glaucous as upper surface). **Inflorescences** with 3–12 flowers, lateral (us.) or terminal (clearly some on a cultivated plant from Landcare Research garden, G2278), unbranched, (0.5–)0.7–1.5(–2.3) cm; peduncle 0.2–0.6(–1.2) cm; rachis 0.3–1.2 cm. **Bracts** lowermost pair opposite, then subopposite or alternate above, broadly oblong or ovate or lowermost sts lanceolate, obtuse (us.) or subacute (sts lowermost pair). **Flowers** Dec–Mar, ⚥. **Pedicels** absent (us.) or if present then always shorter than bracts. **Calyx** 2.3–3(–3.4) mm; *lobes* ovate to oblong, subacute to obtuse, rarely hairy outside. **Corolla** tube glabrous, 1–1.9 × 1.5–1.8 mm, funnelform, shorter than calyx; *lobes* white at anthesis, ovate to lanceolate or elliptic, obtuse, suberect to patent, longer than corolla tube. **Stamen filaments** 4–4.7 mm; anthers magenta, *c.* 0.8–1.3 mm. **Ovary** broadly ovoid to globose, hairy (hairs often quite long), *c.* 0.6–0.8 mm, apex (in septum view) obtuse or slightly didymous; ovules *c.* 10–11 per locule; style 2.5–5 mm, hairy (esp. toward base). **Capsules** Feb–Apr(–Nov), obtuse or subacute, (2–)2.7–3.7 × 1.9–2.5 mm, hairy, loculicidal split extending ¼–½-way to base. **Seeds** weakly flattened, ovoid-ellipsoid to irregular, ± smooth, pale brown, 1–1.5 × 0.6–1 mm, MR 0.3–0.4 mm. $2n = 80$ or 40.

DISTRIBUTION AND HABITAT Mountains of SI, mostly east of the Main Divide, from the Malte Brun Ra., Aoraki/Mt Cook NP, to the Longwood Ra. It grows in open penalpine/subalpine areas on rocks (e.g. Figs 69 and 74A), debris slopes, in low shrubland, or sometimes in grassland.

NOTES Distinguished from most species by the combination of small, glaucous leaves, leaf buds closely surrounded by pairs of recently diverged leaves, sessile flowers, bracts about the same length as calyces, and white corollas. Differences from *H. pinguifolia* are not clear-cut, and are discussed under that species.

ETYMOLOGY Honours John Buchanan (1819–98), botanist, artist and one of the collectors of the type specimen.

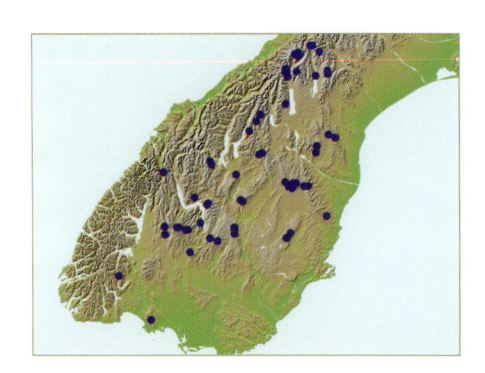

Hebe buchananii   SUBCARNOSAE

FIG. 74   A plant on Garvie Mts. **B** sprig. **C** leaf bud with no sinus. **D** lower and upper leaf surfaces. **E** inflorescence, showing lateral view of flowers. **F** young inflorescences, showing large bracts overlapping calyces. **G** frontal view of flower. **H** apical view of leaf bud, showing that it is surrounded by several imbricate, partly separated leaf pairs. **I** capsule in septicidal (left) and loculicidal view; magnified inset shows hairs. **J** seeds.

22. *Hebe pimeleoides* (Hook.f.) Cockayne et Allan

DESCRIPTION Spreading low or bushy shrub (subsp. *pimeleoides* us. very low-growing and spreading, sts mat-like; subsp. *faucicola* more upright) to 0.5(–0.9) m tall. **Branches** prostrate or decumbent or erect, old stems brown or grey; branchlets brown or red-brown or black, glabrous or pubescent, hairs bifarious or uniform; internodes (0.5–)2–10(–14.5) mm; leaf decurrencies obscure. **Leaf bud** distinct, or tightly surrounded by recently diverged leaves; sinus absent, or small and acute. **Leaves** erecto-patent to patent; *lamina* elliptic (narrowly to broadly) or lanceolate or ovate to sub-circular (rarely), subcoriaceous or coriaceous, flat to slightly concave, (2–)3.5–12(–15.5) × (0.7–)2–5(–8.7) mm; *apex* acute to subacute or occasionally obtuse; *midrib* not thickened or slightly thickened below, sts evident in fresh leaves; *margin* glabrous or rarely minutely ciliolate, sts tinged red; *upper surface* glaucous and light green (rarely mottled red), dull, with many stomata, glabrous (us.) or uniformly egl. pubescent; *lower surface* glaucous and light green (rarely mottled red), glabrous (us.) or uniformly egl. pubescent. **Inflorescences** with 4–12(–24) flowers, lateral, unbranched, (0.8–)1.5–5.5(–7) cm; peduncle (0.2–)0.4–2(–2.8) cm; rachis (0.2–)0.4–1.7(–3) cm. **Bracts** opposite and decussate, ovate to lanceolate or elliptic, acute to subacute, outside glabrous or hairy (on hairy-leaved plants). **Flowers** Dec–Mar, ⚥. **Pedicels** absent, or if evident then always shorter than bracts (us. only present on lowermost flowers), 0–1.2(–3.5) mm. **Calyx** (2–)3.5–4.5(–5.5) mm; *lobes* ovate to lanceolate (rarely linear-lanceolate) or elliptic, acute to subacute, egl. ciliate to ciliolate, sts hairy outside (on hairy-leaved plants). **Corolla** tube glabrous, (1–)1.5(–2) × 1.7–2(–2.5) mm, funnelform and sts contracted at base, shorter than calyx; *lobes* violet or blue to pale mauve at anthesis, violet or blue to pale mauve or sts fading to pale pink or almost white with age, obovate or ovate or elliptic or sts lanceolate, acute to obtuse, patent to recurved (with age), longer than corolla tube; corolla throat mauve to white. **Stamen filaments** mauve, fading to almost white, 3–4 mm; anthers pale pink or mauve or occasionally magenta, *c.* 1.9–2.1 mm. **Ovary** sts hairy, (0.8–)1–1.2(–1.5) mm; ovules 10–22 per locule, in 1–2 layers; style (2–)3–4.5 mm, sts hairy. **Capsules** (Jan–)Feb–May(–Dec), subacute (us.) or acute or obtuse, 3.5–5 × 2.2–3.2 mm, sts hairy, loculicidal split extending ¼–½-way to base. **Seeds** flattened, ellipsoid (sts broadly), pale brown, 0.8–1.3 × 0.6–1 mm, MR 0.2–0.5 mm. $2n = 40$ or 80.

Subsp. *pimeleoides*
Very low-growing shrub, up to *c.* 30 cm tall, branches prostrate or sprawling to decumbent, sometimes forming a dense mat. Leaves narrowly elliptic to elliptic or ovate, lamina (2–)3.5–8.9(–12.1) × (0.7–)1.5–4.5(–5.2) mm, usually glabrous, but sometimes with one or both surfaces covered in short egl. hairs. Inflorescences with 4–12 flowers. **Flowers blue or violet to mauve, fading to mauve after pollination.** Calyces and bracts ciliolate or ciliate on the margins and, on hairy-leaved plants, covered in egl. hairs. $2n = 40$ or 80.

Subsp. *faucicola* Kellow et Bayly
Small bushy shrub, up to *c.* 70(–90) cm tall, branches ascending to erect. Leaves lanceolate or elliptic to subcircular, 7.5–15.5 × (1.8–)3.1–8.7 mm, glabrous. Inflorescences with 4–12(–24) flowers. **Flowers mauve, fading to pale pink or almost white after pollination.** Calyces and bracts ciliolate or ciliate only. $2n = 40$ or 80.

DISTRIBUTION AND HABITAT Subsp. *pimeleoides* occurs on SI, on drier mountains east of the Main Divide, from the Inland Kaikoura Ra. to near Lk. Wakatipu, mostly on terraces, slopes or embankments near lakes and rivers. Subsp. *faucicola* occurs in central Otago, SI, in the Manuherikia, Clutha and Kawarau river valleys, in exposed positions on rocky outcrops and cliff faces, often in gorges. The species probably occurs on mountains between Lk. Aviemore and the Ida Ra. (not shown on map because we have not seen specimens); subsp. *pimeleoides* would most likely occur in this area, but Lovis (1990) recorded plants of subsp. *faucicola* (as var. *rupestris*) from near Lk. Aviemore.

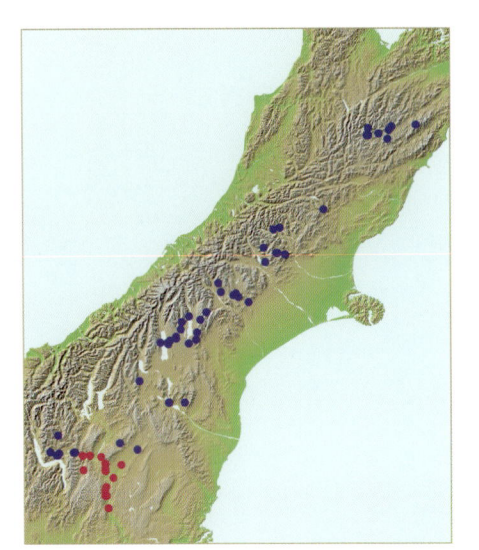

NOTES A distinctive but highly variable species, distinguished from other "Subcarnosae" by its leaf shape, habit and flower shape and colour.

Subsp. *pimeleoides* varies in stature, from small mat-forming plants (Fig. 75A) to sprawling plants with long trailing stems. It also varies in leaf shape and size (Appendix 4); internode length; inflorescence length; the indumentum of leaves, stems and ovaries; the presence/absence, or prominence, of a leaf bud sinus (Figs 75E and H); leaf colour; and in chromosome number (Table 5). These characters may vary both within and between populations,

• Subsp. *pimeleoides*
• Subsp. *faucicola*

Hebe pimeleoides   **SUBCARNOSAE**

FIG. 75   **A** subsp. *pimeleoides* at Lk. Sedgemere. **B** subsp. *faucicola* near Lk. Dunstan. **C** sprigs, subsp. *pimeleoides* (left) and subsp. *faucicola*. **D** lower and upper leaf surfaces, subsp. *pimeleoides* (top) *and* subsp. *faucicola*. **E–H** subsp. *pimeleoides*: E, leaf bud with a conspicuous sinus; F, lateral view of flower; G, frontal view of flower; H, leaf bud without an obvious sinus. **I–J** subsp. *faucicola*: I, lateral view of flower; J, frontal view of flower. **K** capsule in septicidal (left) and loculicidal view. **L** seeds.

TAXONOMIC TREATMENT   141

with some morphological traits varying on individual plants. Some traits may be related to environment; aspects of morphological variation are discussed by Kellow et al. (2003*a*).

Subsp. *faucicola* has generally been known to New Zealand botanists as *H. pimeleoides* var. *rupestris*; nomenclatural reasons for a change in name are outlined by Kellow et al. (2003*a*). It is distinguished from subsp. *pimeleoides* by its height, its stouter, ascending to erect branches, its generally paler flowers, and by its habitat, growing exclusively on rocky outcrops or cliff faces, in river valleys and in gorges. Like subsp. *pimeleoides*, the presence of a leaf bud sinus, stem and ovary indumentum and leaf shape (Appendix 4) are all variable, and two chromosome numbers are recorded (Table 5).

ETYMOLOGY *Pimeleoides* means "resembling *Pimelea*", a genus in the family Thymelaeaceae (Gk, *-oides* = resembling, like); *faucicola* means "gorge dweller" (L. *fauces* = gorge; *-cola* = dweller, only exists on), and refers to the usual habitat of the subspecies.

23. *Hebe biggarii*   (Cockayne) Cockayne

DESCRIPTION Small bushy shrub (sparingly branched, mostly near base) to 0.3(–0.5) m tall. **Branches** spreading or decumbent or ascending or erect, old stems grey; branchlets green or brown or red-brown or purplish, puberulent or pubescent or glabrous, hairs bifarious; internodes 3–10(–12) mm; leaf decurrencies evident. **Leaf bud** distinct; sinus absent. **Leaves** decussate to weakly subdistichous, erecto-patent to patent; *lamina* narrowly to broadly elliptic, coriaceous, concave or flat, (5–)7–15(–20) × (2–)4–8(–10) mm; *apex* acute to rounded; *midrib* not thickened, only sts evident in fresh leaves; *margin* glabrous, often tinged red (may vary in a population); *upper surface* glaucous, with many (but obscure) stomata, glabrous; *lower surface* glaucous. **Inflorescences** with 8–25 flowers, lateral, unbranched, (2–)2.5–4(–5) cm; peduncle (0.5–)1–1.5(–1.8) cm; rachis 1.5–2(–2.5) cm. **Bracts** alternate, linear to narrowly deltoid, subacute to acute. **Flowers** (Oct–)Nov–Jan[–Jun], ♀. **Pedicels** shorter than or equal to bracts, 1–3 mm. **Calyx** 1.8–2.5(–3) mm, 4–5-lobed (5th lobe small, posterior); *lobes* lanceolate, subacute or acute, with mixed gl. and egl. cilia (often more gl. than egl.). **Corolla** tube glabrous, 1.5–1.8 × 2–2.5 mm, contracted at base, equal to or slightly longer than calyx; *lobes* white or tinged pink at anthesis, elliptic, obtuse or subacute, erect to recurved, longer than corolla tube. **Stamen filaments** slightly diverging with age, 3–4 mm; anthers magenta, 1.6–1.8 mm. **Ovary** 1.4–1.8 mm; ovules *c*. 12 per locule; style (3.5–)4–5 mm. **Capsules** [Nov–]Jan–Apr[–May], often (together with calyx, rachis and peduncle) tinged pinkish-red when immature (but colour may vary dramatically within a single population), acute, 3.5–4 × 2–3 mm, loculicidal split extending ¼–⅓-way to base. **Seeds** flattened (sts weakly), ellipsoid to oblong-ellipsoid to irregular, straw-yellow to pale brown, 0.8–1.3 × 0.6–0.8 mm, MR 0.3–0.4 mm. $2n = 40$.

DISTRIBUTION AND HABITAT SI, endemic to Southland, where it is known from the Eyre Mts, Thomson Mts and Mid Dome. It grows primarily on rock outcrops (cliffs and bluffs), but also in open areas of snow tussock grassland.

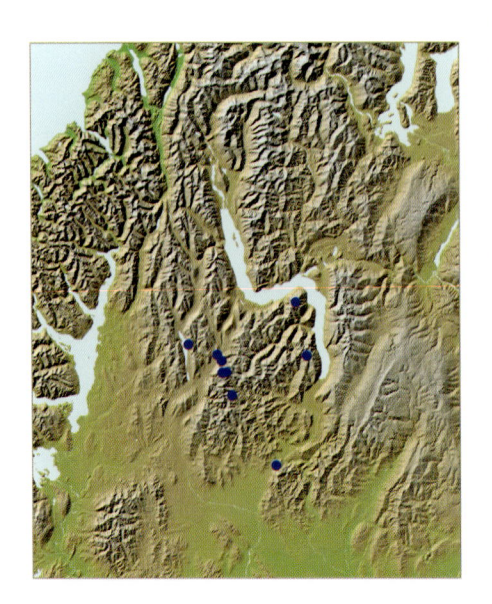

NOTES Differs from similar species of "Subcarnosae" (e.g. *H. pinguifolia* and *H. buchananii*) in having an erect habit, flowers (at least lowermost) that are conspicuously pedicellate, usually glabrous ovaries and capsules, and developing fruit with acute apices. It is readily distinguished from *H. pimeleoides* subsp. *faucicola*, which occurs in areas to the east and north, by its flower colour. Young infructescences are often bright red throughout – that is, including the peduncle, rachis, bracts, pedicels, calyx and developing capsules. Such extensive red pigmentation has not been seen in other *Hebe* species, but pigmentation of both leaves and inflorescences (Figs 76B, I and J) of *H. biggarii* may vary within a population. Moore (in Allan 1961) placed the species *incertae sedis*, but it has subsequently been generally accepted by New Zealand botanists (e.g. Parsons et al. 1998).

ETYMOLOGY Honours George Biggar (1885–1952) of Gore, who accompanied D. L. Poppelwell, collector of the type, on many botanical excursions.

Hebe biggarii   SUBCARNOSAE

FIG. 76   **A** plant on Eyre Mts. **B** sprigs of two individuals from the same population, showing variation in leaf colour. **C** leaf bud with no sinus. **D** lower and upper leaf surfaces. **E** flowering shoot. **F** lateral view of flower. **G** frontal view of flower. **H–J** young infructescences showing a range of variation in pigmentation. H also shows the comparatively long peduncle, and the condensed and opposite-decussate arrangement of the flowers/fruits on the inflorescence. **K** capsule in septicidal (left) and loculicidal view.

"Occlusae"

CRITICAL FEATURES Shrubs or small trees, prostrate to erect, 0.05–13 m tall; **leaf buds distinct**; **sinus absent**; leaf bases free; leaves not appressed or scale-like, entire or shallowly toothed, **rarely glaucous** (except *H. albicans*, *H. topiaria*, *H. glaucophylla*), abscising at nodes; inflorescences lateral; capsule latiseptate.

NUMBER OF SPECIES 31

DISTRIBUTION NI (including immediately surrounding islands), SI, Chatham Ids, and Rapa

NOTES "Occlusae" was divided into two groups by Moore (in Allan 1961). These were: "(a)", a group of large-leaved, chiefly lowland species, occurring mostly on NI and the Chatham Ids; "(b)", a group of small-leaved, subalpine and lowland species, occurring mostly on SI. These groups are retained here in the arrangement of species, and are referred to as "large-leaved" (species 39–54) and "small-leaved" (species 24–38) "Occlusae", respectively.

Despite a general difference in leaf size between these groups, variation in this and other characters is such that it was not possible to write a useful key to separate them clearly. A key to the two groups, and then to species, would involve much redundancy, with many species included more than once, and would increase the number of steps required to obtain identifications. For this reason, three geographically based keys are provided here: one to species of NI; one to species of SI; and one to species of the Chatham Ids and Rapa. If you do not know the geographic origin of a specimen, you will need to use all three keys and from them determine which gives a more likely result.

KEY TO SPECIES OF NORTH ISLAND (INCLUDING IMMEDIATELY SURROUNDING ISLANDS)

1. Flowers magenta to deep violet [North Cape area] .... **50. *Hebe brevifolia***
   Flowers white, violet, mauve, pink or blue 2

2. Inflorescences all branched [Three Kings Ids] .... **76. *Hebe insularis***
   (treated under "Apertae")
   Inflorescences all or mostly simple (some branched only in aberrant specimens) [not Three Kings Ids] 3

3. Stamen filaments of mature flower buds conspicuously incurved at apex (e.g. Fig. 43); leaves linear, linear-lanceolate or lanceolate (sometimes elliptic around Kaweka and Ruahine ranges), ≤ 10 mm wide 4
   Stamen filaments of mature flower buds straight or only very slightly hooked at apex; leaves of various shapes and widths 5

4. Shrub or small tree; leaf margins usually hairy; leaf surfaces smooth or faintly pitted with small depressions; corolla tube hairy inside .... **37. *Hebe parviflora***
   Shrub; leaf margins usually glabrous; leaf surfaces (at least below) conspicuously pitted with small depressions, each containing a twin-headed hair; corolla tube glabrous inside .... **35. *Hebe stenophylla***

5. Leaves slightly concave above, not prominently furrowed along midvein; plants of subalpine shrubland [Ruahine and Tararua ranges] 6
   Leaves flat or m-shaped in TS, or if concave above, then distinctly furrowed along midvein; plants of lowlands to subalpine zone [various areas of NI] 7

6. Plant with dense canopy (e.g. Fig. 80A); leaves often broadest above the midpoint, 12–28 mm long [Tararua Ra.] .... **26. *Hebe evenosa***
   Plant with more open canopy; leaves often broadest below the midpoint, (13–)18–42 mm long [Ruahine Ra.] .... **27. *Hebe truncatula***

7. Leaves linear or linear-lanceolate; plants of riverbanks or the rocky walls of river gorges 8
   Leaves of various shapes, if linear then plants not from riverbanks or walls of gorges 9

8	Corolla tube 1.3–2.5(–2.8) mm long, almost always shorter than calyx on posterior side; calyx lobes at least sparsely hairy outside [North Auckland]	44. *Hebe acutiflora*
	Corolla tube 2.2–4 mm long, longer than calyx; calyx lobes often glabrous outside [Gisborne and Wellington]	43. *Hebe angustissima*
9[1]	Youngest portion of branchlets strongly tinged maroon, purple or red; if green, then outsides of leaf buds tinged purple [near west coast, Muriwai Beach to Kawhia]	10
	Youngest portion of branchlets green, yellow or orange; outside of leaf buds (apart from leaf margins and midribs) not tinged purple [various localities]	11
10	Usually ± prostrate (e.g. Fig. 98A), < 0.5 m tall; leaves usually obovate or elliptic, (6–)20–55 mm long, margins usually conspicuously pubescent (e.g. Fig. 98D)	41. *Hebe obtusata*
	Semi-erect or sprawling shrub (e.g. Fig. 97A), to 1 m tall, leaves often lanceolate or narrowly elliptic, (20–)40–90 mm long, margins usually minutely pubescent	40. *Hebe bishopiana*
11	Ovary hairy; capsules ≥ 4.8 mm long; corolla lobes acute to subacute; posterior side of corolla tube ≤ calyx); free portion of stamen filaments ≥ 7 mm long; [western Northland, from Ahipara to Maungaraho Rock]	48. *Hebe perbella*
	Ovary glabrous or, if hairy, then either capsules < 4.8 mm long, or *both* corolla lobes obtuse (e.g. Fig. 99G) *and* posterior side of corolla tube > calyx [various localities]	12
12	Corolla tube ≥ 2.8 mm wide [Whangarei to Kawhia]	49. *Hebe macrocarpa*
	Corolla tube < 2.8 mm wide	13
13	Plants with leaves broadest at or above the midpoint, mostly ≥ 15 mm wide, apex shortly apiculate to obtuse or subacute; outside of calyx lobes glabrous (only very rarely hairy) [Poor Knights and Hen and Chickens Ids; nearby areas of the NI coast]	47. *Hebe bollonsii*
	Plants with *any of*: leaves broadest below midpoint, *or* leaves mostly < 15 mm wide, *or* leaf apex acute or acuminate, *or* outside of calyx lobes hairy (or if glabrous then plants of Wellington region) [various localities]	14
14	Corolla tube of male-fertile flowers ≤ 3 mm long, usually ≤ calyx [Dargaville northwards]	15
	Corolla tube of male-fertile flowers ≥ 3 mm long (except some plants of *H. stricta* var. *macroura* and var. *egmontiana*), always > calyx [various localities]	16
15	Leaves (30–)50–100(–135) × 6–20(–29) mm, often conspicuously narrowed toward apex (Appendix 4); calyx lobes hairy outside [mostly upland areas of central and western Northland]	45. *Hebe flavida*[2]
	Leaves (12–)26–50(–100) × (4.2–)6–10(–20) mm, often more evenly tapering toward apex (Appendix 4); calyx lobes often glabrous outside [mostly near coastal areas of eastern Northland, and North Cape to Cape Reinga]	46. *Hebe ligustrifolia*[2]
16	Leaves linear or narrowly lanceolate and tapering evenly from a broad base [Gisborne and northern Hawke's Bay]	42. *Hebe tairawhiti*
	Leaves of various shapes; if linear to narrowly lanceolate, then not from Gisborne or northern Hawke's Bay [throughout NI]	39. *Hebe stricta*

---

1 This couplet will not give correct identifications for specimens from Hikurangi Swamp, Northland (see notes under *H. stricta*).
2 Differences between these species are not clear-cut; see notes under *H. flavida*.

KEY TO SPECIES OF SOUTH ISLAND

1 Leaves conspicuously glaucous (at least on lower surface) 2
  Leaves glossy or dull, but not glaucous 4

2 Ovaries (always) and capsules (usually) minutely hairy [southeast Marlborough to
    mid-Canterbury] .... **32.** *Hebe glaucophylla*
  Ovaries and capsules glabrous (rarely hairy in *H. albicans*) [NW Nelson to Lewis Pass] 3

3 Corolla tubes ≤ 2.5 mm; erect, bushy shrub with a dense, rounded habit (the product of
    extensive branching toward shoot tips) [subalpine shrubland] .... **33.** *Hebe topiaria*
  Corolla tubes > 2.5 mm; plant sprawling or erect, usually not dense or rounded (usually
    sparingly branched toward shoot tips) [rocky places, lowland to subalpine] .... **34.** *Hebe albicans*

4 Ovaries and capsules minutely hairy 5
  Ovaries and capsules glabrous 7

5 Upper surfaces of fresh leaves dull; leaves frequently ± linear, often narrowed to a fine
    point or mucro at apex [Banks Peninsula and Port Hills] .... **38.** *Hebe strictissima*
  Upper surface of fresh leaves glossy; leaves usually not linear, often rounded or shortly
    apiculate at apex [not Banks Peninsula or Port Hills] 6

6 Mature leaves mostly > 2 cm [NW Nelson] .... **31.** *Hebe calcicola*
  Mature leaves mostly < 2 cm [eastern Marlborough to Southland] .... **30.** *Hebe rakaiensis*

7 Young branchlets purplish-red to almost black .... **29.** *Hebe decumbens*
  Young branchlets green, yellow, orange, light brown, or only faintly tinged with red 8

8 Leaf margins smooth and glabrous 9
  Leaf margins hairy (at least very minutely ciliate toward the apices of young leaves) 12

9 Plants decumbent, sprawling or ascending, < 30 cm tall; leaves obovate, oblanceolate
    or elliptic; corolla tube glabrous inside [subalpine shrubland or penalpine
    grassland] .... **28.** *Hebe treadwellii*
  Plants usually erect, only rarely creeping with ascending branches, usually > 30 cm tall;
    leaves linear, lanceolate, oblong or elliptic; corolla tube hairy or glabrous inside
    [lowland to penalpine] 10

10 Lower surfaces of leaves (and sometimes upper surfaces) pitted with small depressions,
    each containing a twin-headed hair; corolla tubes (1.8–)3–4.9 mm long, always > calyx;
    margins of calyx lobes usually with egl. hairs only .... **35.** *Hebe stenophylla*
  Lower surfaces of leaves not pitted; corolla tubes 1–2.2 mm long, sometimes *c.* equalling
    or only just longer than calyces; margins of calyx lobes with a mixture of egl. and gl. hairs 11

11 Leaves elliptic, (3–)11–18(–24) mm long; inflorescences 10–31 mm long; calyx lobes
    obtuse to subacute [mostly lowland, D'Urville Id to Bryant Ra.] .... **25.** *Hebe urvilleana*
  Leaves lanceolate, oblanceolate, linear or elliptic, (7–)12–31(–51) mm long; inflorescences
    (11–)19–60 mm long; calyx lobes acute to subacute [montane to subalpine, southwards
    from Wairau Va.] .... **24.** *Hebe subalpina*

12 Corolla tube *c.* equalling or only very slightly longer than surrounding calyx lobes
    (e.g. Fig. 93F) [Banks Peninsula and Port Hills] .... **38.** *Hebe strictissima*
  Corolla tube clearly > surrounding calyx lobes (e.g. Figs 92F and 90G)
    [not Banks Peninsula or Port Hills] 13

13	Leaves > both 7 mm wide and 30 mm long (commonly much larger), usually m-shaped in TS, often conspicuously tapered in apical third .... **39. *Hebe stricta* var. *atkinsonii***
	Leaves mostly either < 7 mm wide or < 30 mm long; if larger, then slightly concave or ± flat in TS; usually evenly tapered from mid-point or near base, or not strongly tapering to apex **14**

14	Leaf surfaces (at least below) conspicuously pitted with small depressions, each containing a single twin-headed hair; margins of calyx lobes usually with egl. hairs only; inside of corolla tubes glabrous (except in near-coastal areas of NW Nelson) .... **35. *Hebe stenophylla***
	Leaf surfaces not, or only faintly, pitted; margins of calyx lobes with a mixture of egl. and gl. hairs; inside of corolla tubes hairy **15**

15	Capsules 2.5–3.5 mm long; a small tree (to 12 m) when mature, with a compact canopy or dome-like shape when young (e.g. Fig. 92A) .... **37. *Hebe parviflora***
	Capsules (3.5–)4–5(–5.5) mm long; usually an openly-branched shrub (e.g. Fig. 91A) .... **36. *Hebe traversii***

KEY TO SPECIES OF THE CHATHAM ISLANDS AND RAPA

1	Corolla tube ≤ 2 mm long; at least slightly shorter than surrounding calyx 2
	Corolla tube ≥ 2.5 mm long; often *c.* equal to or longer than surrounding calyx 3

2	Calyx lobes hairy outside, at least sparsely; small tree [Chatham Ids] .... **51. *Hebe barkeri***
	Calyx lobes hairy on margins only; bushy shrub [Rapa] .... **54. *Hebe rapensis***

3	Plants usually ≤ 30 cm tall; inflorescences usually < 4.5 cm long .... **53. *Hebe chathamica***
	Plants usually > 30 cm tall; inflorescences usually > 4.5 cm long .... **52. *Hebe dieffenbachii***

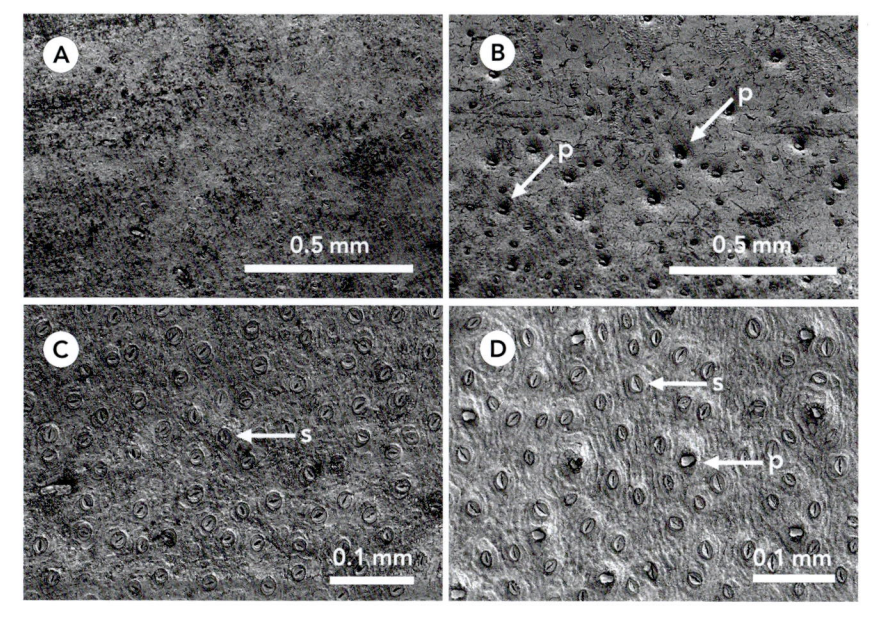

FIG. 77 Scanning electron micrographs of upper (**A**, **B**) and lower (**C**, **D**) leaf surfaces of *H. parviflora* (A, C) and *H. stenophylla* var. *stenophylla* (B, D). Shallow pits (p), each containing a glandular hair, are evident on the leaves of *H. stenophylla*. Stomata (s) are also evident on the leaves of both species, especially on the lower surfaces.

24. *Hebe subalpina* (Cockayne) Andersen

DESCRIPTION Bushy or spreading low shrub to 1.4 m tall. **Branches** erect or spreading or ascending or decumbent, old stems brown or red-brown or black (on drying); branchlets green (with dark bands at nodes), pubescent (often minutely), hairs bifarious; internodes (0.5–)3–13(–16) mm; leaf decurrencies evident. **Leaf bud** distinct; sinus absent. **Leaves** erecto-patent to patent; *lamina* lanceolate or elliptic or oblong-elliptic or almost linear (e.g. some plants in forest rather than above tree-line), subcoriaceous, concave, (7–)12–31(–51) × (3–)5–11 mm; *apex* subacute to obtuse; *margin* very narrowly cartilaginous, glabrous; *upper surface* light to dark green, glossy, with many stomata, hairy along midrib; *lower surface* light green. **Inflorescences** with (4–)8–32 flowers, lateral, unbranched, (1.1–)1.9–6 cm, us. longer than or sts about equal to subtending leaves; peduncle (0.3–)0.6–1.8 cm; rachis (0.8–)1.3–4.2 cm. **Bracts** alternate, ovate or lanceolate, subacute to acute. **Flowers** [Nov–]Dec–Feb, ♂ or ♀ (on different plants). **Pedicels** 0.8–4(–5.5) mm (can vary considerably within one inflorescence, us. longer near base). **Calyx** 1.7–3.4 mm; *lobes* elliptic or lanceolate, subacute or acute. **Corolla** tube glabrous or hairy inside; tube of ♂ flowers 1–2.2 × *c.* 1.7–1.8 mm, funnelform, *c.* equalling or longer than calyx; tube of ♀ flowers *c.* 1–2.2 × 1.7–2.1 mm, funnelform, equalling or longer than calyx; *lobes* white or faintly tinged pink or mauve at anthesis, broadly to narrowly ovate or elliptic or deltoid, obtuse or subacute, suberect to patent, longer than corolla tube, us. papillate inside. **Stamen filaments** 3–6 mm (staminodes 0.7–1 mm); anthers pale pink or mauve, (1.5–)1.8–2.3 mm; sterile anthers of ♀ flowers pink or mauve or white, 0.7–1.1 mm. **Ovary** 0.8–1.2 mm; ovules *c.* 10–15 per locule; style 2–9.5 mm. **Capsules** [Dec–]Jan–May, obtuse or subacute, 3–4.2 × 2.4–2.9 mm, loculicidal split extending up to ⅓-way to base. **Seeds** flattened, ellipsoid (sts broadly), not or only weakly winged, pale brown, 1.2–2 × 0.9–1.4 mm, MR 0.4–0.8 mm. $2n = 80$.

DISTRIBUTION AND HABITAT Widespread on SI, chiefly on wetter mountains on and west of the Main Divide, from the Wairau Va. to the Cameron Mts. It grows in subalpine shrubland, penalpine grassland, and sometimes in beech forest close to the tree-line or along streams.

NOTES A variable species (Appendix 4), similar to a number of other small-leaved "Occlusae" (see notes under *H. rakaiensis*, *H. treadwellii*, *H. urvilleana*, *H. calcicola* and *H. truncatula*). Plants in beech forest tend to have longer internodes and leaves than those in open places. Plants cultivated at Lincoln (e.g. CHR 103130; Table 5) and at Otari-Wilton's Bush (illustrated by Eagle 1982, pers. comm. 2003) were reportedly from the Garvie Mts. The species is not otherwise recorded there, and the locality is omitted from the distribution map.

The circumscription adopted here includes *H. fruticeti* G.Simpson & J.S.Thomson (see Brandon 1995 for further discussion). Aspects of reproductive biology are discussed by Delph (1988, 1990*a*, 1990*c*) and Delph & Lloyd (1991, 1996).

ETYMOLOGY L. (*sub-* = below, under, almost; *alpinus* = alpine), refers to the species' habitat.

Hebe subalpina   OCCLUSAE

FIG. 78   **A** plant at Lk. Tennyson. **B** sprig. **C** leaf bud with no sinus. **D** lower and upper leaf surfaces; magnified inset shows glabrous leaf margin. **E** inflorescences of hermaphrodite (left) and female plants. **F** top view of inflorescence, hermaphrodite. **G** frontal view of female flower. **H** lateral view of hermaphrodite flower. **I** capsule in septicidal (left) and loculicidal view.

25. *Hebe urvilleana* W.R.B.Oliv.

DESCRIPTION Bushy shrub to 1.5 m tall. **Branches** erect (us.) or prostrate, old stems grey; branchlets green, pubescent, hairs bifarious; internodes (0.5–)2–6(–10) mm; leaf decurrencies evident. **Leaf bud** distinct; sinus absent. **Leaves** erect to erecto-patent; *lamina* elliptic (us.) or lanceolate or oblanceolate, coriaceous, concave, (3–)11–18(–24) × (2–)4–6(–8) mm; *apex* acute to subacute; *midrib* slightly thickened below and depressed to grooved above; *margin* glabrous and sts minutely papillate; *upper surface* green, us. glossy, with many stomata, glabrous or hairy along midrib; *lower surface* green, dull or glossy. **Inflorescences** with 5–23 flowers, lateral, unbranched, 1–3.1 cm; peduncle (0.3–)0.4–0.5(–1) cm; rachis 0.7–2.5 cm. **Bracts** alternate, elliptic, subacute or acute. **Flowers** Jan–Feb, ♂ or ♀ (on different plants). **Pedicels** (1–)2–3(–4) mm. **Calyx** 2.2–2.9 mm; *lobes* elliptic or ovate, obtuse or subacute. **Corolla** tube sparsely hairy inside; tube of ♂ flowers 1.8–2.2 × 1.8–2.2 mm, contracted at base, longer than calyx; *lobes* white at anthesis, elliptic (sts broadly), obtuse (posterior sts emarginate), suberect to patent, longer than corolla tube. **Stamen filaments** slightly incurved at apex in bud, 3.7–4.7 mm; anthers dark magenta or purple, 1.7–2 mm. **Ovary** 1–1.3 mm; ovules *c.* 8–9 per locule; style 5–8 mm. **Capsules** Feb–Apr(–Nov), subacute, (3.5–)4–5.3 × 2.2–3 mm, loculicidal split extending ¼-way to base. **Seed** characters not recorded. $2n = 120$.

DISTRIBUTION AND HABITAT D'Urville Id and nearby areas of eastern Nelson, SI, possibly as far south as Mt Starveall, Bryant Ra. It grows chiefly in mānuka scrub and open shrubland.

NOTES Distinguished from most other small-leaved "Occlusae" by the combination of a shrubby habit, leaves that are glossy above, glabrous leaf margins, glabrous ovaries and fruit, and corolla tubes that are hairy within and longer than the surrounding calyces. It has sometimes been confused with *H. stenophylla* vars *stenophylla* and *oliveri* (which also occur on D'Urville Id and/or nearby parts of the outer Marlborough Sounds), but these, in comparison, have leaves that are duller, and very prominently pitted on the lower surface, longer corolla tubes (2.5–4.9 mm) that are usually glabrous within, and calyces that usually have only egl. cilia. The affinities of *H. urvilleana* are not immediately apparent, but it has a general similarity to *H. subalpina* and, among the small-leaved "Occlusae", the chromosome number $2n = 120$ is shared only with *H. evenosa* (although this number could have developed independently in both species). The species shows considerable variation in leaf shape and size, even within one population (e.g. Appendix 4, WELT 81684). Other hebes of D'Urville Id are *H. stricta* var. *atkinsonii*, *H. rigidula* var. *sulcata*, *H. elliptica*, *H. stenophylla* and *H. divaricata*.

The occurrence of the species in the Bryant Ra. is supported by two specimens collected by A. P. Druce in Jan 1981 (CHR 387363, 387364). These are sterile, and identified only on the basis of vegetative features. Flowering specimens from this area might allow more confident identification of these plants.

ETYMOLOGY Refers to D'Urville Id, which is where the species was first collected and originally described from.

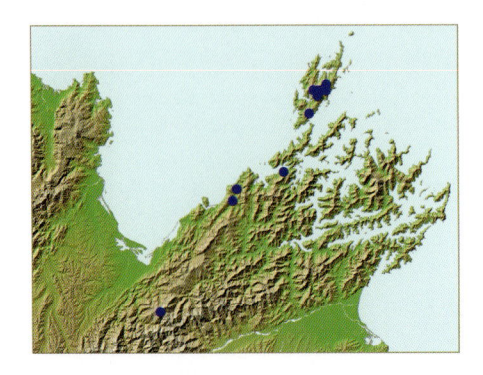

Hebe urvilleana   OCCLUSAE

FIG. 79   **A** plant on D'Urville Id. **B** sprig. **C** leaf bud with no sinus. **D** lower and upper leaf surfaces. **E** inflorescence showing frontal view of flowers. **F** lateral view of flower. **G** dorsal view of flower. **H** capsule in septicidal (left) and loculicidal view.

TAXONOMIC TREATMENT   151

26. *Hebe evenosa* (Petrie) Cockayne et Allan

DESCRIPTION Bushy shrub (us. with a rounded habit) to 2 m tall. **Branches** erect, old stems grey to brown; branchlets green, puberulent to pubescent, hairs bifarious to uniform (hairs on decurrencies, when present, often somewhat shorter); internodes 2–9 mm; leaf decurrencies evident. **Leaf bud** distinct; sinus absent. **Leaves** us. erecto-patent to patent or recurved (with age); *lamina* obovate or oblanceolate to elliptic, concave, 12–28 × 4–9 mm; *apex* obtuse (us.) or subacute (rarely) or shortly apiculate; *margin* ciliolate; *upper surface* dark green, ± glossy, glabrous or hairy along midrib; *lower surface* dull or glossy. **Inflorescences** with 15–40 flowers, lateral, unbranched, 1.4–5 cm; peduncle 0.5–1.1 cm; rachis 0.8–4 cm. **Bracts** alternate, narrowly deltoid or lanceolate or elliptic, subacute or acute. **Flowers** Jan–Feb, ⚥ or ♀ (on different plants). **Pedicels** longer than or equal to bracts, 0.5–3.3 mm. **Calyx** 2–2.5 mm, 4–5-lobed (5th lobe small, posterior); *lobes* ovate or elliptic, obtuse (us.) or subacute. **Corolla** tube hairy inside; tube of ⚥ flowers *c.* equalling calyx; tube of ♀ flowers 1.5–2 mm, shorter than calyx; *lobes* white at anthesis, rhomboid or ovate to elliptic, obtuse, suberect to patent (us.) or recurved, longer than corolla tube, sts with a few hairs toward base on inner surface. **Stamen filaments** 2.5–5 mm; anthers mauve or pink, *c.* 0.9–1.2 mm. **Ovary** *c.* 0.9–1.1 mm; ovules 10–12 per locule; style 3–7.2 mm. **Capsules** Feb–Apr(–Jun), obtuse, *c.* 3–3.7 × 3–4 mm, loculicidal split extending ⅓–¾-way to base. **Seeds** flattened (sts strongly), broad ellipsoid to discoid, weakly winged, ± smooth, brown, 1.2–1.5 × 1.1–1.3 mm, MR 0.2–0.3 mm. $2n = 120$.

DISTRIBUTION AND HABITAT Endemic to the Tararua Ra., NI, in the areas around Mitre, Mt Holdsworth, Field Hut, Dennan and Mt Hector. It grows in subalpine shrubland, often close to the tree-line.

NOTES Differs from *H. truncatula*, the other small-leaved subalpine member of "Occlusae" from NI, in: habit, with branches highly branched toward the apex, producing a dense canopy (Fig. 80A); usually smaller, oblanceolate to elliptic (rather than lanceolate to elliptic) leaves (Appendix 4); chromosome number (Table 5); and flavonoid profile (Mitchell et al. in prep.). It has not been determined whether differences in anther colour shown in Figs 80 and 81 are consistent. The species resembles, and is possibly closely related to, *H. topiaria*, from which it differs most obviously by its non-glaucous leaves.

A few collections – for example, by Petrie and Aston (WELT 16842, 16843, upper piece of 47655) and by Sneddon (WELT 83173), from the Tararua Ra. – have larger leaves than generally seen in *H. evenosa*. These specimens could be hybrids with *H. stricta*; this hybrid combination was reported for the area by Druce (1968). However, the extent of variation in leaf size in *H. evenosa* has not been thoroughly assessed, since many herbarium specimens come from the same small populations.

The locality, "base of Mt Ruapehu", given on WELT 16845 is not likely to be correct. There are no other records from this well-collected area, and it is not represented on the distribution map.

ETYMOLOGY *Evenosa* means "without veins" or "veinless", and refers to the appearance of the leaves.

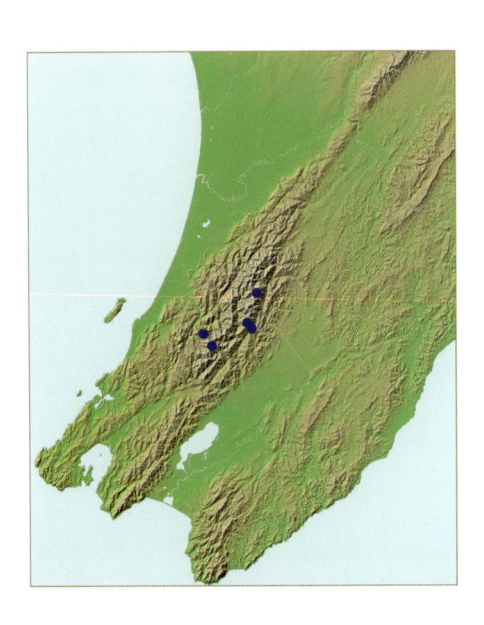

Hebe evenosa   OCCLUSAE

FIG. 80   **A** plants on Tararua Ra., near Powell Hut. **B** sprig. **C** leaf bud with no sinus. **D** lower and upper leaf surfaces; magnified inset shows minute hairs on margin. **E** inflorescence. **F** frontal view of flower. **G** lateral view of flower. **H** septicidal view of capsule. **I** loculicidal view of capsule.

TAXONOMIC TREATMENT   153

27. *Hebe truncatula* (Colenso) L.B.Moore

DESCRIPTION Bushy shrub to 2 m tall. **Branches** erect, old stems brown; branchlets green, pubescent or puberulent, hairs uniform or bifarious; internodes 2–10 mm; leaf decurrencies evident. **Leaf bud** distinct; sinus absent. **Leaves** erecto-patent; *lamina* lanceolate or narrowly elliptic or narrowly oblong, coriaceous, slightly concave, (13–)18–42 × (3.7–)5.2–8.5(–10.3) mm; *apex* truncate or slightly apiculate or subacute; *midrib* conspicuously depressed to grooved above and thickened below; *margin* ciliate or pubescent; *upper surface* dark green, glossy, without evident or with few stomata, hairy along midrib; *lower surface* light green. **Inflorescences** with 28–68 flowers, lateral, almost always unbranched but rarely (only seen on a cultivated specimen) tripartite, 3–7.5 cm, with all flowers (including those near the apex) generally developing to maturity (although sts with a few undeveloped flowers); peduncle 0.8–1.8 cm; rachis 2.4–6 cm. **Bracts** alternate, linear or deltoid or elliptic or oblanceolate, acute or obtuse. **Flowers** [Nov–]Jan–Feb[–Mar], ☿ or ♀ (on different plants). **Pedicels** 0.8–1.8 mm. **Calyx** 2–3 mm; *lobes* lanceolate or elliptic, obtuse or subacute. **Corolla** tube hairy inside; tube of ☿ flowers 2–3.2 × *c.* 1.7–2.2 mm, funnelform or cylindric, slightly shorter to longer than calyx; *lobes* white or tinged mauve at anthesis, ovate or elliptic, obtuse, suberect to patent, longer than or equalling corolla tube. **Stamen filaments** 4.5–5.5 mm; anthers purple, *c.* 1.8–2.1 mm. **Ovary** *c.* 1.2–1.5 mm; ovules 15–28 per locule, in 1–2 layers; style 5–7 mm. **Capsules** Apr–May, obtuse or truncate to subacute, 3–4 × 2–3 mm, loculicidal split extending from ¼ to all the way to base (although often barely splitting at all). **Seed** characters not recorded. $2n = 80$.

DISTRIBUTION AND HABITAT Ruahine Ra., from near Reporoa Bog to near Maharahara, in subalpine shrubland. A specimen labelled "Tararua Range" (CHR 132237, *N. L. Elder*) possibly also belongs to this species.

NOTES Generally resembles *H. subalpina*, and Druce (1993), Wilson & Galloway (1993) and Brandon (1995) considered it should be recognised as a subspecies or variety of that species. However, it can be distinguished from *H. subalpina* by its hairy leaf margins, longer corolla tubes and by its flavonoid profile (Mitchell et al. in prep.), and it is at least as distinct from *H. subalpina* as are some other species (e.g. *H. calcicola*). Because there is no compelling evidence that *H. subalpina* is necessarily its closest relative, *H. truncatula* is retained here as a distinct species.

Differences from *H. evenosa*, which also occurs on NI, are discussed under that species. *H. truncatula* differs from *H. calcicola*, which has similar glossy leaves with hairy leaf margins, by its glabrous ovaries and fruit, and longer corolla tubes. Leaves are illustrated in Appendix 4.

ETYMOLOGY L. (*truncatus* = truncate, that is ending very abruptly as if cut straight across; *-ulus* is a diminutive suffix), refers to the leaf apex.

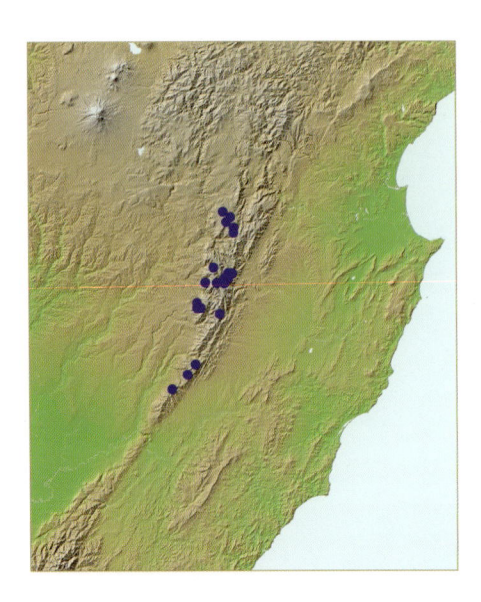

Hebe truncatula   OCCLUSAE

FIG. 81   **A** plant near Rangiwahia Hut, Whanahuia Ra. **B** sprig. **C** leaf bud with no sinus. **D** lower and upper leaf surfaces; magnified inset shows hairs on margin. **E** inflorescence. **F** frontal view of flowers. **G** lateral view of flower. **H** septicidal view of capsule. **I** loculicidal view of capsule.

TAXONOMIC TREATMENT   155

28. *Hebe treadwellii* Cockayne et Allan

DESCRIPTION Spreading low shrub to 0.3 m tall. **Branches** decumbent or ascending, old stems dark brown or red-brown; branchlets green or red-brown, pubescent to puberulent or glabrous, hairs bifarious; internodes 1.5–10(–11.5) mm; leaf decurrencies evident (sts weakly). **Leaf bud** distinct; sinus absent. **Leaves** decussate to somewhat subdistichous, erecto-patent; *lamina* elliptic to oblanceolate or obovate, subcoriaceous to fleshy, concave, (10.7–)13–29(–31.6) × (4.2–)5.5–12(–14.6) mm; *apex* subacute to obtuse; 2 lateral *secondary veins* sts evident at base of fresh leaves; *midrib* slightly depressed to grooved above; *margin* us. cartilaginous, glabrous; *upper surface* light to dark green, glossy, with many stomata (but these not always readily visible), glabrous or hairy along midrib; *lower surface* green or light green, dull or glossy. **Inflorescences** with (5–)8–34 flowers, lateral, unbranched, 1.5–3.1 cm, about equal to or slightly longer than subtending leaves; peduncle 0.5–1.4 cm; rachis (0.7–)0.9–2.2 cm. **Bracts** alternate (sts with lowermost pair ± opposite), linear or lanceolate or ovate or oblong or deltoid, obtuse to acute. **Flowers** Dec–Feb, ☿ or ♀ (on different plants). **Pedicels** 0–2.5 mm. **Calyx** 2.2–3.3 mm; *lobes* oblong to lanceolate or deltoid, acute to subacute. **Corolla** tube glabrous; tube of ☿ flowers 1.9–3.5 × 1.3–2 mm, cylindric or funnelform and sts contracted at base, equalling or longer than calyx; tube of ♀ flowers 1.3–1.8 × 1.5–1.7 mm, contracted at base, shorter than or equalling calyx; *lobes* white at anthesis, ovate (anterior sts narrowly) or elliptic or oblong-elliptic, obtuse (but sts appearing subacute because of margin inrolling), suberect to patent, longer than or equalling corolla tube. **Stamen filaments** 1.5–4.5 mm; anthers pink (sts pale) or cream; sterile anthers of ♀ flowers yellow, 0.9–1.2 mm. **Ovary** conical to ovoid; ovules *c.* 12–17 per locule, in 1–2 layers; style 3.5–10.5 mm; stigma larger in ♀ flowers. **Capsules** [Dec–]Feb(–Oct), acute or obtuse, (2.5–)2.9–5 × 1.6–3 mm, loculicidal split extending ¼–½-way to base. **Seeds** flattened, ellipsoid to discoid, not winged to only weakly winged, straw-yellow or pale brown, 0.9–1.5 × 0.8–1.1(–1.3) mm, MR 0.3–0.6 mm. $2n = 40$.

DISTRIBUTION AND HABITAT Mountains of Nelson, Canterbury and Westland, on or west of the Main Divide, from Bald Knob Ridge to the Selbourne Ra. It grows in subalpine shrubland and penalpine grassland.

NOTES Placed *incertae sedis* by Moore (in Allan 1961), but generally recognised as a distinct species by subsequent authors (e.g. Wardle 1975; Druce 1980; Heads 1993; Wilson & Galloway 1993; Wilson 1996; Parsons et al. 1998). It is similar to, and often confused with, *H. subalpina*. It is distinguished from that species primarily on the basis of habit, but also tends to have relatively broader, dish-like, obovate or oblanceolate to elliptic leaves (Appendix 4), and sometimes longer corolla tubes. It also has a different flavonoid profile (Mitchell et al. in prep.).

Near the type locality, Sealy Ra. (Aoraki/Mt Cook area), and presumably at other locations where *H. treadwellii* and *H. subalpina* occur in close proximity (e.g. Fig. 82), the two species are readily distinguished in the field, on the basis of habit and leaf shape. Reliable identification of herbarium specimens is more problematic, because of variation in leaf shape in both species, and because details of habit are often lost when plants are pressed. For this reason, the map shown here may underestimate the distribution of *H. treadwellii*.

Included are specimens from near Amuri Pass described by Simpson & Thomson (1942) as *H. brockiei*, and those from Bald Knob Ridge identified by Druce & Courtney (1989) as *H. matthewsii* (a name placed *incertae sedis* here).

FIG. 82 *Hebe subalpina* (left) and *H. treadwellii* (indicated by arrow) growing together on slopes above Amuri Pass (the type locality for *H. brockiei*).

Hebe treadwellii   OCCLUSAE

FIG. 83   **A** plant at Amuri Pass. **B** sprig of plant from Sealy Tarns (top) and plant from Amuri Pass. **C** leaf bud with no sinus. **D** lower and upper leaf surfaces; magnified inset shows glabrous leaf margin. **E** lateral view of hermaphrodite flower. **F** lateral view of female flower. **G** frontal view of hermaphrodite flowers. **H** inflorescences of hermaphrodite (left) and female plants. **I** septicidal view of capsule. **J** loculicidal view of capsule.

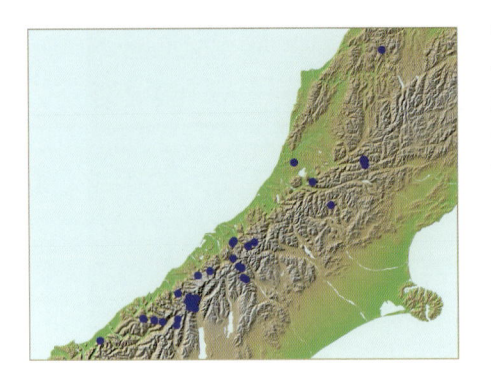

These specimens have the lax habit and dish-like leaves of *H. treadwellii* and, where examined, similar flavonoid profiles (Mitchell et al. in prep.). Specimens of *H. brockiei* (e.g. Appendix 4, WELT 82196, 81709) differ substantially from the type of *H. treadwellii* in the size of their leaves but, given the variation in leaf size among other specimens, no clear grounds have been found to separate these taxa. Plants from Bald Knob Ridge are available in some New Zealand garden centres under the name *Hebe* 'Bald Knob'.

Specimens from Mt Cuthbert, identified by Garnock-Jones et al. (2000) as *H. treadwellii*, are included under *H. subalpina*.

ETYMOLOGY Honours Charles H. Treadwell (1862–1936).

29. *Hebe decumbens* (J.B.Armstr.) Cockayne et Allan

DESCRIPTION Openly branched, spreading low shrub to 0.35 m tall. **Branches** decumbent to erect; branchlets purplish-red to almost black, pubescent, hairs bifarious (us.) or uniform; internodes 1.5–16.5 mm; leaf decurrencies obscure to evident. **Leaf bud** distinct; sinus absent. **Leaves** erecto-patent; *lamina* elliptic (often broadly) or obovate or oblanceolate, concave, 6.5–23.5 × 2–13 mm; *apex* subacute or acute; *midrib* not thickened or depressed to grooved above and thickened below, only sts faintly evident in fresh leaves; *margin* ciliate (with very short, stiff hairs) or glabrous (only very rarely, with hairs us. present at least near apices of youngest leaves), almost always tinged red; *upper surface* dark green, glossy, with many stomata, glabrous. **Inflorescences** with 2–25 flowers, lateral, unbranched, 0.6–3 cm, longer than (us.) to shorter than (rarely) subtending leaves; peduncle 0.15–1.6 cm; rachis 0.4–2.4 cm. **Bracts** alternate (although lowermost bracts may be in a ± opposite pair, or whorl of three), deltoid (sts narrowly) or ovate, acute (us.) to obtuse, glabrous outside. **Flowers** (Nov–)Dec–Feb, ⚥. **Pedicels** 0.2–1.7 mm, hairy or sts glabrous. **Calyx** 1–3 mm, 4–5-lobed (5th lobe small, posterior); *lobes* deltoid or ovate, acute or acuminate or obtuse, with mixed gl. and egl. cilia (gl. cilia sts greatly outnumbering egl. cilia). **Corolla** tube glabrous, 3–6 × 1.8–2 mm, cylindric and contracted at base, much longer than calyx; *lobes* white at anthesis, elliptic (sts broadly) or ovate, obtuse (us.) to acute (posterior sts slightly emarginate), suberect to patent, shorter than corolla tube. **Stamen filaments** at least slightly incurved at apex in bud, 4.2–7 mm; anthers magenta, 1.6–2 mm. **Ovary** 0.9–1.1 mm; ovules 17–20 per locule; style 6–10 mm. **Capsules** Jan–May(–Dec), subacute, 2.5–5.5 × 2–3.5 mm, loculicidal split extending ¼–½-way to base. **Seeds** flattened, ovoid to ellipsoid to oblong, ± finely papillate, brown, (1–)1.2–1.7 × 0.8–1.3 mm, MR 0.3–0.7 mm. $2n = 40$.

DISTRIBUTION AND HABITAT Primarily on drier mountains of Marlborough, SE Nelson and north Canterbury, SI, from the Awatere Va. in the north to the Waiau Va. in the south, and as far west as the Matakitaki Va. (near Mt Baldy). It grows mostly in grassland or low shrubland in rocky, open areas.

NOTES A distinctive species recognised by the combination of a decumbent habit (Fig. 84A), dark branches that are often bifariously hairy in broad bands (Fig. 84B, inset), glossy leaves, leaf margins that are usually red (except in plants from the Black Birch Ra.) and minutely ciliate, especially toward the apex (Fig. 84D), long corolla tubes (Fig. 84F), and bracts not extending beyond tips of calyces (Fig. 84G). It shows some variation in the shape of the leaves (from broadly to narrowly elliptic) and leaf apices (from barely subacute to acute). Herbarium specimens have frequently been misidentified as *H. pinguifolia* (with which it sometimes co-occurs), and vice versa. *H. pinguifolia* can be distinguished by its usually glaucous leaves, usually shorter pedicels, larger bracts, and relatively shorter corolla tubes (Fig. 73G).

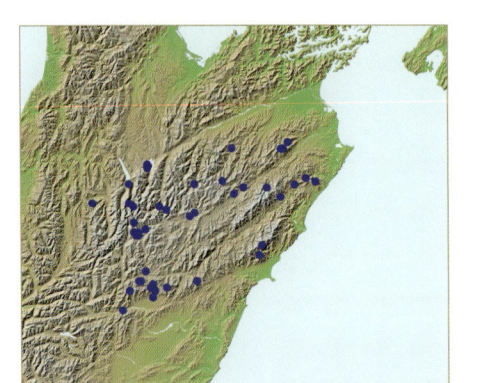

ETYMOLOGY Refers to the decumbent habit of the species.

Hebe decumbens OCCLUSAE

FIG. 84 **A** plant at Lk. Tennyson. **B** sprig; inset shows bifarious bands of stem hairs. **C** leaf bud with no sinus. **D** lower and upper leaf surfaces; magnified inset shows minute hairs on leaf margin. **E** shoot with inflorescences. **F** lateral view of flower. **G** close-up showing relative sizes of bracts and calyces. **H** capsule in septicidal (left) and loculicidal view.

30. *Hebe rakaiensis* (J.B.Armstr.) Cockayne

DESCRIPTION Bushy shrub (often with a rounded habit) to 2 m tall. **Branches** erect, old stems dark grey or red-brown; branchlets green (with dark bands at nodes), puberulent to pubescent, hairs bifarious; internodes (1.7–)2– 6(–7.3) mm; leaf decurrencies evident. **Leaf bud** distinct; sinus absent. **Leaves** erecto-patent to patent; *lamina* oblanceolate or obovate or elliptic, subcoriaceous, concave, (6.4–)8–20(–24) × (2.9–)4.8–7(–8.8) mm; *apex* subacute; *margin* cartilaginous, minutely ciliolate (esp. toward apex); *upper surface* light to dark green, glossy, with few to many stomata, glabrous or hairy along midrib; *lower surface* light green, glossy, almost always glabrous or very rarely hairy along midrib (seen only in CHR 401841, Glenroy Va.). **Inflorescences** with 12–48 flowers, lateral, unbranched, 1.7–4.5 cm; peduncle 0.47–0.87 cm; rachis (1–)2–3.8 cm. **Bracts** alternate, ovate, subacute. **Flowers** [Nov–]Jan–Mar, ⚥ or ♀ (on different plants). **Pedicels** longer than or equal to bracts, (0.5–)1–2(–4.3) mm. **Calyx** (1.3–)1.5–2 mm; *lobes* ovate, subacute, rarely hairy outside (e.g. CHR 401841, 386298). **Corolla** tube hairy inside; tube of ⚥ flowers 0.4–1.4 mm, funnelform, shorter than or equalling calyx; *lobes* white at anthesis, obovate or elliptic (anterior only), obtuse (posterior sts emarginate), patent to recurved (with age), longer than corolla tube, papillate inside and us. ciliolate (often minutely or sparsely). **Stamen filaments** straight or slightly incurved at apex in bud, 1.7–2.5 mm (sterile *c.* 1.7–2 mm; fertile *c.* 2.2–2.5 mm); anthers mauve, *c.* 1.2–1.4 mm; sterile anthers *c.* 0.9–1 mm. **Ovary** hairy, *c.* 0.6–0.8 mm; ovules *c.* 10–13 per locule; style (2–)3–4 mm, hairy (often sparsely) or apparently glabrous. **Capsules** Jan–May(–Nov), obtuse or subacute, 3–3.8 × *c.* 1.9–2.1 mm, hairy, loculicidal split extending ¼–½-way to base (us. about ¼). **Seeds** flattened (sts strongly), ± broad ellipsoid, ± smooth, brown, 0.8–1.6 × 0.7–1 mm, MR 0.2–0.4 mm. $2n = 80$.

DISTRIBUTION AND HABITAT Widespread on SI, chiefly on drier mountains east of the Main Divide, probably from the Inland Kaikoura Ra. in the north (see notes below) to the Blue Mts in the southeast and the Takitimu Mts in the southwest. It grows mostly in subalpine shrubland/scrub, often by streams.

NOTES Similar to a number of small-leaved "Occlusae", in particular *H. calcicola*, *H. subalpina*, *H. traversii*, *H. strictissima* and *H. glaucophylla*. It is recognised (Fig. 85) by the combination of its hairy ovaries and capsules (usually distinguishing it from most similar species, except *H. calcicola* and *H. glaucophylla*), short corolla tubes that are shorter than or equal to calyces (distinguishing it from *H. traversii*), and minutely hairy leaf margins (distinguishing it from *H. subalpina*). It is distinguished from *H. glaucophylla* by its light green, glossy (rather than glaucous) leaves, and usually from *H. calcicola* by having leaves that are mostly < 20 mm long and ciliate corolla margins.

The disjunct, northernmost distribution record rests on a single specimen (CHR 386977, Hodder Va., Marlborough). This is identified as *H. rakaiensis* on the basis of its leaf size and shape, hairy leaf margins, hairy ovaries and ± sparsely ciliate corolla lobes. Unlike other *H. rakaiensis*, the specimen has corolla tubes longer than calyx lobes, although corollas are present only on unopened buds that are probably infested by insects, and may not be properly formed. Further flowering specimens could help to verify or refute the occurrence of *H. rakaiensis* in this area, particularly given that the sometimes vegetatively similar *H. traversii* (which has longer corolla tubes) also occurs there.

Illustrations showing variation in leaf shape and size are provided by Bayly et al. (2001; but note that the distribution map in that paper contained minor errors, which are corrected here).

It is possible that *H. rakaiensis* sometimes hybridises with *H. subalpina* – for example, in the Forbes Mts (WELT 79939, 79942, 80885, 80894, 80899, 80905 and 80907).

ETYMOLOGY L. (*-ensis* = an adjectival suffix implying origin or place), implies that the species occurs near the Rakaia R. in Canterbury, the only province from which it was known when originally described.

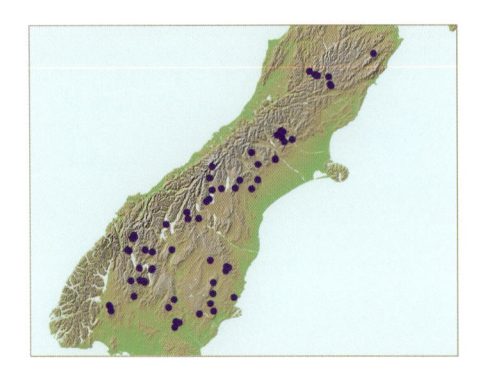

Hebe rakaiensis OCCLUSAE

FIG. 85 **A** plant at Wye Ck, The Remarkables. **B** sprig. **C** leaf bud with no sinus. **D** lower and upper leaf surfaces; magnified inset shows minute hairs on leaf margin. **E** inflorescence showing frontal views of flowers, and magnified inset showing hairs on corolla lobe margin. **F** inflorescence showing lateral view of flowers. **G** part of infructescence, with magnified inset showing hairs on capsule.

31. *Hebe calcicola*   Bayly et Garn.-Jones

DESCRIPTION Bushy shrub to *c.* 1.4 m tall. **Branches** spreading or ascending or erect, old stems mottled grey or brown; branchlets brown or red-brown or green (esp. very youngest), hairs bifarious or uniform (hairs on decurrencies often less densely distributed and somewhat finer); internodes 1.5–15 mm; leaf decurrencies evident (us. at least weakly). **Leaf bud** distinct; sinus absent. **Leaves** decussate, erecto-patent (mostly) to patent or recurved (with age); *lamina* oblong-elliptic (mostly) or lanceolate or oblanceolate (rarely), thin to subcoriaceous, (13–)20–38(–45) × 3.5–9 mm; *apex* subacute or obtuse; 2 lateral *secondary veins* sts evident at base of fresh leaves; *margin* ciliolate (us., esp. toward apex) or glabrous; *upper surface* dark green, glossy, with many stomata (but these are often not esp. evident in fresh leaves), us. hairy along midrib; *lower surface* green, glabrous or sts covered with minute gl. hairs (when young). **Inflorescences** with 25–45 flowers, lateral, unbranched, 2.5–7(–8.5) cm; peduncle (0.4–)1–2(–3) cm; rachis 2.2– 5.6 cm. **Bracts** alternate, deltoid or ovate, obtuse to acute. **Flowers** Jan–Apr, ♀. **Pedicels** 0.5–3 mm. **Calyx** *c.* 1.5–2.5 mm, 4–5-lobed (5th lobe small, posterior); *lobes* oblong or lanceolate (posterior us. shorter and broader), obtuse or more rarely subacute. **Corolla** tube hairy inside, *c.* 0.7–1.2 mm, shortly funnelform, shorter than calyx; *lobes* white at anthesis, elliptic or ovate or obovate, obtuse, erect to slightly recurved, longer than corolla tube, papillate inside and sts with a few hairs toward base or more widespread on inner surface, margins of corolla lobes glabrous. **Stamen filaments** slightly incurved at apex in bud, (3–)3.5–5 mm; anthers magenta, 1–2 mm. **Ovary** hairy, *c.* 0.8–1.1 mm; ovules 10–14 per locule, in 1 layer (although some slightly overlapping); style 3.5–5 mm. **Capsules** Mar–Jun, subacute, 2–3.5 × 2.5–3 mm, hairy, loculicidal split extending ⅓–¾-way to base. **Seeds** flattened, ± broad ellipsoid, brown, 1.2–1.9 × 0.8–1.2 mm, MR 0.4–0.6 mm. $2n = 80$.

DISTRIBUTION AND HABITAT Endemic to NW Nelson, SI, where it is known from the Peel, Lockett, Douglas and Arthur ranges. It probably also occurs in the Turks Cap Ra. (Marino Mts; Williams 1993), but we have seen no specimens from that locality, and it is not represented on the distribution map. All known populations occur on outcropping marble rocks of the Mt Arthur Group. Given the geological complexity of the region (Grindley 1961), the substrate specificity of the species is striking.

NOTES Most similar to *H. rakaiensis*, *H. subalpina*, *H. truncatula*, *H. traversii* and *H. strictissima*, none of which occurs in NW Nelson. It is distinguished from *H. rakaiensis* in having leaves that are mostly > 2 mm long, and corolla lobe margins that are glabrous. From the other species it can usually be distinguished by having ovaries and fruit that are minutely hairy (Fig. 86H). It is further distinguished from *H. subalpina* by its usually ciliate leaf margins (Fig. 86D), and from *H. traversii* by possessing corolla tubes that are shorter than their surrounding calyces (cf. Figs 86F and 91F).

ETYMOLOGY L. (a combination of *calcium* and *-cola* = dweller; exists only on), refers to the apparent restriction of wild populations to areas of calcium-rich rocks.

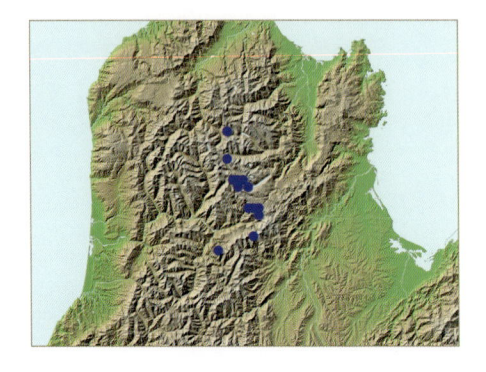

Hebe calcicola OCCLUSAE

FIG. 86   A plant on Peel Ra., near Mt Mytton. B sprig. C leaf bud with no sinus. D lower and upper leaf surfaces; magnified inset shows minute hairs on leaf margin. E inflorescence showing frontal view of flowers. F close-up of inflorescence showing lateral view of flowers. G close-up light micrograph showing papillae on inner surface of corolla lobe. H capsule in septicidal (left) and loculicidal view; magnified inset shows capsule hairs.

TAXONOMIC TREATMENT   163

32. *Hebe glaucophylla* (Cockayne) Cockayne

DESCRIPTION Bushy shrub to 2 m tall. **Branches** erect, old stems brown; branchlets green or red-brown, puberulent or pubescent, hairs bifarious (us.) or uniform; internodes (1–)2–7(–9) mm; leaf decurrencies evident to obscure (sts with narrow ridges along margins and medial line of each decurrency, and sts also with a slight bulge immediately below each leaf scar). **Leaf bud** distinct; sinus absent. **Leaves** erecto-patent (mostly) to patent or recurved (with age); *lamina* oblong or elliptic or lanceolate, subcoriaceous, flat or slightly concave, (7–)9–25 × (2–)3–7(–8) mm; *apex* subacute or acute; 2 lateral *secondary veins* very faintly evident at base of fresh leaves; *midrib* thickened below and depressed to grooved above (weakly, and not evident near leaf apex); *margin* sts minutely cartilaginous, minutely papillate and either ciliolate (with minute antrorse hairs) or glabrous; *upper surface* glaucous, with many stomata, glabrous or hairy along midrib (with minute egl. and/or gl. hairs); *lower surface* glaucous. **Inflorescences** with 15–31 flowers, lateral, unbranched, (1.3–)1.9–3.9(–4.6) cm, longer than (us.) or about equal to subtending leaves (rarely); peduncle 0.4–0.8(–1) cm; rachis (0.9–)1.3–3(–3.6) cm. **Bracts** alternate (although often with an opposite/subopposite pair at base), lanceolate or deltoid (sts narrowly), obtuse to acute, rarely hairy outside. **Flowers** (Dec–)Jan–Feb(–Mar), all ⚥ or possibly ⚥ or ♀ (on different plants). **Pedicels** 0.5–2(–3) mm, hairy (us.) or glabrous (rarely). **Calyx** 1.5–2.2 mm; *lobes* ovate or lanceolate or oblong, obtuse (us.) or subacute, very rarely hairy outside. **Corolla** tube hairy inside; tube of ⚥ flowers 1.1–2.3 × *c.* 1.7 mm, funnelform or contracted at base, shorter than to longer than calyx (mostly shorter than or equal to calyx, but some specimens from the Hanmer area have corolla tubes slightly longer than the calyx); *lobes* white at anthesis, circular or rhomboid or oblong (broadly) or obovate, obtuse (sts emarginate), patent, longer than corolla tube, sts with a few hairs toward base on inner surface. **Stamen filaments** incurved at apex in bud, 3–4.6 mm; anthers pink or yellow, 1.7–2 mm. **Ovary** hairy, *c.* 1–1.2 mm; ovules 8–10 per locule; style 3–5.3 mm, often hairy. **Capsules** Jan–May(–Dec), obtuse or subacute, 2.5–4 × 1.9–3.1 mm, us. hairy, loculicidal split extending ¼–¾-way to base. **Seeds** flattened, broad ellipsoid, brown, 1.6–2.2 × 1.3–1.5 mm, MR *c.* 0.6 mm. $2n = 80$.

DISTRIBUTION AND HABITAT Mountains east of the Main Divide, northern SI, from the Clarence Va. (Kaikoura ranges, Marlborough) in the north to near Castle Hill (mid-Canterbury) in the south. It grows in shrubland and scrub.

NOTES Distinguished from other small-leaved "Occlusae" by the combination of: glaucous leaves (otherwise found only in *H. topiaria* and *H. albicans*); hairy ovaries and capsules (otherwise common only in *H. calcicola* and *H. rakaiensis*); corolla tubes that are usually shorter than or equal to calyces (only sometimes – for example, Fig. 87F – are they slightly longer than calyces). *H. glaucophylla* is distinguished from most glaucous-leaved species of "Subcarnosae" by its small floral bracts and pedicellate flowers. The name "*glaucophylla*" has sometimes, erroneously, been associated with plants of *H. albicans* from NW Nelson (see notes under that species).

ETYMOLOGY Gk (*glaucus* = glaucous, bluish- or grey-green; *phyllon* = leaf, foliage), refers to the characteristically glaucous leaves of the species.

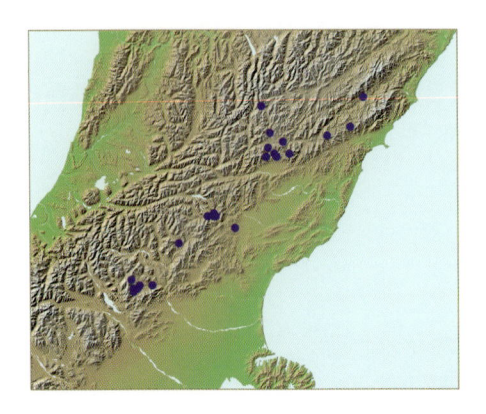

Hebe glaucophylla   OCCLUSAE

FIG. 87   **A** plant at Jacks Pass. **B** sprig. **C** leaf bud with no sinus. **D** lower and upper leaf surfaces. **E** inflorescence. **F** lateral view of flower. **G** close-up of capsule showing conspicuous hairs. **H** capsule in septicidal (left) and loculicidal view.

33. *Hebe topiaria* L.B.Moore

DESCRIPTION Bushy shrub (often of neat, rounded habit) to 1.2 m tall. **Branches** erect, old stems brown or grey (mottled); branchlets green or red-brown or brown or black (when dry), pubescent (mostly with coarse white-golden hairs that are often upward-facing), hairs us. bifarious or rarely uniform; internodes (1–)2–6(–10) mm; leaf decurrencies evident. **Leaf bud** distinct; sinus absent. **Leaves** erect to erecto-patent; *lamina* elliptic or obovate, concave, (5–)8–21(–23) × (3–)3.5–8 mm; *apex* obtuse to acute or apiculate; 2 lateral *secondary veins* sts evident at base of fresh leaves; *margin* us. minutely papillate (on lower surface, esp. toward apex) and rarely ciliolate; *upper surface* glaucous or glaucescent, dull or slightly glossy, with many stomata, us. hairy along midrib and rarely covered with minute gl. hairs; *lower surface* glaucous. **Inflorescences** with 9–33 flowers, lateral, unbranched, 1–3(–4) cm; peduncle 0.2–0.8(–1.1) cm; rachis 0.6–2.4(–3.2) cm. **Bracts** alternate (but often with an opposite basal pair), deltoid (sts narrowly) or elliptic (basal bracts often larger, lanceolate to elliptic), acute to obtuse. **Flowers** [Dec–]Jan–Feb(–Apr), ☿ or ♀ (on different plants). **Pedicels** 0.5–2.5 mm. **Calyx** 1.5–2.8 mm; *lobes* elliptic or ovate, obtuse to acute. **Corolla** tube hairy inside; tube of ☿ flowers 1.5–2.5 mm, longer than calyx; tube of ♀ flowers *c.* 2 mm, longer than calyx (but sts barely); *lobes* white at anthesis, elliptic (often broadly) or circular, obtuse, suberect to recurved, *c.* equalling corolla tube. **Stamen filaments** incurved at apex in bud, 2.5–4.5 mm; anthers magenta, 1.25–1.5 mm. **Ovary** 1.2–1.5 mm; ovules *c.* 12–14 per locule; style 3.5–6.5 mm. **Capsules** Feb–May, subacute, 4–5 × 2.5–3.5 mm, loculicidal split extending less than ¼-way to base. **Seeds** flattened (sts strongly), ± ellipsoid to discoid or irregular, not winged or only weakly winged, brown, 1.2–2 × 0.9–1.3(–1.8) mm, MR 0.2–0.5 mm. $2n = 122$.

DISTRIBUTION AND HABITAT Northern SI, from near Boulder Lk. in the northwest to Amuri Pass and the Poplars Ra. in the south, and as far east as the Richmond Ra. It grows in shrubland and tussock grassland above the tree-line. It can be one of the dominant species in some shrubland associations.

NOTES The combination of usually glaucous leaves and compact habit distinguishes *H. topiaria* from most other species of small-leaved "Occlusae". Its leaves can be similar to those of *H. glaucophylla*, from which it is distinguished in having glabrous ovaries, and corolla tubes that are always longer than calyces.

The species shows considerable variation in the shape and size of the leaves (Appendix 4). Leaves range from *c.* 8 × 4 mm to *c.* 22 × 6 mm, from narrowly to broadly elliptic or obovate, and with apices that are rounded to acute.

The four easternmost and somewhat disjunct collections represented on the distribution map are from Mt Richmond (CHR 286388), Mt Starveall (CHR 76140), Motueka River Left Branch (CHR 401652) and the Red Hills (CHR 387478). Some of these specimens are sterile and/or difficult to identify with certainty, but they have most of the critical features of *H. topiaria*. Since some of these areas (e.g. Mt Richmond) have been intensively botanised, the relative paucity of collections might suggest that *H. topiaria* is not as plentiful or conspicuous in these areas as it is in wetter mountains to the west and south.

ETYMOLOGY Refers to the habit of the species, which is often a neat, globular bush giving the appearance of topiary.

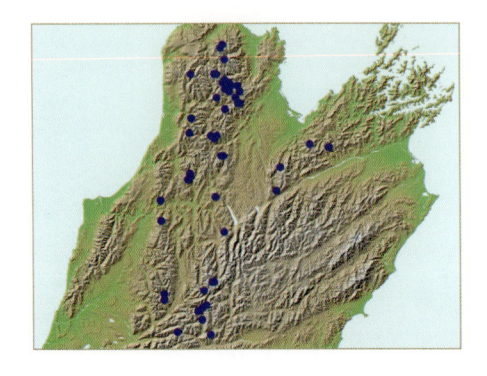

Hebe topiaria   OCCLUSAE

FIG. 88   **A** plants on Mt Arthur. **B** sprig. **C** leaf bud with no sinus. **D** lower and upper leaf surfaces; inset shows glabrous leaf margins. **E** inflorescence of hermaphrodite plant (lateral view). **F** frontal view of hermaphrodite flower. **G** inflorescence of female plant (apical view). **H** lateral view of hermaphrodite flower. **I** capsule in septicidal (left) and loculicidal view; magnified inset shows that it is glabrous. **J** seeds.

34. *Hebe albicans*   (Petrie) Cockayne

DESCRIPTION Openly branched, spreading low or small bushy shrub to 1 m tall. **Branches** decumbent or pendent or erect, old stems brown; branchlets green to red-brown, pubescent, hairs bifarious (sts restricted to tufts at nodes) or uniform; internodes (1.9–)2.5–9(–15.5) mm; leaf decurrencies evident or obscure. **Leaf bud** distinct; sinus absent. **Leaves** erecto-patent to recurved; *lamina* oblong or elliptic or ovate or lanceolate or linear-lanceolate, fleshy or coriaceous or subcoriaceous or thin, flat or concave, 11–42 × 3.4–16.3 mm; *apex* obtuse to acute; *base* prominently amplexicaul to truncate or cuneate; 2 lateral *secondary veins* evident at base of fresh leaves; *midrib* slightly thickened below and depressed to grooved above; *margin* sts cartilaginous, glabrous; *upper surface* glaucous or glaucescent, dull, with many stomata, glabrous or hairy along midrib (near base); *lower surface* glaucous or glaucescent. **Inflorescences** with 15–45 flowers, lateral, unbranched, 1.7–5.1 cm, longer than or about equal to subtending leaves; peduncle 0.4–1.4(–1.7) cm; rachis 0.7–3.9 cm. **Bracts** alternate, lanceolate or elliptic or ovate, acute or subacute. **Flowers** (Nov–)Dec–Mar(–May), ⚥. **Pedicels** (0.5–)0.8–2.5(–3.5) mm. **Calyx** 1.5–3(–4) mm; *lobes* lanceolate or elliptic, acute to subacute. **Corolla** tube hairy inside or glabrous, (2.5–)3–6 × 1.2–1.8 mm, cylindric, longer than calyx; *lobes* white at anthesis, ovate or elliptic (sts broadly), obtuse, suberect to recurved, shorter than corolla tube. **Stamen filaments** sts incurved at apex in bud, 3–5.5 mm; anthers magenta, 1.5–2.4 mm. **Ovary** ovoid (sts narrowly), rarely sparsely hairy, 0.8–1.2 mm; ovules 7–16 per locule; style 4.5–9 mm. **Capsules** [Jan–]Feb–Aug(–Nov), subacute or obtuse or slightly emarginate, 2.5–4.5 × 2–3.2 mm, rarely hairy, loculicidal split extending ¼–½-way to base. **Seeds** flattened (sts strongly), ellipsoid to discoid, brown, 1–1.8(–2) × 0.8–1.1 mm, MR 0.2–0.6 mm. $2n = 40$ or 80.

DISTRIBUTION AND HABITAT Endemic to Nelson, SI, from Mt Burnett in the north to the Glasgow Ra. in the south, and as far east as the Bryant Ra. It grows in a range of habitats, usually on rocks (often calcareous), from just above sea-level (e.g. on the banks of the Aorere R.) to subalpine and penalpine situations.

NOTES An extremely variable species, distinguished from most others by the combination of glaucous leaves, conspicuously pedicellate flowers, short bracts, long corolla tubes, and glabrous ovaries and fruit. It includes plants previously placed in *H. recurva* (e.g. Moore, in Allan 1961) and *H.* "*glaucophylla* NW Nelson" (*sensu* Druce 1993). The species includes both sprawling, broad-leaved plants of subalpine areas (e.g. Fig. 89A), and erect, narrow-leaved plants of lowlands (e.g. Fig. 89C). Two chromosome numbers are recorded (Table 5). Leaf variation is illustrated in Appendix 4. Given this variation it is possible that further taxonomic subdivision is warranted, but many plants have characteristics intermediate between the morphological extremes, and a study by Kellow et al. (2005) could not identify clear grounds on which a subdivision could be based.

H. albicans hybridises with *H. calcicola* (Bayly et al. 2001), and probably also with *H. salicifolia* (based on observations in the Cobb Va.; WELT 80687, 81577, 81747).

ETYMOLOGY L. (*albicans* = becoming white, whitish, not quite perfect white), presumably refers to the glaucous leaves.

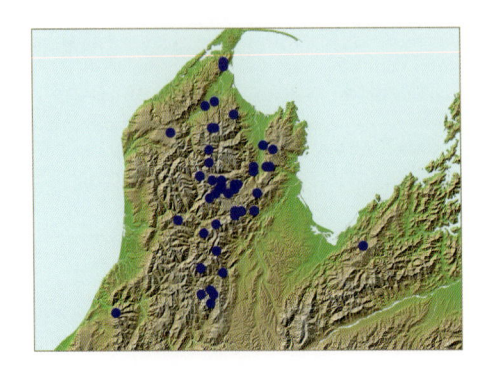

Hebe albicans OCCLUSAE

FIG. 89 **A–C** plants: A, on Peel Ridge; B, on Mt Burnett; C, by Aorere R. **D** sprigs: cultivated plant (of unknown origin, left) and plant from Aorere R. **E–F** leaf buds without sinuses, plants from: E, Mt Burnett; F, Aorere R. **G** upper leaf surfaces, plants from (left to right) Mt Burnett, Takaka Hill, Aorere R. **H–I** lateral views of flowers, plants from: H, Mt Burnett; I, Aorere R. **J** capsules in septicidal (left) and loculicidal view, plant from Mt Burnett. **K–L** inflorescences, plants from: K, Mt Arthur; L, Aorere R.

35. *Hebe stenophylla*   (Steudel) Bayly et Garn.-Jones

DESCRIPTION Spreading low or bushy shrub (often openly branched) to 2 m tall. **Branches** erect, old stems grey; branchlets olive-green or red-brown, glabrous or puberulent, hairs bifarious or uniform (mostly in var. *hesperia*); internodes (1–)4–13(–24) mm; leaf decurrencies evident (us. weakly), or obscure. **Leaf bud** distinct; sinus absent. **Leaves** patent to recurved; *lamina* linear or lanceolate or elliptic (esp. var. *oliveri* or some NI populations of var. *stenophylla*), subcoriaceous, flat or concave, (16–)23–53(–87) × 2.5–6.5(–10) mm; *apex* acute; 2 lateral *secondary veins* sts evident at base of fresh leaves; *margin* mostly glabrous but occasionally pubescent; *upper surface* dark to light green, dull, with few or many stomata, sparsely (often only along margins) to conspicuously pitted with small depressions that each contain a twin-headed gl. hair, hairy along midrib; *lower surface* green (often paler than upper), conspicuously pitted with small depressions that each contain a twin-headed gl. hair. **Inflorescences** with (35–)55–130(–170) flowers, lateral, unbranched, (2.5–)4.5–6.5(–9.5) cm; peduncle (0.5–)1–1.5(–2.1) cm; rachis (2–)4.5–6(–9) cm. **Bracts** alternate, ovate or deltoid, acute (mostly) or obtuse. **Flowers** Dec–Apr(–Sep), ☿ or ♀ (on different plants). **Pedicels** longer than or equal to bracts, (0.5–)1–3(–5) mm, hairy or glabrous. **Calyx** 1.5–2.5 mm; *lobes* ovate or oblong, obtuse to acute, egl. ciliate (us.) or with mixed gl. and egl. cilia (gl. hairs often with a single, rounded cell at the apex; twin-headed hairs, when present, us. sparse). **Corolla** tube us. glabrous, sts hairy inside (esp. var. *hesperia*); tube of ☿ flowers (1.8–)3–4.9 × 1.3–2 mm, contracted at base and cylindric or expanded in lower half, longer than calyx; *lobes* white or tinged mauve at anthesis, ovate (often broadly) or circular or elliptic, obtuse (posterior sts emarginate), suberect to recurved (mostly patent), shorter than corolla tube. **Stamen filaments** incurved at apex in bud, 2.5–4.4 mm; anthers magenta, 1–1.5 mm. **Ovary** c. 0.8–0.9 mm; ovules 4–10 per locule; style 3–7 mm. **Capsules** Jan–Jul(–Dec), acute or obtuse, (2–)2.5–3.5 × (0.8–)1.5–3 mm, loculicidal split extending ¼–¾-way to base. **Seeds** flattened, ± ellipsoid to oblong, straw-yellow to pale brown, 0.9–1.5(–2) × 0.7–0.9(–1.1) mm, MR 0.2–0.3 mm. $2n = 40$.

KEY TO VARIETIES

1. Corolla tube 1.8–3(–3.5) mm long, hairy within; branchlets bifariously to uniformly puberulent; upper surface of leaves with few stomata .... **var. *hesperia***
   Corolla tube usually > 3 mm long, glabrous or (very rarely) hairy within; branchlets glabrous or bifariously to (rarely) uniformly puberulent; upper surface of leaves with few to many stomata 2

2. Leaves 3–6 times as long as broad, upper surface with few stomata .... **var. *oliveri***
   Leaves > 6 times as long as broad, or, if not, then upper leaf surface with many stomata .... **var. *stenophylla***

Var. *stenophylla*
Branchlets glabrous or bifariously to uniformly puberulent. *Leaves* linear, or lanceolate, or elliptic, 19–87 × 2.5–6.5(–9.5) mm; upper surface conspicuously pitted to apparently smooth (except along margins), stomata sparse to dense; margins broad, glabrous (us.) or minutely hairy (especially toward the apex). Inflorescences 26–95 mm long. Corolla tube usually glabrous within, 3–4.9 mm long.

Var. *hesperia* Bayly et Garn.-Jones
Branchlets bifariously to uniformly puberulent. Leaves oblong-elliptic, linear-lanceolate or lanceolate, (16–)25–45(–62) × (3.5–)4–8 mm; upper surface usually smooth, mostly pitted only toward the margin but sometimes with obvious pits toward middle of lamina, stomata sparse except at apex and sometimes along either side of midrib; margins broad, glabrous (us.) or with very fine hairs toward the apex. Inflorescences 32–64 mm long. Corolla tube hairy within, 1.8–3(–3.5) mm long.

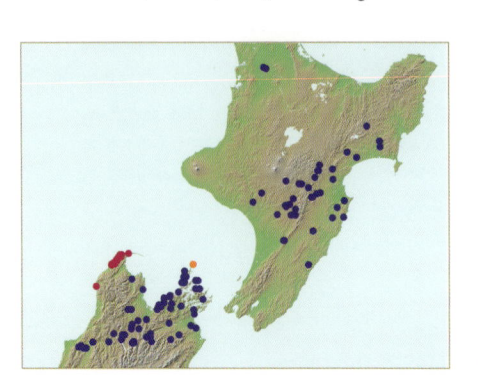

Var. *oliveri* Bayly et Garn.-Jones
Branchlets glabrous or bifariously puberulent. Leaves lanceolate or elliptic, 22–37 × 5–10 mm; upper surface mostly smooth, not conspicuously pitted except toward margins, stomata sparse except near apex; margins broad, glabrous (us.) or with very fine hairs. Inflorescences 26–60 mm long. Corolla tube glabrous within, 2.5–3.8 mm long.

- Var. *stenophylla*
- Var. *hesperia*
- Var. *oliveri*

Hebe stenophylla   OCCLUSAE

FIG. 90   **A** var. *stenophylla* ("mudstone" variant) near Mangaweka. **B** sprig, var. *stenophylla*. **C** leaf bud with no sinus. **D** lower and upper leaf surfaces, var. *stenophylla*; magnified inset shows glabrous leaf margin and pits on upper leaf surface. **E** lower and upper leaf surfaces, var. *oliveri*, at same scale as D; inset shows glabrous leaf margin and pits on lower leaf surface. **F** low-growing, small-leaved plant of var. *stenophylla* growing on alpine scree on Makahu Spur, Kaweka Ra. **G** lateral views of flowers, var. *hesperia* (top) and var. *stenophylla*. **H** frontal view of flowers, var. *stenophylla*. **I** capsule of var. *stenophylla* in septicidal (left) and loculicidal view.

DISTRIBUTION AND HABITAT Var. *stenophylla* occurs in central and eastern NI (from near Gisborne to Wanganui and northern Wairarapa, with outlying populations near Hamilton) and northern SI (as far south as a line from Westport to Cape Campbell), growing mostly on bluffs, terraces or rocky areas, often along streams and roadsides. Var. *hesperia* occurs in near-coastal areas in northwest SI, between Cape Farewell and the Heaphy R., and usually grows on limestone bluffs. Var. *oliveri* is known only from Stephens Id, Cook Strait, and grows mostly in wind-shorn vegetation on exposed bluffs on the western side.

NOTES Similar to *H. traversii*, *H. parviflora* (in which it was included, as var. *angustifolia*, by Moore, in Allan 1961) and *H. strictissima* (see notes under those species). Key distinguishing features include: the presence of small pits (Figs 90D, 77B and D), at least on the lower leaf surface (usually conspicuous under a dissecting microscope, each containing a recessed glandular hair); leaf margins that are usually smooth and glabrous (only occasionally hairy); corolla tubes that are longer than calyces and usually glabrous within (except in var. *hesperia*); and calyx cilia only rarely including twin-headed glandular hairs.

Var. *stenophylla* is variable in habit, and in the shape and size of the leaves. This is partly correlated with geographic distribution and habitat, and several distinctive variants can be identified, but relationships of these are unknown, and some specimens remain difficult to place (hence no further, formal subdivision of the variety is attempted; Bayly et al. 2000). Typical forms occur on SI, are shrubs up to *c.* 2 m high, and have linear-lanceolate leaves that are often strongly recurved or falcate, giving plants an "inelegant appearance" (as described by T. Kirk in the protologue of *Veronica squalida*). On NI, one extreme of the morphological range is an erect, openly branched shrub, usually less than 1 m high, with very narrowly linear-lanceolate leaves that are often somewhat recurved (e.g. Fig. 90A). Plants of this type are apparently restricted to exposed mudstone on cliffs or embankments, are fairly uniform in appearance, and seem to retain their features in cultivation (e.g. CHR 7440). Superficially, they most closely resemble SI plants, from which they differ primarily (though not consistently) in stature, leaf width and the possession of numerous stomata on the upper leaf surface. Plants from other NI populations are generally somewhat broader-leaved than the "mudstone" variant, and occur mostly in rocky areas on cliffs, bluffs or scree (e.g. at Makahu Spur, Kaweka Ra.; Fig. 90F).

Collections of var. *hesperia* vary in leaf size, with small- and larger-leaved plants apparently occurring in close proximity.

ETYMOLOGY *Stenophylla* (Gk *stenos* = narrow; *phyllon* = leaf) means narrow-leaved; *hesperia* (Gk *hesperos* = western, of evening) refers to the distribution of the variety on the West Coast of SI; *oliveri* honours Walter R. B. Oliver (1883–1957), former Director of the Dominion Museum, who collected the type specimen.

36. *Hebe traversii* (Hook.f.) Andersen

DESCRIPTION Bushy shrub to 2.5 m tall. **Branches** erect, old stems grey or brown; branchlets green, puberulent (sts very minutely and sparsely), hairs almost always uniform or rarely bifarious; internodes (2–)3–8(–9.5) mm; leaf decurrencies weakly evident, or obscure. **Leaf bud** distinct; sinus absent. **Leaves** erecto-patent; *lamina* narrowly oblong or oblong-lanceolate, subcoriaceous, ± flat or concave, 16–39(–44) × (2.5–)3–8(–9) mm; *apex* subacute (us.) or obtuse or acute (rarely); 2 lateral *secondary veins* evident at base of fresh leaves; *margin* scabrous or ciliate or pubescent (with short, stiff, antrorse hairs); *upper surface* green or light green, dull, with many stomata (but these not always apparent), hairy along midrib and toward base or glabrous; *lower surface* green or light green. **Inflorescences** with 34–72 flowers, lateral, unbranched, (2.3–)3–5(–7.3) cm, with all flowers (including those near the apex) generally developing to maturity (although sts with a small number of aborted flowers); peduncle (0.35–)0.6–0.9(–1.4) cm; rachis (1.4–)2–4.5(–5.9) cm. **Bracts** alternate, ovate or lanceolate, subacute or acute, sts hairy outside (near base). **Flowers** Dec–Feb(–Mar), ♂ or ♀ (on different plants). **Pedicels** (0.8–)1.2–3 mm. **Calyx** 1.8–2.2 mm, 4–5-lobed (5th lobe small, posterior); *lobes* ovate, subacute, either with mixed gl. and egl. cilia or apparently egl. ciliate. **Corolla** tube hairy inside; tube of ♂ flowers (2.5–)3–4.5 mm, cylindric, longer than calyx; tube of ♀ flowers 2–4 × *c.* 1.3–1.6 mm, cylindric, longer than calyx; *lobes* white or tinged mauve or pink at anthesis, elliptic or obovate, obtuse, suberect to recurved, shorter than corolla tube, sts with a few hairs toward base on inner surface and/or bluntly papillate inside. **Stamen filaments** incurved at apex in bud, 2.5–3.3 mm; anthers magenta or pink, *c.* 1.5–1.6 mm; sterile anthers of

Hebe traversii   OCCLUSAE

FIG. 91   **A** plant on Inland Road, near Whales Back. **B** sprig. **C** leaf bud with no sinus. **D** lower and upper leaf surfaces; inset shows minute hairs on leaf margin. **E** inflorescence. **F** lateral view of flower. **G** seeds. **H** infructescence.

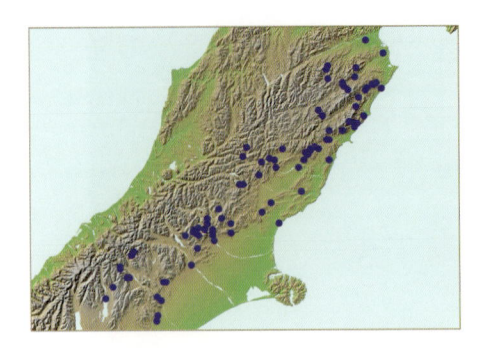

♀ flowers magenta, 0.8–1.4(–1.8) mm. **Ovary** 0.9–1.1 mm; ovules 4–10 per locule; style 4–7 mm. **Capsules** (Jan–)Feb–Jun(–Nov), subacute or obtuse, (3.5–)4–5(–5.5) × (1.8–)2–4 mm, loculicidal split extending ¼-way to base. **Seeds** flattened (sts strongly), ellipsoid or ovoid or oblong, brown, (1.3–)1.5–2.2(–2.4) × 1–1.5(–1.8) mm, MR 0.3–0.6 mm. $2n = 40$.

DISTRIBUTION AND HABITAT Eastern SI, from near Blenheim, Marlborough, to the Four Peaks Ra., south Canterbury. It grows in scrub and at forest margins, often in river valleys, in situations ranging from near-coastal to montane or subalpine.

NOTES Similar to some forms of *H. stenophylla*, from which it is distinguished by the combination of non-pitted leaves (cf. Figs 77B, D and 90D), minutely hairy leaf margins (Fig. 91D), and hairs inside corolla tubes. It is distinguished from most other species of small-leaved "Occlusae" by its large capsules and its corolla tubes, which are markedly longer than the calyces (Fig. 91F). Leaves are illustrated in Appendix 4.

ETYMOLOGY Presumably honours William T. L. Travers (1819–1903).

37. *Hebe parviflora*   (Vahl) Andersen

DESCRIPTION Bushy shrub or small tree (often highly branched toward the tips, and dome-shaped when young) to 7.5(–12) m tall. **Branches** erect, old stems pale grey; branchlets progressing from olive-green to brown or red-brown, puberulent or rarely glabrous, hairs bifarious (mostly) or uniform; internodes (1.5–)3–17(–20) mm; leaf decurrencies obscure. **Leaf bud** distinct; sinus absent. **Leaves** erecto-patent to recurved; *lamina* lanceolate or linear-lanceolate, subcoriaceous, flat or concave, (8–)25–60(–76) × 1.5–7 mm; *apex* whitish, acute or shortly acuminate; 2 lateral *secondary veins* sts evident at base of fresh leaves; *margin* scabrous or minutely pubescent (with short, stiff, basally-swollen, antrorse hairs); *upper surface* light green, dull, us. with many stomata, hairy along midrib; *lower surface* light green, not pitted (although frequently with many small gl. hairs) or sts faintly pitted with small depressions that each contain a twin-headed gl. hair, or often glabrous. **Inflorescences** with (20–)40–80(–130) flowers, lateral, unbranched, (2–)4–10(–12) cm; peduncle (0.35–)0.5–1.9 cm; rachis (1.6–)3–10.2 cm. **Bracts** alternate, ovate to deltoid or oblong, acute to obtuse. **Flowers** [Sep–]Jan–Mar(–Aug), ⚥ or ♀ (on different plants). **Pedicels** (0.3–)0.5–3.5(–4) mm, hairy or glabrous, sts recurved in fruit. **Calyx** 1.5–2.3(–2.7) mm; *lobes* ovate to elliptic (often broadly), obtuse to acute (sts on one inflorescence). **Corolla** tube hairy inside; tube of ⚥ flowers 2.1–3.8 × 1.4–3(–3.8) mm, cylindric or slightly expanded in lower half, longer than calyx; *lobes* white tinged with pink or mauve at anthesis, ovate (sts broadly), obtuse (posterior sts emarginate), suberect to recurved, sts with a few hairs toward base on inner surface. **Stamen filaments** incurved at apex in bud, 3–5 mm; anthers magenta, 1.5–2.2 mm; sterile anthers of ♀ flowers very pale lilac or light brown, *c.* 1 mm (when dry). **Ovary** *c.* 0.8–1 mm; ovules *c.* 8–12 per locule; style 3.5–6 mm. **Capsules** (Jan–)Feb–Jun(–Nov), obtuse or subacute, 2.5–3.5 × 1.4–2.5 mm, loculicidal split extending ¼–¾-way to base. **Seeds** strongly flattened, ellipsoid to discoid, weakly winged, straw-yellow to pale brown, 0.9–1.8 × 0.8–1.4 mm, MR 0.1(–0.5) mm. $2n = 80$.

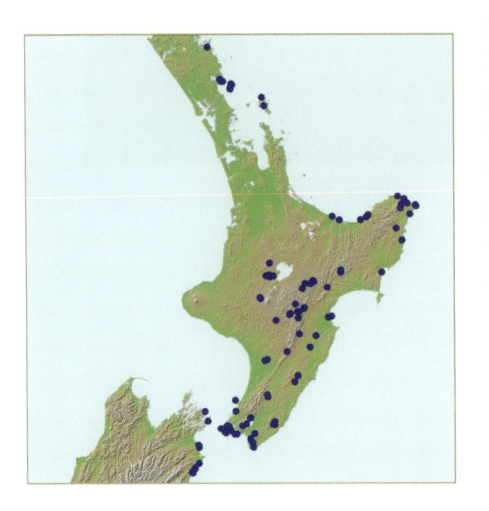

DISTRIBUTION AND HABITAT Eastern, central and southern NI (including the Hen and Chickens and Great Barrier islands), and northeast SI, ranging from near Russell (NI) to near Kekerengu (SI). It generally grows in scrub on hillsides, along streams and at forest margins, from near-coastal to montane situations.

NOTES Similar to *H. stenophylla* and *H. strictissima*. It is distinguished from: the former by often smooth (only sometimes pitted) leaf surfaces (Figs 77A and C), minutely hairy leaf margins, corolla tubes that are hairy within (Fig. 92G) and calyx cilia always including twin-headed glandular hairs; and from the latter by having corolla tubes longer than calyces. It can also be more arborescent than either of these species.

Hebe parviflora   OCCLUSAE

FIG. 92   **A** plants on Hauhangaroa Ra. **B** sprig. **C** leaf bud with no sinus. **D** lower and upper leaf surfaces; magnified inset shows marginal hairs. **E** inflorescence showing frontal view of flowers. **F** lateral view of flowers. **G** close-up of flower showing hairs inside corolla tube. **H** capsule in septicidal (left) and loculicidal view.

Plants are often highly branched toward their extremities, giving a characteristic dome-shaped appearance when young, which may persist to some age in open situations (e.g. Fig. 92A). The species is, however, variable in habit, as it also is in corolla tube length and the degree to which the leaf surface is pitted with recessed glandular hairs. Along the Taruarau and Rangitikei rivers (e.g. WELT 80944, 80945, 81037, 81051) some plants are quite openly branched and similar to *H. stenophylla* in having relatively long corolla tubes and leaves that are distinctly pitted on the lower surface. These plants are included under *H. parviflora* on the basis of their hairy leaf margins, hairy corolla tubes, conspicuous glandular hairs on calyx margins, and flavonoid profile (Bayly et al. 2000).

ETYMOLOGY L. (*parvus* = little, small; *flores* = flowers), means small-flowered.

38. *Hebe strictissima* (Kirk) L.B.Moore

DESCRIPTION Bushy shrub to 2.5 m tall. **Branches** erect, old stems brown or grey; branchlets olive-green or red-brown, minutely puberulent (us.) or glabrous, hairs bifarious; internodes (1–)2–13(–18) mm; leaf decurrencies evident to obscure (often with narrow ridges along medial line and margins). **Leaf bud** distinct; sinus absent. **Leaves** erecto-patent to patent; *lamina* linear-lanceolate or narrowly oblong, subcoriaceous, flat or concave, (9–)16–41(–49) × 3–7(–8) mm; *apex* acute to subacute or shortly apiculate; 2 lateral *secondary veins* sts evident at base of fresh leaves; *margin* sts ± cartilaginous, us. egl. puberulent or sts glabrous; *upper surface* green, dull, with many stomata, hairy along midrib or glabrous; *lower surface* light green. **Inflorescences** with (11–)20–72 flowers, lateral, unbranched, 1.7–7.7(–10.7) cm, almost always longer than or very rarely about equal to subtending leaves; peduncle (0.3–)0.4–1(–1.5) cm; rachis 1.2–6.5(–9.2) cm. **Bracts** alternate, ovate or deltoid (sts narrowly), obtuse (mostly) to acute. **Flowers** Dec–Mar(–Jun), ⚥ or ♀ (on different plants). **Pedicels** 1–4 mm, hairy (indumentum sts sparse when compared with that of peduncle and rachis). **Calyx** 1.5–2(–2.5) mm; *lobes* ovate, obtuse. **Corolla** tube hairy inside; tube of ⚥ flowers 1.4–2.9 × 1.3–1.7 mm, funnelform, equalling or slightly longer than calyx; tube of ♀ flowers *c.* 1.3–1.8 mm, funnelform, equalling or slightly longer than calyx; *lobes* white or tinged deep mauve at anthesis, circular or elliptic (often broadly) or oblong (anterior only), obtuse (posterior sts emarginate), suberect to recurved, longer to shorter than corolla tube (posterior lobe always longer than tube, anterior lobe often shorter than tube). **Stamen filaments** incurved at apex in bud, 3–5 mm; anthers dark magenta or purple, 1.2–1.8 mm; sterile anthers of ♀ flowers purple or light brown. **Ovary** sts hairy, 0.9–1.3 mm; ovules 11–16 per locule, in 1–2 layers; style 2.3–6 mm, sts hairy. **Capsules** [Jan–]Mar–Apr(–Jun), subacute or obtuse, 2.9–4 × *c.* 2.2–2.6 mm, occasionally hairy, loculicidal split extending up to ¾-way to base. **Seeds** flattened, ± broad ellipsoid, not winged to only weakly winged, brown, 1–1.6 × (0.8–)0.9–1.3 mm, MR 0.1–0.4 mm. $2n = 80$.

DISTRIBUTION AND HABITAT Endemic to Banks Peninsula and the Port Hills, SI. It grows mostly in open areas on banks and bluffs, or in scrub.

NOTES Similar to *H. traversii*, *H. stenophylla* and *H. parviflora*, from which it is distinguished by (among other features) having corolla tubes equal to or only slightly exceeding surrounding calyces (Figs 93F and G). The only other hebes on Banks Peninsula are *H. salicifolia* and *H. odora*. Moore (in Allan 1961) suggests the species may form hybrids with *H. salicifolia*, and a specimen potentially matching this parentage is WELT 84066. Aspects of reproductive ecology are discussed by Delph (1990*b*). Leaves are illustrated in Appendix 4.

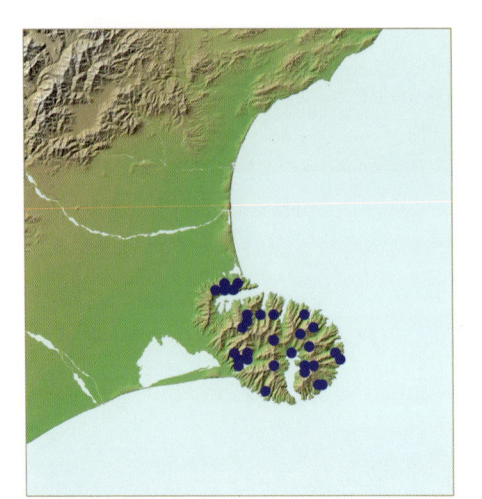

ETYMOLOGY L. (*strictus* = drawn close together, very upright, very straight; *issimus* = most so, to the greatest degree), presumably refers to the inflorescences, which are described in the protologue as "strict, erect".

Hebe strictissima   OCCLUSAE

FIG. 93   **A** plant near Lk. Forsyth, Banks Peninsula. **B** sprig. **C** leaf bud with no sinus. **D** lower and upper leaf surfaces; magnified inset shows hairs on leaf margin. **E** inflorescence showing frontal view of flowers. **F** lateral view of flower. **G** dorsal view of flower. **H** infructescence.

TAXONOMIC TREATMENT   177

39. *Hebe stricta* (Benth.) L.B.Moore

DESCRIPTION Spreading low or bushy shrub (openly to densely branched) to 4 m tall. **Branches** erect or spreading or decumbent, old stems brown (sts dark) or grey; branchlets green or olive-green or yellowish or red-brown or orange-brown, puberulent or pubescent or glabrous, hairs bifarious or uniform; internodes 2.25–49(–58) mm; leaf decurrencies obscure or evident. **Leaf bud** distinct; sinus absent. **Leaves** erecto-patent to recurved; *lamina* linear or lanceolate or oblanceolate or ovate or obovate or oblong or narrowly to broadly elliptic, thin to coriaceous, flat or m-shaped in TS or slightly concave, (19–)26–106(–127) × 5.5–45 mm; *apex* acute or acuminate to obtuse; brochidodromous *secondary veins* sts evident in fresh leaves; *margin* narrowly cartilaginous, ciliolate to pubescent or sts glabrous (with age), entire or distantly denticulate; *upper surface* light to dark green or yellowish-green, dull or slightly glossy, without evident or with few to many stomata, us. hairy along midrib but sts glabrous or covered with a mixture of egl. and gl. hairs (when young); *lower surface* light green, glabrous or hairy along midrib or covered with a mixture of egl. and gl. hairs (when young). **Inflorescences** with 35–300 flowers, lateral, unbranched, 2.6–21.5 cm, sts with a conspicuous number of unopened flowers toward the apex (leaving a protruding "rat's tail"); peduncle 0.7–4.1 cm; rachis (1.6–)2.4–19 cm. **Bracts** alternate or lowermost pair opposite, then subopposite or alternate above (sts with a pair, or whorl of 3, sterile bracts at base), linear or lanceolate or ovate or deltoid (sts narrowly) or oblong or oblanceolate, obtuse to acute, sts hairy outside. **Flowers** Jan–May(–Sep), ⚥ or ♀ (on different plants). **Pedicels** 0.5–4.3 mm, sts recurved in fruit. **Calyx** 1.7–4.2 mm; *lobes* lanceolate or elliptic or deltoid or oblong, acute to obtuse, sts hairy outside. **Corolla** tube hairy inside and sts outside (particularly below clefts between lobes); tube of ⚥ flowers (1.5–)3–5 × 1.1–2.4 mm, cylindric to funnelform, longer than calyx; *lobes* white or tinged mauve at anthesis, ovate or elliptic or obovate or oblanceolate, obtuse, suberect to patent, longer to shorter than corolla tube, sts with a few hairs toward base on inner surface and then sts ciliate (near base) or hairy outside. **Stamen filaments** white or occasionally coloured, 3.2–6.5 mm; anthers mauve or violet or purple or white or yellow, (0.9–)1.2–2.6 mm. **Ovary** sts hairy (esp. near apex and along septal grooves), 0.6–1.3 mm; ovules 7–24 per locule, in 1–2 layers; style 2.3–8 mm, sts hairy. **Capsules** Jan–May(–Sep), obtuse or subacute, 1.7–4(–5.5) × 1.5–2.9 mm, sts hairy, loculicidal split extending ¼–¾-way to base. **Seeds** flattened, ellipsoid to discoid, straw-yellow, 0.9–1.2 × 0.6–1 mm, MR 0.2–0.3 mm. $2n = 40$ or 80.

KEY TO VARIETIES

1. Leaves narrowly lanceolate to linear (Appendix 4); Mt Taranaki and Pouakai Ra. (lowland to subalpine) .... **var. *egmontiana***
   Leaves of various shapes, but rarely narrowly lanceolate or linear; various localities, scarce near base of Mt Taranaki (var. *stricta* only) 2

2. Plants of compact habit, < 1 m tall; leaves firm-textured; Raukumara to Kaimanawa and Kaweka ranges (subalpine) .... **var. *lata***
   Plants usually of open lax habit, mostly > 1 m tall (except at some coastal sites); leaves thin to subcoriaceous; various localities (lowland to montane) 3

3. Calyx lobes with marginal cilia only; south from Manawatu R. .... **var. *atkinsonii***
   Calyx lobes hairy on outer surface; mostly north from Manawatu R. (extending southwards only as far as Otaki R. on east side of Tararua Ra.) 4

4. Leaves broad (Appendix 4) but of various shapes – broadly lanceolate, ovate, broadly elliptic, broadly oblanceolate, obovate, elliptic or oblanceolate, rarely lanceolate; Bay of Plenty to Taranaki and Hawke's Bay (coastal only) .... **var. *macroura***
   Leaves narrower (Appendix 4), of various shapes – lanceolate, oblanceolate, elliptic or oblong; widespread (coastal to montane) .... **var. *stricta***

Var. *stricta*
Small to large bushy shrub, to 4 m tall. Branchlets uniformly hairy or glabrous; internodes (3–)5–35(–45) mm long. Leaves lanceolate, oblanceolate, elliptic (often narrowly), or oblong; apex acuminate or acute or subacute; (19–)30–100(–127) × (5.5–)7–22(–30) mm; thin to subcoriaceous; lower surface us. hairy along midrib but sts glabrous or covered with a mixture of egl. and gl. hairs (when young). Inflorescences (4–)6–21.5 cm long. Calyx lobes ciliate, hairy outside. Corolla tube 3.3–4.5 mm long. Capsules sts hairy. $2n = 40$ or 80.

Hebe stricta var. *stricta* and var. *macroura*   OCCLUSAE

FIG. 94   *Hebe stricta* var. *stricta* and var. *macroura*. **A** var. *stricta* near Rangiwahia. **B** var. *macroura* at Makorori. **C** sprigs of var. *stricta* (left) and var. *macroura*. **D–E** lower and upper leaf surfaces: D, var. *stricta*; magnified inset shows marginal hairs; E, var. *macroura*. **F–G** close-ups of leaf buds with no sinuses: F, var. *stricta*; G, var. *macroura*. **H–I** lateral views of flowers: H, var. *stricta*; I, var. *macroura*; magnified insets show hairs on outer surfaces of calyces. **J–K** inflorescences showing frontal view of flowers: J, var. *stricta*; K, var. *macroura*. **L–M** capsule of var. *macroura* in: L, septicidal view; M, loculicidal view.

TAXONOMIC TREATMENT   179

Var. *macroura* (Benth.) L.B.Moore
Spreading or erect shrub, 0.2–1.4 m tall. Branchlets glabrous or uniformly hairy; internodes 8.6–49(–58) mm long. Leaves broadly lanceolate, ovate, broadly elliptic, broadly oblanceolate, obovate, elliptic or oblanceolate, rarely lanceolate; apex acute, subacute or obtuse; (31–)37–106 × (10.8–)16–44.8 mm; coriaceous; lower surface glabrous or hairy along midrib. Inflorescences (2.6–)5.5–16.7 cm long. Calyx lobes ciliate, almost always hairy outside. Corolla tube (1.5–)2–3.5 mm long. Capsules sts hairy near apex. $2n = 40$.

Var. *atkinsonii* (Cockayne) L.B.Moore
Bushy shrub, to 3 m tall. Branchlets uniformly or bifariously hairy; internodes 3–35 mm long. Leaves lanceolate, oblanceolate, obovate (mostly coastal forms), or elliptic; apex acute, acuminate, subacute or obtuse; (20–)35–80(–95) × (7–)9–19(–28) mm; subcoriaceous; lower surface glabrous or hairy along midrib. Inflorescences 4.5–19 cm long. Calyx lobes ciliate, glabrous outside. Corolla tube 3–5 mm long. Capsules glabrous. $2n = 40$.

Var. *egmontiana* L.B.Moore
Bushy shrub, to 4 m tall. Branchlets bifariously hairy or glabrous; internodes 3–13(–18) mm long. Leaves lanceolate to linear; apex acute or acuminate; 36–100(–106) × 5.5–14.5 mm; subcoriaceous; lower surface glabrous. Inflorescences (4–)7–15.5 cm long. Calyx lobes ciliate, sts sparsely hairy outside. Corolla tube (2.2–)3–4 mm long. Capsules sts hairy. $2n = 80$.

Var. *lata* L.B.Moore
Compact shrub, 0.1–1 m tall. Branchlets usually glabrous or bifariously or uniformly hairy; internodes 2.25–17.2(–25.2) mm long. Leaves lanceolate, narrowly elliptic, or occasionally oblanceolate; apex acute; 25.8–70.5 × 5.7–16.1 mm; firm; lower surface glabrous. Inflorescences 3.2–8.1(–12.4) cm long. Calyx lobes ciliate, rarely sparsely hairy outside. Corolla tube 3.2–5 mm long. Capsules rarely with a few hairs near apex. $2n = 80$.

DISTRIBUTION AND HABITAT Var. *stricta* is widespread on NI, from near Cape Reinga, Northland, to near Otaki, Wellington, including islands of Hauraki Gulf and the Bay of Plenty. It grows in a range of mostly open habitats from the coast to near altitudinal tree-line, including scrub, open areas in forest and on forest margins, riverbanks, slips, bluffs and other rocky areas.

Var. *macroura* occurs in coastal areas of South Auckland, Gisborne, Hawke's Bay and Taranaki. It grows on rocks and in scrub.

Var. *atkinsonii* is found on NI in areas south of the Manawatu R., Wellington. On SI, it occurs in the Marlborough Sounds, and in coastal areas as far west as Golden Bay and as far south as Napenape, Canterbury. It grows in a range of mostly open habitats from the coast to near altitudinal tree-line, including scrub, open areas in forest and on forest margins, riverbanks, slips, bluffs and other rocky areas.

Var. *egmontania* is endemic to Mt Taranaki and the Pouakai Ra., NI. It grows on forest margins and in scrub, from lowland areas to the subalpine zone.

Var. *lata* occurs on the western and central mountains of NI from the Raukumara to Kaimanawa and Kaweka ranges. It grows above the tree-line in tussock grassland, low scrub and rocky places.

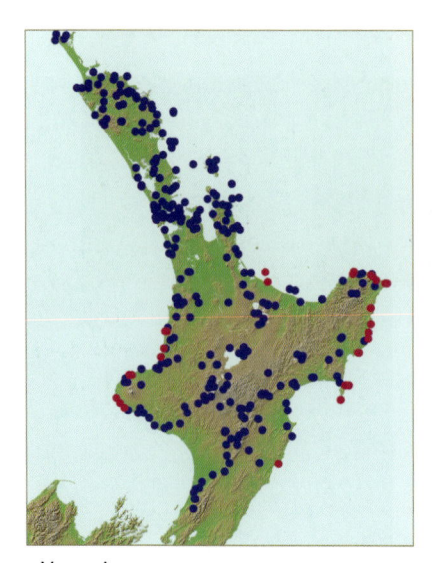

• Var. *stricta*
• Var. *macroura*

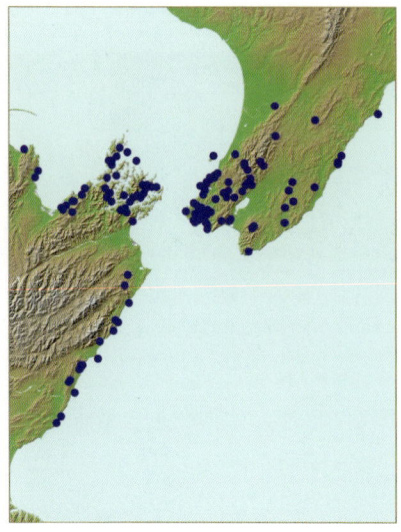

• Var. *atkinsonii*

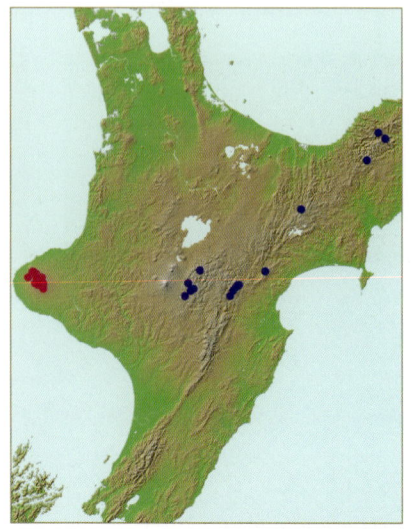

• Var. *lata*
• Var. *egmontiana*

Hebe stricta var. *atkinsonii*   OCCLUSAE

FIG. 95   *Hebe stricta* var. *atkinsonii*. **A** plant near Maitai R., Nelson. **B** sprig. **C** leaf bud with no sinus. **D** lower and upper leaf surfaces. **E** inflorescence. **F** lateral view of flower; magnified inset shows calyx with glabrous outer surface. **G** frontal view of flowers. **H** capsule in septicidal (left) and loculicidal view. **I** seeds.

NOTES *Hebe stricta* is a widespread and morphologically variable species (Figs 94–96; Appendix 4). On central and southern NI, and on northern SI, it is the most common large-leaved hebe of areas below the tree-line. It is generally distinguished from similar species of "Occlusae" in those areas (*H. angustissima, H. tairawhiti*) by its leaf shape and size. On northern NI, var. *stricta* is distinguished from several morphologically similar "Occlusae" species (*H. acutiflora, H. flavida, H. ligustrifolia, H. perbella*) by its usually longer corolla tubes (which are longer than the calyx lobes) and obtuse corolla lobes. Other similar species in that area are: *H. bishopiana*, distinguished by its smaller decumbent stature, and by the colour of its stems and undersides of young leaves; and *H. macrocarpa*, distinguished by its more robust leaves, calyx lobes that are often glabrous outside, broad flowers, long filaments and large fruit.

The varieties recognised here are based, with some modification, on those recognised by Moore (in Allan 1961), and their limits are not clear-cut. Var. *egmontiana* is one of the most distinctive. Var. *stricta* and var. *atkinsonii* represent broad geographic groups that differ primarily in the presence/absence of hairs on the outside of the calyx lobes. Var. *lata* is a subalpine variant; we have hardly seen it in the wild and have not been able to judge its distinctiveness with confidence. Var. *macroura*, which includes broad-leaved coastal plants, might be better viewed as a descriptive group, rather than a biological entity. For example, populations on East Cape and the Taranaki coast are conceivably more closely related to surrounding populations of var. *stricta* than to each other. Similar broad-leaved plants from the Wellington coast included in var. *macroura* by Moore are included here in var. *atkinsonii*. This is because, unlike plants from further north, they lack hairs on the outside of the calyces (as do other members of var. *atkinsonii*), and because field and herbarium observations suggest a complete gradation between inland and coastal forms (also suggested by Cockayne 1916). Development of larger leaves at coastal sites is seen in many unrelated plant species, and can have both an environmental and genetic basis (Blackman et al. 2005).

Variation in chromosome number (Table 5) prompted Druce (1980, 1993) to propose recognition of segregate species. A broad circumscription is retained here for three reasons: because of the morphological similarity and overlap between varieties; because variation in chromosome number in var. *stricta* (Table 5) makes chromosome differences less clear-cut; and because flavonoid chemistry suggests that all varieties are closely related (Mitchell et al. in prep.). Further morphological study, chromosome counts and use of other data sources (e.g. DNA fingerprinting) would aid investigation of the circumscriptions of both the varieties and the species.

Collections from Hikurangi Swamp, Northland (e.g. AK 202263, WELT 81920) are similar to *H. stricta* in most vegetative and reproductive features. They differ most noticeably in that the leaves and stems, at least when young, are tinged maroon-purple. They also differ from *H. stricta* in flavonoid profile (Mitchell et al. in prep.), and are tetraploid (Murray & de Lange 1999, as *H.* aff. *stricta* (B)). The most appropriate taxonomic placement of these specimens has yet to be determined; they are mentioned by de Lange et al. (1999, 2004a) as "*Hebe* aff. *bishopiana*".

ETYMOLOGY *Stricta* (L. *strictus* = drawn close together, very upright, very straight) possibly refers to leaves or habit. *Egmontiana* refers to the distribution of the variety on Mt Taranaki (also known as Mt Egmont). *Lata* (L.) refers to habit, usually broader than tall. *Macroura* (Gk *macro* = long, large; *urus* = tailed), presumably refers to the inflorescences. *Atkinsonii* honours Esmond H. Atkinson (1888–1941), botanical artist, who worked for the New Zealand Department of Agriculture and, later, the Dominion Museum.

FIG. 96   *Hebe stricta* var. *egmontiana* and var. *lata*. **A** var. *egmontiana* on Mt Taranaki. **B** var. *lata* on Makahu Spur, Kaweka Ra. **C** sprigs, var. *egmontiana* (left) and var. *lata*. **D–E** lower and upper leaf surfaces: D, var. *egmontiana*; E, var. *lata*; magnified insets show marginal hairs. **F–G** leaf buds with no sinuses: F, var. *egmontiana*; G, var. *lata*. **H–I** lateral view of flowers: H, var. *egmontiana*; I, var. *lata*; magnified insets show calyces with glabrous outer surfaces. **J–K** inflorescences showing frontal view of flowers: J, var. *egmontiana*; K, var. *lata*. **L–M** capsules of var. *lata* in: L, septicidal view; M, loculicidal view.

40. *Hebe bishopiana* (Petrie) D.Hatch

DESCRIPTION Openly branched, spreading low shrub or bushy shrub to 1 m tall. **Branches** spreading or ascending, old stems red-brown or black; branchlets purplish or sts green, minutely puberulent (us.) or glabrous, hairs uniform (us.) or bifarious; internodes (7.6–)20–24 mm; leaf decurrencies weakly evident, or obscure. **Leaf bud** distinct (and tinged maroon to purple outside); sinus absent. **Leaves** patent to recurved; *lamina* lanceolate or narrowly elliptic or oblanceolate, subcoriaceous, flat or m-shaped in TS, (20–)40–90 × (8–)13–18(–22) mm; *apex* acute or slightly acuminate; *margin* narrowly cartilaginous, pubescent (often minutely), often tinged red; *upper surface* dark green (sts tinged maroon, esp. along midrib), glossy, with few or without evident stomata, hairy along midrib; *lower surface* often maroon when young, otherwise light green. **Inflorescences** with 70–170 flowers, lateral, unbranched, (5–)9–17 cm, sts with a conspicuous number of unopened flowers toward the apex (leaving a protruding "rat's tail"); peduncle 1–2.2 cm; rachis (4.2–)7.5–15 cm. **Bracts** alternate, linear-lanceolate or elliptic, acute, sts sparsely hairy outside. **Flowers** (Oct–)Feb–Jun(–Jul), ⚥. **Pedicels** longer (mostly) to shorter than bracts, 1.5–3 mm, sts recurved in fruit. **Calyx** 2–4 mm, 4–5-lobed (5th lobe small, posterior); *lobes* lanceolate or elliptic, acute to obtuse, hairy outside. **Corolla** tube hairy inside and sts outside, 2–4 × 1.9–2.1 mm, funnelform, equalling or longer than calyx; *lobes* white or tinged mauve at anthesis, narrowly elliptic or lanceolate, subacute or obtuse, suberect or patent, equalling or shorter than corolla tube, with a few hairs toward base on inner surface and often also ciliate (near base). **Stamen filaments** white or coloured, 5.5–8 mm; anthers mauve, 1.9–2.5 mm (flowers sts have an additional 1 or 2 stamens, with ± malformed anthers, inserted at margins of anterior corolla lobes). **Ovary** sts hairy, *c.* 1–1.1 mm; ovules *c.* 10–13 per locule; style 6–8 mm, sts hairy. **Capsules** May–Aug(–Nov), subacute, 3–4.5 × 2.4–3 mm, hairy, loculicidal split extending ¼–½-way to base. **Seed** characters not recorded. $2n = 40$.

DISTRIBUTION AND HABITAT Endemic to the Waitakere Ranges, west of Auckland. It grows on streamsides, shaded cliff faces, seepages on exposed outcrops and sometimes in low scrub by tracksides.

NOTES This was originally treated by Petrie (1926) as a hybrid between *Veronica salicifolia* [*H. stricta* var. *stricta*] and *V. obtusata* [*H. obtusata*], but has been treated as a distinct species by Eagle (1982), Smith-Dodsworth (1991), Druce (1993), and de Lange (1996). It is distinguished from most large-leaved "Occlusae" by the colour of its stems and the undersides of young leaves, and by its open, spreading habit. It most closely resembles *H. obtusata*, from which it differs in its greater stature, leaf shape and size, and lack of conspicuous pubescence on the leaf margin. The parentage suggested by Petrie (1926) is historically feasible, but does not explain the present distribution of *H. bishopiana*, which sometimes occurs with one or other of the supposed parents (more frequently *H. stricta*) but not both. That said, de Lange (pers. comm. 2005) noted that, until road widening in 1993, the three species were sympatric on a roadside bank in the car park on Mt Donald McLean, near the track to the summit. *H. bishopiana* also comes true from seed (Hatch 1966; de Lange 1996). A detailed distribution map, and notes on nomenclature, typification, relationships, recognition and conservation, are provided by de Lange (1996).

ETYMOLOGY Honours John J. Bishop (1865–1933), who collected the species and grew it in his garden, and whose collection (jointly with H. Carse and E. Jenkins) is the type.

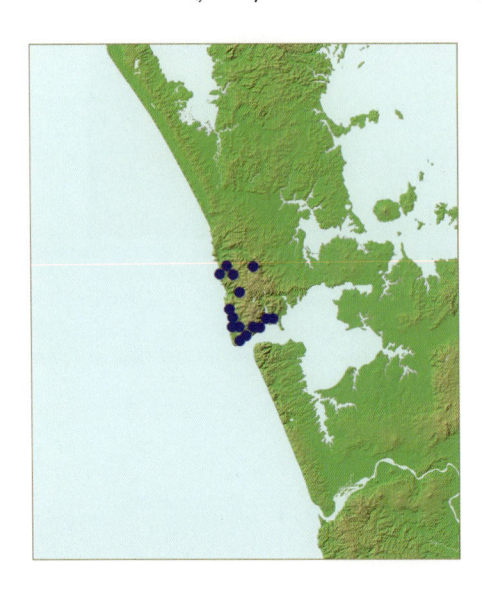

Hebe bishopiana   OCCLUSAE

FIG. 97   **A** plant on Mt Donald McLean, Waitakere Ranges. **B** sprig. **C** leaf bud with no sinus. **D** lower and upper leaf surfaces. **E** inflorescence. **F** inflorescence showing frontal view of flowers. **G** lateral view of flower. **H** capsule in septicidal (left) and loculicidal view.

41. *Hebe obtusata* (Cheeseman) Cockayne et Allan

DESCRIPTION Spreading low shrub to 0.5 m tall. **Branches** prostrate or decumbent, old stems brown or grey; branchlets red-brown, puberulent, hairs bifarious or uniform (sts very sparse); internodes (2–)10–39 mm; leaf decurrencies evident (often with a narrow ridge along medial line) or obscure. **Leaf bud** distinct; sinus absent. **Leaves** decussate to subdistichous, erecto-patent to patent; *lamina* obovate (us.) or elliptic or circular or oblong, coriaceous, flat or slightly m-shaped in TS, (6–)20–55 × (6.5–)11–20(–28.5) mm; *apex* obtuse or rounded or rarely retuse; *margin* cartilaginous, pubescent (us. very conspicuously), often tinged red; *upper surface* dark green, glossy, without evident stomata, us. hairy along midrib; *lower surface* light green, hairy along midrib and sts covered with minute gl. hairs (when young). **Inflorescences** with 34–88 flowers, lateral, unbranched, 3.8–12.6 cm; peduncle 0.7–3.1 cm; rachis 3.2–9.5 cm. **Bracts** alternate, lanceolate or narrowly oblong, acute or subacute, sts hairy outside. **Flowers** (Nov–)Jan–Aug, ⚥. **Pedicels** 1–3.3 mm, sts recurved in fruit. **Calyx** 2.5–3.5 mm; *lobes* narrowly elliptic or oblong, obtuse to acute, often hairy outside. **Corolla** tube hairy inside and sts outside, 2.3–4 × 1.7–2 mm (us. quite asymmetric – longer on anterior side), funnelform, about equalling or longer than calyx; *lobes* white or tinged mauve at anthesis, white or mauve with age, ovate or lanceolate or elliptic, obtuse or subacute, suberect to patent, longer to shorter than corolla tube (varies on one flower because of asymmetric division of corolla), hairy inside or at least with a few hairs toward base on inner surface and sts ciliate (near base) or hairy outside. **Stamen filaments** white or mauve, 4.7–7.5 mm; anthers buff or mauve, 1.5–2.2 mm. **Ovary** sts sparsely hairy, 0.8–1 mm; ovules *c.* 10–13 per locule; style 5–7 mm, sts sparsely hairy. **Capsules** (Nov–)Jan–Sep, acute or subacute, 3.7–4.5 × 2–2.3 mm, sts hairy, loculicidal split extending ¼-way to base. **Seeds** flattened, ellipsoid to discoid, straw-yellow, 0.9–1.3 × 0.7–0.9 mm, MR *c.* 0.2 mm. $2n = 40$.

DISTRIBUTION AND HABITAT Occurs near the west coast of NI, chiefly west of Auckland between Muriwai and Manukau Heads, but also further south, at Kawhia Harbour. It grows mostly on near-coastal slopes, cliffs and rocks.

NOTES Distinguished from most large-leaved "Occlusae" by the combination of its low-growing habit, leaf shape and size, the colour of young branchlets, and the usually conspicuous pubescence on the leaf margin (Fig. 98D). It resembles *H. bishopiana* (see notes under that species). de Lange (1996) suggests that it sometimes hybridises with *H. stricta* var. *stricta*.

ETYMOLOGY L. (*obtusus* = blunt, obtuse; *-atus* is an adjectival suffix for nouns), refers to the leaf apex.

Hebe obtusata   OCCLUSAE

FIG. 98   **A** plant near Takatu Head. **B** sprig. **C** leaf bud with no sinus. **D** lower and upper leaf surfaces, with magnified region showing hairs on leaf margin. **E** inflorescence. **F** lateral view of flower. **G** close-up of calyx, with magnified region showing hairs on the outer surface of calyx lobe. **H** capsule in septicidal (left) and loculicidal view. **I** seeds.

42. *Hebe tairawhiti* B.D.Clarkson et Garn.-Jones

DESCRIPTION Bushy shrub to 3 m tall. **Branches** erect, old stems brown; branchlets green, becoming red-brown to black (with age), puberulent or glabrous, hairs bifarious or uniform; internodes 7–15 mm; leaf decurrencies obscure to evident. **Leaf bud** distinct; sinus absent. **Leaves** recurved; *lamina* linear to lanceolate (tapered evenly from broad base), subcoriaceous, flat or m-shaped in TS, (45–)60–100(–150) × (5–)6–11 mm; *apex* acute; *base* abruptly cuneate; very faint brochidodromous *secondary veins* evident in fresh leaves; *margin* puberulent, entire or distantly denticulate; *upper surface* green (with broad yellow midrib), dull, without evident stomata, hairy along midrib or glabrous; *lower surface* light green (paler than upper). **Inflorescences** with 120–180 flowers, lateral, unbranched, 6.8–15 cm; peduncle 0.6–2.3 cm; rachis 5.5–14 cm. **Bracts** alternate, narrowly deltoid, acute, hairy outside. **Flowers** Jan–Apr, ⚥. **Pedicels** longer (mostly) to shorter than bracts, 1.3–2 mm. **Calyx** 2–2.5 mm; *lobes* lanceolate, acute, rarely hairy outside. **Corolla** tube hairy inside, 3–3.5 × 1.5–1.8 mm, expanded in lower half, longer than calyx; *lobes* white or pale mauve at anthesis, elliptic or oblong, obtuse, erect (mostly) or suberect to patent, shorter than corolla tube. **Stamen filaments** 3–5 mm; anthers mauve or violet, 2–2.3 mm. **Ovary** ovoid or ellipsoid, 1–1.2 mm; ovules 10–12 per locule; style 3.5–4 mm. **Capsules** (Feb–)Apr(–Dec), subacute, 2.7–3.5(–5.2) × 1.8–2.6 mm, loculicidal split extending ¼-way to base. **Seeds** flattened (sts strongly), ellipsoid or obovoid or discoid, straw-yellow to brown, (0.8–)1–1.2 × (0.5–)0.7– 1 mm, MR *c.* 0.2 mm. $2n = 80$.

DISTRIBUTION AND HABITAT Gisborne and northern Hawke's Bay, NI, from the Maraehara R. near East Cape in the north, to near Wairoa, and the Mahia Peninsula in the south. It grows in near-coastal shrub-dominated vegetation and at inland sites, on riverbanks, road cuttings and rock outcrops.

NOTES A recently described species (Clarkson & Garnock-Jones 1996), previously known informally as *Hebe* sp. (n) (Druce 1980; Eagle 1982) and *H.* "Wairoa" (Cameron et al. 1993; Druce 1993). It is distinguished from most "Occlusae" by its narrowly lanceolate to linear leaves, which taper gradually from a broad base. Other similar taxa are *H. stricta* var. *egmontiana*, which, as circumscribed here, does not occur in Gisborne or northern Hawke's Bay; *H. angustissima*, which is usually smaller statured and occurs in different habitats; and *H. parviflora*, which often has smaller leaves, and stamen filaments that are strongly incurved at the apex in bud. The flavonoid profile of *H. tairawhiti* is distinct from those other "Occlusae" (Mitchell et al. in prep.).

ETYMOLOGY Māori-derived, Tairawhiti being the Māori name for the region in which the species mostly occurs.

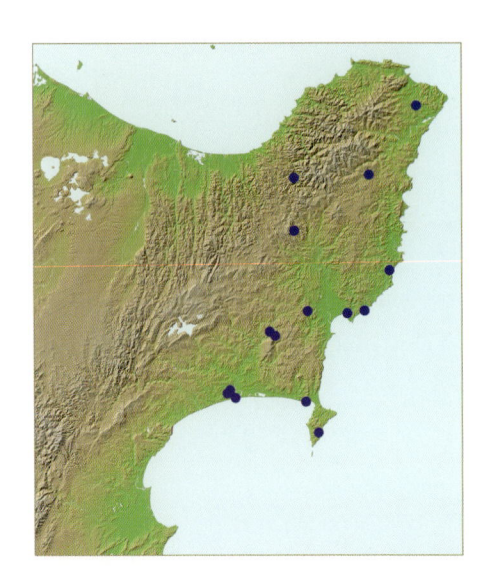

Hebe tairawhiti   OCCLUSAE

FIG. 99   **A** plants near Maraehara River, Gisborne. **B** sprig. **C** leaf bud with no sinus. **D** lower and upper leaf surfaces; magnified inset shows two minute teeth on leaf margin. **E** inflorescence. **F** lateral view of flower. **G** frontal view of flowers. **H** septicidal view of capsule. **I** loculicidal view of capsule.

43. *Hebe angustissima*   (Cockayne) Bayly et Kellow comb. nov.

DESCRIPTION Openly branched bushy shrub to 1(–1.4) m tall. **Branches** erect or ascending, old stems brown; branchlets green, puberulent, hairs bifarious or uniform; internodes 6–14 mm; leaf decurrencies weakly evident (with a narrow ridge along the medial line of each decurrency). **Leaf bud** distinct; sinus absent. **Leaves** patent to recurved; *lamina* linear or linear-lanceolate, thin, ± flat, (22–)30–75 × 3–9 mm; *apex* acute (mostly) or subacute; *margin* narrowly cartilaginous, puberulent (with a mixture of gl. and egl. hairs); *upper surface* dark green, dull, with few stomata, hairy along midrib and/or sts covered with minute gl. hairs; *lower surface* light green, sts hairy along midrib or sts covered with minute gl. hairs. **Inflorescences** with 29–95 flowers, lateral, unbranched, 4.7–11 cm; peduncle 0.5–1.8 cm; rachis 4.1–9.5 cm. **Bracts** alternate, narrowly oblong or oblanceolate or deltoid, acute to obtuse. **Flowers** Feb–Apr(–Jun), ⚥. **Pedicels** longer than or equal to bracts, 1.8–3.5 mm, sts recurved in fruit. **Calyx** 1.9–3.4 mm; *lobes* narrowly oblong or lanceolate or elliptic, acute (mostly) to obtuse, sts hairy outside. **Corolla** tube densely hairy inside, 2.2–4 × 0.7–0.9 mm, cylindric and contracted at base, longer than calyx; *lobes* white or tinged mauve at anthesis, lanceolate or elliptic, obtuse or subacute, suberect to patent, equalling or longer than corolla tube (esp. anterior), sts with a few hairs toward base on inner surface. **Stamen filaments** 4–4.5 mm; anthers mauve or magenta (sts drying purple), 1.4–1.8 mm. **Ovary** 0.6–0.9 mm; ovules 10–12 per locule; style 3.5–6 mm. **Capsules** Mar–Jul(–Jan), obtuse or subacute, 2.2–3 × 1.4–2.6 mm, loculicidal split extending ¼–½-way to base. **Seeds** flattened, ± ellipsoid to discoid, straw-yellow, 0.8–1 × 0.6–0.8 mm, MR 0.1–0.2 mm. $2n = 40$.

DISTRIBUTION AND HABITAT Disjunct between Gisborne (Motu and Waioeka rivers) and Wellington (Otaki R. and tributaries, as well as Takapu Stm). It grows on rocky sites in lowland river gorges, often in the flood zone.

NOTES Originally described by Cockayne (1918), using specimens from Otaki Gorge, as *Veronica salicifolia* var. *angustissima*. It was included in *H. stricta* by Moore (in Allan 1961), but informally accepted as a distinct variety by Druce (1980), and later as a distinct species by Druce (1993).

It differs from *H. stricta* in flavonoid profile (Mitchell et al. in prep.), and in its usually narrower leaves (Appendix 4); *H. stricta* var. *egmontiana* has similarly narrow leaves, but has a denser canopy, occupies different habitats (chiefly montane to subalpine) and is tetraploid (rather than diploid). The two species sometimes co-occur, particularly in Gisborne, but *H. stricta* is more common in adjoining scrub and forest than in the exposed, rocky and often flood-prone areas usually occupied by *H. angustissima*.

Other similar species are: *H. acutiflora*, which usually has shorter corolla tubes; and *H. tairawhiti*, which usually has firmer leaves, a more erect habit and a different habitat range, and is tetraploid. *H. angustissima* also differs from both of these species in flavonoid profile.

The unusual distribution of *H. angustissima* poses questions about whether it occurs at intervening localities, and whether the Gisborne and Wellington populations are truly conspecific. These disjunct populations are included together here because they are similar both morphologically and ecologically, and have a distinctive flavonoid profile (flavonoid variation between populations is less than that within populations).

The record from Takapu Stm (the southernmost on the distribution map) rests on one collection (AK 233642, CHR 490854), which resembles others in leaf shape and habitat, but from which leaf flavonoids were not sampled.

ETYMOLOGY L. (*angustus* = narrow; *-issimus* = most so, to the greatest degree), refers to the leaves.

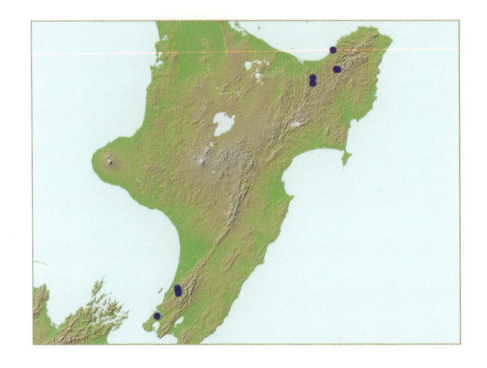

Hebe angustissima   OCCLUSAE

FIG. 100   A plant in Otaki Gorge. **B** sprig. **C** leaf bud with no sinus. **D** lower and upper leaf surfaces; magnified inset shows minute hairs on leaf margin. **E** inflorescence. **F** lateral view of flower. **G** frontal view of flower. **H** septicidal view of capsule. **I** loculicidal view of capsule.

TAXONOMIC TREATMENT   191

44. *Hebe acutiflora* Cockayne

DESCRIPTION Openly branched bushy shrub to 1.2(–1.5) m tall. **Branches** erect, old stems brown; branchlets green, puberulent or rarely glabrous (only on youngest branchlets), hairs bifarious (mostly) or uniform; internodes (1.5–)3–18(–29) mm; leaf decurrencies weakly evident (with a narrow ridge along the medial line of each decurrency), or obscure. **Leaf bud** distinct; sinus absent. **Leaves** erecto-patent to patent; *lamina* linear or linear-lanceolate, thin, ± flat, (15–)25–85(–118) × (3–)4–9(–12) mm; *apex* acute to obtuse; brochidodromous *secondary veins* sts evident in fresh leaves; *margin* very narrowly cartilaginous, ciliolate, entire or distantly denticulate; *upper surface* dark green, dull, without evident stomata, hairy along midrib (us.) and sts covered with minute gl. hairs or occasionally glabrous; *lower surface* light green, faintly pitted with small depressions that each contain a twin-headed gl. hair or not pitted, otherwise glabrous or rarely hairy along midrib. **Juvenile leaves** pinnatifid (near apex), minutely ciliolate (and midrib hairy above). **Inflorescences** with 13–81 flowers, lateral, unbranched (almost always) or tripartite (seen on only one inflorescence on one specimen), 2.7–8.4(–13.6) cm, longer than (mostly) or about equal to subtending leaves; peduncle 0.5–1.9 cm; rachis 2.2–7.2(–11.7) cm. **Bracts** alternate, narrowly deltoid or lanceolate, acute, hairy outside. **Flowers** Jan–May(–Jun), ⚥. **Pedicels** longer than or equal to bracts, (0.5–)2–4(–5) mm, sts recurved in fruit. **Calyx** 2.2–3(–3.5) mm; *lobes* very narrowly deltoid or lanceolate, acute or acuminate, hairy outside (us. densely, but sts sparsely). **Corolla** tube hairy inside and occasionally outside, 1.3–2.5(–2.8) × *c*. 1.6–1.8 mm, shortly funnelform, shorter than (almost always true for posterior portion) to very slightly longer than calyx; *lobes* white or tinged pale mauve at anthesis, lanceolate, acute or subacute, suberect to patent, longer than (mostly) to shorter than corolla tube, sts ciliate and sts hairy outside. **Stamen filaments** (4.2–)5–6.5 mm; anthers mauve, 1.9–2.2 mm. **Ovary** narrowly ovoid, often sparsely and minutely hairy (esp. along septal grooves), *c*. 0.9–1.2 mm; ovules 8–10 per locule; style (3.5–)4–5(–5.5) mm. **Capsules** Jan–Jun(–Dec), obtuse or subacute, (2–)2.5–3.5 × (1.6–)2–3 mm, sts sparsely hairy (mostly along septal margins), loculicidal split extending ¼–½-way to base. **Seeds** strongly flattened, ellipsoid to discoid, weakly winged, straw-yellow, 0.9–1.4 × 0.8–1.1 mm, MR 0.1–0.3 mm. $2n = 40$.

DISTRIBUTION AND HABITAT Endemic to Northland, NI, where it occurs with certainty on the Kerikeri R. and Puketotara Stm (both near Kerikeri), Waipapa R. (Puketi Forest) and Waipoua R. (Waipoua Forest). It grows exclusively near riverbanks, often in the flood zone.

NOTES Distinguished from most large-leaved "Occlusae" by the combination of linear or linear-lanceolate leaves (Appendix 4), calyces that are hairy outside, and corolla tubes that are mostly shorter than or equalling the calyx. It is similar to *H. flavida* and *H. angustissima* (see notes under those species). It sometimes grows with *H. stricta* var. *stricta*, which usually has broader leaves and longer corolla tubes.

H. acutiflora possibly also occurs in Herekino Forest, as recorded by Allan (1939), based on CHR 56644. Flowering specimens and/or more information on the habit/habitat of plants would help to confirm its occurrence there (also see notes under *H. flavida*).

The label of a cultivated specimen of *H. acutiflora* (CHR 189031) indicates that it originally came from shoreline cliffs at "Mangonui Inlet" (Te Puna Inlet, Bay of Islands, on recent maps), where it was "not uncommon". A search of this area in 2001 failed to find any similar plants, and the locality is not represented on the distribution map. *H. ligustrifolia* is common in this area and, although sometimes narrow-leaved, does not have leaves as conspicuously linear as CHR 189031.

Plants from Trounson Kauri Park, Northland, have been included by some (e.g. Murray & de Lange 1999) in *H. acutiflora*. They have similar, linear leaves and occupy riparian habitats, but differ markedly in flavonoid profile from samples of *H. acutiflora* (Mitchell et al. in prep.) and have corolla tubes at least slightly longer than calyces (e.g. AK 235668, WELT 82786). They potentially represent an undescribed taxon, to which a morphologically similar specimen from Kaihu Forest (AK 167579) probably also belongs.

ETYMOLOGY L. (*acutus* = sharp-pointed; *flores* = flowers), probably refers to general flower shape, both calyx and corolla lobes usually being acute.

Hebe acutiflora OCCLUSAE

FIG. 101 **A** plant near Rainbow Falls, Kerikeri. **B** sprig. **C** leaf bud with no sinus. **D** lower and upper leaf surfaces; magnified inset shows hairs on leaf margin. **E** inflorescence and young infructescence. **F** lateral view of flowers. **G** frontal view of flower. **H** capsule in septicidal (left) and loculicidal view; magnified inset shows hairs on outer surface of calyx lobe.

45. *Hebe flavida*   Bayly, Kellow et de Lange sp. nov.

DESCRIPTION Small tree (us.) or bushy shrub to 8 m tall. **Branches** erect, old stems brown or grey; branchlets green, puberulent to pubescent, hairs uniform; internodes (2.5–)6–20(–27) mm; leaf decurrencies evident (sts weakly). **Leaf bud** distinct; sinus absent. **Leaves** erecto-patent to patent; *lamina* linear-lanceolate to narrowly elliptic or oblanceolate, thin or subcoriaceous, flat or slightly m-shaped in TS, (30–)50–100(–135) × (6–)10–20(–29) mm; *apex* acuminate or acute; brochidodromous *secondary veins* evident in fresh leaves; *margin* narrowly cartilaginous, puberulent, entire or distantly denticulate; *upper surface* light to dark green (with midrib and base of lamina us. yellow), dull, with few or without evident stomata, hairy along midrib; *lower surface* light green, hairy along midrib and sts covered with minute gl. hairs (when young) or rarely glabrous. **Inflorescences** with 60–140(–155) flowers, lateral, unbranched (although one small secondary branch seen on one inflorescence of WELT 80664), (4–)7–16(–24.5) cm; peduncle (0.7–)1–3(–4.5) cm; rachis (2.8–)5.5–14(–20.5) cm. **Bracts** alternate, lanceolate or narrowly deltoid, acute, hairy outside. **Flowers** Jan–Jun, ⚥. **Pedicels** 1.5–4.2 mm. **Calyx** 2.1–2.7 mm; *lobes* linear or narrowly deltoid, acute or acuminate, hairy outside. **Corolla** tube hairy inside and sts outside (near base of corolla lobes), 1.5–3 × 1.7–2.7 mm, funnelform, shorter than or equalling calyx (us. asymmetrically divided – anterior side is *c.* equal to calyx, but posterior is shorter); *lobes* white or tinged mauve to pink at anthesis (sts very faintly), lanceolate (sts narrowly) or ovate or elliptic, subacute (us.) or obtuse, suberect to patent, longer than corolla tube, sts sparsely hairy inside. **Stamen filaments** 5.5–6.8 mm; anthers violet or purple or blue, 1.5–2.5 mm. **Ovary** sts hairy, *c.* 0.8–1 mm; ovules *c.* 9–13 per locule; style 4–7.2 mm, sts hairy. **Capsules** Mar–Jun(–Sep), obtuse or subacute, 2.5–4 × (2–)2.5–3.5 mm, sts hairy, loculicidal split extending ½–¾-way to base. **Seeds** flattened (sts strongly), ± broad ellipsoid to discoid, pale brown, (0.8–)0.9–1.4(–1.6) × 0.7–1.2 mm, MR 0.1–0.3 mm. $2n = 40$.

DISTRIBUTION AND HABITAT Endemic to Northland, NI, where it occurs with certainty between Warawara Forest and Waikaraka Va. in the north and Tangihua Forest in the south. The northernmost point on the distribution map is based on historical records (WELT 16511, CHR 328803) from "Okahu" near Kaitaia. It grows mostly in upland areas, above *c.* 250 m asl, often in cloud forest.

NOTES A new species described herein (diagnostic description and type details are given on page 310). It is distinguished from most species by the combination of: a shrub to tree habit (up to *c.* 8 m tall, with trunk 10 cm dbh); no leaf bud sinus (Fig. 102C); corolla tubes shorter than or equalling calyx (Fig. 102E); mostly subacute corolla lobes; leaves with the upper surface of the petiole and base of midvein usually conspicuously yellow (Fig. 102D); and inflorescences that are usually held erect, even in fruit (Fig. 102B).

The three species it most closely resembles are *H. stricta* (of which var. *stricta* occurs in Northland), *H. acutiflora* and *H. ligustrifolia*. *H. stricta* has a similar shrubby habit and superficial appearance (although it doesn't grow as tall as *H. flavida* often does), but is readily distinguished by its flowers (Fig. 94H), which have corolla tubes much longer than the calyces. *H. acutiflora* has flowers similar to *H. flavida*, with short corolla tubes (Fig. 101F), but differs most substantially in having leaves that are generally narrower both in absolute terms and in proportion to their length (they are (3–)4–9(–12) wide, and linear or linear-lanceolate), and in that it is a low-growing shrub, usually not exceeding 1.2 m in height, that apparently grows exclusively on riverbanks, in the zone of vegetation subjected to occasional flooding. Differences from *H. ligustrifolia* are less clear-cut, but *H. flavida* often has longer leaves that are more conspicuously tapered toward the apex (Appendix 4) and a larger habit, and usually occupies more upland habitats. Flavonoid chemistry distinguishes *H. flavida* from all samples of *H. stricta* and *H. acutiflora*, but not consistently from samples of *H. ligustrifolia* (Mitchell et al. in prep.).

Specimens of *H. flavida* are not uniform in appearance. They vary in leaf shape and size (Appendix 4), inflorescence length (Fig. 103), and in the hairiness of stems, ovaries, undersides of leaf midribs and outer surfaces of calyx lobes. On some specimens the leaves are tightly arranged and restricted to the apices of branchlets (Fig. 103A), whereas on others the leaves are more widely spaced along the branchlets (Fig. 103B); these features probably vary with plant age and the degree of exposure of both whole plants and individual branches.

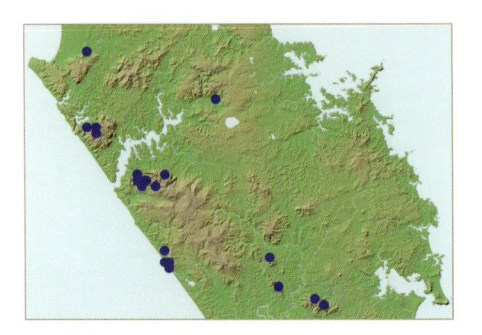

Some herbarium specimens that might be *H. flavida* cannot be identified with certainty without flowers, or without further information on habit or habitat. These include narrow-leaved, lowland specimens from Herekino (e.g. CHR 316527) and Warawara Forest (AK 175866) that might be *H. acutiflora* or *H. flavida*.

Hebe flavida — OCCLUSAE

FIG. 102 **A** plant on Horokaka, Tangihua Ra. **B** sprig. **C** leaf bud with no sinus. **D** lower and upper leaf surfaces; magnified inset shows small tooth and hairs on leaf margin. **E** lateral view of flower. **F** frontal view of flower. **G** inflorescence and infructescence. **H** loculicidal view of capsule. **I** septicidal view of capsule.

Lowland specimens from Waipoua Forest, near the mouth of Ohae Stm (e.g. WELT 83433, AK 153629) and Kararoa Rd (e.g. WELT 81935), are only tentatively identified as *H. flavida*, and their relationships to *H. ligustrifolia* and *H. acutiflora* are worthy of further consideration. These specimens have reasonably long, broad leaves, and flowers (where present) with short corolla tubes. Some from Ohae Stm have flavonoid profiles generally similar to *H. flavida*, but in cultivation in Wellington have different flowering times and paler leaves than any other specimens of *H. flavida*. Vegetation around Ohae Stm is highly modified, and these hebes might not be indigenous there.

ETYMOLOGY L., means yellowish (*flavus* = yellow ; -*idus* indicates a state or action in progress) and refers to the characteristic colour of the petioles.

FIG. 103 Herbarium specimens of *H. flavida* from Warawara Forest, showing variation in form. **A** AK 201784. **B** WELT 82484.

46. *Hebe ligustrifolia* (A.Cunn.) Cockayne et Allan

DESCRIPTION Openly branched, us. a bushy shrub or spreading low shrub, rarely a small tree (near Te Paki), to 2.5(–8) m tall. **Branches** erect to spreading, old stems brown or grey; branchlets olive-green to ± orange or sts purplish, minutely puberulent, hairs uniform; internodes (1.9–)4–10(–17.5) mm; leaf decurrencies evident (often with a narrow ridge along medial line) or obscure. **Leaf bud** distinct; sinus absent. **Leaves** erecto-patent to patent; *lamina* elliptic or oblong-elliptic or linear-lanceolate, subcoriaceous, flat or slightly m-shaped in TS, (12–)26–50(–100) × (4.2–)6– 10(–20) mm; *apex* subacute to obtuse; 2 lateral veins arising from base or brochidodromous *secondary veins* evident in fresh leaves; *midrib* thickened below and either slightly thickened above or depressed to grooved above; *margin* sts narrowly cartilaginous, puberulent or glabrous, rarely tinged red; *upper surface* light to dark green (with midrib and base of lamina often yellow), dull to slightly glossy, without evident stomata, us. minutely hairy along midrib or sts glabrous; *lower surface* light green, glabrous (mostly) or hairy along midrib (only toward base) or rarely covered with minute gl. hairs (on youngest leaves). **Inflorescences** with (15–)20–70 flowers, lateral, unbranched, (2.5–)3–8 cm; peduncle 0.45–1.5(–2.2) cm; rachis 1.5–6.5 cm. **Bracts** alternate or lowermost pair opposite, then subopposite or alternate above, ovate or narrowly lanceolate, acute or subacute, rarely hairy outside. **Flowers** Jan–Dec, ⚥. **Pedicels** us. longer than or equal to bracts, 1–2.5 mm. **Calyx** (1.5–)2–3 mm; *lobes* lanceolate or elliptic, acute or subacute or acuminate, sts hairy outside. **Corolla** tube hairy inside and sts outside, (1.2–)1.6–3 × 1.8–2.2 mm, funnelform, shorter than (us.) or equalling or sts slightly longer than calyx; *lobes* white or tinged mauve at anthesis, ovate to deltoid or lanceolate or elliptic (last two states mostly in anterior lobes), acute or subacute, suberect to recurved, longer than corolla tube, bluntly ciliate (often) or with a few hairs toward base on inner surface and sts hairy outside. **Stamen filaments** white or mauve,

Hebe ligustrifolia   OCCLUSAE

FIG. 104   **A** plant on Unuwhao, Northland. **B** sprig. **C** leaf bud with no sinus. **D** lower and upper leaf surfaces. **E** inflorescence showing frontal view of flowers. **F** lateral view of flowers. **G** faded (white) flowers, from same plant as in E and F. **H** capsule in septicidal (left) and loculicidal view.

5–6.5 mm; anthers mauve or purple, (1.5–)1.7– 2.5 mm. **Ovary** 0.75–1 mm; ovules *c.* 6–15 per locule; style 4–6 mm. **Capsules** Jan–Dec, acute or subacute, 2.5–4(–6) × 1.7–3(–3.7) mm, loculicidal split extending ½–¾-way to base. **Seeds** flattened, broad ellipsoid to ± discoid, straw-yellow, 0.9–1.5 × 0.7–1.1 mm, MR 0.2–0.4 mm. $2n = 40$.

DISTRIBUTION AND HABITAT Endemic to Northland, NI, from North Cape to Whangarei Heads, mostly on the eastern and northern coasts (between North Cape and Cape Reinga). It might also occur in western Northland (see notes below). It grows chiefly in near-coastal sites in scrub, in forest, on cliffs or on slips.

NOTES Distinguished from most large-leaved "Occlusae" by the combination of: leaf shape and size; having corolla tubes mostly shorter than calyces; and broad, acute, to subacute corolla lobes that are longer than the corolla tube. The leaves are generally less robust than those of *H. perbella* and broader than those of *H. acutiflora*, both of which have similar flowers. The outsides of calyces are frequently, though not always, glabrous; this can distinguish plants with only fruit or buds from *H. stricta* var. *stricta*, with which it may co-occur. It is probably most similar to *H. flavida* (see notes under that species); both often have midribs that are conspicuously yellow above.

H. ligustrifolia is variable in habit (from sprawling to, more commonly, erect) and leaf size (Appendix 4), and two informal segregates have been proposed: *Hebe* sp. "m" of Druce (1980) and Eagle (1982), also called *H.* "Whangarei" by Druce (1993); *H.* aff. *ligustrifolia* of de Lange & Murray (2002), databased at AK as *H. ligustrifolia* "var. Surville", and also listed, without an informal name, by Druce (1993). Neither is considered sufficiently distinct for recognition here.

A suite of specimens from Matai Bay, Karikari Peninsula (e.g. WELT 81894–81896), representing the only *Hebe* seen there, may have some affinity to *H. ligustrifolia*, but they are difficult to identify with any certainty. They have: long hairs on the branchlets, undersides of midribs, leaf margins and outsides of calyx lobes (all uncommon in *H. ligustrifolia*); corolla tubes slightly to conspicuously longer than calyces; and leaf flavonoids roughly intermediate between those of *H. stricta* var. *stricta* and *H. ligustrifolia* (Mitchell et al. in prep.).

It is possible that *H. ligustrifolia* also occurs in western Northland, particularly around and south of Hokianga Harbour, including Waima (P. J. de Lange pers. comm. 2005) and Waipoua forests. Some similar specimens from that area are identified here as *H. flavida* (but see notes under that species). We have had few opportunities for fieldwork in these areas, and cannot identify all herbarium specimens with confidence.

ETYMOLOGY L., suggests that the leaves (*folium* = leaf) resemble those of privet (genus *Ligustrum*, family Oleaceae).

47. *Hebe bollonsii* (Cockayne) Cockayne et Allan

DESCRIPTION Bushy shrub to 2.5 m tall. **Branches** erect, old stems brown; branchlets green, puberulent (often minutely), hairs uniform; internodes (1–)4–35(–45) mm; leaf decurrencies evident (often with a narrow ridge along medial line). **Leaf bud** distinct; sinus absent. **Leaves** erect to patent; *lamina* oblanceolate to obovate or oblong or elliptic, coriaceous, ± flat, (14–)20–95(–130) × (8–)15–31(–42) mm; *apex* shortly apiculate to subacute or obtuse; brochidodromous *secondary veins* evident in fresh leaves; *margin* narrowly cartilaginous, glabrous or minutely ciliate; *upper surface* dark green, dull or glossy, without evident or with few stomata, hairy along midrib; *lower surface* light green. **Inflorescences** with (24–)35–120(–125) flowers, lateral, unbranched, (3.5–)4.5–12(–15) cm; peduncle (0.6–)1.4–3 cm; rachis (2.7–)3.8–12.4(–12.6) cm. **Bracts** alternate or lowermost pair opposite, then subopposite or alternate above, lanceolate to linear-lanceolate or ovate, acute. **Flowers** (Oct–)Nov–Feb(–Sep), ⚥. **Pedicels** (1–)1.5–6(–6.5) mm. **Calyx** 2.5–4.5(–5.5) mm; *lobes* lanceolate, acute, very rarely hairy outside. **Corolla** tube hairy inside and often outside (near the points at which lobes diverge), 3–5 × 1.9–2.4 mm, funnelform, about equalling or longer than calyx; *lobes* tinged mauve at anthesis, mauve to white with age, lanceolate or narrowly elliptic, subacute or obtuse (anterior lobe), patent to recurved, equalling or longer than corolla tube, with a few hairs toward base on inner surface and often shortly ciliate (near base). **Stamen filaments** 3.8–7 mm; anthers mauve or purple, 2–2.7 mm. **Ovary** 0.6–1.2 mm; ovules *c.* 10–15 per locule, in 1 layer (us.) or in 2 layers (when ovule number is high); style 5.5–8.5 mm. **Capsules** [Nov–]Jan–Feb(–Sep),

Hebe bollonsii OCCLUSAE

FIG. 105 **A** plant at Crater Bay, Aorangi Id, Poor Knights Ids. **B** sprig. **C** leaf bud with no sinus. **D** lower and upper leaf surfaces. **E** close-up of inflorescence. **F** lateral view of flower (corolla lobes tinged mauve). **G** lateral view of flower (corolla lobes white). **H** capsule in septicidal (left) and loculicidal view. **I** seeds.

TAXONOMIC TREATMENT 199

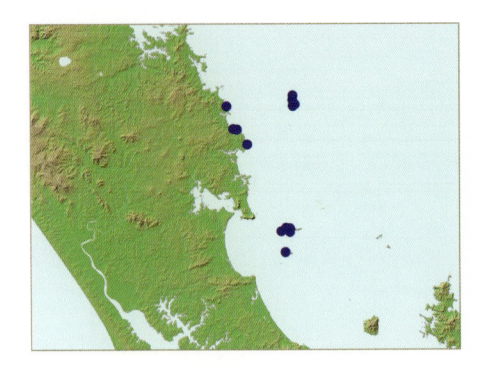

subacute, 2.5–5.5 × (1.8–)2.4–4 mm, loculicidal split extending ¼–½-way to base. **Seeds** strongly flattened, broad ellipsoid, ± winged, straw-yellow, 1–1.7 × 0.9–1.3 mm, MR 0.2–0.4 mm. $2n = 40$.

DISTRIBUTION AND HABITAT Poor Knights Ids, Hen and Chickens Ids, and nearby areas of the NI coast, between Tutukaka and Mimiwhangata Bay. It grows in near-coastal scrub and forest.

NOTES Distinguished from other northern members of large-leaved "Occlusae" by its broad, robust leaves. It may resemble some coastal forms of *H. stricta* (e.g. var. *macroura*) and forms of *H. pubescens* subsp. *sejuncta* on the Mokohinau Ids. It differs from the former by its mostly subacute corolla lobes and corolla tubes that are not much longer than calyces, and from the latter in that it lacks a leaf bud sinus. Some aspects of variation and relationships are discussed by Bayly et al. (2003).

ETYMOLOGY Honours Capt. John P. Bollons (1862–1929).

48. *Hebe perbella* de Lange

DESCRIPTION Bushy shrub to 1.8 m tall. **Branches** erect, old stems brown or grey; branchlets green (when fresh) or sts purplish (when dry), glabrous or puberulent (hairs very short and stubbly), hairs bifarious; internodes 4–26 mm; leaf decurrencies evident (with a prominent narrow ridge along the medial line). **Leaf bud** distinct (often strongly flattened); sinus absent. **Leaves** patent or erecto-patent; *lamina* lanceolate or oblanceolate or oblong-elliptic, coriaceous, slightly m-shaped in TS, (10–)40–90(–110) × (9–)14–18(–25) mm; *apex* obtuse or subacute; faint brochidodromous *secondary veins* evident in fresh leaves; *margin* thickly cartilaginous, glabrous or minutely papillate (with occasional gl. or egl. hairs), sts tinged red; *upper surface* dark green, glossy or dull, without evident stomata (using a dissecting microscope) or with few stomata (seen on epidermal peels), minutely hairy along midrib (with a mixture of egl. and gl. hairs, esp. toward base); *lower surface* light green, sparsely covered with minute gl. hairs (these often inconspicuous and sitting in shallow depressions) and sts sparsely hairy along midrib. **Inflorescences** with (20–)50–85 flowers, lateral, unbranched (us.) or tripartite (occasionally, according to protologue), 4–10(–15) cm, longer than (mostly) or about equal to subtending leaves; peduncle 0.7–5 cm, minutely hairy or glabrous (or very nearly so); rachis 2.5–11 cm. **Bracts** alternate, lanceolate or oblong, acute to obtuse. **Flowers** [Jun–Nov], ⚥. **Pedicels** 2–5 mm. **Calyx** 2.6–3.3 mm, 4–5-lobed (5th lobe small, posterior); *lobes* lanceolate or deltoid, acuminate. **Corolla** tube hairy inside and sts sparsely hairy outside, 1.5–3 × 1.8–2.8 mm, shortly and broadly funnelform, shorter than or equalling calyx (longest on anterior side); *lobes* deep mauve or violet at anthesis, lanceolate or narrowly ovate, subacute or acute, suberect, longer than corolla tube, ciliolate (near base) and with a few hairs toward base on inner surface. **Stamen filaments** coloured (mauve to purple when young) or white (with age), 7–9 mm; anthers blue to purple, 2–2.7 mm. **Ovary** narrowly ovoid, sparsely hairy (esp. along septal grooves), 1–1.4 mm; ovules 9–11 per locule; style 4.5–7 mm, often sparsely hairy (mostly toward base). **Capsules** Nov–?, acute, *c.* 4.8–5.8 × *c.* 3–3.5 mm, often hairy, loculicidal split extending ⅓–½-way to base. **Seed** characters not recorded. $2n = 40$.

DISTRIBUTION AND HABITAT Western Northland, NI, mostly upland areas, where it occurs in the Ahipara gumlands, Herekino Forest, Warawara Forest and Waima Forest, and with a somewhat disjunct southernmost occurrence at Maungaraho Rock, near Tokatoka. It grows in open areas on rock outcrops, bluffs and slips.

NOTES Distinguished from most large-leaved "Occlusae" by the combination of the shape and size of its leathery leaves, and by having: peduncle and rachis often reddish when young; flower buds pointed and well spaced on the inflorescence; corolla tubes shorter than or equalling calyces; coloured corolla lobes fading dramatically with age; and ovaries sparsely hairy. It is most similar to *H. adamsii* (see notes under that species). Detailed notes on

Hebe perbella   OCCLUSAE

FIG. 106   **A** plant at Hauturu Trig, Waima Forest. **B** sprig. **C** leaf bud with no sinus. **D** lower and upper leaf surfaces. **E** young inflorescence. **F** inflorescence showing frontal view of flowers. **G** lateral view of flower. **H** capsule in septicidal (left) and loculicidal view.

recognition, distribution, variation and cultivation are provided by de Lange (1998). It was previously known informally as *Hebe* sp. "x" (Eagle 1982) and *H.* "Bartlett" (Druce 1993).

ETYMOLOGY L. (*per* = extra, very; *bellus* = delightful, beautiful, pleasing), refers to the attractive and colourful flowers of the species.

49. *Hebe macrocarpa* (Vahl) Cockayne et Allan

DESCRIPTION Bushy shrub to 3 m tall. **Branches** erect, old stems brown or grey; branchlets green, pubescent or glabrous, hairs bifarious or uniform; internodes (2–)5–41 mm; leaf decurrencies obscure or weakly evident (with a faint ridge along medial line). **Leaf bud** distinct; sinus absent. **Leaves** erecto-patent to recurved; *lamina* lanceolate or linear or oblong or oblanceolate or elliptic (often narrowly), coriaceous, m-shaped in TS, (23–)45–110(–163) × (5–)9–22(–32) mm; *apex* acute to obtuse or apiculate or sts acuminate; *base* cuneate or truncate; brochidodromous *secondary veins* sts evident in fresh leaves; *margin* narrowly cartilaginous, ciliolate or glabrous; *upper surface* green or dark green, us. glossy, without evident or rarely with few stomata, hairy along midrib (us.) or glabrous; *lower surface* light green. **Juvenile leaves** crenate, ciliolate (and with scattered hairs above midrib). **Inflorescences** with (13–)25–85 flowers, lateral, unbranched, 3–13.2 cm, shorter to longer than subtending leaves (almost always shorter than leaves in var. *latisepala*; variable in var. *macrocarpa*); peduncle 0.6–1.9(–3.6) cm; rachis (2–)3–11.3 cm. **Bracts** alternate (apart from lowermost pair in most cases), lanceolate or deltoid (sts narrowly) or oblong, obtuse to acute or acuminate. **Flowers** Apr–Nov(–Dec)[–Jan], ⚥. **Pedicels** 1.5–5.5 mm, sts recurved in fruit. **Calyx** (2–)2.5–3.7(–4.2) mm; *lobes* lanceolate or elliptic or ovate or deltoid, acute to obtuse, very rarely hairy outside. **Corolla** tube hairy inside, (2.2–)3.2–5.5 × 2.8–4.2 mm, funnelform and contracted at base, at least slightly longer than calyx; *lobes* violet (var. *latisepala*) or white or tinged with pink or mauve at anthesis, violet (var. *latisepala*) or white with age, ovate or elliptic, obtuse, erect to patent (us. only posterior lobe patent), shorter to longer than corolla tube, sts ciliolate or hairy inside; corolla throat white or violet. **Stamen filaments** white or coloured (var. *latisepala*), 5.5–12.2 mm; anthers mauve or pink or violet or yellow, 2.3–3 mm. **Ovary** very rarely hairy, 1–1.6 mm; ovules *c.* 8–10 per locule; style 5–11.5 mm. **Capsules** Jan–Dec, acute or subacute, 3.8–10 × 3–6.5 mm, loculicidal split extending ¼–½-way to base. **Seeds** flattened (sts strongly), broad ellipsoid to discoid, winged, ± smooth, brown (sts pale), 1–2.5(–3.2) × 0.9–1.7 mm, MR 0.2–0.6 mm. $2n = 80$ or 120.

DISTRIBUTION AND HABITAT NI, from near Whangarei to near Kawhia, including islands of Hauraki Gulf and the Mercury Ids. It occurs in coastal to upland areas, in scrub, at forest margins or in open areas in forest, and on rocky sites.

NOTES Distinguished from most large-leaved "Occlusae" by its: leathery leaves; large, broad flowers, with corolla tubes longer than calyces, and long filaments; and large fruit.

Moore (in Allan 1961) recognised three varieties of *H. macrocarpa*. One of these is here treated as a distinct species, *H. brevifolia*. The other two, var. *macrocarpa* and var. *latisepala*, are probably worthy of recognition (and are treated by some authors, e.g. Druce 1980, 1993, as distinct species). It has not, however, been possible to delimit these varieties to an extent that they could each be mapped and described separately here. Var. *macrocarpa*, as traditionally defined, has white flowers, $2n = 80$ chromosomes and occurs on the NI mainland, including the Coromandel Peninsula. Var. *latisepala*, as traditionally defined, has violet flowers, $2n = 120$ chromosomes and occurs on Great Barrier and Little Barrier islands, and possibly near Whangarei Harbour and on Coromandel Peninsula. The two varieties are difficult to discriminate, particularly using herbarium specimens because: they are variable in leaf shape and size (Appendix 4); specimens do not always have flowers (and colour may not always be retained); and the true limits of the chromosome races are unknown. Their discrimination is further confused because the geographic and morphological boundaries are not as clear-cut as outlined by Moore (in Allan 1961). For example, white-flowered plants occur on Great Barrier Island (personal observation), some of which have $2n = 80$ chromosomes (de Lange & Murray 2002); plants with $2n = 120$ occur on Coromandel

Hebe macrocarpa   OCCLUSAE

FIG. 107   **A** var. *macrocarpa* at Shakespeare Cliff, Coromandel Peninsula. **B** var. *latisepala* at Windy Canyon, Great Barrier Id. **C** sprigs of var. *macrocarpa* (left) and var. *latisepala*. **D** lower and upper leaf surfaces, var. *macrocarpa* (left) and var. *latisepala*. **E** inflorescence and infructescences, var. *macrocarpa*; shows large fruits and flowers. **F** leaf bud with no sinus. **G** frontal view of flower, var. *macrocarpa*. **H** lateral view of flower, var. *macrocarpa*. **I** lateral view of flower, var. *latisepala*.

Peninsula (de Lange & Murray 2002) and the Hunua Ranges (Hair 1967), the former at least with white flowers; both white-flowered and violet-flowered plants with $2n =120$ are sympatric with white-flowered plants with $2n = 80$ on Bream Head; violet-flowered plants with $2n =120$ are sympatric with white-flowered plants with $2n = 80$ on [Mt] Manaia (P. J. de Lange pers. comm. 2005). The limits of the two varieties are worthy of further investigation.

H. macrocarpa is possibly closely related to, and may grade into *H. corriganii* (see notes under that species). It probably hybridises with *H. stricta* var. *stricta* at a range of sites, and the name *H.* ×*affinis* probably applies to this hybrid combination. Included here under *H. macrocarpa* are narrow-leaved plants from Great Barrier Island considered by Druce (1980, as *H.* "sp. (w)"; 1993, as *H.* "Great Barrier") to constitute an undescribed species, and by de Lange & Murray (2002) possibly to be hybrids between *H. macrocarpa* and *H. pubescens* subsp. *rehuarum*. There are no clear grounds for either treating them as a distinct species or as hybrids.

ETYMOLOGY Gk (*macros* = long, large, great; *carpos* = fruit), refers to the large capsules.

50. *Hebe brevifolia*   (Cheeseman) de Lange

DESCRIPTION Spreading low shrub to 0.7 m tall. **Branches** decumbent to erect, old stems grey or brown; branchlets green or yellowish (drying grey to black), puberulent, hairs uniform; internodes (1.5–)2–23 mm; leaf decurrencies mostly obscure (but with a thin medial ridge extending ± the entire length of the internode). **Leaf bud** distinct; sinus absent. **Leaves** patent or erecto-patent; *lamina* narrowly or broadly elliptic (sts oblong-elliptic) or obovate, coriaceous, flat or m-shaped in TS, (16–)20–75 × (6–)10–25 mm; *apex* often apiculate and obtuse or subacute; *margin* ciliolate (us. with very fine hairs, esp. toward the apex or base) or occasionally glabrous; *upper surface* with few stomata (but not very evident) or without evident stomata, us. hairy along midrib; *lower surface* glabrous or sts covered with minute gl. hairs. **Inflorescences** with 16–57 flowers, lateral, unbranched, 2–9.6 cm, sts with a conspicuous number of unopened flowers toward the apex; peduncle 0.6–3.5 cm; rachis (1.5–)3–7.5 cm. **Bracts** alternate (2 lowermost bracts can be subopposite), ovate (mostly) or elliptic or narrowly deltoid, obtuse or subacute. **Flowers** (Oct–)Nov–May(–Aug), ⚥. **Pedicels** always longer than bracts, (1–)2–5(–7) mm. **Calyx** (1.9–)3–4 mm; *lobes* ovate, obtuse (us.) or subacute. **Corolla** tube hairy inside or glabrous, 3–5.5 mm, longer than calyx; *lobes* magenta to deep violet at anthesis, rose pink or violet with age, ovate (sts broadly) or elliptic, obtuse, erect to suberect, longer to shorter than corolla tube, occasionally hairy inside; corolla throat magenta or violet. **Stamen filaments** magenta to violet, 5.5–10.5 mm; anthers red-purple, 0.9–1.6 mm. **Ovary** 0.9–1.5 mm; ovules *c.* 7–8 per locule; style 7.5–11.8 mm. **Capsules** Jan–May(–Dec), subacute, (4.5–)5.5–7.5(–8.5) × 4–5 mm, hairy, loculicidal split extending ⅓–¾-way to base. **Seeds** strongly flattened, broad ellipsoid to discoid, finely papillate, pale to dark brown, 1.7–2.4 × 1.2–1.8 mm, MR 0.4–0.6 mm. $2n = 118$.

DISTRIBUTION AND HABITAT Endemic to the Surville Cliffs and surrounding plateau of the North Cape area, northern NI. It occurs predominantly in low-growing shrubland on ultramafic substrates.

NOTES Distinguished from other species by, among other features, its magenta flowers, low, spreading habit, and lack of sinus in the leaf bud (Fig. 108C). It has previously been included, at variety rank, in both *H. speciosa* and *H. macrocarpa*, but is readily distinguished from both of those species (from the former, by lacking a sinus, and having usually narrower leaves; and from the latter by its habit and flower colour). Detailed notes on morphology, distribution and conservation status are provided by de Lange (1997).

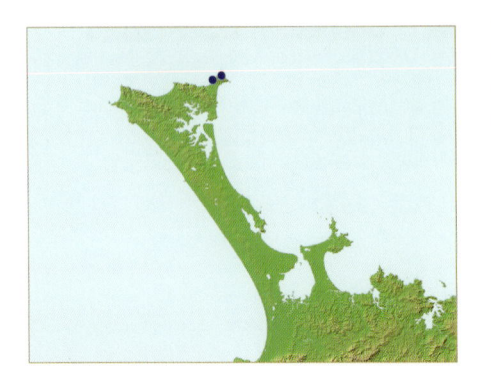

ETYMOLOGY L., refers to the size of the leaves (*brevi-* = short; *-folius* = leaved). The species is not particularly short-leaved when compared with many hebes, but the name was originally used by Cheeseman (1906) when distinguishing these plants, at variety rank, from *H. speciosa*.

Hebe brevifolia   OCCLUSAE

FIG. 108   A plant on the Surville Cliffs. **B** sprig. **C** leaf bud with no sinus. **D** lower and upper leaf surfaces. **E** inflorescence (left) and young infructescence. **F** inflorescence showing frontal view of flowers. **G** lateral view of flower. **H** septicidal view of capsule. **I** loculicidal view of capsule.

51. *Hebe barkeri*   (Cockayne) Cockayne

DESCRIPTION Small tree (often with a dense, rounded canopy when young) to 13 m tall. **Branches** erect, old stems brown; branchlets green or purplish or red-brown, pubescent, hairs uniform; internodes (3–)6–17(–29) mm; leaf decurrencies mostly obscure or weakly evident. **Leaf bud** distinct; sinus absent. **Leaves** erecto-patent to patent; *lamina* linear-lanceolate or lanceolate, subcoriaceous, mostly flat, (24–)36–79 × (4–)8–15(–22) mm; *apex* acute; faint brochidodromous *secondary veins* sts evident in fresh leaves; *margin* narrowly cartilaginous, minutely pubescent (with short, multicellular, egl. hairs); *upper surface* light green, dull, without evident stomata, hairy along midrib and sts covered with minute gl. hairs or rarely uniformly egl. pubescent (apparently only when lower surface is also hairy); *lower surface* light green, conspicuously or faintly pitted with small depressions that each contain a twin-headed gl. hair, hairy along midrib (mostly) and sts uniformly egl. pubescent or glabrous. **Juvenile leaves** minutely denticulate, ciliate (with long multicellular hairs) and lower surface pubescent (on midrib). **Inflorescences** with 23–39 flowers, lateral, unbranched, 2.8–7.6 cm, shorter to longer than subtending leaves; peduncle 0.6–1.4 cm; rachis 2.2–6.2 cm. **Bracts** alternate, deltoid (sts narrowly) or oblong, obtuse to acute, sts hairy outside. **Flowers** Oct–Dec, ⚥. **Pedicels** always longer than bracts, 1–5.5 mm. **Calyx** (1.5–)2–4.2 mm; *lobes* deltoid (us.) or lanceolate (sts broadly), acute to obtuse, hairy outside (sts sparsely). **Corolla** tube hairy inside, 1.4–2 × 1.6–1.9 mm, broadly funnelform, shorter than calyx; *lobes* white tinged with pale blue or mauve at anthesis, elliptic or lanceolate or rhomboid or ovate, obtuse, suberect to recurved, longer than corolla tube. **Stamen filaments** white or mauve, straight or slightly incurved at apex in bud, 4–5 mm; anthers purple, 1.5–2 mm. **Nectarial disc** ciliate (hairs egl., 1–4 cells long, with tapered ends). **Ovary** hairy, 1.1–1.3 mm; ovules c. 4–6 per locule; style 2.5–4.5 mm, hairy. **Capsules** Dec–Jan, subacute, 4–5 × 2.8–3.3 mm, hairy, loculicidal split extending ⅓–¾-way to base. **Seeds** strongly flattened, ellipsoid-oblong to broad ellipsoid, winged, pale to dark brown, 1.1–2 × 1–1.4 mm, MR 0.3–0.6 mm. $2n = 40$ or 80.

DISTRIBUTION AND HABITAT Endemic to the Chatham Ids, where it occurs on Chatham Id (Rekohu, Wharekauri). It is also reported from Pitt Id (Rangiauria) and South East Id (Rangatira) (Crisp et al. 2000; Walls et al. 2003), but we have not seen specimens from these localities and they are not represented on the distribution map. It grows chiefly in forest, often near streams, and sometimes in scrub, or on coastal scarps.

NOTES One of the tallest-growing species in the genus, and one of the few, including *H. parviflora* and *H. flavida*, that are truly arborescent. Despite striking differences in form, it is probably most closely related to the other Chatham Ids species (*H. chathamica*, *H. dieffenbachii*), and to *H. rapensis*. These species all have: sessile, sometimes ± amplexicaul leaves; often coarse branchlet pubescence (although the nature of the hairs vary, and they are sometimes absent); similar flowers usually with acute calyx lobes; and nectarial discs that are often ciliate (a feature uncommon in *Hebe*, seen otherwise only in some specimens of *H. elliptica*).

Specimens have often been confused with *H. dieffenbachii*. It can be distinguished from that species by the combination of: an arborescent habit; adult leaves that are often brighter green and tend to be broadest below the midpoint; leaf margins that are minutely pubescent with several rows of short hairs, at least when young; youngest leaves that often have long, ± flattened hairs on the underside of the midrib; lower leaf surfaces on which the stomata, although numerous, are not conspicuous; and shorter corolla tubes that do not exceed the calyx and are shorter than corolla lobes. In contrast, *H. dieffenbachii* has: a shrubby habit; adult leaves that are often duller green (sometimes ± greyish), and tend to be broadest at or above the midpoint; leaf margins that are usually glabrous, although sometimes fringed with a row of long hairs, and only occasionally have a minute pubescence of short hairs; youngest leaves on which the midrib is usually glabrous below, only occasionally having short eglandular hairs; lower leaf surfaces on which stomata are usually conspicuous, at least on the youngest leaves of dried specimens; and longer corolla tubes that usually exceed the calyx (despite some variation in flower size, and in calyx length in particular) and are longer than corolla lobes.

A possibly unique feature of *H. barkeri* is the presence of hairs inside the ovary – a single tuft in each locule, on the septum wall, below the placenta. This feature was seen in flowers of all live and spirit-preserved material available for examination, and was not observed in any other species.

ETYMOLOGY Honours Samuel D. Barker (1848–1901).

Hebe barkeri   OCCLUSAE

FIG. 109   **A** plant beside the Tuku a Tamatea R., Chatham Id. **B** sprig. **C** lower and upper leaf surfaces; magnified insets show lower leaf surface without evident stomata (upper inset) and leaf margin with minute hairs. **D** leaf bud with no sinus. **E** capsule in septicidal (left) and loculicidal view. **F** inflorescence showing frontal view of flowers. **G** lateral view of flowers.

52. *Hebe dieffenbachii*   (Benth.) Cockayne et Allan

DESCRIPTION Bushy or spreading low shrub (varying from stunted or pendent in some rocky situations, to large and erect in more favourable sites) to 3 m tall. **Branches** spreading or erect or pendent, old stems grey or brown; branchlets green (sts tinged maroon), puberulent to pubescent or glabrous, hairs bifarious or uniform; internodes (2.9–)6–22(–34) mm; leaf decurrencies obscure to moderately evident (sts giving branchlets a flattened appearance). **Leaf bud** distinct; sinus absent. **Leaves** erecto-patent to recurved; *lamina* elliptic or oblong-elliptic to oblanceolate or obovate, coriaceous, flat, (26–)38–74(–102) × (4.5–)8.5–20(–25) mm; *apex* subacute or obtuse; *base* truncate to subcordate or amplexicaul; 2 lateral *secondary veins* sts evident at base of fresh leaves; *margin* cartilaginous, glabrous or rarely ciliate (with long or short egl. hairs); *upper surface* light to dark green, dull, without evident stomata, hairy along midrib and rarely uniformly egl. pubescent; *lower surface* light green or glaucescent, not pitted (but often with minute gl. hairs when young), sts glabrous or rarely uniformly egl. pubescent. **Inflorescences** with (34–)45–110(–135) flowers, lateral, unbranched, 5–10(–11.5) cm; peduncle 0.9–1.8(–2) cm; rachis 3.5–7.5(–9.7) cm. **Bracts** alternate, lanceolate or linear-lanceolate or deltoid, acute (us.) or subacute, sts hairy outside. **Flowers** Dec–Feb(–Mar), ⚥. **Pedicels** 0.7–2(–3.8) mm. **Calyx** (1.5–)2.2–3.2(–4) mm, 4–5-lobed (5th lobe small, posterior); *lobes* lanceolate or ovate or deltoid, acute to subacute, sts hairy outside. **Corolla** tube hairy inside, *c.* 2.5–3.5 × *c.* 1.5–1.8 mm, shortly cylindric, longer than calyx; *lobes* white or mauve at anthesis, elliptic or ovate, obtuse, patent to recurved, slightly shorter than corolla tube, hairy inside or at least with a few hairs toward base on inner surface. **Stamen filaments** white or faintly mauve coloured, 3.5–4 mm; anthers magenta, 1.5–1.9 mm. **Nectarial disc** ciliate (us.) or glabrous. **Ovary** sts hairy, 0.9–1.1 mm; ovules *c.* 13–15 per locule; style 4–5(–7.3) mm, sts hairy. **Capsules** (Jan–)Feb–Apr(–Dec), obtuse or subacute, 3.5–4.5(–5.6) × (2.7–)3–3.8(–4.3) mm, sts hairy, loculicidal split extending ¼-way to base. **Seeds** flattened, ± discoid, ± smooth, brown (sts pale), 0.8–1.5 × 0.8–1.2 mm, MR 0.2–0.4 mm. $2n = 40$.

DISTRIBUTION AND HABITAT Endemic to the Chatham Ids, where it occurs on Chatham Id (Rekohu, Wharekauri), Pitt Id (Rangiauria) and South East Id (Rangatira). It is also reported from Mangere Id (Crisp et al. 2000), but we have not seen specimens from this locality and it is not represented on the distribution map. It grows on rock outcrops and in near-coastal scrub.

NOTES Similar to *H. barkeri* and *H. chathamica*, which also occur on the Chatham Ids. Notes on recognition are provided in the entries for those two species.

Leaves vary in both size and shape (Appendix 4), from relatively narrow (e.g. oblong-elliptic and oblanceolate) to broad (broadly elliptic or obovate), with broad-leaved plants being most often collected in southern parts of Chatham Id.

ETYMOLOGY Honours German explorer and naturalist J. K. Ernst Dieffenbach (1811–55), who collected the type specimen (his collections of plants and animals from the Chatham Ids were the first made by any European).

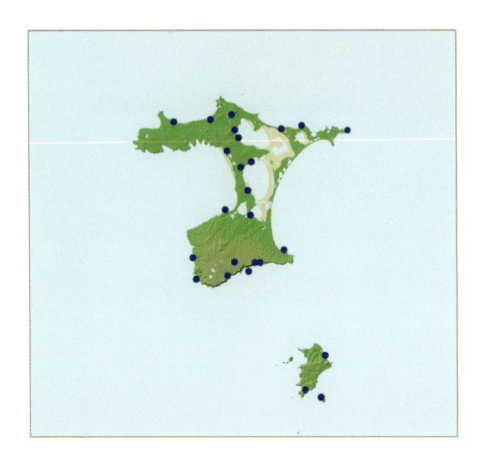

Hebe dieffenbachii OCCLUSAE

FIG. 110 **A** plant near Kaingaroa, Chatham Id. **B** sprig. **C** leaf bud with no sinus. **D** lower and upper leaf surfaces; magnified insets show lower leaf surface with clearly visible stomata (upper inset) and glabrous leaf margin. **E** close-up of broad, slightly amplexicaul leaf base. **F** frontal view of flower. **G** lateral view of flower with corolla tube much longer than calyx. **H** lateral view of flower with corolla tube about equal to calyx. **I** capsule in septicidal (left) and loculicidal view; inset shows hairs on surface.

TAXONOMIC TREATMENT 209

53. *Hebe chathamica* (Buchanan) Cockayne et Allan

DESCRIPTION Spreading low shrub (often ± mat-like) to 0.25 m tall. **Branches** prostrate or decumbent or pendent, old stems brown or grey; branchlets green or red-brown, pubescent, hairs uniform; internodes (1.9–)3–17(–25.5) mm; leaf decurrencies evident. **Leaf bud** distinct; sinus almost always absent but very rarely present and then small and rounded (seen only on some branches of WELT 42721). **Leaves** decussate or subdistichous, erecto-patent to recurved; *lamina* elliptic to obovate or oblanceolate, coriaceous, flat, 8.5–33 × (3.3–)5.5–16.5 mm; *apex* subacute to obtuse; *margin* narrowly cartilaginous, glabrous (us.) or pubescent (only because leaf surfaces are hairy), sts slightly red-tinged; *upper surface* green to dark green, dull, with few or without evident stomata, glabrous (rarely) or hairy along midrib and often uniformly egl. pubescent; *lower surface* light green, glabrous (sts) or hairy along midrib and often uniformly egl. pubescent. **Inflorescences** with (20–)25–40 flowers, lateral, unbranched, 1.3–4.1 cm; peduncle 0.5–2 cm; rachis 0.2–1.8 cm. **Bracts** alternate or lowermost pair opposite, then subopposite or alternate above, lanceolate to linear-lanceolate, acute, margins glabrous (rarely) or hairy, often hairy outside. **Flowers** Dec–Feb[–Mar], ⚥. **Pedicels** 1–2.6 mm. **Calyx** 2.5–4 mm; *lobes* linear-lanceolate or deltoid, acute, egl. ciliate (or apparently so), often hairy outside. **Corolla** tube hairy inside and sts outside, 2.5–4 × 2–2.3 mm, cylindric, us. equalling or longer than calyx (only sts very slightly shorter than calyx); *lobes* white or tinged purplish mauve at anthesis, elliptic or ovate, obtuse, patent, shorter than corolla tube, hairy inside or at least with a few hairs toward base on inner surface. **Stamen filaments** *c.* 4–4.5 mm; anthers pale brown or pale mauve, 2–2.4 mm. **Nectarial disc** ciliate or glabrous. **Ovary** sts hairy, 1.2–1.5 mm; ovules 10–29 per locule, in 1–2 layers; style 5–6 mm, sts hairy. **Capsules** Jan–Mar(–Dec), subacute, 3.5–4.5(–5) × 2.5–3.5 mm, sts hairy, loculicidal split extending up to ¼-way to base. **Seeds** flattened, broad ellipsoid to sub-discoid, brown, 1.2–1.6 × 0.9–1.3 mm, MR 0.4–0.5 mm. $2n = 40$.

DISTRIBUTION AND HABITAT Endemic to the Chatham Ids, where it occurs on The Sisters (Rangitatahi), Chatham Id (Rekohu, Wharekauri), Pitt Id (Rangiauria) and South East Id (Rangatira). It grows on coastal rocks and cliffs, and in very low vegetation at beach margins.

NOTES Most similar to *H. dieffenbachii*, from which it is distinguished by a low-growing habit and smaller leaves that are generally broader relative to their length. Where the two species co-occur near Kaingaroa, Chatham Id (and probably at other sites), plants display a range of habits and leaf forms, suggesting hybridisation or introgression between them. At this site, low-growing plants of *H. chathamica* occur in situations closest to the coast, most exposed to salt spray, whereas shrubby plants of *H. dieffenbachii* are more common in positions further away from the beach margin.

H. chathamica is variable in leaf size (Appendix 4), and highly variable in the distribution and density of hairs on stems, leaves and calyces. Some larger-leaved plants match the description and type of *Veronica coxiana*, but patterns of variation are complex, and there are no compelling grounds for recognising that or any other segregates.

ETYMOLOGY Refers to the species' distribution.

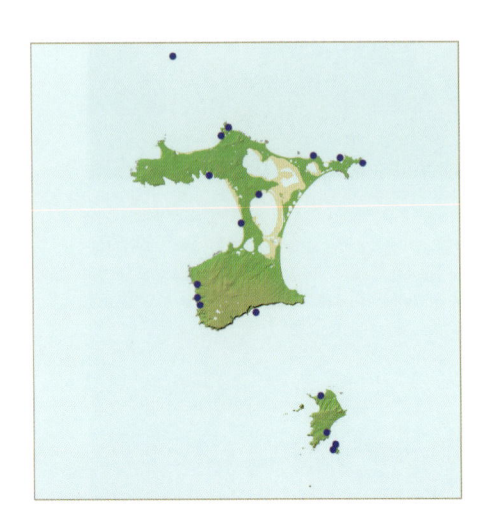

Hebe chathamica   OCCLUSAE

FIG. 111   A plant near Kaingaroa, Chatham Id. B sprig. C leaf bud with no sinus. D lower and upper leaf surfaces. E inflorescence. F frontal view of flower. G lateral view of flower. H capsule in septicidal (left) and loculicidal view; inset shows capsule hairs. I seeds.

54. *Hebe rapensis* (F.Br.) Garn.-Jones

DESCRIPTION Bushy shrub to *c.* 1.5 m tall. **Branches** erect, old stems brown; branchlets pubescent, hairs bifarious or uniform; internodes 2.5–12 mm; leaf decurrencies obscure or very weakly evident (as a narrow ridge below leaf). **Leaf bud** distinct; sinus absent. **Leaves** erecto-patent to patent; *lamina* oblanceolate or oblong-elliptic, subcoriaceous, slightly m-shaped in TS (at least when dry), (27–)40–55(–67) × (6.5–)8.5–12(–14.5) mm; *apex* subacute or acute; *base* slightly amplexicaul; *midrib* conspicuously thickened below and depressed to grooved above; *margin* cartilaginous, us. glabrous but sts pubescent (esp. toward base); *upper surface* dark green, dull, with few or without evident stomata, conspicuously hairy along midrib; *lower surface* light green, glabrous or covered with minute gl. hairs (seen only on a few young leaves). **Inflorescences** with 20–51 flowers, lateral, unbranched, (2.5–)3–6.5 cm, about equal to or longer than subtending leaves; peduncle 0.35–1.5 cm; rachis (1.7–)3–5.3 cm. **Bracts** alternate, linear or narrowly deltoid, acute or acuminate. **Flowers** Jun–Oct, ⚥. **Pedicels** longer than or equal to bracts, (1–)2–3(–4.5) mm, sts recurved in fruit. **Calyx** (1.7–)2.5–4 mm; *lobes* narrowly deltoid or lanceolate, acute to acuminate or obtuse. **Corolla** tube hairy inside or glabrous, 1.4–1.7 × *c.* 0.9–1.5 mm, funnelform, shorter than calyx; *lobes* white or tinged pale blue or mauve at anthesis, ovate or elliptic, obtuse, longer than corolla tube, sts with a few hairs toward base on inner surface. **Stamen filaments** *c.* 2.5–3 mm; anthers *c.* 0.7–0.9 mm (when dry). **Nectarial disc** ciliate or possibly sts glabrous. **Ovary** sts sparsely hairy, *c.* 0.6–0.8 mm; ovules *c.* 6–8 per locule; style 2–4.5 mm. **Capsules** Jul–Oct(–Jan), acute or subacute, 2.5–4.2 × 1.7–3 mm, occasionally sparsely hairy, loculicidal split extending ¼–¾-way to base. **Seeds** flattened, ovoid or oblong, brown, *c.* 1 × 0.6–0.7 mm, MR 0.2–0.3 mm. Chromosome number unknown.

DISTRIBUTION AND HABITAT Endemic to Rapa in the Austral Ids, French Polynesia. Notes on herbarium specimens suggest it commonly occurs on cliffs.

NOTES Probably most closely related to the three Chatham Ids species (see notes under *H. barkeri*) and particularly similar to *H. barkeri* and *H. dieffenbachii*. It is distinguished from the former by its shrubby habit, calyx lobes that are glabrous on the outer surface, and tendency for leaf margins (at least toward the apex) to be glabrous; from the latter by its flowers with shorter corolla tubes (shorter than both calyces and corolla lobes); and from both these species by having leaves that are slightly m-shaped in transverse section, with midribs strongly depressed and margins slightly recurved (at least when dry). It is the only *Hebe* on Rapa, and the only one endemic to an area wholly outside of New Zealand.

ETYMOLOGY L. (*-ensis* = an adjectival suffix implying origin or place), refers to the species' distribution.

Hebe rapensis   OCCLUSAE

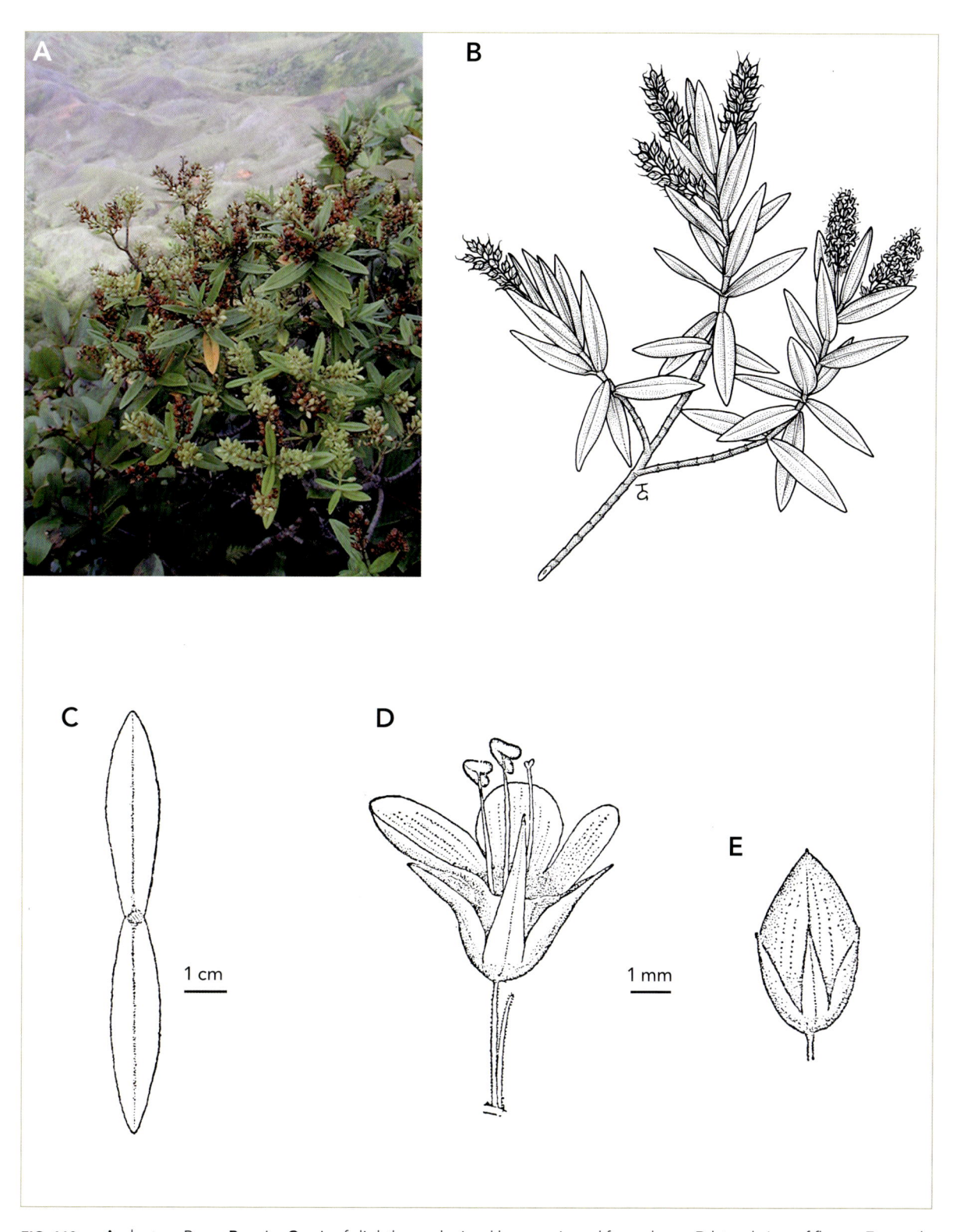

FIG. 112   **A** plant on Rapa. **B** sprig. **C** pair of slightly amplexicaul leaves, viewed from above. **D** lateral view of flower. **E** capsule in loculicidal view. [C, D and E are reproduced from the original description of the species by Brown (1935).]

FIG. 113 *Hebe odora* shrubland, Garvie Mts, Southland.

FIG. 114 Low-growing plant of *Hebe odora* on Mt Anglem, Stewart Id.

"Buxifoliatae"

CRITICAL FEATURES Shrubs, prostrate to erect, up to 1(–2) m tall; **leaf buds distinct, or tightly surrounded by recently diverged leaves**; **sinus ± shield-shaped**; **leaf bases free**; leaves not appressed or scale-like, entire or minutely crenulate, not glaucous, **usually abscising just above base, leaving a fragment of petiole attached to stem**; inflorescences terminal or lateral or both; capsule latiseptate.

NUMBER OF SPECIES 4

DISTRIBUTION Mountains of NI (south from Lk. Waikaremoana), SI, and Stewart Id; also the Auckland Ids

KEY TO SPECIES[1]

1 Upper surface of leaves with no stomata (e.g. Fig. 115D) 2
  Upper surface of leaves with obvious stomata (e.g. Fig. 117D) 3

2 Inflorescences lateral only (e.g. Fig. 116B) .... 56. *Hebe mooreae*
  Inflorescences terminal spikes, beneath which there are usually lateral spikes in the axils
    of the uppermost leaf pairs (e.g. Figs 115E, H and I) .... 55. *Hebe odora*

3 Inflorescence bracts large, mostly obscuring calyces (e.g. Figs 117I and J);
    inflorescences strictly terminal; leaves sharply keeled beneath (along the midrib)
    throughout their length (e.g. Fig. 117D) .... 57. *Hebe masoniae*
  Inflorescence bracts distinctly shorter than calyces (e.g. Figs 118H and I);
    inflorescences strictly lateral; leaf keel (along the underside of the midrib)
    characteristically flattened just below the tip (e.g. Fig. 118D) .... 58. *Hebe pauciramosa*

---

[1] This key will not give correct identifications for specimens of *H. mooreae* from the Caswell Sound and Denniston Plateau areas, or specimens of *H. odora* from near Arthur's Pass (which often have stomata on the upper surfaces of their leaves – a feature otherwise unusual in these species). If trying to identify specimens from the Caswell Sound area, you should compare the notes and illustrations provided for *H. mooreae*, *H. pauciflora* and *H. odora*. Likewise, for specimens from near Denniston you should compare notes and illustrations for *H. mooreae* and *H. odora*, and for specimens from near Arthur's Pass you should compare notes and illustrations for *H. odora* and *H. pauciramosa*.

55. *Hebe odora* (Hook.f.) Cockayne

DESCRIPTION Spreading low or bushy shrub (varies from a rounded bush to very lax and open, e.g. Figs 115A and 114) to 0.9(–1.7) m tall. **Branches** spreading or decumbent or ascending or erect, old stems brown or red-brown or grey or black (at least when dry); branchlets green, puberulent or pubescent, hairs bifarious; internodes (0.9–)1.3–4.5 mm; leaf decurrencies evident and us. swollen; leaves abscising above nodes and lower part of petioles remaining attached to stem. **Leaf bud** distinct; sinus broad and shield-shaped. **Leaves** erect to patent; *lamina* ovate or lanceolate to elliptic or obovate or sts almost circular, rigid and coriaceous, concave, (3.6–)4.5–11.5 × 2.3–5.4 mm; *apex* subacute; *base* truncate; *midrib* thickened below; *margin* glabrous, us. entire or sts minutely crenulate; *upper surface* dark green, glossy, without evident stomata (us.) or with many stomata (on many plants from Arthur's Pass area), glabrous; *lower surface* green (paler than upper); *petiole* 0.5–1.5(–2.2) mm, glabrous. **Inflorescences** mostly terminal and lateral but sts only terminal, unbranched, (0.6–)1–2.8 cm; peduncle 0.13–0.36 cm, bifariously hairy or glabrous; rachis 0.5–1.7 cm, hairy (us. bifariously). **Bracts** opposite and decussate, free, ovate, subacute. **Flowers** (Nov–)Dec–Jan(–Mar), ⚥ (although E. M. Low (pers. comm. 2005) suggests that some populations include ♀ plants). **Pedicels** absent. **Calyx** 3.5–5 mm; *lobes* elliptic, subacute to obtuse. **Corolla** tube hairy inside, *c.* 3–3.5 × *c.* 1.5 mm, narrowly cylindric, *c.* equalling or longer than calyx; lobes white at anthesis, narrowly to broadly elliptic, obtuse, patent to recurved, equalling or longer than corolla tube, sts sparsely hairy inside. **Stamen filaments** 2–3.2 mm; anthers pink, 1.9–2.4 mm. **Ovary** ovoid or globose, 0.7–1.2 mm; ovules *c.* 8–13 per locule; style *c.* 5.5–7 mm. **Capsules** Dec–Apr(–Nov), subacute or obtuse, 3.9–4.5 × *c.* 3.4–3.6 mm, loculicidal split extending ¼–½-way to base. **Seeds** flattened, ellipsoid (sts broadly), not winged to only weakly winged, straw-yellow to pale brown, 1.2–1.8 × 0.9–1.3 mm, MR 0.3–0.6 mm. 2n = 42 or 84.

DISTRIBUTION AND HABITAT Widespread, south from the Huiarau Ra., Lk. Waikaremoana, on mountains of NI, SI, Stewart Id and the Auckland Ids. It grows in montane to penalpine grassland, shrubland (Fig. 113), bogs and flushes. Its distribution may extend, in the north, to the Raukumara Ra. (as implied by Druce 1980; Eagle 1982), but there are no specimens from this area at WELT, CHR or AK.

NOTES Distinguished from similar species of "Buxifoliatae" by the combination of: bracts not extending beyond tips of calyces (Fig. 115E); inflorescences that consist of a terminal spike, beneath which there are usually also lateral spikes in the axils of the uppermost leaf pairs (Figs 115E, H and I); no stomata on the upper leaf surface (Fig. 115D), except in many specimens from the Arthur's Pass area; leaves that are sharply keeled beneath (along the midrib) throughout their length (Fig. 115D); leaf buds that are usually not closely surrounded by several imbricate leaf pairs (particularly when compared with *H. masoniae* and *H. pauciramosa*); free anterior calyx lobes; and corolla lobes that are comparatively narrow relative to their length (particularly when compared with *H. masoniae* and *H. mooreae*). Leaves are variable in shape and size (Appendix 4).

Two chromosome races in *H. odora* were treated by Druce (1980, 1993) and Eagle (1982) as distinct species. Diploid plants, assumed to be typical *H. odora*, are recorded from NI, northern SI and the Auckland Ids, while tetraploids are recorded at a range of SI localities (from Island Pass, Nelson, southwards) and on Stewart Id (Table 5; Dawson & Beuzenberg 2000). Although there is possibly a correlation between ploidy and flavonoid profile (Markham et al. 2005), no consistent morphological differences between the two chromosome races have been identified.

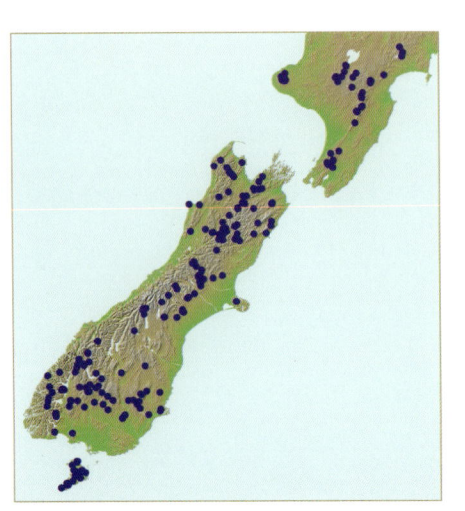

Auckland Islands

Hebe odora   BUXIFOLIATAE

FIG. 115   **A** plant at Blue Lk., Garvie Mts. **B** sprig; magnified inset shows a branchlet from which the leaves have been shed, leaving a portion of petiole (yellow-coloured) attached to the branchlet. **C** leaf bud with shield-shaped sinus. **D** lower and upper leaf surfaces; magnified inset shows upper surface, on which stomata are absent. **E** flowering shoot with an apical cluster of inflorescences; this includes a terminal spike, beneath which there are lateral spikes in the axils of the uppermost leaf pairs. **F** frontal view of flower. **G** lateral view of flower. **H** and **I** young and old infructescences, respectively. **J** apical view of leaf bud, showing that it is not closely surrounded by several imbricate leaf pairs.

TAXONOMIC TREATMENT   217

H. odora is sometimes confused with *H. venustula* and *H. brachysiphon*, with which it may co-occur in subalpine shrubland on NI and SI, respectively. It is readily distinguished from both these species by: its shield-shaped leaf bud sinus (Fig. 115C); terminal clusters of inflorescences (Figs 115E, H and I); and flowers and fruits that are sessile and subtended by coriaceous and comparatively larger bracts (Fig. 115E).

ETYMOLOGY L. (*odorus* = having a smell, usually sweet-smelling), presumably refers to the flowers, which Hooker (1844), when naming the species, described as having a "delicious fragrance" (this feature is apparently not, however, always found in the species).

56. *Hebe mooreae* (Heads) Garn.-Jones

DESCRIPTION Spreading low or bushy shrub to 1.2(–2) m tall. **Branches** erect, old stems brown; branchlets green, pubescent (with somewhat strap-like, white, multicellular hairs often ± appressed and upward-facing), hairs bifarious; internodes 1–4(–8) mm; leaf decurrencies swollen and extended for length of internode (us. somewhat saddle-shaped); leaves abscising above nodes and lower part of petioles remaining attached to stem. **Leaf bud** distinct; sinus broad and shield-shaped. **Leaves** erecto-patent; *lamina* oblong or oblong-elliptic or oblong-lanceolate, rigid, slightly concave, (7–)14–18(–28) × (3–)4–6(–8) mm; *apex* acute or subacute; *base* cuneate (mostly) or truncate; *midrib* thickened below and depressed to grooved above (but not necessarily prominent on upper surface); *margin* glabrous, minutely crenulate; *upper surface* dark green, glossy, without evident stomata (us.) or with many stomata (on plants from Caswell Sound or Denniston Plateau), glabrous; *lower surface* dark green; *petiole* 1–3 mm, glabrous. **Inflorescences** with 3–13 flowers, lateral, unbranched, 0.8–2.9 cm, shorter to longer than subtending leaves; peduncle 0.1–0.5 cm; rachis 0.7–2.6 cm. **Bracts** opposite and decussate, connate, ovate or deltoid, obtuse or subacute. **Flowers** Nov–Feb(–Jun), ⚥. **Pedicels** absent or when present always shorter than bracts, 0–1 mm. **Calyx** 3.3–4.5 mm, with anterior lobes free for most of their length or united to ⅓-way to apex; *lobes* lanceolate or ovate, subacute or obtuse, egl. ciliolate or with mixed gl. and egl. cilia (can vary on calyces from one plant). **Corolla** tube hairy inside or glabrous, *c.* 4 × 2.4–2.6 mm, funnelform (narrowly) and contracted at base, equalling or longer than calyx; *lobes* white at anthesis, elliptic (often broadly) or lanceolate (anterior only), obtuse, patent to recurved, longer to shorter than corolla tube, sparsely hairy inside. **Stamen filaments** 3–4 mm; anthers pink (often faintly). **Ovary** ovoid or ellipsoid or globose, 1–1.3 mm; ovules 8–15 per locule; style 5.5–8.5 mm. **Capsules** Jan–Jun(–Dec), subacute or obtuse, 3.5–4.5 × 2–2.8 mm, loculicidal split extending ¼-way to base. **Seeds** flattened, ± broad ellipsoid, ± winged, ± smooth, pale brown, 1.2–1.8 × 0.9–1.2 mm, MR *c.* 0.4 mm. $2n = 126$.

DISTRIBUTION AND HABITAT Widespread on SI, from the Wakamarama Ra. in the north to the Longwood Ra. in the south, chiefly on wetter mountains west of the Main Divide. It grows mostly in penalpine grassland and subalpine shrubland. The two anomalous eastern localities shown on the distribution map are based on WELT 17242 ("Cameron R. Canterbury alps", *R. M. Laing*, undated), and CHR 4138 ("East Dome", no collector or date given).

NOTES Can be distinguished from similar species of "Buxifoliatae" by having a combination of: bracts not extending beyonds tips of calyces (Figs 116I and J); inflorescences that are strictly lateral (Fig. 116B); no stomata on the upper leaf surface (Fig. 116D), except in some specimens from the Caswell Sound (Fiordland) and Denniston Plateau (Nelson) areas (e.g. Fig. 116F); leaves that are sharply keeled beneath (along the midrib) throughout their length (Fig. 116D); leaf buds that are usually not closely surrounded by several imbricate leaf pairs (particularly when compared with *H. masoniae* and *H. pauciramosa*); anterior calyx lobes that are usually free (Fig. 116I), but may be united up to ⅓ the way to the apex; and corolla lobes that are comparatively broad relative to their length (particularly when compared with *H. odora* and *H. pauciramosa*). An often conspicuous feature of *H. mooreae* is possession of strongly crenulate leaf margins (Fig. 116E). This feature is not, however, evident on all specimens, and is also seen on some specimens of *H. odora*.

ETYMOLOGY Honours Lucy B. Moore (1906–87) (M. Heads pers. comm. 1996), former botanist at DSIR Botany Division, who prepared most of the last comprehensive treatment of *Hebe* (in Allan 1961).

Hebe mooreae BUXIFOLIATAE

FIG. 116 **A** plant on Hump Ridge, Southland. **B** sprig, including lateral infructescence; magnified inset shows a branchlet from which the leaves have been shed, leaving a portion of petiole (yellow-coloured) attached to the branchlet. **C** leaf bud with shield-shaped sinus. **D** lower and upper leaf surfaces; magnified inset shows no evident stomata on the upper surface (the most common condition). **E** close-up of minutely crenulate leaf margin. **F** leaf; magnified inset shows stomata on upper surface (common only in plants from near Denniston and Caswell Sound). **G** top view of inflorescences. **H** lateral view of flower. **I** young infructescence. **J** mature infructescence.

57. *Hebe masoniae* (L.B.Moore) Garn.-Jones

DESCRIPTION Spreading low or bushy shrub to 0.5 m tall. **Branches** decumbent or ascending or erect, old stems brown; branchlets green or red-brown, pubescent (hairs multicellular, ± appressed, us. upward-facing), hairs bifarious; internodes (1–)2–4(–5) mm; leaf decurrencies extended for length of internode and often ± swollen; leaves us. abscising above nodes with a small portion of lower part of petioles remaining attached to stem. **Leaf bud** tightly surrounded by recently diverged leaves; sinus broad and shield-shaped. **Leaves** appressed to patent; *lamina* oblong (often broadly) or elliptic or sub-circular, rigid, concave, (3–)6–9(–10) × 4–8 mm; *apex* obtuse; *base* truncate (often abruptly); *midrib* evident in fresh leaves (below), forming a thickened keel throughout the length of leaf; *margin* glabrous or ciliolate (with very short, stiff hairs) or ciliate; *upper surface* dark green, glossy, with many stomata, glabrous; *lower surface* dark green, glossy; *petiole* 1–1.5(–2) mm, glabrous. **Inflorescences** with 2–10(–14) flowers, terminal, unbranched, 0.8–1.8 cm; rachis glabrous or hairy (but not evident without removal of flowers and bracts). **Bracts** opposite and decussate, connate, large and almost obscuring calyx, elliptic or sub-circular, obtuse (us.) or subacute, margins hairy (cilia us. longer than those of *H. pauciramosa*). **Flowers** (Oct–)Dec–Feb(–Apr), ⚥. **Pedicels** absent. **Calyx** 5.5–7 mm; *lobes* lanceolate to elliptic, obtuse or subacute, us. egl. ciliate or very rarely with mixed gl. and egl. cilia. **Corolla** tube hairy inside, 4.5–6 × *c.* 1.5–2 mm, cylindric to funnelform, *c.* equalling calyx; *lobes* white (us.) or tinged mauve at anthesis, ovate (often broadly) or elliptic, obtuse (posterior sts emarginate), suberect to recurved, longer to shorter than corolla tube, sts sparsely hairy inside. **Stamen filaments** 3–4.2 mm; anthers magenta, 1.8–2.6 mm. **Ovary** 1–1.4 mm, apex (in septum view) truncate or emarginate; ovules 8–14 per locule; style 7–9 mm. **Capsules** Jan–May(–Nov), obtuse or subacute, 4–5 × *c.* 4 mm, loculicidal split extending ¼–⅓-way to base. **Seeds** strongly flattened, broad ellipsoid or obovoid, weakly winged, pale brown, 1.5–2.1 × 1–1.4 mm, MR 0.3–0.5 mm. $2n = 118$.

DISTRIBUTION AND HABITAT Mountains of western Nelson, SI, from near Boulder Lk. in the north to the Braeburn Ra. in the south. It grows in *Chionochloa australis* grassland (Fig. 117A), tussock grassland, or scrub, sometimes in wet sites.

NOTES Distinguished from similar species of "Buxifoliatae" by the combination of: large ciliolate bracts that largely obscure the calyx (Fig. 117I); strictly terminal inflorescences; stomata on the upper leaf surface (Fig. 117D); leaves that are sharply keeled beneath (along the midrib) throughout their length (Fig. 117D); leaf buds that are closely surrounded by several imbricate leaf pairs (Fig. 117E); free anterior calyx lobes; and corolla lobes that are almost as broad as they are long (Fig. 117F and G). This was originally described by Moore (in Allan 1961) as a subspecies of *H. pauciramosa*, but has been treated as a separate species for some time (e.g. by Heads 1987, 1992; Garnock-Jones 1993a). Heads (1987), mostly on the basis of leaf shape and size, divided *H. masoniae* into two varieties with ± overlapping distributions; these are not recognised here as distinct.

ETYMOLOGY Honours Ruth Mason (1913–90), former botanist at DSIR Botany Division (and co-worker of L. B. Moore), who recognised some of the distinguishing features of the species.

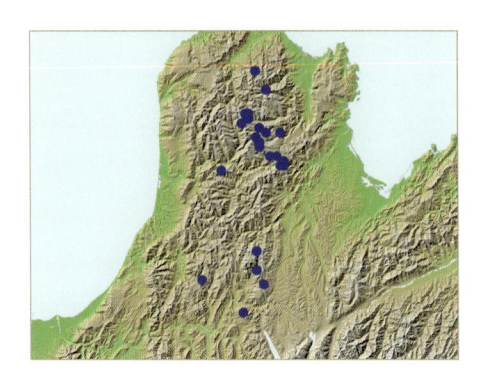

Hebe masoniae   **BUXIFOLIATAE**

FIG. 117   **A** plants on Peel Ra., NW Nelson. **B** sprig; magnified inset shows a branchlet from which the leaves have been shed, leaving a portion of petiole (yellow-coloured) attached to the branchlet. **C** leaf bud with shield-shaped sinus. **D** lower and upper leaf surfaces; magnified inset shows margin, and stomata visible on upper surface. **E** apical view of leaf buds that are closely surrounded by several imbricate leaf pairs. **F** lateral view of inflorescence. **G** inflorescence showing frontal view of flower. **H** seeds. **I** close-up of young infructescence showing that large bracts almost completely obscure the calyces. **J** close-up of mature infructescence.

TAXONOMIC TREATMENT   221

58. *Hebe pauciramosa* (Cockayne et Allan) L.B.Moore

DESCRIPTION Spreading low or bushy shrub to 0.5 m tall. **Branches** erect or ascending, old stems brown or black; branchlets green or yellowish, pubescent (hairs multicellular, us. upward-facing), hairs bifarious; internodes (1–)1.5–4.5(–5.5) mm; leaf decurrencies extended for length of internode and often ± swollen; leaves abscising above nodes with a small portion of lower part of petioles remaining attached to stem. **Leaf bud** tightly surrounded by recently diverged leaves; sinus broad and shield-shaped. **Leaves** erect to patent; *lamina* broadly oblong or elliptic to sub-circular, rigid, concave, (3–)3.5–7(–9) × (1.5–)3.5–6(–11.5) mm; *apex* obtuse; *base* abruptly truncate; *midrib* evident in fresh leaves (below), forming a thickened keel that is characteristically flattened toward leaf apex; *margin* glabrous or ciliolate (with short, stiff hairs); *upper surface* dark green, glossy, with many stomata, glabrous; *lower surface* dark green, glossy; *petiole* (0.5–)1–1.5(–3) mm, glabrous. **Inflorescences** with 2–10 flowers, lateral, unbranched, 0.7–1.7(–2) cm; rachis coarsely hairy. **Bracts** opposite and decussate, connate, deltoid, obtuse, margins minutely hairy or glabrous. **Flowers** Nov–Jan(–Mar)[Oct–Jan in Allan], ♀. **Pedicels** absent or if evident then always shorter than bracts. **Calyx** 3.5–4.5 mm, 3–4-lobed (i.e. depending on extent of fusion of anterior lobes), with anterior lobes united from ⅔ to all the way to apex (sts splitting secondarily as fruit matures); *lobes* narrowly to broadly oblong, obtuse or emarginate (the latter in anterior lobes when fused to just below apex), minutely egl. ciliolate. **Corolla** tube glabrous or sparsely hairy inside, 3.7–5 × 1.3–1.7 mm, cylindric, longer than calyx; *lobes* white at anthesis, elliptic or ovate, obtuse (posterior sts emarginate), patent to recurved, shorter than or equalling corolla tube. **Stamen filaments** 3.7–5.5 mm; anthers magenta, 1.4–1.7 mm. **Ovary** 1–1.4(–1.6) mm; ovules 10–15 per locule; style 3.8–8 mm. **Capsules** Jan–May(–Nov), obtuse, 4–5.5 × 2.5–3.5 mm, loculicidal split extending to *c.* ⅓-way to base (although sts splitting further when very old). **Seeds** flattened (sts strongly), ellipsoid (sts broadly), weakly winged, pale brown, 1.1–1.8 × 0.8–1.1 mm, MR 0.3–0.5 mm. $2n = 42$.

DISTRIBUTION AND HABITAT Mountains of SI, from the Allen Ra. southward, mostly on or west of the Main Divide, but with a few records from drier mountains of the east. It also occurs on Stewart Id, where it has been collected from Mt Anglem and Mt Rakeahua. It usually grows in moist or boggy areas in tussock grassland (Fig. 118A).

NOTES Distinguished from similar species of "Buxifoliatae" by the combination of: bracts distinctly shorter than calyces (Figs 118H and I); strictly lateral inflorescences; stomata on the upper leaf surface (Fig. 118D); leaves in which the keel (along the underside of the midrib) is characteristically flattened just below the tip (Fig. 118D); leaf buds that are closely surrounded by several imbricate leaf pairs (Fig. 118F); anterior calyx lobes fused between ⅔ (Fig. 118H) and all the way to the apex; and corolla lobes (Fig. 118G) that are comparatively narrow relative to their length (particularly when compared with *H. masoniae* and *H. mooreae*).

As noted by Moore (in Allan 1961), there is one collection at WELT (17269) labelled "Mt Egmont, 4000 ft" (*D. Petrie*, 6 Feb. 1901), and another (WELT 17268) labelled "Port Ross" (*T. Kirk*, 10 Jan. 1981). No other records are known from either NI or Auckland Id, and these specimens are not represented on the distribution map.

ETYMOLOGY L. (*pauci-* = few; *ramus* = branch), refers to the habit of the species. Having lateral inflorescences and indeterminate growth of individual shoots, *H. pauciramosa* is generally not as highly branched as the similar species *H. masoniae* and *H. odora* (in which inflorescences are terminal and upward vegetative growth after flowering is continued by lateral shoots).

Hebe pauciramosa — BUXIFOLIATAE

FIG. 118 **A** plants at Lk. Tennyson. **B** sprig; magnified inset shows a branchlet from which the leaves have been shed, leaving a portion of petiole (yellow-coloured) attached to the branchlet. **C** leaf bud with shield-shaped sinus. **D** lower and upper leaf surfaces; magnified inset shows stomata on upper surface. The keel of the leaf (lower surface) is characteristically broad and flattened toward the apex (cf. *H. masoniae*, Fig. 117D). **E** flowering shoot. **F** apical view of leaf buds, showing that they are closely surrounded by several imbricate leaf pairs. **G** frontal view of flower. **H** close-up anterior view of basal part of flower showing the size of the bract relative to the calyx, and that the anterior calyx lobes are fused for *c.* ⅔ their length. **I** young infructescence showing the opposite-decussate arrangement of the sessile flowers/fruits, and the relative sizes of bracts and calyces.

Small-leaved "Apertae"

CRITICAL FEATURES Shrubs, decumbent to erect, us. < 2 m tall; **leaf buds distinct; sinus small and rounded to broad and acute**; leaf bases free; leaves not appressed or scale-like, **usually < 4 cm long**, entire or toothed, sometimes glaucous, **abscising at nodes; inflorescences lateral** (rarely also terminal in some cultivated specimens); capsule latiseptate.

NUMBER OF SPECIES 20

DISTRIBUTION Three Kings Ids, NI, SI, Stewart Id, Snares Ids, the Auckland Ids, Campbell Id, southern South America, Falkland Ids; naturalised in Tasmania (Rozefelds et al. 1999) and Europe (Webb 1972)

KEY TO SPECIES

1 Leaf bud strongly tetragonous in transverse section (e.g. Fig. 127F); leaves glaucous
   (at least below); rupestral plants of central NI (from Kawhia southwards) 2
   Leaf bud terete, flattened, or only weakly tetragonous in transverse section
   (e.g. Fig. 125D); leaves glaucous or dull or glossy; if rupestral, then not from central NI 3

2 Corolla tubes longer than calyces (Fig. 127G); margins of calyx lobes ciliolate; branchlets
   hairy (at least bifariously); leaves entire, upper surface usually not glaucous (Awaroa Va.,
   southeast of Kawhia) .... **67.** *Hebe scopulorum*
   Corolla tubes shorter than or about equal to calyces (e.g. Fig. 126E); margins of calyx lobes
   and branchlets glabrous (only occasionally with short sparse hairs); leaves toothed or
   entire, upper surface usually glaucous, at least when young (southeast of central NI) .... **66.** *Hebe colensoi*

3 Bracts usually equal to or longer than calyx, surrounding and obscuring calyx in anterior
   view (e.g. Fig. 124F); small shrubs of rocky places of eastern Marlborough .... **64.** *Hebe rupicola*
   Bracts usually shorter than calyx, not surrounding and not, or only partly, obscuring calyx
   in anterior view; plants of various habit, habitat and distribution 4

4 Inflorescences all, or mostly, unbranched, with flowers in the axils of lowermost bracts 5
   Inflorescences all, or mostly, branched (at least 1–2 branches at base; e.g. Fig. 134F),
   or if all unbranched, then basal bracts comparatively large and sterile (e.g. Fig. 121H, right) 17

5 Leaf margin conspicuously white pubescent except at plicate-mucronate tip (e.g. Fig. 139F)
   and well-defined petiole; calyx (3.5–)4–6.5 mm long .... **77.** *Hebe elliptica*
   Leaf margin glabrous or minutely or sparsely hairy and without such a strong contrast of
   indumentum at apex and petiole; calyx usually < 4 mm long (except *H. townsonii*) 6

6 Leaf bud sinus square to rounded (e.g. Fig. 143C) or elliptic (e.g. Fig. 140C), sometimes
   filled with hairs or scarcely visible (e.g. Figs 144E and F) 7
   Leaf bud sinus more elongate, acute (e.g. Figs 119C and 135C) 9

7 Outer surface of calyx lobes hairy (e.g. Fig. 144G); underside of leaf lamina often with long,
   shaggy hairs; [NI, Coromandel Peninsula and nearby islands] .... **81.** *Hebe pubescens*
   (treated under large-leaved "Apertae")
   Outer surface of calyx lobes glabrous apart from marginal cilia; underside of leaves glabrous
   except along midrib [SI] 8

8 Leaves tapering to a fine point, often conspicuously narrowed *c.* ¾ of the way to apex
   [Westland, northwest Otago] .... **80.** *Hebe paludosa*
   (treated under large-leaved "Apertae")
   Leaves either: not tapering to a fine point; or gradually tapering from about or just past
   the midpoint [Nelson, Marlborough, Canterbury, north Westland] .... **78.** *Hebe leiophylla*

Small-leaved
APERTAE

9   Lower surface of leaves with two rows of domatia (e.g. Fig. 149E) .... **85.** *Hebe townsonii*
(treated under large-leaved "Apertae")
    Lower surface of leaves without domatia 10

10 Leaves glaucous or glaucescent on both surfaces (at least when young) 11
    Upper surface of leaves dull or glossy, but not glaucous 12

11 Decumbent and sparsely branched (e.g. Figs 122A and B) .... **62.** *Hebe societatis*
    Low, bushy shrub (e.g. Figs 123A and B) .... **63.** *Hebe carnosula*

12 Leaves mostly > 2 cm long, anthers pale yellow or white [Southland] .... **71.** *Hebe arganthera*
    Leaves mostly < 2 cm long or, if mostly longer, then anthers magenta, purple, mauve or
        dark pink [Wellington to Southland] 13

13 Corolla tube < 1.5 mm long, shorter than calyx; anthers usually pale pink or white .... **59.** *Hebe vernicosa*
    Corolla tube usually >1.5 mm, *c.* equalling or longer than calyx; anthers usually more
        strongly coloured 14

14 Plants low-growing, < 1 m tall, often spreading and open (e.g. Fig. 120A); branchlets
        uniformly and densely hairy (rarely bifarious); margins of petiole, and often rest of
        leaf margin, prominently hairy .... **60.** *Hebe canterburiensis*
    Plants of various habit, but if low-growing, then: branchlets bifariously hairy, or weakly
        uniformly hairy on leaf decurrencies; petiole and leaf margins often glabrous 15

15 Lower surface of leaves usually glaucous, at least when young 16
    Lower surface of leaves dull or glossy, but not glaucous .... **74.** *Hebe brachysiphon* and **73.** *Hebe venustula*[1]

16 Upper surface of leaf midribs finely pubescent (hairs up to *c.* 0.075 mm long;
        many < 0.05 mm) or glabrous; longest leaves usually < 2 cm long [south Westland,
        Otago, Southland] .... **70.** *Hebe cockayneana*
    Either upper surface of leaf midribs coarsely pubescent (hairs 0.075–0.175 mm long),
        or longest leaves > 2 cm long [Nelson, Marlborough, north Canterbury]
        .... **68.** *Hebe crenulata* and **69.** *Hebe cryptomorpha*[2]

17 Low-growing shrub (< 50 cm), often creeping, sometimes ± mat-forming; corolla tube
        shorter than or about equal to calyx [Southland] .... **61.** *Hebe dilatata*
    Various habits, but if low-growing (< 50 cm) then corolla tube longer than calyx
        [mid-Canterbury and northwards] 18

18 Sinus very small (e.g. Fig. 138C), leaves often ± glaucescent above [Three Kings Ids] .... **76.** *Hebe insularis*
    Sinus more prominent, leaves not glaucescent above [NI, SI] 19

19 Leaves glaucous and waxy below, flowers often ± sessile .... **65.** *Hebe rigidula*
    Leaves often dull but not glaucous and waxy below, flowers usually conspicuously pedicellate 20

20 With either some leaves minutely toothed (e.g. Fig. 134D), or anterior calyx lobes of
        some flowers fused for most of their length (e.g. Fig. 134G) [north of Auckland] .... **72.** *Hebe diosmifolia*
    Leaves entire; anterior calyx lobes free for most of their length [south of Auckland]
        .... **74.** *Hebe brachysiphon*, **73.** *Hebe venustula* and **75.** *Hebe divaricata*[1]

---

1 See notes under these species, particularly *H. brachysiphon*, for information on recognition.
2 See notes under these species for information on recognition.

59. *Hebe vernicosa* (Hook.f.) Cockayne et Allan

DESCRIPTION Spreading low shrub to 0.8 m tall. **Branches** spreading or ascending, old stems dark grey or brown; branchlets green or brown, hairs bifarious to uniform (hairs on leaf decurrencies often finer and shorter than those between); internodes 1.5–7 mm; leaf decurrencies evident or obscure. **Leaf bud** distinct; sinus broad and acute or almost shield-shaped. **Leaves** subdistichous, patent or erecto-patent; *lamina* elliptic or obovate, very slightly concave, 5–20 × (2.5–)3.5–7.5(–8) mm; *apex* apiculate and obtuse or subacute; *midrib* depressed to grooved above and thickened below (at least slightly); *margin* sts cartilaginous, glabrous or ciliolate; *upper surface* dark green, glossy, with few or without evident stomata, hairy along midrib (us.) or glabrous; *lower surface* dull or slightly glossy (but less so than upper surface). **Inflorescences** with (9–)12–35(–43) flowers, lateral, unbranched, (1.6–)2.5–5(–7.2) cm; peduncle (0.2–)0.5–1.2(–1.4) cm; rachis (1.2–)1.6–5.5(–6) cm. **Bracts** mostly opposite and decussate below and becoming alternate above or alternate, ovate or deltoid, obtuse to acute, glabrous outside. **Flowers** [Oct–]Nov–Jan, ⚥. **Pedicels** 0–3.5 mm. **Calyx** (1.2–)1.4–1.8(–2.5) mm; *lobes* ovate to elliptic, almost always obtuse or rarely subacute (e.g. in some material from Picton and Pelorus Sound). **Corolla** tube glabrous, 0.6–1.5 × 0.8–1.3 mm, shorter than calyx; *lobes* white at anthesis, ovate (sts narrowly) or elliptic, obtuse, suberect to patent, longer than corolla tube. **Stamen filaments** 3–5(–5.5) mm; anthers pale pink or white, *c.* 1.25–1.75 mm. **Ovary** 0.7–1.4 mm; ovules *c.* 9–10 per locule; style 2.7–5.5 mm. **Capsules** (Dec–)Feb–May(–Sep), subacute or obtuse, 2.8–4.2 × 1.6– 2.5 mm, loculicidal split extending ¼–½-way to base. **Seeds** flattened, ellipsoid or obovoid or oblong, pale brown, 1.3–1.5 × 0.9–1.1 mm, MR 0.3–0.5 mm. $2n = 42$.

DISTRIBUTION AND HABITAT Northern SI, from Gouland Downs in the northwest to the St James Ra. in the south, and as far east as the Robertson Ra., near Picton. It grows in beech forest at a range of altitudes, often at or near the tree-line.

NOTES Most similar to *H. canterburiensis*, with which it often co-occurs. Specimens with flowers or fruits are readily separated, but sterile specimens can be difficult to identify. *H. vernicosa* differs by having (Figs 119E–G): corolla tubes shorter than calyces; calyces (1.2–)1.4–1.8(–2.5) mm long; usually longer, tapering inflorescences of (1.6–)2.5–5 (–7.2) cm; pale pink or white anthers. It has, in comparison to *H. canterburiensis*, a more restricted ecological range, occurring in or at the margins of beech forest and (although it can be common at or near tree-line) is not widespread in open, subalpine habitats.

The locality "Arthur's Pass" given on CHR 63350 (*G. Simpson*, undated) is probably erroneous, and is not represented on the distribution map.

ETYMOLOGY L. (*vernicosus* = glossy), refers to the glossy upper surface of the leaves.

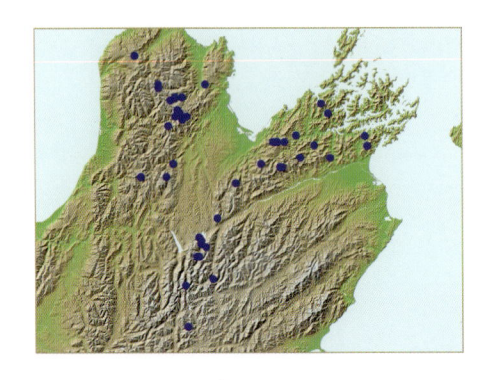

Hebe vernicosa

Small-leaved
APERTAE

FIG. 119 A plant on Mt Arthur. B sprig. C leaf bud with shield-shaped sinus. D lower and upper leaf surfaces; magnified inset shows hairs on leaf margin. E inflorescences. F lateral view of flower. G dorsal view of flower. H septicidal view of capsule. I loculicidal view of capsule.

TAXONOMIC TREATMENT   227

60. *Hebe canterburiensis* (J.B.Armstr.) L.B.Moore

DESCRIPTION Openly branched, spreading low shrub to 1 m tall. **Branches** spreading or decumbent or ascending; branchlets green (with dark bands at nodes) or red-brown or brown, puberulent, hairs uniform (us.) or bifarious (rarely); internodes (1–)2–6(–7.5) mm; leaf decurrencies obscure, or evident and extended for length of internode. **Leaf bud** distinct; sinus narrow and acute, or broad and acute. **Leaves** subdistichous, erecto-patent or patent; *lamina* elliptic or ovate or obovate, rigid, concave, 8–15(–18.5) × (3.2–)3.5–6.5(–7.8) mm; *apex* subacute or obtuse; *base* cuneate (us.) or truncate; *margin* sts cartilaginous, ciliolate to puberulent (almost always some hairs) or rarely glabrous; *upper surface* green, glossy (us.) or dull, with few or without evident stomata, minutely hairy along midrib (esp. toward base); *lower surface* green, us. dull; *petiole* 1–2.5(–3) mm. **Juvenile leaves** entire, minutely ciliolate and midrib minutely puberulent above. **Inflorescences** with 5–12 flowers, lateral, unbranched, 1–3 cm; peduncle 0.2–1 cm; rachis 1–2(–2.5) cm. **Bracts** alternate or opposite and decussate below and becoming alternate above, ovate or elliptic or deltoid, obtuse to acute. **Flowers** [Oct–]Nov–Jan(–Apr), ⚥. **Pedicels** (0–)0.3–3(–4) mm. **Calyx** 2–4.1 mm; *lobes* ovate (mostly) or deltoid or elliptic, subacute or obtuse or acute (rarely). **Corolla** tube glabrous, 1.4–3.5 × 1–1.5(–2) mm, contracted at base, longer than or equalling calyx; *lobes* white at anthesis, ovate or obovate, obtuse or subacute, suberect to recurved, longer than corolla tube. **Stamen filaments** remaining erect or slightly diverging with age, 3–5 mm; anthers purple or magenta, (0.85–)1–1.5(–1.65) mm. **Ovary** 1.1–1.8 mm; ovules 13–22 per locule, in 1–3 layers; style (3.5–)4–7.2 mm. **Capsules** [Nov–]Dec–Apr(–Oct), subacute or acute, (2.6–)3–4.9 × 2.1–3.8 mm, loculicidal split extending up to ½-way to base. **Seeds** flattened (sts strongly), broad ellipsoid to discoid, ± weakly winged, brown, 1.3–1.7 × 1–1.4 mm, MR 0.2–0.5 mm. $2n = 40$.

DISTRIBUTION AND HABITAT Southern NI (Tararua Ra., near Mt Holdsworth) and mountains of SI, mostly on or west of the Main Divide (an exception being Mt Riley, Richmond Ra., Marlborough), from NW Nelson to Arthur's Pass, with apparent southern disjunctions to south Westland and possibly to Southland (see below). It grows in beech forest at or close to tree-line, and in subalpine grassland and shrubland.

NOTES Most similar to *H. vernicosa*, with which it co-occurs on the mountains of Nelson. Specimens with flowers or fruits are readily separated, but sterile specimens may be difficult to identify. *H. canterburiensis* differs by having: corolla tubes equal to or longer than calyces (Figs 120E and F); calyces 2–4.1 mm long; usually shorter inflorescences of 1–3 cm; and magenta or purple anthers (Figs 120E–G). It has, in comparison to *H. vernicosa*, a wider ecological range, occurring not only in or at the margins of beech forest (especially close to the tree-line), but also more widely in open habitats of the subalpine and penalpine zones.

The three disjunct, southernmost distribution records are each represented only by single collections. The southernmost of these, based on a specimen labelled "The Hump, Fiord[land] Co[unty]", *J. C. Smith* [undated], WELT 5358, in particular, requires confirmation.

ETYMOLOGY L. (-*ensis* = an adjectival suffix implying origin or place), implies that the species occurs in Canterbury, which, when the name was first published (Armstrong 1879), was the extent of its known distribution.

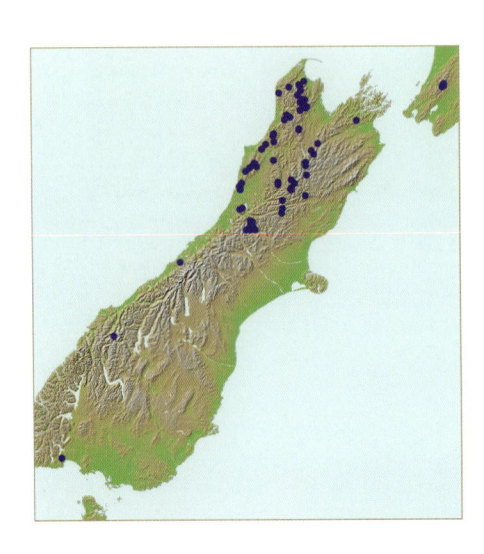

Hebe canterburiensis

Small-leaved
APERTAE

FIG. 120   **A** plants on Mt Arthur. **B** sprig. **C** leaf bud with narrow, acute sinus. **D** lower and upper leaf surfaces; magnified inset shows cilia on leaf margin. **E** flowering shoot. **F** lateral view of flower. **G** frontal view of flowers. **H** capsule in septicidal (left) and loculicidal view.

TAXONOMIC TREATMENT   229

61. *Hebe dilatata*   G.Simpson et J.S.Thomson

DESCRIPTION Spreading low shrub (either openly or densely branched, sts ± mat-like) to 0.4 m tall. **Branches** decumbent, old stems dark brown or black; branchlets green or red or brown, pubescent to puberulent, hairs bifarious; internodes (1–)2–10(–18) mm; leaf decurrencies evident. **Leaf bud** distinct; sinus small and acute (narrow). **Leaves** decussate or slightly subdistichous, erect to recurved; *lamina* obovate or spathulate or sts elliptic, coriaceous, concave, (4–)8–20(–25) × (4–)6–8(–11) mm; *apex* obtuse to rounded; 2 lateral *secondary veins* occasionally faintly evident at base of fresh leaves; *margin* sts narrowly cartilaginous, often minutely papillate or glabrous; *upper surface* green to glaucescent, glossy or dull, with many stomata, glabrous or hairy along midrib (esp. toward base); *lower surface* green to glaucescent or glaucous, glossy or dull; *petiole* (0.5–)1–2(–4) mm, often hairy below and/or hairy above. **Inflorescences** with 5–52 flowers, lateral, unbranched or tripartite (can vary on one plant; when unbranched, lowermost bracts often sterile), (0.5–)2–3.5(–4.2) cm, longer than (almost always) or about equal to subtending leaves; peduncle (0.1–)0.2–0.9 cm; rachis (0.5–)1–2.3 cm. **Bracts** alternate or opposite and decussate, elliptic or ovate or narrowly deltoid or lanceolate, subacute or obtuse. **Flowers** Nov–Jan(–Mar), ⚥ or ♀ (on different plants). **Pedicels** 0.5–2 mm. **Calyx** 2.4–3.8 mm, 3–4(–5)-lobed (5th lobe small, posterior), with anterior lobes united from ⅓ to all the way to apex; *lobes* ovate, subacute or obtuse or emarginate (compound anterior lobe). **Corolla** tube glabrous; tube of ⚥ flowers 1.8–2.5 × 2.1–2.5 mm, funnelform, shorter than or *c.* equalling calyx; tube of ♀ flowers 1.2–1.8 × 1.8–2.2 mm, funnelform, shorter than calyx; *lobes* white at anthesis, ovate, obtuse, suberect or patent, longer than (mostly) or about equalling corolla tube. **Stamen filaments** *c.* 3.8–4.2 mm (filaments of staminodes *c.* 1–2.5 mm); anthers magenta, 1.9–2.2 mm; sterile anthers of ♀ flowers light brown, 0.9–1.7 mm. **Ovary** 0.7–1.5 mm; ovules *c.* 6–10 per locule; style 3.4–4.2 mm; stigma generally larger in ♀ flowers. **Capsules** Nov–Mar, obtuse or subacute, 3.4–4.5 × 2.1–2.8 mm, rarely hairy, loculicidal split extending ¼–⅓-way to base. **Seeds** flattened, broad ovoid to sub-discoid, straw-yellow to pale brown, 0.9–1.1 × 0.6–1 mm, MR 0.2–0.4 mm. $2n = 120$.

DISTRIBUTION AND HABITAT Southern SI, where it is known from the Umbrella, Garvie, Eyre and Takitimu mountains, with one historical record (WELT 82011) from the Blue Mts near Tapanui. It grows in rocky areas on boulders or scree, sometimes in tussock grassland, and occasionally in boggy areas.

NOTES Not readily confused with other species of southern SI, from which it is generally distinct in habit and leaf bud sinus (also see notes for *H. cockayneana*). It is similar in many respects to *H. societatis* (differences are discussed under that species)

H. dilatata is not considered here to be distinct from *H. crawii* Heads (*cf.* Heads 1987, 1992). It is variable in: habit (open and more sparsely branched in grassland or on moist ground, more closely branched and mat-forming in rockier sites), size and degree of glaucousness of leaves, and the degree of inflorescence branching. The type of *H. dilatata*, from the Garvie Mts, has ± glossy leaves and simple inflorescences and is matched well by only a few specimens from the same area. Specimens matching the type of *H. crawii*, from the Takitimu Mts, have more glaucous leaves and often compound (tripartite) inflorescences. Further specimens suggest there is a continuum of forms between typical *H. dilatata* and *H. crawii*, with a study of inflorescence variation (Bayly et al. 2002) showing specimens to vary from consistently simple (9.1 per cent), to a mixture of tripartite and simple (54.5 per cent) or consistently tripartite (36.4 per cent). On the small number of specimens with consistently simple inflorescences, at least the lowermost pair of bracts on each inflorescence (e.g. Fig. 121H) is sterile, with aborted buds in their axils (suggesting that the simple condition may, at least in these cases, result from a failure of the basal inflorescence branches to develop, rather than an inherently different inflorescence architecture). Only occasionally (on specimens with mixed simple and tripartite inflorescences) are flowers borne in axils of the lowermost inflorescence bracts.

A limited study of leaf flavonoids (Bayly et al. 2002) highlighted differences between samples from the Garvie and Takitimu Mts (two samples from each locality), and this might indicate that the limits of *H. dilatata* and *H. crawii* are worthy of further investigation. However, without further flavonoid data from these and intermediate localities, and without clear-cut morphological distinctions, there are no compelling reasons for (or ways of) recognising the two as distinct species.

ETYMOLOGY L. (*dilatatus* = broadened, expanded, widened); the meaning of the name is not known with certainty, but it might refer to the sometimes broad, spreading, mat-like habit of the species.

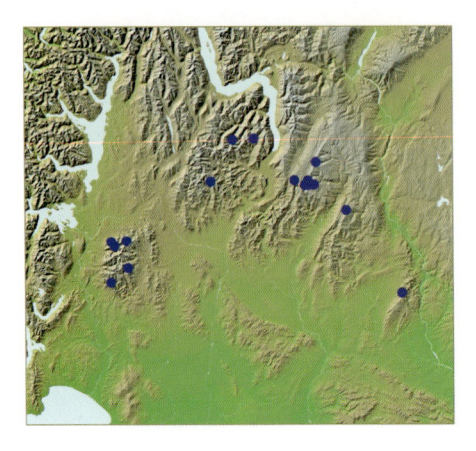

Hebe dilatata — Small-leaved APERTAE

FIG. 121 **A** plant in the Garvie Mts. **B** sprig. **C** leaf bud with small acute sinus. **D** lower and upper leaf surfaces. **E** frontal view of female flower. **F** lateral view of female flower. **G** lateral view of hermaphrodite flower. **H** tripartite inflorescence (left) and simple inflorescence with sterile lower bracts.

62. *Hebe societatis* Bayly et Kellow

DESCRIPTION Sparsely branched, spreading low shrub to 0.3 m tall. **Branches** decumbent or ascending, old stems brown; branchlets green or red-brown, pubescent, hairs bifarious; internodes (2.8–)3.5–6.5(–9.7) mm; leaf decurrencies evident. **Leaf bud** distinct; sinus broad and acute. **Leaves** subdistichous, erect or erecto-patent; *lamina* elliptic to obovate, coriaceous, concave, (5–)9–24(–37) × (2–)4–8.5(–10) mm; *apex* obtuse to subacute; *midrib* slightly thickened below and slightly depressed to grooved above; *margin* glabrous or minutely ciliolate (esp. toward apex); *upper surface* glaucescent to glaucous, dull, with many stomata, hairy along midrib and/or toward base; *lower surface* glaucescent to glaucous; *petiole* 1–2 mm, hairy above and/or along margins. **Inflorescences** with 10–25 flowers, lateral, unbranched, 1.5–3.5 cm; peduncle 0.45–0.7 cm; rachis 0.65–2.4 cm. **Bracts** with lowermost pair opposite, then subopposite or alternate above, sub-circular to broadly deltoid, obtuse to subacute. **Flowers** Jan–Feb, ♀. **Pedicels** always shorter than bracts, 0.2–1(–1.7) mm. **Calyx** (2–)2.5–3.5 mm, with anterior lobes free for most of their length or united to ⅔-way to apex; *lobes* elliptic (sts broadly), subacute to obtuse. **Corolla** tube glabrous, 2–2.5 × *c.* 1.5 mm, funnelform, *c.* equalling calyx; *lobes* white at anthesis, elliptic to circular or ovate, obtuse, suberect to patent, slightly longer than corolla tube. **Stamen filaments** *c.* 2.8–3.5 mm; anthers purple to magenta, 1.7–2 mm. **Ovary** 1–1.4 mm; ovules *c.* 11–14 per locule; style 5–5.5 mm. **Capsules** Feb–Mar, subacute or acute, 3.7–5 × 2.4–3.2 mm, loculicidal split extending up to ⅓-way to base. **Seed** characters not recorded. $2n = 42$.

DISTRIBUTION AND HABITAT Known only from Mt Murchison in the Braeburn Ra., SI, where it occurs on a steep northeast-facing slope in low subalpine herbfield dominated by *Chionochloa australis* (Fig. 122A).

NOTES The combination of a decumbent and sparsely branched habit (with ascending terminal branches), leaves that are glaucous to glaucescent on both surfaces, an acute leaf bud sinus, shortly pedicellate flowers, bracts that are shorter than calyces, and corolla tubes that are about equal to the calyces, distinguishes *H. societatis* from most other hebes. It resembles *H. dilatata*, which has similar flowers and leaves, and a sometimes similar habit, but it can be distinguished from that species by its usually broader leaf bud sinus (cf. Figs 122C and 121C, which are reproduced at the same scale), by its chromosome number, and by its consistently simple inflorescences that bear flowers in the axils of all bracts (Fig. 122E; cf. Fig. 121H, and see notes on inflorescence structure for *H. dilatata*). *H. societatis* also has a general morphological similarity to *H. vernicosa* (which also has $2n = 42$ chromosomes, an unusual number in the "Apertae", as defined here), but can be distinguished from that species by its habit, glaucous or glaucescent leaves, longer calyces ((2–)2.5–3.5 mm long), longer corolla tubes (2–2.5 mm long), comparatively smaller corolla lobes, usually shorter (2.8–3.5 mm) and stouter filaments, and purple anthers. *H. canterburiensis*, which also occurs on Mt Murchison, is generally similar to *H. societatis*, but never has glaucous leaves.

ETYMOLOGY L., means "of the society" (*societas* = society), and refers to the Nelson Botanical Society, whose members discovered the species in February 2000.

Hebe societatis

Small-leaved APERTAE

FIG. 122   A plant on Mt Murchison, Braeburn Ra. B sprigs, the two showing variation in leaf size. C leaf bud with acute sinus. D lower and upper leaf surfaces. E inflorescence. F frontal view of flower. G lateral view of flower. H capsule in septicidal (left) and loculicidal view.

63. *Hebe carnosula* (Hook.f.) Cockayne

DESCRIPTION Spreading low or bushy shrub to 0.6 m tall. **Branches** erect, old stems dark grey or brown or black; branchlets brown or red-brown, pubescent, hairs bifarious; leaf decurrencies evident and extended for length of internode. **Leaf bud** distinct; sinus small and acute. **Leaves** erect to patent; *lamina* elliptic or obovate, coriaceous, concave, 6.2–12.3 × 4.4–7.4(–9.1) mm; *apex* obtuse or subacute; *base* cuneate or truncate; *midrib* thickened below (sts weakly), sts evident in fresh leaves (below); *margin* minutely papillate; *upper surface* light to dark olive-green or glaucous or glaucescent, dull, with many stomata, glabrous; *lower surface* olive-green or glaucous or glaucescent; *petiole* 1–1.5 mm, glabrous (us.) or hairy along margins (esp. near base). **Inflorescences** with 6–22 flowers, lateral (almost always) or terminal (seen on one specimen only), unbranched, 1.2–3 cm; peduncle 0.6–0.9 cm; rachis 1.3–2.5 cm. **Bracts** alternate or opposite and decussate (and sts becoming alternate above), ovate or lanceolate, obtuse or subacute or acuminate (esp. lowermost). **Flowers** [Nov–]Dec–Jan(–Apr), ☿ or ♀ (on different plants). **Pedicels** absent or when present always shorter than bracts, 0–0.8 mm. **Calyx** 1.8–2.5 mm, 4–5-lobed (5th lobe small, posterior); *lobes* elliptic, obtuse. **Corolla** tube glabrous; tube of ☿ flowers 2.1–2.7 × *c.* 1.9–2.1 mm, shortly cylindric, equalling or longer than calyx; tube of ♀ flowers 1.2–1.7 × *c.* 1.4–1.6 mm, funnelform, shorter than or equalling calyx; *lobes* white at anthesis, elliptic (sts narrowly) or ovate, obtuse, patent to recurved, longer than corolla tube. **Stamen filaments** (1.7–)2.5–5.2 mm; anthers magenta, 1.7–1.8 mm; sterile anthers *c.* 1 mm. **Ovary** narrowly ovoid, sts sparsely hairy (toward apex), *c.* 1.2–1.3 mm; ovules 10–16 per locule; style 5–5.7 mm, occasionally hairy (esp. toward base). **Capsules** Jan–Apr, obtuse or subacute, 3–4.5 × 2.6–3 mm, glabrous (or possibly sts hairy), loculicidal split extending ¼–¾-way to base. **Seed** characters not recorded. Chromosome number unknown.

DISTRIBUTION AND HABITAT Known with certainty only from ultramafic areas on Dun Mtn and Red Hills Ridge, northern SI. It grows in grassland and sparse shrubland.

NOTES Distinguished from most species by the combination of: leaves usually glaucous (or glaucescent) on both surfaces; an acute leaf bud sinus; simple inflorescences; and inflorescence bracts not overtopping calyces. It is distinguished from *H. societatis*, which shares these features, by its more erect and more branching habit. Its affinities and best infrageneric placement are not clear. It is placed here with other "Apertae" because it consistently has a leaf bud sinus and small inflorescence bracts, but it is also similar to some glaucous-leaved plants of "Subcarnosae". The name *carnosula* has sometimes, probably incorrectly, been applied to some specimens included here under *H. pinguifolia* (see notes under that species). The species possibly hybridises with *H. divaricata* and/or *H. brachysiphon* on the Red Hills Ridge (chiefly at the margins of ultramafic areas), where plants display a range of variation in leaf shape, glaucousness and pedicel length.

ETYMOLOGY L. (*carnosulus* = slightly fleshy), refers to leaf texture.

Hebe carnosula | **Small-leaved APERTAE**

FIG. 123 A plant on Red Hills Ridge, Wairau Va. **B** sprig. **C** leaf bud with acute sinus. **D** lower and upper leaf surfaces. **E** inflorescence. **F** lateral view of flower. **G** frontal view of flower. **H** capsule in loculicidal view. **I** young infructescence with opposite-decussate arrangement of bracts.

64. *Hebe rupicola* (Cheeseman) Cockayne et Allan

DESCRIPTION Low-growing bushy shrub to 0.8(–1.5) m tall. **Branches** decumbent or ascending or erect, old stems brown; branchlets green (but dark when dry), pubescent or puberulent, hairs bifarious (us. in broad bands between leaf decurrencies); internodes (1–)2–9(–18) mm; leaf decurrencies extended for length of internode and often swollen. **Leaf bud** distinct; sinus narrow and acute. **Leaves** erecto-patent to patent; *lamina* elliptic (mostly) or oblanceolate, coriaceous, slightly concave, (4–)7–21(–24) × (2–)5–7(–9) mm; *apex* acute to obtuse; *midrib* thickened below and depressed to grooved above (but evident above to varying extents); *margin* minutely papillate and sts ciliolate or gl.-ciliate; *upper surface* light green or yellowish-green, dull, with many stomata, glabrous or hairy along midrib; *lower surface* light green or sts glaucous or glaucescent; *petiole* (1–)2–4(–5) mm, hairy along margins (us.) and hairy above (often) or glabrous. **Inflorescences** with (5–)9–37 flowers, lateral, unbranched or tripartite, 1.3–3(–4.7) cm, longer than (us.) to shorter than subtending leaves; peduncle (0.3–)0.6–1.7(–2) cm; rachis (0.7–)1–2.1(–2.9) cm. **Bracts** large, opposite and decussate, free (us.) or connate (sts, and then very shortly), ovate or lanceolate, acute (mostly) or subacute or obtuse. **Flowers** Dec–Feb(–Mar), ☿ or ♀ (on different plants). **Pedicels** us. absent but if evident then always shorter than bracts, 0–1 mm. **Calyx** 3–4(–5) mm, with anterior lobes free for most of their length or united ⅓–⅔-way to apex (degree of fusion can vary in one inflorescence); *lobes* ovate or lanceolate, subacute or obtuse. **Corolla** tube glabrous; tube of ☿ flowers 3–4.9 × 1.5–2.3 mm, funnelform (narrowly) or cylindric, longer than (us.) or equalling calyx; *lobes* white at anthesis, ovate (sts narrowly) or lanceolate, subacute or obtuse, suberect to recurved, shorter to longer than corolla tube. **Stamen filaments** white, 3.3–4.2 mm; anthers white or buff or pink or mauve or violet, 1.3–1.7 mm; sterile anthers 0.6–0.8 mm (on herbarium specimens). **Ovary** c. 1–1.3 mm; ovules 8–9 per locule; style 5.5–8.3 mm. **Capsules** (Jan–)Feb–May(–Oct), obtuse, 4–4.5 × 2.3–3.2 mm, loculicidal split extending ¼–½-way to base. **Seeds** flattened, ovoid to obovoid-oblong, straw-yellow to pale brown, 1.4–2.2 × 0.9–1.4 mm, MR 0.5–0.7 mm. $2n = 40$.

DISTRIBUTION AND HABITAT Endemic to Marlborough and north Canterbury, SI, between the Chalk Ra. in the northeast and the Mason R. in the southwest. It grows mostly on cliffs or other rocky areas, often near rivers.

NOTES Has most commonly been confused with *H. rigidula*, which is a similar small shrub of rocky areas and also has a sinus in the leaf bud. It can be distinguished from this and other similar species by the possession of large bracts that surround and almost completely obscure the calyces (Fig. 124F). It also differs from *H. rigidula* in having longer calyces (3–4(–5) mm), and leaves that are usually duller green above.

The localities "Key Summit" (Livingstone Mts, Southland, given on CHR 33033) and "Aorere River ford" (NW Nelson, given on CHR 468465) are not likely to be correct.

ETYMOLOGY L. (*rupes* = rock, cliff; *-cola* = dweller, only exists on), refers to the usual habitat of the species.

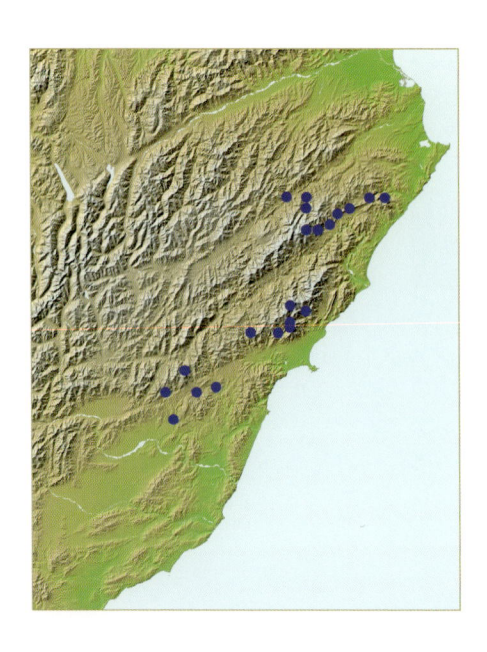

Hebe rupicola — **Small-leaved APERTAE**

FIG. 124 **A** plant by Inland Road, near Whales Back Station. **B** sprig. **C** leaf bud with acute sinus. **D** lower and upper leaf surfaces. **E** top view of inflorescence. **F** flower, with large bract peeled back to illustrate its length relative to calyx. **G** lateral view of flower. **H** capsule in septicidal (left) and loculicidal view.

65. *Hebe rigidula* (Cheeseman) Cockayne et Allan

DESCRIPTION Low-growing, openly branched, bushy shrub to 0.3(–0.6) m tall. **Branches** erect or ascending, old stems brown; branchlets green, pubescent, hairs bifarious; internodes (1–)2–5(–5.5) mm; leaf decurrencies evident and extended for length of internode. **Leaf bud** distinct, terete to weakly tetragonous in TS; sinus narrow and acute. **Leaves** erect to patent; *lamina* narrowly to broadly elliptic or oblong or lanceolate or oblanceolate, thin to coriaceous, m-shaped in TS or concave, (7.1–)12.5–25(–31.4) × (2.1–)3–7.5(–8.7) mm; *apex* acute or subacute; *margin* minutely papillate; *upper surface* green, dull or often slightly glossy, with few or without evident stomata, glabrous (mostly) or hairy along midrib (esp. toward base); *lower surface* glaucous or glaucescent; *petiole* 2–3.2 mm, glabrous or hairy along margins (esp. youngest leaves) and/or hairy above (rarely). **Inflorescences** with 10–40 flowers, lateral, tripartite (us.) or with more than 3 branches (occasionally) or unbranched (rarely), 1.4–2.2(–3.2) cm, longer than (us.) to shorter than subtending leaves (rarely); peduncle 0.3–1(–1.2) cm; rachis 0.7–2.3(–2.8) cm. **Bracts** opposite and decussate, or opposite and decussate below and becoming alternate above, ovate or deltoid, obtuse or subacute. **Flowers** [Nov–]Dec–Feb, ♀. **Pedicels** absent or if evident then always shorter than bracts, (0–)0.3–2 mm. **Calyx** (1.5–)2–2.5 mm; *lobes* elliptic or oblong or ovate, obtuse or subacute, with mixed gl. and egl. cilia or apparently egl. ciliate. **Corolla** tube glabrous, (2.5–)3–4 × *c.* 1.7 mm, cylindric, longer than calyx; *lobes* white at anthesis, elliptic or ovate or oblong, obtuse (posterior sts emarginate), suberect to recurved, shorter than corolla tube. **Stamen filaments** 3–4 mm; anthers buff or slightly tinged pink, 1.7–2 mm. **Ovary** 0.9–1 mm; ovules 9–14 per locule; style 5–7 mm. **Capsules** (Dec–)Jan[–Mar](–Oct), obtuse or subacute, 2.4–3.5(–4) × 1.8– 2.5 mm, loculicidal split extending ¼–½-way to base. **Seeds** flattened, ellipsoid to discoid or irregular, brown (sts pale), 0.8–1.6 × 0.7–1.1 mm, MR 0.3–0.7 mm. $2n = 40$.

Var. *rigidula*
Leaves narrowly elliptic (usually), or elliptic or oblanceolate, **concave above**, (7.7–)12.5–20(–31.4) × (2.1–)3.4–5.6(–7.4) mm, **ratio of leaf length to width** (2.7–)3.6–4.4(–4.8). Pedicels (0.1–)0.3–0.8 mm long.

Var. *sulcata* Bayly et Kellow
Leaves elliptic or broadly elliptic, **m-shaped in transverse section** (9.4–)13.5–25(–28.2) × (3.2–)5–7.5(–8.7) mm, **ratio of leaf length to width** (2.5–)2.8–3.3(–3.5). Pedicels (0.3–)0.4–2 mm long.

DISTRIBUTION AND HABITAT Var. *rigidula* occurs on northern SI, in the Pelorus and Maitai valleys, on the Bryant and Richmond ranges, and on Serpentine Hill in the Lee Va. area, in open habitats including sparse scrub, rock outcrops, crevices, boulder falls and the sides of river gorges. Var. *sulcata* occurs on D'Urville Id and on Editor Hill and Lookout Peak, northern SI, in open rocky areas and short scrub, especially associated with the mineralised zone of D'Urville Id. An historical collection of var. *rigidula*, not represented on the distribution map, is labelled with the vague locality "Awatere Valley" (CHR 331953, *T. F. Cheeseman*, date unknown). The locality given on CHR 132109 ("Gouland Downs") is probably incorrect, and is also not represented on the distribution map.

NOTES Similar to *H. rupicola* and *H. scopulorum* (see notes under both species). Variation in the species is discussed by Bayly et al. (2002). The two varieties are narrowly geographically separated, and differ only in minor morphological and flavonoid characters.

Var. *rigidula* shows considerable variation in leaf shape and size but, in comparison with var. *sulcata*, the leaves are usually more concave above and/or more narrowly elliptic, tending to be narrower in absolute terms, and in the ratio of leaf length to width. Specimens from Mt Duppa, Bryant Ra., have leaves that are quite short and broad when compared with those from other populations. In terms of the ratio of leaf length to width, these specimens are similar to var. *sulcata*, but are placed here under var. *rigidula* on the basis of their small overall leaf length, (7.1–)11.3–16.5(–19.7) mm, and generally concave leaves.

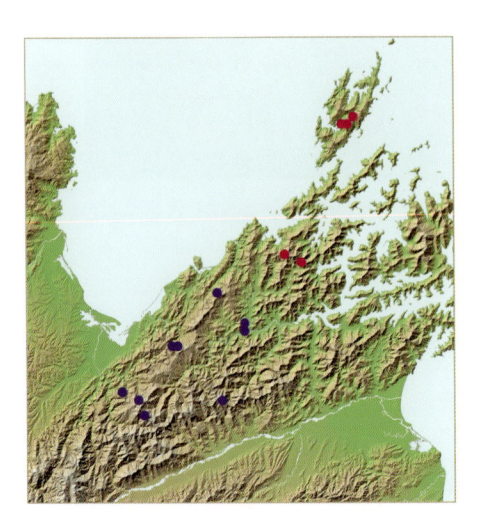

ETYMOLOGY *Rigidula* (L. *rigidulus* = somewhat rigid) refers to the leaves; *sulcata* (L. *sulcatus* = furrowed or grooved) also refers to the leaves, which in this variety are characteristically furrowed along the midvein.

- Var. *sulcata*
- Var. *rigidula*

Hebe rigidula — Small-leaved APERTAE

FIG. 125 A var. *rigidula* at Pelorus Bridge, Nelson. B sprig of var. *rigidula* (left) and var. *sulcata*. C lower and upper leaf surfaces of var. *rigidula* (above) and var. *sulcata*. D top view of leaf buds (to same scale) of var. *rigidula* (D1; terete) and var. *sulcata* (D2; weakly tetragonous). E leaf bud with acute sinus. F branched inflorescence (left) and infructescence. G frontal view of flower. H inflorescence showing lateral view of flowers.

66. *Hebe colensoi* (Hook.f.) Cockayne

DESCRIPTION Openly branched, small bushy shrub or spreading low shrub to 0.4(–0.75) m tall. **Branches** erect, old stems brown or grey; branchlets initially green, becoming brown, glabrous or very sparsely puberulent, hairs bifarious; internodes (1.5–)2–5(–8) mm; leaf decurrencies evident. **Leaf bud** distinct, tetragonous in TS; sinus narrow and acute. **Leaves** decussate or subdistichous, erecto-patent; *lamina* obovate or elliptic (narrowly to broadly), coriaceous, shallowly m-shaped in TS (the margins being slightly revolute) or flat, (10–)14–27(–42) × (2–)4.5–9(–15.5) mm; *apex* subacute or obtuse; *margin* sts very narrowly cartilaginous, glabrous and minutely papillate (to the inside of outer cartilaginous portion), entire or shallowly toothed (may vary on one plant); *upper surface* glaucous (often less so than lower surface), with many stomata, glabrous or hairy along midrib; *lower surface* glaucous; *petiole* glabrous or hairy above. **Inflorescences** with (11–)15–21(–29) flowers, lateral and sts also terminal (e.g. WELT 14900; see comments by Elder (1939)), tripartite and/or unbranched, only sts with more than three branches, (1.7–)2.5–4.5 cm; peduncle 0.5–1.3 cm, glabrous (us.) or hairy; rachis (1.2–)1.9–3.3 cm. **Bracts** alternate (lowermost pair may be subopposite or opposite), lanceolate or deltoid or oblong, acute or subacute, margins glabrous (us.) or hairy (very rarely, and only with a few cilia near base). **Flowers** (Aug–)Sep–Nov(–Jan), ⚥ or ♀ (on different plants). **Pedicels** longer than or equal to bracts, 0.5–2(–3) mm. **Calyx** (1.5–)2–2.5(–3) mm, 4–5-lobed (5th lobe small, posterior), with anterior lobes free for most of their length or united to ⅓–⅔-way to apex; *lobes* deltoid or lanceolate, acute or subacute, margins glabrous (us.) or egl. ciliolate (only ever with sparse, short hairs). **Corolla** tube glabrous; tube of ⚥ flowers 1.8–2.3 × 1–1.5 mm, funnelform, shorter than (mostly) or equalling calyx; *lobes* white at anthesis, lanceolate or ovate, subacute or obtuse, patent to recurved, longer than corolla tube. **Stamen filaments** 2–3.5 mm; anthers yellow or buff or pink or mauve or violet, 1.2–1.9 mm. **Ovary** ovoid (sts very narrowly), 0.8–1.3 mm; ovules 4–8 per locule; style 2.2–4.5 mm. **Capsules** Dec–Apr(–Oct), subacute, (2.5–)2.8–3.5(–3.8) × 1.9–2.5 mm, loculicidal split extending ¼–½-way to base. **Seeds** flattened, ± ellipsoid-oblong, ± smooth, pale brown (with orange component), 1.1–1.5 × 0.7–0.9 mm, MR 0.2–0.3 mm. $2n = 40$.

DISTRIBUTION AND HABITAT Central NI, in the upper catchments of the Moawhango, Mohaka, Rangitikei, Taruarau and Ngaruroro rivers. It grows on rock outcrops on bluffs, gorges and riverbanks.

NOTES Distinguished from most species by the combination of: low-growing, rupestral habit; leaf bud sinus; leaves glaucous on both surfaces, m-shaped in TS; and corolla tubes shorter than or equal to calyces. It is most similar to *H. scopulorum* (see notes under that species). Calyces without marginal cilia (the usual condition in the species) are almost unique in the genus, being otherwise seen only on some specimens of *H. pareora* and *H. macrocalyx*.

Elder (1939, 1971) presented a study of variation in *H. colensoi*, and Moore (in Allan 1961) recognised two varieties, distinguished on the basis of leaf size, leaf shape and the number of teeth on the leaf margin. There is substantial variation in these features (Appendix 4), no doubt representing both environmental and genetic differences (perhaps likely between small, isolated populations), but strong grounds for distinguishing two distinct biological entities are lacking, and no varieties are recognised here. Some populations include mixtures of plants representing each of Moore's varieties, and Elder (1939) suggested that many plants are of intermediate form (i.e. between what he called "jordanon 1" and "jordanon 2").

ETYMOLOGY Honours William Colenso (1811–99), upon whose specimen the original description of the species was (in part) based.

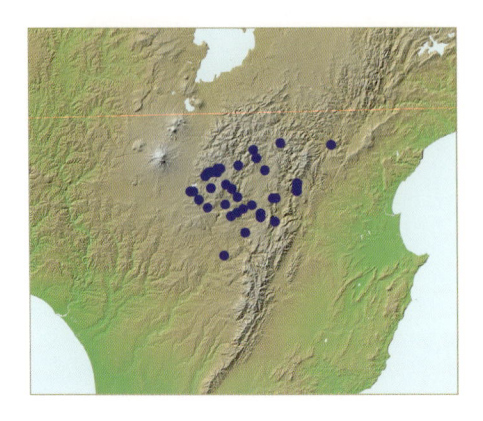

Hebe colensoi — Small-leaved APERTAE

FIG. 126 **A** plant near Taruarau R. **B** sprig. **C** leaf bud with acute sinus. **D** lower and upper leaf surfaces. **E** lateral view of flower; magnified inset shows glabrous margins of calyx lobes. **F** frontal view of flower. **G** young infructescences. **H** capsule in septicidal (left) and loculicidal view. **I** seeds.

TAXONOMIC TREATMENT 241

67. *Hebe scopulorum* Bayly, de Lange et Garn.-Jones

DESCRIPTION Bushy shrub to 0.7 m tall. **Branches** erect or ascending, old stems black or grey (producing thick corky bark with age); branchlets green to brown, pubescent, hairs bifarious to uniform; internodes 2–5(–8) mm; leaf decurrencies evident and extended for length of internode. **Leaf bud** distinct, tetragonous in TS; sinus narrow to broad, acute. **Leaves** decussate to subdistichous, erecto-patent to patent; *lamina* linear-elliptic or elliptic to narrowly oblanceolate, subcoriaceous, m-shaped in TS, (14–)20–44(–55) × (4–)6–11(–16) mm; *apex* plicate and subacute or acute or subapiculate or obtuse; *midrib* thickened below and strongly depressed to grooved above; *margin* cartilaginous, minutely papillate; *upper surface* green to dark green, glossy, without evident stomata, hairy along midrib; *lower surface* glaucous; *petiole* 2–4(–5) mm, hairy above. **Juvenile leaves** incised. **Inflorescences** with (7–)15–30(–40) flowers, lateral, tripartite and some rarely also unbranched or with more than 3 branches, (1.5–)2–3.5(–4.6) cm, shorter than or about equal to subtending leaves; peduncle (0.5–)0.8–1.5(–2.5) cm; rachis 1–1.8(–2.5) cm. **Bracts** us. opposite and decussate below and becoming alternate above, lanceolate to linear or rarely ovate, acute to acuminate or rarely obtuse. **Flowers** (Sep–)Oct–Nov, ⚥. **Pedicels** shorter than or equal to bracts, (0.5–)1–3(–5) mm, sts recurved in fruit. **Calyx** 2.3–3.5 mm, with anterior lobes free for most of their length or rarely united to ⅔-way to apex (seen on one specimen only); *lobes* lanceolate (us.) to ovate or elliptic, acute (us.) to obtuse, egl. ciliolate or with mixed gl. and egl. cilia. **Corolla** tube glabrous, 3–4.2 × 1.5–2 mm, shortly cylindric, longer than calyx; *lobes* white or pale mauve at anthesis, elliptic or lanceolate or oblong, subacute, suberect to patent, equalling or shorter than corolla tube. **Stamen filaments** remaining erect or slightly diverging with age, 3–4.5(–5) mm; anthers pale mauve or violet or white, 1.6–1.8(–2.2) mm. **Ovary** ovoid, 1–1.5 mm; ovules 6–14 per locule; style 5–7.5 mm. **Capsules** (Sep–)Nov– Feb(–Apr), acute, 3.2–4.5 × 2–3 mm, loculicidal split extending ½–¾-way to base. **Seeds** flattened, ellipsoid or ovoid or oblong, pale brown, 1.1–1.4 × 0.7–1.1 mm, MR 0.3–0.4 mm. $2n = 40$.

DISTRIBUTION AND HABITAT Southeast of Kawhia, NI, in the upper Awaroa Va. and northern Taumatatotara Ra., where it grows on remnant tors and mesa of the Orahiri Limestone Formation of the Te Kuiti Group (Kear & Schofield 1959). All known populations are within *c.* 10 km of each other.

NOTES Morphologically most similar to *H. colensoi* and *H. rigidula*. It differs from the former species in that the upper surfaces of leaves are usually not glaucous (Figs 127B and D), corolla tubes are longer than calyces (Fig. 127G), margins of calyx lobes are ciliolate, and branchlets are hairy (at least bifariously). In *H. colensoi*, upper surfaces of leaves are often glaucous (at least when young; Figs 126B–D), corolla tubes are usually shorter than calyces (Fig. 126E; or sometimes about equal), and margins of calyx lobes and branchlets are usually glabrous (with short, sparse hairs only occasionally present on either). *H. scopulorum* differs from *H. rigidula* in that the leaf buds are strongly tetragonous (Fig. 127F; rather than terete or weakly tetragonous, Fig. 125D), pedicels are generally longer ((0.5–)1–3(–5) mm; cf. Figs 127H, 125F and H), and leaves tend to be longer (in terms of mean leaf length, or the length of the largest leaves on individual specimens). Detailed notes on distribution, habitat and variation in *H. scopulorum* are given by de Lange (1986) and Bayly et al. (2002).

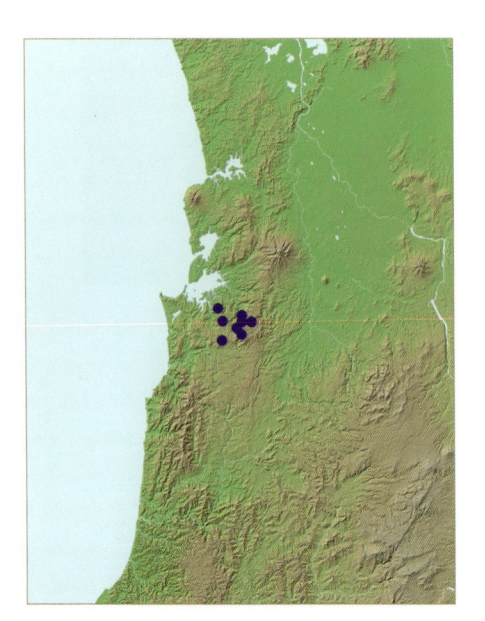

ETYMOLOGY L., means "of the crags" (*scopulus* = pointed rock, cliff, crag), and refers to the characteristic habitat of the species.

Hebe scopulorum — Small-leaved APERTAE

FIG. 127 **A** plant at Rock Peak, upper Awaroa Va. **B** sprig. **C** leaf bud with acute sinus. **D** lower and upper leaf surfaces. **E** close-up of an inflorescence showing a flower in frontal view. **F** top view of a leaf bud, which is strongly tetragonous. **G** lateral view of flower. **H** infructescence, which is tripartite and has fruits on obvious pedicels.

TAXONOMIC TREATMENT 243

68. *Hebe crenulata* Bayly, Kellow et de Lange

DESCRIPTION Densely branched, spreading low or bushy shrub to 1 m tall. **Branches** spreading or ascending or erect, old stems dark brown (mostly) or grey; branchlets brown or red-brown or green, pubescent, hairs bifarious; internodes (1–)2–6(–7.5) mm; leaf decurrencies evident. **Leaf bud** distinct; sinus broad and acute. **Leaves** decussate, erecto-patent to patent; *lamina* us. oblanceolate or obovate or less commonly elliptic, coriaceous, concave, (6.4–)7–16(–24.2) × (3.5–)5–7(–7.9) mm; *apex* obtuse to subacute or acute; *margin* minutely papillate and sts gl.-ciliate (with clusters of hairs at bases of marginal teeth), entire or shallowly toothed (may vary on one plant); *upper surface* green, slightly glossy or dull, with or without evident stomata, hairy along midrib (hairs 0.075–0.175 mm); *lower surface* glaucous or glaucescent; *petiole* 1–2.7 mm, hairy above (along midrib). **Inflorescences** with 4–10(–16) flowers, lateral, unbranched (mostly) or sts tripartite, 0.9–3.1 cm, longer than (mostly) or about equal to subtending leaves; peduncle (0.1–)0.4–0.65 cm; rachis (0.4–)1–1.5(–2.1) cm. **Bracts** opposite and decussate, us. free or sts connate (and then only connected by a very narrow ridge), lanceolate or deltoid, subacute or acute or acuminate. **Flowers** (Dec–)Jan–Feb, ⚥ or ♀ (on different plants). **Pedicels** absent or when present always shorter than bracts, 0–1.5 mm. **Calyx** 2–3 mm; *lobes* lanceolate to elliptic, obtuse or subacute. **Corolla** tube glabrous; tube of ⚥ flowers 1.8–3 × 2.2–2.5 mm, broadly funnelform and contracted at base, ± equalling calyx; tube of ♀ flowers 1.4–1.6 × c. 1.5–1.8 mm, contracted at base, shorter than or equalling calyx; *lobes* white at anthesis, ovate or elliptic, obtuse, suberect to recurved, longer than corolla tube. **Stamen filaments** straight at apex in bud, 2.5–6.5 mm (lower value represents staminodes only); anthers magenta, c. 2–2.2 mm; sterile anthers c. 0.7–1.1 mm. **Ovary** 1.5–1.7 mm; ovules 7–18 per locule (in ⚥ plants); style 5–7 mm. **Capsules** (Jan–)Feb–Apr(–Dec), subacute or obtuse, 3.2–4 × (2.2–)2.5–3.2 mm, loculicidal split extending ¼–½-way to base. **Seed** characters not recorded. $2n = 80$.

DISTRIBUTION AND HABITAT Mountains of western and central areas of northern SI, between the Douglas Ra. in the northwest and the Poplars Ra. in the southeast. It grows in subalpine shrubland and tussock grassland, often in shallow soils on rocky sites.

NOTES Similar to *H. cryptomorpha* and *H. cockayneana*, in which it was previously included (Moore, in Allan 1961; Druce 1980, 1993; Eagle 1982; Heads 1992). All share a similar bushy habit, an acute sinus in the leaf bud, leaves that are mostly glaucous below and dark green above, similar coarse bifarious pubescence on their branchlets, simple inflorescences with an opposite-decussate flower arrangement, and similar flowers with short corolla tubes and us. obtuse corolla lobes. Separation of *H. crenulata* from these species rests largely on its distinct chromosome number (Table 5) and consistent and substantial differences in leaf flavonoids, which are of the extent that grouping them together under one species would not adequately reflect their relationships and diversity (Bayly et al. 2002). It differs morphologically from *H. cockayneana* in having the combination of: coarse pubescence (hairs 0.075–0.175 mm long) along the midrib on upper surfaces of leaves; usually oblanceolate leaves; and sometimes shallowly toothed leaf margins. Specimens cannot always be distinguished morphologically from *H. cryptomorpha*, but those with highly toothed leaf margins are distinct. *H. crenulata* usually also differs in having the combination of: 7–18 ovules per ovary locule in hermaphrodite plants; longest leaves (including petioles) on individual plants that are (11.9–)14–22(–25.4) mm long (although the two species overlap in this character).

Leaves show considerable variation in size and the degree of marginal incision (Appendix 4). In the northern part of the species' range highly toothed forms are generally more common, and plants generally have at least some toothed leaves. In the southern part of its range (e.g. around Ada Pass, Spenser Mts) plants with more sparsely toothed leaves, a greater proportion of entire leaves, or only entire leaves, are more common.

Plants from southern areas, particularly those with large (longer than c. 20 mm), entire or only faintly toothed leaves, are the most difficult to distinguish from *H. cryptomorpha*. This area is geographically close to the southern distributional limit of *H. cryptomorpha*, but there is no evidence of introgression, or of overlap in the two distributions. More chemical and cytological investigation of plants throughout this area would be worthwhile, and would allow a more precise assessment of the geographic limits of the two species.

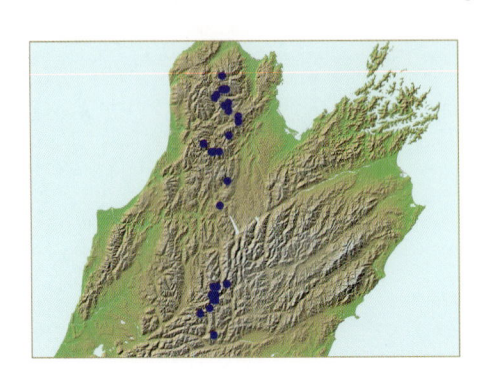

ETYMOLOGY L. (*crenulatus* = crenulate, having small rounded teeth), refers to the often shallowly toothed leaf margins. Although not found in all specimens, this feature can be conspicuous and, when present, is helpful in distinguishing *H. crenulata* from similar species.

Hebe crenulata — Small-leaved APERTAE

FIG. 128 **A** plant near Lk. Peel, Nelson. **B** sprig. **C** leaf bud with acute sinus. **D** lower and upper views of leaf with shallowly toothed margins; magnified inset shows hairs above midrib. **E** old inflorescence showing opposite-decussate arrangement of flowers. **F** frontal view of flower. **G** lateral view of flower. **H** leaf with entire margins. **I** capsule in septicidal (left) and loculicidal view.

TAXONOMIC TREATMENT 245

69. *Hebe cryptomorpha* Bayly, Kellow, G. Harper et Garn.-Jones

DESCRIPTION Densely branched, spreading low or bushy shrub (often of rounded habit) to 1.2 m tall. **Branches** spreading or ascending or erect, old stems light to very dark brown or red-brown; branchlets brown (us.) or green, pubescent, hairs bifarious (sts sparsely); internodes 1–8(–10) mm; leaf decurrencies evident. **Leaf bud** distinct; sinus broad and acute. **Leaves** erecto-patent (us.) to recurved (with age); *lamina* oblanceolate or obovate or elliptic (less commonly), coriaceous, concave or slightly m-shaped in TS (on large-leaved specimens), (7–)13–28(–33) × (3–)4–8(–9) mm; *apex* acute or subacute; *margin* minutely papillate, entire (almost always) or shallowly toothed (rarely); *upper surface* green, dull or slightly glossy, with many or with few stomata (us.) or without evident stomata, hairy along midrib; *lower surface* glaucous or glaucescent; *petiole* 1–4(–6) mm, hairy above (along midrib). **Inflorescences** with 5–17 flowers, lateral, unbranched (almost always) or tripartite (very rarely, and with lateral branches short and not well developed), 1.2–3.7 cm, about equal to or longer than subtending leaves; peduncle (0.2–)0.3–0.6(–1.3) cm, hairy (sts sparsely); rachis (0.8–)1.2–2.6(–3.1) cm. **Bracts** almost always opposite and decussate, free (us.) or connate (sts very shortly, esp. near base of inflorescence), lanceolate or ovate or deltoid, acute or subacute or obtuse (infrequently). **Flowers** Dec–Feb, ☿ or ♀ (on different plants). **Pedicels** absent or when present always shorter than bracts, 0–3(–5) mm. **Calyx** 2.4–3(–3.5) mm, (3–)4-lobed, with anterior lobes free for most of their length (us.) or united between ⅓ and all the way to apex; *lobes* lanceolate to elliptic, obtuse to acute. **Corolla** tube glabrous; tube of ☿ flowers 1.5–3 × 2–3 mm, shortly cylindric, equalling or slightly longer than calyx; *lobes* white at anthesis, ovate to elliptic, obtuse, suberect to recurved, longer than corolla tube. **Stamen filaments** 3.5–5.5 mm; anthers dark pink or purplish mauve, 1.9–2.4 mm. **Ovary** sts sparsely hairy, 1.5–2.1 mm; ovules 15–33 per locule (in ☿ plants), in 1–3 layers; style 3–6.5 mm, sts sparsely hairy (esp. toward base). **Capsules** Jan–Mar(–Dec), obtuse or subacute, 3–4 × 2.4–2.8 mm, sts sparsely hairy, loculicidal split extending ¼–½-way to base. **Seeds** oblong, straw-yellow, *c.* 1.1 × *c.* 0.7 mm, MR *c.* 0.3 mm. 2*n* = 40.

DISTRIBUTION AND HABITAT Northern SI, primarily on mountains of the Wairau R. catchment (e.g. the Richmond, Raglan, St Arnaud and Crimea ranges), extending west to Mt Robert and east to Mt Severn and the upper Saxton Va. It grows in subalpine grassland and shrubland, sometimes among rocks and boulder fields.

NOTES Most similar to *H. cockayneana* and some forms of *H. crenulata* (see notes under those species). It is morphologically distinguished from *H. rigidula*, with which it was allied by Druce (1980) and Eagle (1982), by having almost always simple inflorescences, corolla tubes about equalling calyces in length, and a rounded habit. Leaves are illustrated in Appendix 4.

ETYMOLOGY Gk (*crypto-* = hidden, concealed; *-morphus* = -shaped), refers to the fact that morphological differences between this species and some specimens of *H. crenulata* are obscure or unknown.

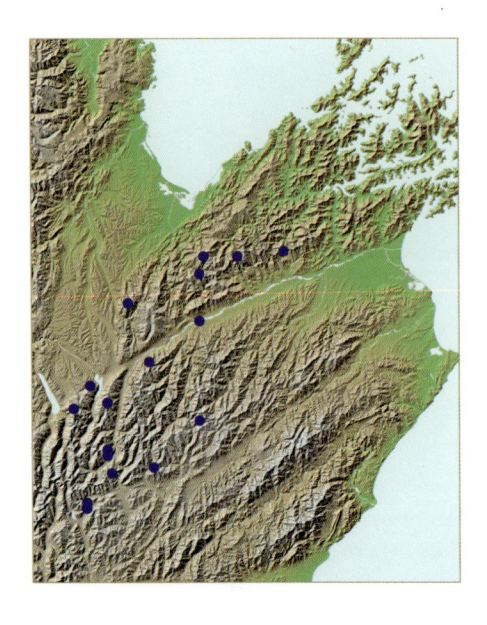

Hebe cryptomorpha

Small-leaved
APERTAE

FIG. 129 **A** plant near Mt Patriarch, Marlborough. **B** sprig. **C** leaf bud with acute sinus. **D** lower and upper leaf surfaces. **E** young infructescence showing opposite-decussate arrangement of bracts. **F** lateral view of flower. **G** frontal view of flower. **H** inflorescence showing lateral view of a pair of opposite flowers.

TAXONOMIC TREATMENT 247

70. *Hebe cockayneana* (Cheeseman) Cockayne et Allan

DESCRIPTION Densely branched, bushy shrub or spreading low shrub (in exposed situations) to 1.2 m tall. **Branches** erect; branchlets pubescent (with long, multicellular hairs, us. of 4 or more cells, golden or white), hairs bifarious; internodes 2.5–6.5 mm; leaf decurrencies obscure to evident and extended for length of internode. **Leaf bud** distinct; sinus narrow and acute. **Leaves** decussate or subdistichous, erecto-patent to patent; *lamina* us. elliptic (sts narrowly) or less commonly obovate, rigid or coriaceous, concave, (5–)10–21.5 × (3–)3.5–7.5(–9) mm; *apex* subacute; *margin* minutely papillate and glabrous or minutely gl.-ciliate; *upper surface* green, glossy, with few or without evident stomata, glabrous or hairy along midrib (hairs to *c.* 0.075 mm, many < 0.05 mm); *lower surface* glaucous (except on midrib or margin); *petiole* 0.5–2 mm. **Inflorescences** with (2–)6–16(–23) flowers, lateral, unbranched (almost always) or with 3 or more branches, 1–3.2 cm; peduncle (0.15–)0.2–1.1 cm; rachis 0.6–2.5 cm. **Bracts** us. opposite and decussate, free or connate (rarely and only very shortly), ovate or deltoid, subacute or acute or rarely obtuse. **Flowers** Dec–Feb(–Mar), ☿ or ♀ (on different plants). **Pedicels** (0.5–)1–5.5 mm. **Calyx** 1.5–4 mm, with anterior lobes free for most of their length or united ⅓–⅔-way to apex; *lobes* ovate, subacute or obtuse. **Corolla** tube glabrous; tube of ☿ flowers 1.5–2.3 × *c.* 1.8–2.5 mm, funnelform, shorter than or equalling calyx; tube of ♀ flowers 1–2 mm, shorter than or equalling calyx; *lobes* white at anthesis, ovate or elliptic, obtuse or subacute, suberect to patent, longer than corolla tube. **Stamen filaments** 1.4–4.3 mm (fertile 3.5–4.3 mm; sterile 1.4–1.8 mm); anthers magenta, 1.6–2 mm; sterile anthers 0.8–1 mm. **Nectarial disc** very broad and glabrous. **Ovary** 1.1–1.3 mm; ovules 9–13 per locule; style 2.5–5.2 mm; stigma noticeably larger in ♀ flowers. **Capsules** Dec–Apr, subacute or obtuse, (2–)3–5.5 × 2.2–3.3 mm, loculicidal split extending ⅓–¾-way to base (frequently less than ½-way). **Seeds** flattened, ± broad ellipsoid, ± smooth, brown (sts pale), 1.1–1.2 × 0.9–1.7 mm, MR 0.3–0.5 mm. $2n = 120$.

DISTRIBUTION AND HABITAT Mountains of southern SI, mostly on or west of the Main Divide, from Lk. Sweeney and the Mataketake Ra. to near Centre Pass, north of Dusky Sound. It grows in alpine shrubland and grassland, sometimes in rocky places.

NOTES Most similar to *H. cryptomorpha* and *H. crenulata* (see notes under that species), neither of which occurs in Fiordland. Flavonoid and chromosome differences (Table 5) are important in separating the three species (Bayly et al. 2002). Morphological differences from *H. crenulata* include: consistently entire leaf margins; leaf midribs that are glabrous or more finely pubescent above (hairs up to *c.* 0.075 mm long); and mostly elliptic leaves. Morphological differences from *H. cryptomorpha* include: mostly elliptic leaves that are generally broader in relation to their width, and smaller in terms of mean leaf length, length of the longest leaves, and the mean distance from a leaf base to its widest point (Fig. 130; Appendix 4).

Similar species from southern SI (also with an acute sinus in the leaf bud) are *H. arganthera* and *H. dilatata*. *H. cockayneana* is distinguished from the former by its darker anthers, usually smaller leaves, leaf margins that are glabrous or have minute glandular hairs (Fig. 132B), and leaves that are usually glossy above and glaucous below. This last feature, together with a bushier habit, usually also distinguishes it from *H. dilatata* (in which the leaves vary in glaucousness, but usually not strikingly between the upper and lower surfaces).

ETYMOLOGY Honours Leonard C. Cockayne (1855–1934), pioneering New Zealand botanist.

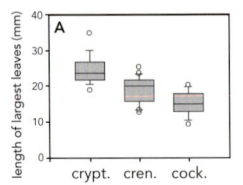

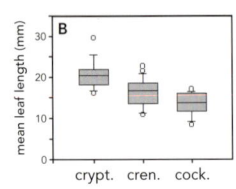

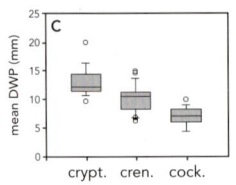

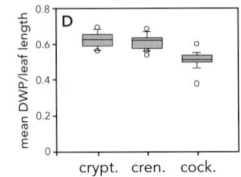

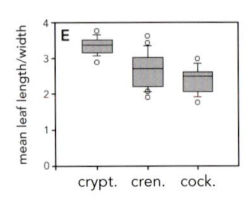

FIG. 130 Box plots showing variation in: **A** maximum leaf length; **B** mean leaf length; **C** mean DWP; **D** mean DWP/leaf length; **E** mean leaf length/leaf width for *Hebe cryptomorpha* (crypt.), *H. crenulata* (cren.) and *H. cockayneana* (cock.). Plots indicate the median, 25th and 75th percentiles, with the 10th and 90th percentiles shown as error bars, and outlying data points as circles. DWP = distance from a leaf base to its widest point. Data from herbarium specimens listed by Bayly et al. (2002).

Hebe cockayneana — Small-leaved APERTAE

FIG. 131 **A** plant near Lk. Harris, head of Routeburn Va. **B** sprig. **C** leaf bud with acute sinus. **D** lower and upper leaf surfaces. **E** frontal view of hermaphrodite flower. **F** lateral view of hermaphrodite flower. **G** frontal view of female flower. **H** lateral view of female flower. **I** inflorescences/infructescences, flowers are opposite-decussate, those on left inflorescence are sessile, those on right pedicellate. **J** capsule in loculicidal (left) and septicidal view.

TAXONOMIC TREATMENT 249

71. *Hebe arganthera*   Garn.-Jones, Bayly, W.G.Lee et Rance

DESCRIPTION Bushy shrub to 1.5 m tall (or hanging branches to 2 m long). **Branches** ascending to erect, old stems brown; branchlets green, pubescent, hairs bifarious; internodes (2–)4–8 mm; leaf decurrencies evident; leaves abscising at (us.) or above nodes (rarely; leaving lower part of petioles attached to stem). **Leaf bud** distinct; sinus narrow and acute. **Leaves** erecto-patent to patent; *lamina* oblong to elliptic or oblanceolate, coriaceous or subcoriaceous, m-shaped in TS, (12–)15–30(–38) × (5–)6–11 mm; *apex* subacute and mucronate; *margin* cartilaginous, minutely papillate and ciliolate to ciliate (with egl. hairs only); *upper surface* green or yellowish-green, dull, with many stomata, hairy along midrib; *lower surface* light green; *petiole* 3–4 mm, hairy above and below. **Inflorescences** with 9–20(–25) flowers, lateral, unbranched, (2–)4–5 cm, about equal to or longer than subtending leaves; peduncle 0.8–1.5 cm; rachis 1.5–3.5 cm. **Bracts** opposite and decussate or opposite and decussate below and becoming alternate above, linear to lanceolate, subacute to acute. **Flowers** Dec–Jan(–May), ⚥. **Pedicels** (0.5–)1.5–3(–5) mm. **Calyx** 2.5–3.5 mm; *lobes* lanceolate to elliptic, obtuse, apparently egl. ciliolate or with mixed gl. and egl. cilia (gl. hairs few). **Corolla** tube glabrous, *c.* 1.8–2.5 × *c.* 2.5–3.5 mm, cylindric, slightly shorter to slightly longer than calyx; *lobes* white at anthesis, elliptic or ovate, obtuse, suberect to recurved, longer than corolla tube. **Stamen filaments** 4–6.5 mm; anthers white or pale yellow, *c.* 1.5 mm. **Ovary** 1.5–2 mm; ovules 20–25 per locule, in 1–2 layers; style 5–7 mm. **Capsules** Mar, acute, 3–4 × 2.5–3.5 mm, loculicidal split extending ¼–⅓-way to base (capsule valves with pronounced midrib). **Seeds** flattened, ellipsoid to discoid, straw-yellow to pale brown, 0.6–1.1 × 0.5–0.9 mm, MR *c.* 0.2 mm. $2n = 40$.

DISTRIBUTION AND HABITAT Known from four widely separated sites in eastern Fiordland NP, SI: Lk. Wapiti; Takahe Va., Murchison Mts; Doubtful Sound above Kellard Point; and Lk. Monk, Cameron Mts. It grows around bluffs of calcium-rich rocks near to the tree-line.

NOTES Can be distinguished from other similar species of small-leaved "Apertae" by the combination of the short eglandular hairs on its leaf margins (Figs 132A and 133D), its pale anthers (Figs 133E–G), and its comparatively large, dull green leaves.

ETYMOLOGY Gk (*argos* = white), refers to the pale anthers, which are a distinctive feature of the species and unusual in the genus.

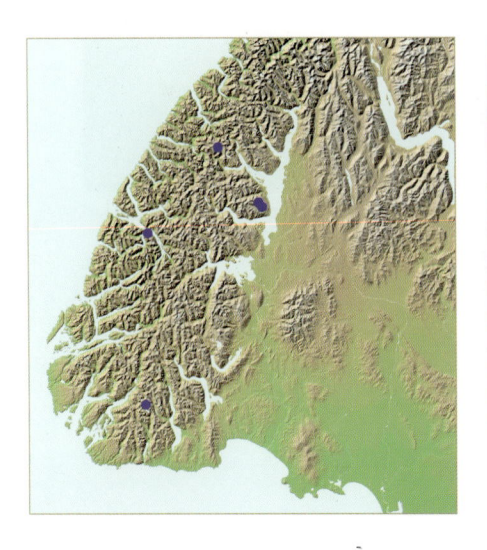

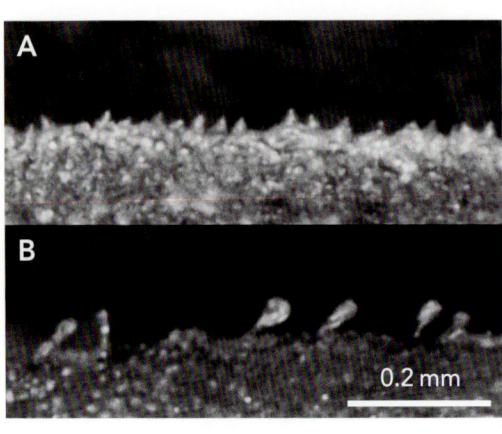

FIG. 132 Leaf margins. **A** *Hebe arganthera* with short eglandular cilia. **B** *H. cockayneana* with longer glandular cilia.

Hebe arganthera — Small-leaved APERTAE

FIG. 133 A plant in Takahe Va., Murchison Mts. B sprig. C leaf bud with acute sinus. D lower and upper leaf surfaces; magnified inset shows minute hairs on leaf margin. E inflorescence. F frontal view of flower. G lateral view of flower. H infructescence.

TAXONOMIC TREATMENT 251

72. *Hebe diosmifolia* (A.Cunn.) Andersen

DESCRIPTION Bushy shrub (us.) or small tree (according to herbarium notes) to 2.5(–6) m tall. **Branches** erect or spreading, old stems light brown to grey; branchlets green, puberulent, hairs us. uniform or sts tending bifarious; internodes (0.5–)1–9(–16) mm; leaf decurrencies evident. **Leaf bud** distinct; sinus narrow to broad, acute. **Leaves** subdistichous, patent; *lamina* narrowly oblong-elliptic or oblong or linear-lanceolate, rigid, flat or slightly concave, (3–)8–20(–30) x (2–)3–6 mm; *apex* obtuse to acute or somewhat acuminate; *margin* ciliate (with egl. and/or gl. hairs) and often minutely papillate, shallowly toothed (us.) or entire; *upper surface* green, dull, without evident stomata (us.) or with few stomata, hairy along midrib (us.) or glabrous (rarely); *lower surface* light green; *petiole* (0.3–)1–4 mm, almost always hairy above and us. hairy along margins. **Inflorescences** with (4–)10–54 flowers, lateral, with three or more branches (almost always) or unbranched (rarely, and never all inflorescences on a plant), (1–)1.5–3.5(–5.5) cm; peduncle (0.3–)0.4–1.8(–2.3) cm; rachis (0.4–)1–2–3.1 cm. **Bracts** opposite and decussate or subopposite to alternate, ovate or lanceolate, subacute or acute. **Flowers** (Aug–)Sep–Jan[–Jul], ⚥. **Pedicels** longer than or equal to bracts, 1.5–4.2 mm. **Calyx** 1.5–2.6 mm, 3–4(–5)-lobed (5th lobe small, posterior), with anterior lobes free for most of their length or united between ⅓ and all the way to apex; *lobes* ovate to deltoid or elliptic, subacute to obtuse or emarginate (fused anterior lobes), us. with mixed gl. and egl. cilia. **Corolla** tube glabrous, 2–2.5 x 1.5–2 mm, funnelform, longer than (us.) or equalling calyx; *lobes* pinkish-mauve or blue or white at anthesis, ovate to deltoid, subacute, patent, longer than corolla tube; corolla throat white or mauve. **Stamen filaments** 4.5–6 mm; anthers mauve or pink or buff, 1.2–1.7 mm. **Ovary** 0.8–1.1 mm; ovules *c.* 9–11 per locule; style 4–8 mm. **Capsules** (Sep–)Oct–May(–Aug), acute or subacute, 3.5–5.4 x (1.8–)2.3–3.7 mm, loculicidal split extending ¼–½-way to base. **Seeds** strongly flattened, broad ovoid or obovoid to discoid, not winged to only weakly winged, straw-yellow, 1.2–2 x 0.9–1.3 mm, MR 0.2–0.6 mm. $2n$ = 40 or 80.

DISTRIBUTION AND HABITAT Northern NI, from Cape Reinga to Woodhill Forest. It grows in lowland scrub, and at forest margins, often in near-coastal situations or near riverbanks.

NOTES A distinctive but variable species that is widely cultivated. It most closely resembles *H. divaricata*, from which it is geographically separated and can be distinguished by having anterior calyx lobes (Fig. 134G) that are partly or wholly fused (at least on some flowers on all specimens). The leaves are often minutely toothed and, when present, these teeth also distinguish specimens from *H. divaricata*. Both of these features, together with the size and colour of the leaves, acute leaf bud sinus, and branched inflorescences, readily distinguish the species from all others of northern New Zealand.

Both diploid and tetraploid populations exist, and there is marked variation in flowering time, habit (Figs 134A and E), shape and size of leaves (Appendix 4), number and size of marginal incisions on leaves, the degree of inflorescence branching, pedicel length (affecting the general appearance of inflorescences) and the degree of fusion of anterior calyx lobes. Two segregate varieties have been proposed (Kirk 1896; Carse 1929), and some authors (e.g. Druce 1980; Eagle 1982) suggest the current circumscription potentially includes two distinct species, but the results of a detailed morphological and cytological study (Newman 1988; Murray et al. 1989) do not support any of these proposals. Plants from more northern localities tend to flower later (December–January) than those from southern localities (September–October), even in cultivation under uniform conditions. The known tetraploids are also all from northern populations. There are, however, populations with intermediate flowering times, some populations with anomalous flowering times (e.g. the southernmost population, at least sometimes, flowers in February), and there is geographic overlap in the distribution of chromosome races. Variation in other characters shows no consistent patterns, is not well correlated with flowering time or chromosome number, and provides no clear grounds for the recognition of segregate taxa.

Colenso (1883) suggested that the species occurs in the Kaweka Ra., Hawke's Bay, based on a collection attributed to Augustus Hamilton (WELT 5352, WELT 79807, K; distributed by Colenso under the name *Veronica trisepala*). This is an unlikely locality for the species, and it is not included on the distribution map (also see comments by Moore, in Allan 1961).

ETYMOLOGY L., implies that the leaves (*folium* = leaf) are similar to those of *Diosma* (a South African plant genus, family Rutaceae).

Hebe diosmifolia

Small-leaved APERTAE

FIG. 134 **A** low-growing, bushy plant, in coastal heathland on Cape Reinga (flowering plant at lower right is *H. ligustrifolia*). **B** sprig. **C** leaf bud with acute sinus. **D** lower and upper leaf surfaces. **E** erect, open plant with slightly weeping branches, on banks of Waipoua R. **F** branched inflorescence and infructescence. **G** flower showing fused anterior calyx lobe. **H** inflorescence showing frontal view of flowers. **I** seeds.

73. *Hebe venustula*   (Colenso) L.B.Moore

DESCRIPTION Bushy shrub to 1.4(–1.8) m tall. **Branches** erect, old stems dark brown or grey; branchlets green, puberulent, hairs bifarious; internodes (1–)2–6(–11) mm; leaf decurrencies evident and often swollen. **Leaf bud** distinct; sinus narrow to broad, acute. **Leaves** decussate to somewhat subdistichous, erect to erecto-patent; *lamina* elliptic to obovate, rigid or coriaceous, concave, (4–)8–20(–29) × (3–)5–8(–10) mm; *apex* subacute to obtuse; *midrib* thickened below and at least slightly depressed to grooved above; *margin* sts cartilaginous, us. minutely papillate and sts also ciliate; *upper surface* dark green, glossy, with few or without evident stomata, glabrous or hairy along midrib or hairy toward base; *lower surface* green; *petiole* (0.5–)1–3(–6) mm, glabrous or hairy along margins or above. **Inflorescences** with 7–75 flowers, lateral, unbranched (often exclusively) or with 3 or more branches, 1.4–5(–6.8) cm; peduncle 0.4–1.4 cm; rachis 1–3.5(–5.7) cm. **Bracts** opposite and decussate or lowermost pair opposite, then subopposite or alternate above, lanceolate (us.) or ovate, subacute or acute. **Flowers** (Dec–)Jan–Feb(–Mar), ⚥. **Pedicels** 0.5–3(–7) mm, hairy (us.) to almost glabrous. **Calyx** 2.3–2.8(–4) mm; *lobes* ovate or oblong or rarely lanceolate, subacute. **Corolla** tube hairy inside, 3–4.2 × 2–2.6 mm, cylindric or funnelform, longer than calyx; *lobes* white tinged with mauve at anthesis, often white with age, ovate (sts broadly) or elliptic or lanceolate, obtuse, patent to recurved, longer to shorter than corolla tube, sts with a few hairs toward base on inner surface. **Stamen filaments** white or faintly coloured, 4–7 mm; anthers mauve to magenta, 2.1–2.6 mm. **Ovary** *c.* 0.9–1.3 mm; ovules *c.* 8–10 per locule, in 1(–2) layers; style 6.5–8(–11) mm. **Capsules** Feb–Apr, subacute, 3.5–5 × 2.3–3.5 mm, loculicidal split extending ⅓–½-way to base. **Seeds** flattened (sts strongly), ellipsoid or ovoid or oblong, not winged to ± winged, brown, (1.3–)1.5–2.2 × 1–1.3(–1.5) mm, MR 0.3–0.6 mm. $2n = 120$.

DISTRIBUTION AND HABITAT Mountains of NI, including the Raukumara Ra., volcanoes of central NI, Mt Taranaki, and the Pouakai, Kaimanawa, Kaweka, Ruahine and Aorangi ranges. It occurs chiefly in subalpine shrubland/penalpine grassland, but apparently occurs at lower altitudes in the eastern Wairarapa.

NOTES Distinguished from most other species, especially on NI, by the combination of: acute leaf bud sinuses; non-glaucous, entire leaves; pedicellate flowers; small bracts; and corolla tubes longer than surrounding calyces. Specimens are sometimes misidentified as *H. odora* (and vice versa), but the species is probably most closely related to, and not clearly morphologically separated from, *H. brachysiphon* (see notes under those species). Specimens from eastern Wairarapa Taipos often have narrow, acute leaves, and highly branched inflorescences. In these respects they resemble *H. divaricata*, and their relationship to that species is worthy of closer scrutiny.

A specimen resembling *H. venustula* is labelled "Kapiti Island" (WELT 13295). Although *Veronica laevis* (a synonym of *H. venustula*) was included in a list of plants introduced to Kapiti Id between 1924 and 1943 (Wilkinson & Wilkinson 1952), no similar specimens are known from this island, or recorded in vegetation surveys (e.g. Fuller 1985), and the specimen is not represented on the distribution map. Another specimen, labelled "Mount Holdsworth, beside Powell Hut" (AK 51014), is also, in the absence of further evidence for the species' occurrence in that well-collected area, not represented on the distribution map.

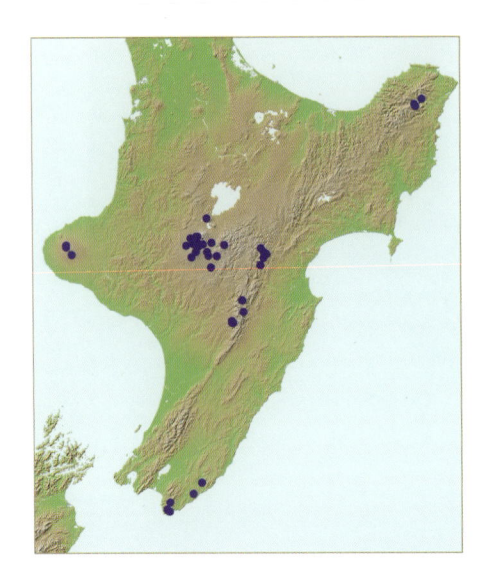

ETYMOLOGY L. (*venustulus* = charming, lovely).

Hebe venustula — Small-leaved APERTAE

FIG. 135 A plant on Makahu Spur, Kaweka Ra. B sprig. C leaf bud with acute sinus. D lower and upper leaf surfaces. E inflorescence, showing pedicellate flowers. F frontal view of flower with narrow corolla lobes. G frontal view of flower with broader corolla lobes. H lateral view of flower. I capsule in septicidal (left) and loculicidal view.

74. *Hebe brachysiphon*   Summerh.

DESCRIPTION Bushy shrub (often closely branched with a compact habit, but sts more open) to 1.8 m tall. **Branches** erect, old stems brown or grey; branchlets green (quickly becoming brown with age), puberulent, hairs bifarious (us.) or uniform; internodes 2–7(–10.5) mm; leaf decurrencies evident. **Leaf bud** distinct; sinus narrow and acute, small and acute, or small and rounded (rarely). **Leaves** decussate to subdistichous, erecto-patent to patent; *lamina* elliptic (sts narrowly) or oblanceolate or ovate or obovate or oblong, rigid or subcoriaceous, concave, (5.5–)8.5–25.5 × 3.3–8(–10) mm; *apex* acute or subacute to occasionally obtuse; 2 lateral *secondary veins* sts evident at base of fresh leaves; *midrib* thickened below and slightly depressed to grooved above; *margin* ciliolate or ciliate (with egl. and/or gl. hairs; occasionally glabrous on older leaves); *upper surface* light to dark green, glossy or dull, with many stomata, hairy along midrib; *lower surface* lighter green than upper surface; *petiole* (0.5–)0.8–3(–3.5) mm, hairy along margins and sts hairy above. **Inflorescences** with (9–)14–36 flowers, lateral, unbranched (us.) or tripartite, (1.2–)1.7–4.1 cm; peduncle 0.5–1.3 cm; rachis 0.6–3.1 cm. **Bracts** lowermost pair opposite, then subopposite or alternate above (us.) or alternate, ovate to lanceolate or deltoid (lowermost), obtuse or subacute (us.) or acute (lowermost). **Flowers** (Oct–)Dec–Feb(–Mar), ⚥ or ♀ (on different plants). **Pedicels** 0.6–2.5(–3) mm, hairy or sts almost glabrous. **Calyx** 1.7–3 mm, 4–5-lobed (5th lobe small, posterior); *lobes* elliptic or ovate, subacute or obtuse or occasionally emarginate. **Corolla** tube hairy inside; tube of ⚥ flowers 2.2–3.5(–4) × 1.7–2.2 mm, cylindric to funnelform, longer than (us.) or equalling calyx; tube of ♀ flowers 1.6–3 × 1.4–1.8 mm, funnelform, equalling or longer than calyx; *lobes* white at anthesis (but sts pale mauve in bud), elliptic to oblong or circular or ovate or obovate (occasionally), obtuse, patent, longer than (us.) to shorter than corolla tube. **Stamen filaments** 1–4.5 mm (1–1.3 mm on ♀ flowers, 3–4.5 mm on ⚥ flowers); anthers mauve or purple, 2.2–2.3 mm; sterile anthers 1–1.3 mm. **Ovary** sts hairy, 0.9–1.2 mm; ovules 8–13 per locule; style 4.2–7.2 mm. **Capsules** Jan–May(–Dec), subacute, 3–6 × 2.3–4.5 mm, sts hairy, loculicidal split extending ¼–½-way to base. **Seed** characters not recorded. $2n = 120$.

DISTRIBUTION AND HABITAT Mountains of Marlborough and Canterbury, from the Red Hills Ridge to near Mt Hutt. It grows in subalpine shrubland and in beech forest close to the tree-line.

NOTES Distinguished from most species by the combination of: a bushy, often rounded habit; us. acute leaf bud sinuses; non-glaucous leaves; pedicellate flowers; small bracts; and corolla tubes us. longer than surrounding calyces.

H. brachysiphon is probably most closely related to *H. venustula*, in which it was included by Druce (1980), Smith-Dodsworth (1991) and Wilson & Galloway (1993). The two are morphologically similar and share the same hexaploid chromosome number. They are retained here as distinct species because differences in flavonoid chemistry (Mitchell et al. in prep.), together with their geographic separation (NI/SI), suggest genetic differentiation and reproductive isolation. Plants of *H. brachysiphon* often have more conspicuous stomata on the upper leaf surface than *H. venustula* (cf. Figs 136D and 135D) and a relatively broader petiole, but these differences are not consistent; reliable characters that morphologically distinguish specimens of the two species are unknown.

The morphological and distributional limits of *H. brachysiphon* and *H. divaricata* also require clarification. In general, *H. brachysiphon* is a rounded shrub (Fig. 136A) of subalpine areas and has simple (Figs 136B and E), or predominantly simple, inflorescences, whereas *H. divaricata* is a plant of variable, though usually more open, habit (Figs 137A and F) that occurs chiefly in montane to lowland areas, and has inflorescences that are mostly branched (Figs 137E and H). *H. brachysiphon* often also has leaves that are comparatively broader (relative to their length) and less acute at their apices (cf. Figs 136D and 137D; Appendix 4). However, despite this, and the fact that a study of chromosome numbers (de Lange et al. 2004c; Table 5) suggests consistent differences between the species, some specimens are difficult to identify with certainty and are omitted from both distribution maps. These specimens are either vegetative (i.e. they do not provide inflorescence characters for comparison), or have a mixture of branched and simple inflorescences. These unmapped specimens are mostly from eastern Marlborough, or from the area between Hanmer Springs and Nelson Lakes. It is likely that one or both species are more widespread in these areas than the distribution maps suggest.

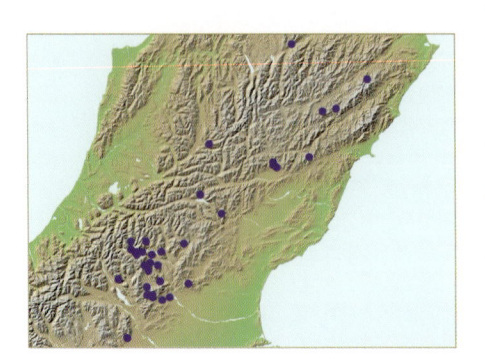

Specimens have sometimes been misidentified as *H. odora* and vice versa (see notes under that species).

ETYMOLOGY Gk, means short-tubed (*brachy-* = short, little; *siphon* = reed, straw, tube); refers to the corolla, particularly in comparison to *H. traversii* with which it was previously confused.

Hebe brachysiphon — Small-leaved APERTAE

FIG. 136   A plant near Lk. Lyndon. B sprig. C leaf bud with acute sinus. D lower and upper leaf surfaces; insert shows stomata on upper surface. E inflorescence. F frontal view of flowers. G lateral view of flower. H septicidal view of capsule. I loculicidal view of capsule.

75. *Hebe divaricata* (Cheeseman) Cockayne et Allan

DESCRIPTION Bushy shrub to 1.8 m tall. **Branches** erect and spreading, old stems brown or grey; branchlets green (often with reddish or dark bands at nodes), puberulent, hairs bifarious (us.) or uniform; internodes 1.7–11 mm; leaf decurrencies obscure or evident. **Leaf bud** distinct; sinus narrow to broad, acute. **Leaves** decussate or subdistichous, erecto-patent or patent; *lamina* elliptic to lanceolate or oblanceolate (sts), rigid or subcoriaceous, concave, (6.3–)9–32(–38.2) × (2.8–)3.5–6(–7.5) mm; *apex* acute to subacute; *midrib* depressed to grooved above and thickened below (slightly, often keeled near apex); *margin* minutely papillate and/or ciliolate (with short, stiff hairs) and/or gl.-ciliate; *upper surface* green, glossy or dull, without evident stomata (or, occasionally, with a few scattered stomata apparently in sunken pits), hairy along midrib; *lower surface* lighter green than upper surface; *petiole* (0.8–)1.5–3.5(–4.5) mm, hairy along margins and above. **Inflorescences** with 24–88 flowers, lateral, with 3 or more branches (occasionally with undeveloped branches in lowest bracts) or unbranched (rarely, and never all inflorescences on one plant), 1–3.6(–5.9) cm; peduncle 0.35–1.4 cm; rachis 0.7–2.5(–4.8) cm. **Bracts** alternate (us.) or lowermost pair opposite, then subopposite or alternate above, ovate to lanceolate or oblong or lowermost sts linear, subacute to acute. **Flowers** Dec–Feb(–Mar), probably ⚥. **Pedicels** 0.5–2(–4) mm. **Calyx** 1.9–3.4 mm, 4–5-lobed (5th lobe small, posterior); *lobes* elliptic to ovate or lanceolate, subacute to obtuse or emarginate (sts). **Corolla** tube hairy inside, (2.1–)2.5–4.3 × 1.2–2.6 mm, cylindric or funnelform (narrowly), longer than calyx; *lobes* white or pink at anthesis and with age, elliptic to ovate, obtuse, suberect to patent, longer to shorter than corolla tube. **Stamen filaments** 3.5–5.5 mm; anthers pink or mauve, 1.6–2.6 mm. **Ovary** 1.1–1.5 mm; ovules 7–11 per locule; style 4.5–8.5 mm. **Capsules** Jan–May(–Oct), subacute, 2.9–4.5 × 1.9–3.2 mm, loculicidal split extending ¼–½-way to base. **Seeds** flattened, broad ellipsoid to discoid, brown (sts with an orange component), 1.2–2 × 0.8–1.4 mm, MR 0.3–0.5 mm. $2n = 80$.

DISTRIBUTION AND HABITAT Nelson and Marlborough, SI, from the Aorere R. to near Nelson Lks and also on D'Urville Id. It grows in a range of habitats, including riverbanks, rock outcrops of various substrates (e.g. ultramafic or calcareous), scrub and beech forest margins, chiefly in montane to lowland situations, but sometimes subalpine.

NOTES Highly variable in terms of habit (e.g. Figs 137A and F) and leaf size. It resembles *H. diosmifolia*, *H. brachysiphon* and some populations of *H. venustula* (see notes under those species). It can be difficult to distinguish from *H. brachysiphon* in particular, and for this reason the southern and eastern limits of the species are unclear (i.e. it may extend further south and east than indicated on the distribution map). It frequently grows with *H. leiophylla*, from which it is distinguished by the usual presence of branched inflorescences (Figs 137E and H), a more elongated sinus (cf. Figs 137C and 140C) and less hairy branchlets (cf. Fig. 140B).

ETYMOLOGY L. (*divaricatus* = spreading at a wide angle), probably refers to the plant habit and the arrangement of branches.

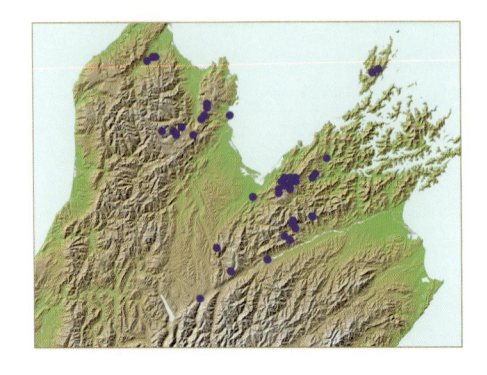

Hebe divaricata — Small-leaved APERTAE

FIG. 137 **A** plant in a sheltered site, Richmond Ra. **B** sprig. **C** leaf bud with acute sinus. **D** lower and upper leaf surfaces. **E** flowering sprig. **F** plant in an exposed site, Richmond Ra. **G** lateral view of flower. **H** young inflorescence and infructescence showing branching pattern. **I** frontal view of recently opened flower with stigma enfolded in anterior corolla lobe. **J** frontal view of older flower with stigma exposed.

TAXONOMIC TREATMENT 259

76. *Hebe insularis* (Cheeseman) Cockayne et Allan

DESCRIPTION Spreading low shrub to 0.5(–1) m tall. **Branches** decumbent or erect or sts pendent, old stems brown; branchlets red-brown, densely pubescent, hairs uniform to bifarious; internodes (1–)4–10.5 mm; leaf decurrencies evident (us. with ± faint ridges running along medial line and margins, with a shallow groove forming between the decurrencies of a leaf pair). **Leaf bud** distinct; sinus absent, or small and rounded to narrow and acute. **Leaves** erecto-patent to patent (us.) or recurved; *lamina* elliptic (us.) or slightly obovate, coriaceous or fleshy, flat or slightly concave, 7.5–32.5 × 3.7–13.5 mm; *apex* subacute or obtuse; *margin* ciliolate or glabrous; *upper surface* green to dark green and often slightly glaucous or glaucescent, dull, with many stomata, glabrous (us.) or hairy along midrib; *lower surface* green, glabrous (us.) or hairy along midrib; *petiole* glabrous (us.) or hairy along margins or above. **Inflorescences** with 7–46 flowers, lateral, with 3 or more branches, 2.2–4 cm, longer than (us.) or about equal to subtending leaves (rarely); peduncle 0.7–1.7 cm; rachis 1.2–2.8 cm. **Bracts** opposite and decussate below and becoming alternate above, deltoid (sts narrowly) or ovate or lanceolate, obtuse to acute. **Flowers** Nov–Dec(–Feb), ⚥. **Pedicels** 0.5–5.5 mm. **Calyx** 2–4 mm, with anterior lobes free for most of their length (mostly) or united ⅓–⅔-way to apex (degree of fusion of anterior lobes varies on one inflorescence); *lobes* lanceolate or ovate, obtuse to acute, egl. ciliate (almost always) or with mixed egl. and occasional gl. cilia. **Corolla** tube glabrous, 2.9–4 × 2.2–3 mm, funnelform and contracted at base, longer than calyx; *lobes* white or tinged mauve at anthesis (often almost purple when young), broadly ovate, obtuse, suberect to recurved, longer than or equalling corolla tube. **Stamen filaments** 4.8–6.5 mm; anthers magenta, 1.8–2.2 mm. **Ovary** *c.* 0.9–1.1 mm; ovules 12–29 per locule, marginal on a flattened placenta (but us. recurved and appearing somewhat scattered), in 1–3 layers; style 3.5–7.2 mm. **Capsules** Dec–Mar[–Aug], subacute or obtuse, 2.5–4.5 × 2–3 mm, septicidal split sts extending only ¾-way to base, loculicidal split extending ¼–½-way to base. **Seeds** flattened, ± ellipsoid, finely papillate, pale brown, 1–1.2 × 0.6–0.8 mm, MR 0.2–0.3 mm. $2n = 40$ (+ fragment).

DISTRIBUTION AND HABITAT Endemic to the Three Kings Ids, northern New Zealand, where it is known from West, South West, Great and North East islands. It grows on rock outcrops and cliffs, especially near the coast.

NOTES A distinctive species recognised by the combination of: leathery, elliptic to obovate leaves, which are often ± glaucous or glaucescent; and branched inflorescences (Figs 138H and I). The leaf bud sinus is either small or absent (Figs 138C and D). Specimens show some variation in the degree of hairiness of leaves and branchlets. It is the only *Hebe* recorded from the Three Kings Ids.

ETYMOLOGY L. (*insularis* = pertaining to islands, insular), refers to the distribution of the species.

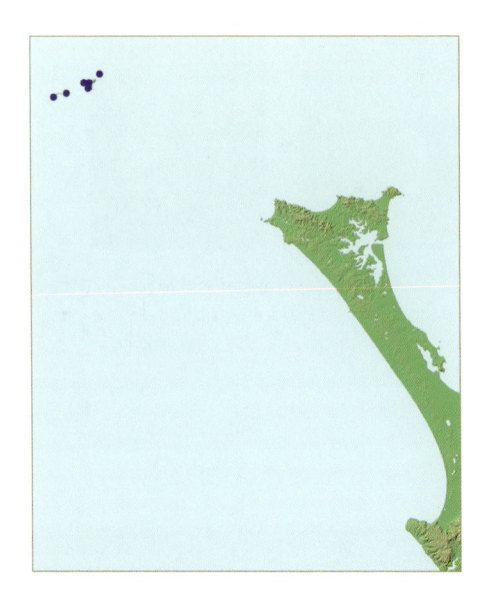

Hebe insularis | Small-leaved APERTAE

FIG. 138   **A** plants on Great Id, Three Kings Ids. **B** sprig. **C** leaf bud with small sinus. **D** leaf bud with no sinus. **E** lower and upper leaf surfaces. **F** lateral view of flower. **G** close-up of inflorescence showing frontal view of flowers. **H** flowering sprig. **I** branched infructescence.

77. *Hebe elliptica* (G.Forst.) Pennell

DESCRIPTION Bushy shrub to 2 m tall. **Branches** erect, old stems brown; branchlets green or red-brown or reddish-black (initial cork formation often in regions between decurrencies), pubescent, hairs strictly bifarious or uniform; internodes (1–)4–13(–17.5) mm; leaf decurrencies evident (and often with a narrow ridge along medial line). **Leaf bud** distinct; sinus square to oblong. **Leaves** decussate or sts ± subdistichous (with petioles twisted so that leaves face in ± one direction), erecto-patent to patent; *lamina* broadly to narrowly elliptic or oblong or obovate or oblanceolate, coriaceous, flat or m-shaped in TS, (5–)12–31(–42) × (3–)6–12(–18) mm; *apex* plicate and mucronate or acute; *base* cuneate to truncate; *margin* sts cartilaginous, conspicuously long-pubescent (with dense, tangled hairs; except at apex), entire or minutely crenulate; *upper surface* green or dark green, dull or slightly glossy, with many stomata, minutely hairy along midrib; *lower surface* light green; *petiole* 1–4(–8.5) mm, glabrous or sts hairy along margins (but hairs much shorter and more sparse than those on rest of leaf margin). **Inflorescences** with (3–)6–14 flowers, lateral, unbranched, 1.5–5.1 cm, shorter to longer than subtending leaves; peduncle 0.4–1.7 cm; rachis 1.1–3.6 cm. **Bracts** alternate (lowermost often a ± subopposite pair or a slightly offset "whorl" of three), deltoid, acute or subacute. **Flowers** (Aug–)Nov–Mar(–Jun), ⚥. **Pedicels** (1.5–)3–8(–9) mm. **Calyx** (3.5–)4–6.5 mm; *lobes* lanceolate or ovate or elliptic, obtuse to acute, with mixed gl. and egl. cilia (egl. most conspicuous, often long and tangled). **Corolla** tube hairy inside or glabrous, 3–4 × 3.5–4 mm, shortly and broadly funnelform, shorter than or equalling calyx; *lobes* mauve or blue at anthesis, ovate or elliptic, obtuse or subacute, patent to recurved, longer than corolla tube. **Stamen filaments** white or mauve, 4.5–5.5 mm; anthers mauve, 2.4–3.2 mm. **Nectarial disc** glabrous or densely ciliate. **Ovary** 1.7–2 mm; ovules 45–61 per locule, in 2–3 layers; style 4–6.5 mm. **Capsules** Nov–Apr(–Oct), subacute, 5.5–8.5 × (3.5–)4–5.5 mm, loculicidal split extending ¼–½-way to base (mostly ¼–⅓). **Seeds** flattened, broad ellipsoid to discoid, winged or not winged, straw-yellow to brown, 0.9–2 × 0.9–1.5 mm, MR 0.3–0.5 mm. $2n = 40$.

DISTRIBUTION AND HABITAT Occurs naturally on NI, SI, Stewart Id, Solander Id, Snares Ids, the Auckland Ids, Campbell Id, Falkland Ids and in southern South America (South American distribution is mapped separately on Fig. 141); naturalised in France (Webb 1972) and Tasmania (Rozefelds et al. 1999). On NI it occurs on the west coast south from Taranaki; on SI it occurs on the north, west and south coasts, and on the east coast as far north as Oamaru. It grows in coastal areas, often in exposed places on rocks.

NOTES Distinguished from other species by the combination of: large flowers; a prominent leaf bud sinus; robust, elliptic, oblong, obovate or oblanceolate leaves; and leaf margins conspicuously pubescent except on petioles and plicate-mucronate apices (Fig. 139F). Plants vary throughout the species' range in terms of overall size, leaf shape and size (Appendix 4), leaf thickness, internode length, branchlet pubescence and flower colour. Moore (in Allan 1961) also reported that some cultivated specimens from South America have terminal, as well as lateral, inflorescences.

Plants with broad, fleshy leaves (e.g. Fig. 139E, lower leaves) from Kapiti Id and Titahi Bay, Wellington, were described by Cockayne & Allan (1926c) as a distinct variety, var. *crassifolia*. That variety is not considered sufficiently distinct, given variation in the species, to be formally recognised here.

There are specimens labelled "Lyttleton" in the Armstrong Herbarium at CHR, and two in WELT (44110, Herb. T. Kirk; 5298, *G. Mair*) apparently from the Chatham Ids. It is not unreasonable that the species occurs/

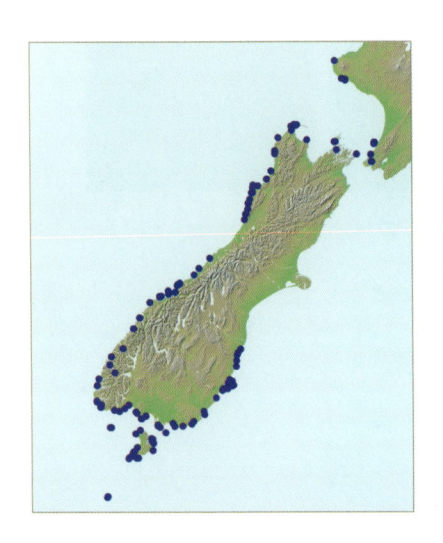

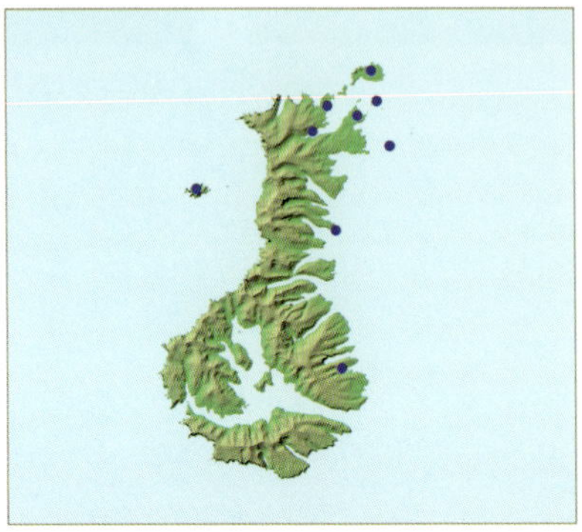

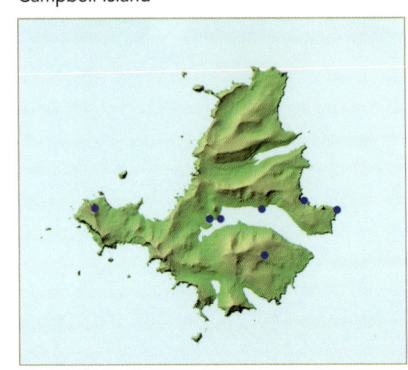

Hebe elliptica — Small-leaved APERTAE

FIG. 139 **A** plant at Pancake Rocks, Punakaiki. **B** sprig. **C** leaf bud with oblong sinus. **D** leaf bud with square sinus. **E** lower and upper surfaces of leaves from cultivated plants originally from: Okarito, Westland (top); Titahi Bay, Wellington. **F** close-up of mucronate-plicate leaf apex; this contrasts with surrounding margins in terms of hairiness. **G** lateral view of flower. **H** frontal view of young flower (anthers not dehisced). **I** frontal view of older, faded flower from same plant as H. **J** septicidal view of capsule. **K** loculicidal view of capsule.

occurred naturally in either area, but if so, it is surprising that its presence in these well-collected localities has not been confirmed by subsequent wild collections, and they are omitted from the distribution map. According to P. J. de Lange (pers. comm. 2005), there are plantings of *H. elliptica* on Chatham Id and the species is currently naturalised there around Waitangi.

The species hybridises with *H. salicifolia* at some sites where they co-occur, particularly on southern SI. It is also one parent of a range of ornamental hybrid cultivars, including the widely grown *H.* ×*franciscana* (Heenan 1994*a*; Metcalf 2001).

ETYMOLOGY L. (*ellipticus* = elliptic), refers to the leaves.

78. *Hebe leiophylla*   (Cheeseman) Andersen

DESCRIPTION Bushy shrub to 3 m tall. **Branches** erect, old stems brown or grey; branchlets green, puberulent, hairs uniform; internodes (2–)4–15(–30) mm; leaf decurrencies evident or obscure. **Leaf bud** distinct; sinus small and rounded, or narrow and acute. **Leaves** erecto-patent to recurved; *lamina* linear-lanceolate or oblong-elliptic, thin or subcoriaceous, flat or slightly m-shaped in TS, (8–)15–40(–56) × (2–)4–8(–10.5) mm; *apex* acute to obtuse; *margin* sts cartilaginous, puberulent; *upper surface* green, dull or glossy, without evident or with few stomata, hairy along midrib and sts covered with minute gl. hairs (when young); *lower surface* green or light green, glabrous or hairy along midrib; *petiole* 0–2(–3) mm, hairy along margins and above and sts below. **Inflorescences** with 14–150 flowers, lateral, unbranched, 3.5–16.5 cm; peduncle 0.5–2.6 cm; rachis 2.5–14.1 cm. **Bracts** alternate, lanceolate to elliptic, subacute to acute. **Flowers** (Dec–)Jan–Apr(–Jun), ♀. **Pedicels** (0.9–)1.5–3 mm. **Calyx** 1.5–2.5(–2.9) mm; *lobes* elliptic or lanceolate or more rarely deltoid or oblong, subacute to obtuse or rarely acute. **Corolla** tube hairy inside and sts sparsely hairy outside, (1.5–)2–3 × 1.7–2.5 mm, funnelform, shorter to longer than calyx; *lobes* white or tinged mauve at anthesis, ovate or elliptic, obtuse, suberect to patent, longer than corolla tube, sts with a few hairs toward base on inner surface. **Stamen filaments** (4.5–)5.5–7.5 mm; anthers mauve or pink, 1.8–2.1 mm. **Ovary** very rarely hairy, *c.* 0.8–1.2 mm; ovules *c.* 11–12 per locule; style 4.7–7 mm, very rarely hairy. **Capsules** (Dec–)Feb–Jun(–Nov), obtuse or subacute, 2.7–4.5(–5) × (1.8–)2.3–3(–3.5) mm, loculicidal split extending ¼–¾-way to base (most *c.* ¼-way). **Seeds** strongly flattened, ± discoid, winged, pale brown, 1.2–1.7 × 1–1.4 mm, MR 0.2–0.4 mm. $2n = 80$.

DISTRIBUTION AND HABITAT Widespread on northern SI, north from the Organ Ra. and Greymouth. It grows in scrub and at forest margins in a range of situations, sometimes in swampy sites, from sea-level to the tree-line.

NOTES A variable species, particularly in leaf shape (Appendix 4), distinguished from most others by the combination of its: leaf size and shape; often small sinus; densely and uniformly hairy branchlets; and simple inflorescences. It sometimes resembles *H. paludosa*, and differences between them are not always clear-cut (see notes under *H. paludosa*). It often grows with *H. salicifolia* (which generally has much larger leaves) and *H. divaricata* (see notes under that species).

This species was included as *H. gracillima* in the treatment of Moore (in Allan 1961), where the name *H. leiophylla*, which has historically been applied to a range of species (including *H. traversii*, *H. strictissima* and *H. rakaiensis*), was placed *incertae sedis*. Examination of type specimens of the two names suggests that they are conspecific and *H. leiophylla*, having priority over *H. gracillima*, is the name that must be adopted under the rules of the ICBN. Any specimens previously identified by us as *H. gracillima* are *H. leiophylla* as treated here.

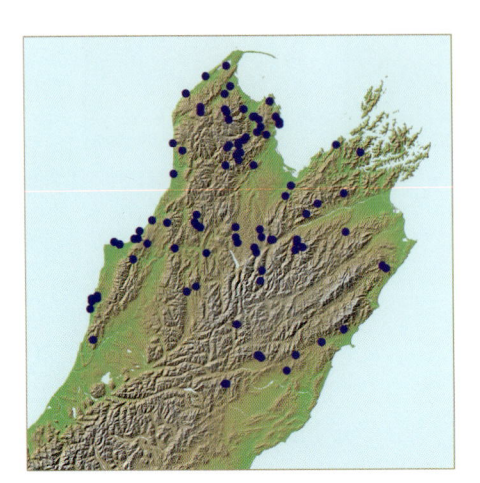

ETYMOLOGY Gk, means smooth-leaved (*leios* = smooth; *phyllon* = leaf).

Hebe leiophylla — **Small-leaved APERTAE**

FIG. 140 **A** plant beside Buller R., near outlet of Lk. Rotoiti, Nelson Lakes. **B** sprig; inset shows close-up of minutely pubescent branchlet. **C** leaf bud with sinus. **D** lower and upper leaf surfaces; magnified inset shows hairs on leaf margin. **E** inflorescence. **F** frontal view of flower. **G** lateral view of flower. **H** septicidal view of capsule. **I** loculicidal view of capsule.

Large-leaved "Apertae"

CRITICAL FEATURES Shrubs, spreading to erect, up to 2.5(–5) m tall; **leaf buds distinct; sinus rounded to square or acute**; leaf bases free; leaves not appressed or scale-like, **usually > 4 cm long**, entire or minutely denticulate, not glaucous, **abscising at nodes; inflorescences lateral**; capsule latiseptate.

NUMBER OF SPECIES 8

DISTRIBUTION NI (including islands of Hauraki Gulf), SI, Stewart Id, Auckland Id (see notes under *H. salicifolia*), Chile; naturalised in Hawai'i, California (de Lange & Cameron 1992) and Europe (Webb 1972)

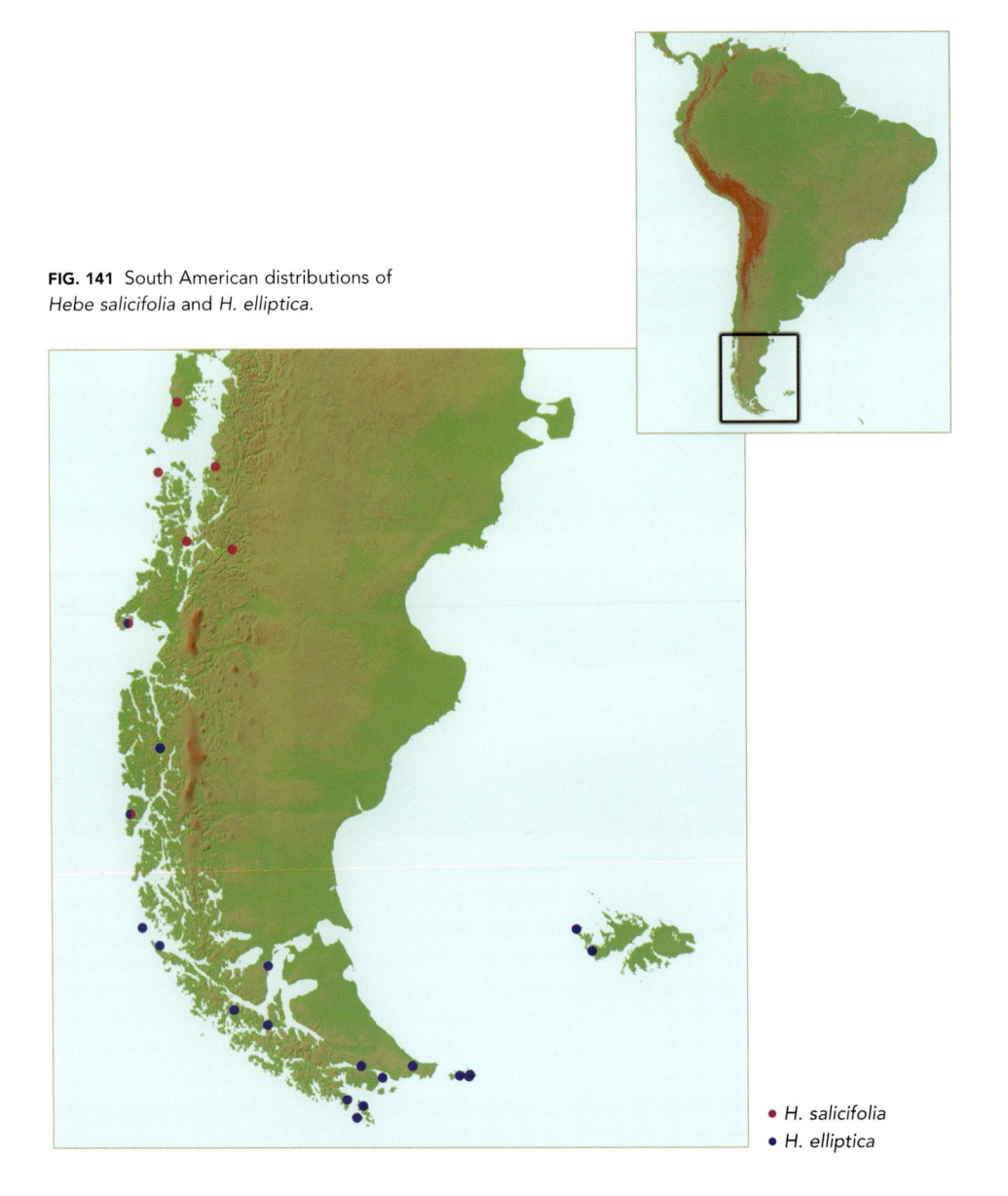

FIG. 141 South American distributions of *Hebe salicifolia* and *H. elliptica*.

- *H. salicifolia*
- *H. elliptica*

Large-leaved APERTAE

KEY TO SPECIES

1 Underside of leaves glaucous (e.g. Fig. 127D); inflorescences mostly branched (e.g. Fig. 127H) .... **67. *Hebe scopulorum***
   (treated under small-leaved "Apertae")
   Underside of leaves not glaucous; inflorescences simple 2

2 Lower surface of leaves with two rows of domatia (e.g. Fig. 149E) .... **85. *Hebe townsonii***
   Lower surface of leaves without domatia 3

3 Corolla magenta (e.g. Fig. 150); leaves mostly > 25 mm wide .... **86. *Hebe speciosa***
   Corolla white, mauve, or violet; leaves often < 25 mm wide 4

4 Inflorescences usually shorter than subtending leaves (e.g. Figs 146B and G); corolla tube shorter than calyx lobes (e.g. Fig. 146F), ≤ 2.3 mm; free portion of stamen filaments ≤ 3 mm [Kermadec Ids] .... **82. *Hebe breviracemosa***
   Inflorescences usually equalling or longer than subtending leaves; corolla tube *either* longer than calyx lobes *or* > 2.3 mm (except some specimens of *H. pubescens* subsp. *pubescens*); free portion of stamen filaments > 4 mm [rest of NZ excluding Kermadec Ids, South America] 5

5 Outer surface of calyx lobes (not just margins of lobes) hairy; lower surface of leaf lamina often hairy .... **81. *Hebe pubescens***
   Calyx lobes ciliate on margins only; lower surface of leaf lamina not hairy (or only with minute, glandular hairs) 6

6 Lateral corolla lobes shorter than corolla tube; leaves linear-lanceolate [Hunua Ra. to Ruahine Ra.] .... **83. *Hebe corriganii***
   Lateral corolla lobes longer than or equalling corolla tube; leaves of various shapes (sometimes linear-lanceolate) [either north of Hunua Ra. or south of Ruahine Ra.] 7

7 Plant of NI 8
   Plant of SI, Stewart Id, Auckland Id or South America 9

8 Corolla tube > 2 mm wide; free portion of stamen filaments > 6 mm long [far north of NI, near Spirits Bay] .... **84. *Hebe adamsii***
   Corolla tube < 2 mm wide; free portion of stamen filaments < 6 mm long [Coromandel Peninsula and islands of outer Hauraki Gulf] .... **81. *Hebe pubescens***

9 Leaves tapering to a fine point, often conspicuously narrowed (± acuminate) *c.* ¾ of the way to apex 10
   Leaves either not tapering to a fine point, or gradually tapering from about or just past the midpoint .... **78. *Hebe leiophylla***
   (treated under small-leaved "Apertae")

10 Leaves mostly < 1 cm wide [Westland] .... **80. *Hebe paludosa***
   Leaves mostly > 1 cm wide [throughout SI (including Westland), Stewart Id, Auckland Id, South America] .... **79. *Hebe salicifolia***

79. *Hebe salicifolia* (G.Forst.) Pennell

DESCRIPTION Openly branched bushy shrub to 2.5 m tall. **Branches** erect, old stems brown or grey; branchlets green or orange, glabrous (often) or puberulent, hairs bifarious to uniform; internodes (1–)6–18(–34) mm; leaf decurrencies evident or obscure. **Leaf bud** distinct; sinus square to oblong. **Leaves** erecto-patent; *lamina* narrowly lanceolate or oblanceolate, coriaceous or subcoriaceous, shallowly m-shaped in TS, (34–)60–106(–132) × (6–)11– 18(–28) mm; *apex* acuminate; brochidodromous *secondary veins* evident in fresh leaves; *margin* cartilaginous, pubescent or ciliate, distantly denticulate or entire; *upper surface* green, dull, with few or many stomata, hairy along midrib; *lower surface* light green, glabrous or hairy along midrib or sts covered with minute gl. hairs; *petiole* (1–)2–4(–5) mm, hairy along margins and above and below. **Inflorescences** with 100–250 flowers, lateral, unbranched, (5–)7–18(–23) cm; peduncle (0.7–)1.3–4.5(–6) cm; rachis (3.5–)5.5–17.5 cm. **Bracts** alternate, lanceolate or linear, acute or subacute. **Flowers** (Oct–)Dec–Jun(–Jul), ♀ (or possibly some ♂-sterile). **Pedicels** (0.7–)1.3–3(–4.7) mm, sts recurved in fruit. **Calyx** 1.5–3 mm; *lobes* ovate or lanceolate, acute or subacute. **Corolla** tube hairy inside and often outside, 2.5–3.2 × 1.6–1.8 mm, contracted at base, longer than calyx; *lobes* white or tinged mauve at anthesis, lanceolate, acute to subacute, suberect or erect, longer than corolla tube, sts with a few hairs toward base on inner surface and/or ciliate (e.g. WELT 16280). **Stamen filaments** 5–8.5 mm; anthers mauve, 1.5–1.9 mm. **Ovary** 0.9–1.1 mm; ovules 14–19 per locule, in 1–2 layers; style 4–6 mm, sts sparsely hairy. **Capsules** (Nov–)Jan– Jun(–Jul), subacute or obtuse, 2.5–3.5 × 2.5–3 mm, loculicidal split extending ¼–¾-way to base (most c. ⅓). **Seeds** flattened, broad ellipsoid to discoid, straw-yellow, (0.6–)0.7–1.1 × 0.6–0.9 mm, MR 0.1–0.2 mm. $2n = 40$.

DISTRIBUTION AND HABITAT Throughout SI (except for Marlborough Sounds) and Stewart Id, and also on Auckland Id (not mapped, see below) and in Chile (mapped separately on Fig. 141); naturalised in western Europe (Webb 1972). In New Zealand it occurs from sea-level to close to the tree-line, mostly in open sites, and in forest. The distribution map, based only on herbarium specimens, under-represents the extent of the species' SI and Stewart Id distribution.

NOTES The most common hebe of lowland and montane areas of SI, distinguished from most others by the size of its leaves (Appendix 4). Similar SI species are *H. stricta*, from which it differs in the presence of a leaf bud sinus, and *H. paludosa* (see notes under that species). It hybridises with *H. elliptica*, *H. calcicola* (Bayly et al. 2001), probably *H. albicans* and *H. strictissima* (see notes under those species), and potentially other species with which it co-occurs.

It is not known whether *H. salicifolia* occurs naturally on Auckland Id or was introduced there. Two confirmed specimens (WELTU 11157, WELT 83266!) are from a single plant growing close to a ruined house site near Lindley Pt (Johnson & Campbell 1975). A more recent specimen from the same general area, CHR 437295, "nr. Deas Head", *W. R. Sykes 25/87*, 13 Feb. 1987, resembles *H. salicifolia*, but is not identified with certainty (in the size of the flowers and inflorescences, and in leaf margin pubescence, it resembles some *H. salicifolia* × *H. elliptica* hybrids).

ETYMOLOGY L., suggests that the leaves (*folium* = leaf) resemble those of willows (genus *Salix*, family Salicaceae).

Hebe salicifolia — **Large-leaved APERTAE**

FIG. 142 **A** plant near the Red Hills Ridge, Wairau Va. **B** sprig; inset shows glabrous stem. **C** leaf bud with small sinus. **D** lower and upper leaf surfaces; magnified inset shows hairs on leaf margin and one obscure tooth. **E** inflorescence. **F** lateral view of flower. **G** frontal view of flowers. **H** dorsal view of flowers. **I** capsule in septicidal (left) and loculicidal view.

TAXONOMIC TREATMENT 269

80. *Hebe paludosa*   (Cockayne) D.A.Norton et de Lange

DESCRIPTION Bushy shrub to 5 m tall. **Branches** erect, old stems red-brown; branchlets green or red-brown, puberulent, hairs uniform (mostly) or bifarious; internodes 6–26 mm; leaf decurrencies obscure. **Leaf bud** distinct; sinus square to oblong, or small and rounded. **Leaves** erecto-patent (mostly) to patent (with age); *lamina* linear-lanceolate (mostly) or linear-elliptic, subcoriaceous, slightly m-shaped in TS, 30–86 × 5.5–11(–14) mm; *apex* acuminate; brochidodromous *secondary veins* evident in fresh leaves; *margin* cartilaginous, minutely pubescent (with egl. and sts gl. hairs), distantly denticulate or entire; *upper surface* dark green or yellowish-green, dull, with few stomata (not readily visible in fresh leaves), hairy along midrib and sts covered with minute gl. hairs (but these hairs us. not readily visible, even at high magnification); *lower surface* light green, hairy along midrib; *petiole* (1–)2–3(–4) mm, hairy along margins and above and below. **Inflorescences** with *c.* 100–150 flowers, lateral, unbranched (one aberrant inflorescence seen with a single branch at base), 9.3–19 cm; peduncle 1.8–4 cm, hairy (sts very sparsely); rachis 7.5–15 cm. **Bracts** alternate, linear or narrowly deltoid, subacute or acuminate, sts hairy outside (esp. lowermost). **Flowers** Jan(–Mar), ⚥. **Pedicels** longer than or equal to bracts, (1–)2–4(–5) mm, sts recurved in fruit. **Calyx** (1.5–)2–3(–3.5) mm; *lobes* ovate or lanceolate or oblong or deltoid, obtuse to acute or acuminate. **Corolla** tube hairy inside and sts outside (around base of corolla lobes), 2.3–3.5 × 1.9–2.1 mm, shortly cylindric, longer than (us.) or equalling calyx; *lobes* white or tinged mauve at anthesis, lanceolate, obtuse or subacute, erect to suberect, longer than corolla tube, sts with a few hairs toward base on inner surface. **Stamen filaments** diverging with age (but probably erect for some time after anthesis), 6.5–8.5 mm; anthers purple or blue, 1.7–2.2 mm. **Ovary** sts sparsely hairy (esp. near apex), 0.8–1.1 mm; ovules 12–16 per locule, in 1–2 layers; style 7–7.5 mm, sts sparsely hairy. **Capsules** Mar–May, obtuse or truncate or subacute, 3–4 × 2–2.5 mm, sts hairy, loculicidal split extending ¼–½-way to base. **Seeds** flattened (sts strongly), ellipsoid to discoid, not winged to only weakly winged, straw-yellow to pale brown, 0.8–1.3 × (0.6–)0.7–1.1 mm, MR 0.2–0.3 mm. $2n = 80$.

DISTRIBUTION AND HABITAT Endemic to Westland and northwest Otago, SI, from the Grey Va. in the north to Big Bay in the south. It usually grows in lowland, mesotrophic wetlands.

NOTES A recently recognised species (Norton & de Lange 1998) based on *Veronica salicifolia* var. *paludosa* (Cockayne 1916). It was included in *H. salicifolia* by Moore (in Allan 1961), but with the suggestion that it might be a hybrid between that species and *H. gracillima* (= *H. leiophylla* here). It is generally distinguished from *H. salicifolia*, with which it co-occurs, by its narrower, more evenly tapered leaves (Appendix 4), conspicuously hairy branchlets (although branchlets of *H. salicifolia* vary from glabrous to bifariously or uniformly hairy), usual wetland habitat, and chromosome number ($2n = 80$, rather than 40). Differences from *H. leiophylla* are less clear-cut. *H. paludosa* generally has longer and more tapered leaves than *H. leiophylla*, but that species sometimes has similar leaves (Appendix 4), and some short-leaved specimens of *H. paludosa* (e.g. CANU 35236) are scarcely different from many specimens of *H. leiophylla*. Because of the similarity between *H. paludosa* and *H. leiophylla*, defining the northern limit of the former (and southwest limit of the latter) is difficult, and requires further assessment. Here we give the same northern limit for *H. paludosa* as Norton & de Lange (1998), but refer plants from some nearby localities to *H. leiophylla*. Limited study of leaf flavonoids (Mitchell et al. in prep.) suggests that *H. paludosa* has a profile distinguished from, but roughly intermediate between, those of *H. leiophylla* and *H. salicifolia*. The possibilities that *H. paludosa* is recently derived from *H. leiophylla* or *H. salicifolia*, might be a stable hybrid between them, or may intergrade with one or other of them (particularly *H. leiophylla*, as suggested by Wardle 1975, using the name *H. gracillima*), are worthy of further investigation.

ETYMOLOGY L. (*paludosus* = marshy, boggy), refers to the species' habitat.

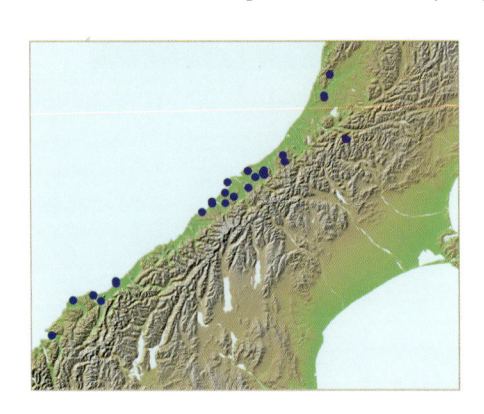

Hebe paludosa — **Large-leaved APERTAE**

FIG. 143 A plant near Lk. Wahapo. B sprig; inset shows minute hairs on branchlet. C leaf bud with small sinus. D lower and upper leaf surfaces; magnified inset shows hairs on margin. E inflorescence. F lateral view of flowers. G dorsal view of flower. H capsule in septicidal (left) and loculicidal view.

81. *Hebe pubescens*   (Benth.) Cockayne et Allan

DESCRIPTION Openly branched bushy shrub to 2 m tall. **Branches** erect (mostly) to spreading (in some situations), old stems brown to red-brown; branchlets green to red, pubescent (hairs varying from relatively short to very long and woolly, on different plants) or glabrous, hairs uniform; internodes (1–)9–21(–39) mm; leaf decurrencies obscure. **Leaf bud** distinct; sinus often filled with hairs, small and rounded, or square to oblong (sts subacute at the apex). **Leaves** erecto-patent to recurved; *lamina* oblong or lanceolate or elliptic or linear or oblanceolate or ovate, coriaceous or sub-coriaceous, ± flat, (15–)30–95(–125) × (3.5–)8–25(–31) mm; *apex* subacute or acute or obtuse; *base* truncate or cuneate; faint brochidodromous *secondary veins* sts evident in fresh leaves; *margin* narrowly cartilaginous, pubescent to sparsely ciliate (us. with at least some hairs toward base); *upper surface* dark green to yellowish-green, dull to somewhat glossy, without evident or with few stomata, hairy along midrib and sts hairy toward base and apex or rarely covered with minute gl. hairs; *lower surface* green or light green, not pitted (but sts with scattered gl. hairs in ± shallow depressions), indumentum variable between plants and populations with lower surfaces being either glabrous or uniformly egl. pubescent (us. on Coromandel Peninsula) and/or hairy along midrib (most populations); *petiole* 0.5–3(–4) mm, hairy along margins and often hairy above and below. **Inflorescences** with (20–)30–130(–190) flowers, lateral, unbranched, (2–)6–15(–20) cm, longer than (mostly) or about equal to subtending leaves, rarely with a conspicuous number of unopened flowers toward the apex (leaving a protruding "rat's tail"); peduncle (0.3–)1–2.8 cm; rachis (1.7–)4–12(–17.5) cm. **Bracts** alternate, narrowly deltoid or lanceolate, acute or subacute, hairy outside. **Flowers** Jan–Dec, ☿ or ♀ (on different plants) (the commonness of ☿ sterility is not known – only a few possibly ♀ specimens are known). **Pedicels** sts recurved in fruit. **Calyx** (1.7–)2.3–4 mm; *lobes* deltoid (sts narrowly) or lanceolate, acute or acuminate or subacute, sts hairy outside. **Corolla** tube hairy inside and often outside (esp. near apex, toward clefts between lobes); tube of ☿ flowers (1.9–)2.5–5.5 × 1.3–1.7(–1.9) mm, narrowly funnelform to shortly cylindric and contracted at base, us. *c.* equalling or longer than or sts slightly shorter than calyx; tube of ♀ flowers 2–2.5 mm; *lobes* white or mauve or violet at anthesis, lanceolate or elliptic, acute to obtuse, suberect to patent, longer to shorter than corolla tube (last condition only for anterior lobe), us. hairy inside or at least with a few hairs toward base on inner surface and sts hairy outside. **Stamen filaments** 4.5–6 mm; anthers mauve or purple or purplish-magenta, 1.1–1.5 mm; sterile anthers 0.8–1 mm (when dry). **Ovary** sts hairy (may vary on one plant), 0.9–1.1 mm; ovules *c.* 6–11 per locule; style 3.5–7(–10.5) mm, sts hairy (may vary on one plant). **Capsules** Jan–Oct(–Dec), obtuse or subacute, 2.5–4(–5) × 2–3.4 mm, sts hairy, loculicidal split extending ¼–½-way to base. **Seeds** flattened, ellipsoid (sts broadly), winged, straw-yellow to pale brown, 1–1.9(–2.5) × 0.9–1.5(–1.7) mm, MR 0.2–0.6(–0.8) mm. $2n = 40$.

KEY TO SUBSPECIES

1 Hairs on underside of leaf midrib > 0.2 mm long; similar hairs usually present on underside of lamina [Coromandel Peninsula and nearby islands] .... **subsp. *pubescens***
  Hairs on underside of leaf midrib < 0.2 mm long or absent; hairs absent from underside of lamina 2

2 Leaves mostly lanceolate or linear-lanceolate (leaves usually broadest below midpoint) [Great Barrier Id and immediately surrounding islands] .... **subsp. *rehuarum***
  Leaves mostly obovate or oblanceolate (leaves usually broadest above midpoint) [Mokohinau and Little Barrier islands] .... **subsp. *sejuncta***

Subsp. *pubescens*

Shrub to 2 m tall, usually erect. Branchlets uniformly hairy, with hairs varying from relatively short to very long and woolly. Leaf bud sinus small, rounded to square or oblong, often filled with hairs. Leaves lanceolate, oblong, elliptic or linear, subcoriaceous to coriaceous, (15–)30–70(–87) × (3.5–)7–14(–18) mm; upper surface dark green to yellowish-green; underside of lamina usually covered with long egl. hairs, rarely glabrous; underside of midrib hairy, with many hairs usually > 0.2 mm long; margins pubescent. Inflorescences (2–)6–13(–20) cm long. Outer surface of calyx lobes hairy. Corolla lobes at least faintly mauve when young, fading to white after anthesis, covered outside with egl. hairs; corolla tubes (1.9–)2.4–3.9 mm long. Ovaries usually hairy (on any one plant at least some ovaries are hairy; hairs, when present, are usually egl. and scattered over entire surface of ovary).

Hebe pubescens subsp. *pubescens* and subsp. *rehuarum*

Large-leaved
APERTAE

FIG. 144 *Hebe pubescens* subsp. *pubescens* and subsp. *rehuarum*. **A** subsp. *pubescens*, Cooks Beach. **B** sprig, subsp. *pubescens*. **C–D** lower and upper leaf surfaces: C, subsp. *pubescens*; D, subsp. *rehuarum*. **E–F** leaf buds with small round sinuses: E, subsp. *pubescens*, sinus almost obscured by leaf hairs; F, subsp. *rehuarum*. **G–H** lateral view of flowers: G, subsp. *pubescens*, magnified inset shows close up of hairs on calyx lobe; H, subsp. *rehuarum*. **I–J** frontal view of flowers: I, subsp. *pubescens*; J, subsp. *rehuarum*. **K–L** capsule of subsp. *rehuarum* in: K, septicidal view; L, loculicidal view.

TAXONOMIC TREATMENT 273

Subsp. *rehuarum*   Bayly et de Lange

Shrub with habit varying from low and spreading (e.g. *c.* 30 cm tall, 1 m wide) to erect (to *c.* 1.5 m tall). Branchlets uniformly and minutely puberulent, or glabrous. Leaf bud sinus small and rounded. Leaves lanceolate, narrowly elliptic, or linear-lanceolate, subcoriaceous to coriaceous, (25–)40–65 × 7–15(–19) mm; upper surface dark green to yellowish-green; underside of lamina glabrous; underside of midrib glabrous or hairy, hairs when present < 0.2 mm long; margins pubescent to puberulent (at least toward base). Inflorescences 5.5–10 cm long. Outer surface of calyx lobes usually glabrous, but sometimes hairy. Corolla lobes at least faintly mauve when young, fading to white after anthesis, glabrous outside (but only a few flowering specimens available for examination); corolla tubes 2.7–3.9 mm long. Ovaries glabrous or sparsely hairy (especially along septal grooves).

Subsp. *sejuncta*   Bayly et de Lange

Bushy shrub, often heavily branched, to *c.* 1.6 m tall. Branchlets either uniformly and minutely puberulent, or glabrous. Leaf bud sinus, rounded to subacute, usually conspicuous, but sometimes (on Mokohinau Ids) very small. Leaves obovate, oblanceolate or narrowly elliptic, subcoriaceous to very robust and coriaceous, (30–)45–95(–125) × (7–)10–25(–31) mm; upper surface dark to very dark green; underside of lamina glabrous; underside of midrib glabrous or hairy, hairs when present < 0.2 mm long; margins pubescent to sparsely ciliate or nearly glabrous. Inflorescences 5.5–14.5 mm long. Outer surface of calyx lobes glabrous or hairy (always glabrous on Mokohinau Ids). Corolla lobes faint mauve to violet when young, usually fading to white after anthesis, outer surface either glabrous (but with margins often ± ciliate near base) or hairy (always glabrous on Mokohinau Ids); corolla tube (2.5–)3–5.5 mm long (longest corolla tubes on Mokohinau Ids). Ovaries usually sparsely hairy (especially along septal grooves), sometimes glabrous.

DISTRIBUTION AND HABITAT   Subsp. *pubescens* occurs on NI on the Coromandel Peninsula, surrounding offshore islands (including the Motukawao Group, Ngamotukaraka, Waimate, Motuoruhi, and Motukaramea, Shoe and the Mercury islands), mostly in coastal sites under open pōhutukawa forest, on cliff faces, rock-strewn ground, slip scars and offshore rock stacks, or sometimes inland (e.g. Tararu and Kauaeranga valleys), especially in open, seral vegetation in historically disturbed areas. Subsp. *rehuarum* is endemic to Great Barrier (Aotea) Id and immediately surrounding islands (including the Broken Ids, and an island between Great Barrier Id and Aiguilles Id), growing in coastal situations, or inland chiefly in rocky places. Subsp. *sejuncta* occurs primarily on the Mokohinau and Little Barrier (Hauturu) islands, with a single plant (e.g. WELT 80844) known on Maungapiko, Great Barrier Id. It grows in flaxland and in scrub, on cliff faces, rock outcrops and slip scars, or sometimes in coastal forest.

NOTES   Distinguished from other hebes of northern New Zealand by its leaf shape and size (Appendix 4), and by possession of a leaf bud sinus; distinguished from *H. corriganii* (which occurs on southern Coromandel Peninsula and also has a leaf bud sinus) by generally flat (rather than m-shaped) leaves, which are often broader relative to their length, and by flower shape, with corolla lobes usually equalling or longer than corolla tubes. It commonly co-occurs with *H. stricta* var. *stricta* and *H. macrocarpa*.

The circumscription used here is broader than that of Moore (in Allan 1961; and corrigenda in Moore & Edgar 1970), and includes plants from islands of the outer Hauraki Gulf, some of which have, in the intervening years, been commonly known by several informal "tag names" (Appendix 1). Variation is described in detail by Bayly et al. (2003), and there is possibly a cline of forms between the Coromandel Peninsula (subsp. *pubescens*) and the Mokohinau Ids (subsp. *sejuncta*), all of which share possession of a leaf bud sinus, similar flowers, basically similar flavonoid chemistry and the same chromosome number.

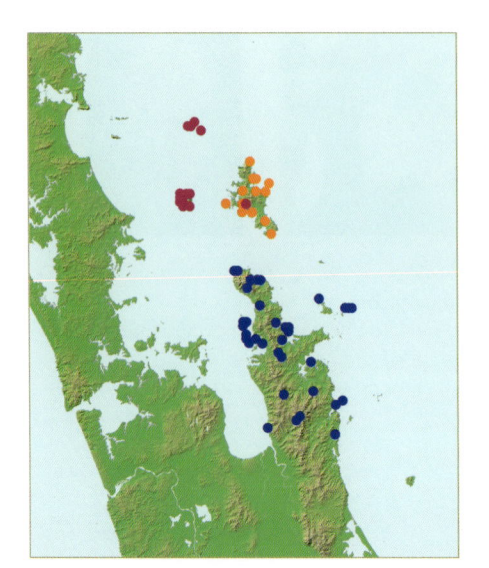

Plants of subsp. *pubescens* on Coromandel Peninsula are typically hairy on the underside of the leaf lamina, although less hairy plants, otherwise matching and included in the subspecies, occur in Kauaeranga Va. (AK 165414), at Lonely Bay (WELT 80516) and on some islands immediately surrounding the Coromandel (AK 167064, 167162, 217291). The locality "Piha Creek: west coast Auckland", given on a specimen of subsp. *pubescens* (AK 22185), is likely to be incorrect (de Lange 1996). The record of subsp. *pubescens* from Papanui Point on the western coast of the Firth of Thames cited by Bayly et al. (2003) was based on the small sprig on CHR 482970. After subsequent examination of the duplicate sheet AK 211025, on which the bases of the leaf buds are more clearly visible, this collection has been

- Subsp. *pubescens*
- Subsp. *rehuarum*
- Subsp. *sejuncta*

Hebe pubescens subsp. sejuncta

Large-leaved
APERTAE

FIG. 145   *Hebe pubescens* subsp. *sejuncta*. **A** plant on Hokoromea Id, Mokohinau Ids. **B** flowering sprig of plant from Little Barrier Id. **C** lower and upper leaf surfaces, plants from Little Barrier Id (top) and Fanal (Motukino) Id. **D** leaf bud with small round sinus. **E–F** flowers of plant from Little Barrier Id in: E, lateral view; F, frontal view. **G–H** flowers of plant from Fanal Id in: G, lateral view; H, frontal view. **I** septicidal view of capsule. **J** loculicidal view of capsule. **K** seeds.

TAXONOMIC TREATMENT   275

reidentified as *H. stricta* var. *stricta* (hairy forms of which are not uncommon in Auckland and the surrounding areas).

Most specimens of subsp. *rehuarum* have either no, or very short, hairs on the undersides of leaf midribs, and on this basis are readily distinguished from specimens of subsp. *pubescens*. One exception is a specimen from Tapuwai Point (WELT 81387), which has midrib hairs considerably longer than those commonly found on Great Barrier Id. The hairs on this specimen are, nonetheless, generally shorter than those seen in subsp. *pubescens* (< 0.2 mm), and the specimen is, therefore, placed, with others from Great Barrier Id, in subsp. *rehuarum*.

The specimen of subsp. *sejuncta* from Great Barrier Id resembles some Little Barrier Id plants, and it seems possible that its occurrence, if truly singular, might result from recent seed dispersal, by wind or with the assistance of birds (Maungapiko is in the flight path of black petrels that move between the islands). Further field sampling would provide additional information/verification of the true extent of geographic overlap between subsp. *sejuncta* and subsp. *rehuarum*.

ETYMOLOGY *Pubescens* (L. *pubescens* = becoming hairy), presumably refers to the leaves (and outside of calyx lobes?); *rehuarum* means "of the Rehua", honouring Ngāti Rēhua, the iwi exercising mana whenua over Great Barrier (Aotea) Id, where the subsp. is endemic; *sejuncta* (L. *sejunctus* = isolated, separated, disunited) refers to both the insular separation from subsp. *pubescens*, and the separation of populations of subsp. *sejuncta* on Little Barrier, Great Barrier and the Mokohinau islands.

82. *Hebe breviracemosa* (W.R.B.Oliv.) Andersen

DESCRIPTION Openly branched bushy shrub to 2 m tall. **Branches** erect, old stems light brown or grey; branchlets green, puberulent, hairs uniform; internodes 3–14 mm; leaf decurrencies evident (with a thin ridge decurrent with the midrib of each leaf and sts, to a lesser extent, with each margin; the latter sts creating a broad ± grooved region between decurrencies). **Leaf bud** distinct; sinus narrow and acute, or small and rounded (not always conspicuous on herbarium specimens). **Leaves** erecto-patent to recurved; *lamina* oblanceolate or obovate or elliptic, subcoriaceous, m-shaped in TS, 23–85(–112) × 6–19(–26) mm; *apex* acute or subacute; faint brochidodromous *secondary veins* evident in fresh leaves; *margin* ciliolate (often sparsely) or glabrous, entire (and sts ± undulate); *upper surface* green, dull or slightly glossy, without evident stomata, minutely hairy along midrib; *lower surface* light green, covered with minute gl. hairs (these scattered and inconspicuous) and often hairy along midrib; *petiole* 1.5–3 mm. **Inflorescences** with 19–45 flowers, lateral, unbranched, 2.5–5 cm, us. shorter than subtending leaves; peduncle 0.5–1.5 cm; rachis 2.1–4.1 cm. **Bracts** alternate, lanceolate or narrowly deltoid, acute to acuminate. **Flowers** [Jan–Jul], ⚥. **Pedicels** shorter than or equal to bracts, 1–3 mm. **Calyx** 2.5–4 mm; *lobes* lanceolate, acute, sts sparsely hairy outside (esp. toward base or along central vein). **Corolla** tube hairy inside and outside, 1.3–2.3 × 1.4–1.7 mm, contracted at base, shorter than calyx; *lobes* white or tinged mauve at anthesis, ovate (sts narrowly), acute to subacute, suberect to recurved, longer than corolla tube. **Stamen filaments** white, 2.4–3 mm; anthers 1.3–1.7 mm. **Ovary** narrowly ovoid or conical, c. 1 mm; ovules 6–8 per locule; style 2.4–4 mm. **Capsules** [Feb–]Aug, acute, 2.5–4.5 × 2–3 mm, loculicidal split extending ¼–⅔-way to base. **Seeds** flattened, ± discoid, straw-yellow, 0.7–1 × 0.6–0.7 mm, MR c. 0.1 mm. 2*n* = 40.

DISTRIBUTION AND HABITAT Endemic to Kermadec Ids, northern New Zealand, where it is known only from Raoul Id, and grows primarily on steep cliffs. It is the only *Hebe* recorded from the Kermadec Ids.

NOTES Considered by Moore (in Allan 1961) to lack a leaf bud sinus and, accordingly, placed in large-leaved "Occlusae" in her informal classification. This assessment was based on only a few specimens, with generally obscured leaf buds, and probably involved some extrapolation (e.g. from leaf base shape). Any fresh or cultivated specimens we have examined have a definite leaf bud sinus (hence placement of the species here), but this may be small, sometimes hair-filled, and not necessarily conspicuous on dried specimens.

The species is similar to other large-leaved species of northern New Zealand (some "Occlusae" and some "Apertae") that have short corolla tubes and acute or subacute corolla lobes – for example, *H. bollonsii*, *H. ligustrifolia*,

Hebe breviracemosa | Large-leaved APERTAE

FIG. 146   **A** plant on Raoul Id. **B** sprig. **C** leaf bud with small acute sinus. **D** lower and upper leaf surfaces. **E** inflorescence. **F** inflorescence (some flowers removed) showing lateral view of flowers. **G** a range of infructescences, with a leaf included for comparison of length. **H** capsule in septicidal (left) and loculicidal view.

H. adamsii and *H. pubescens*. It is distinguished from these by the combination of: inflorescences consistently shorter than subtending leaves; very short and narrow corolla tubes that do not exceed calyces; very short filaments; and the presence of a leaf bud sinus.

When first describing the species, Oliver (1910) noted that it "…was formerly plentiful, but has been almost killed out by goats, and is now found only on cliffs and other places inaccessible to those animals". It was at one stage considered probably extinct (e.g. Sykes 1977; Given 1981; Eagle 1982). In 1983 a single plant was found, and additional plants have since been discovered. Notes on distribution and conservation are provided by de Lange & Stanley (1999).

ETYMOLOGY L. [*brevi-* = short; *racemus* = raceme (inflorescence)], refers to the often short inflorescences.

83. *Hebe corriganii*   Carse

DESCRIPTION Openly branched, small to large bushy shrub to 2.5 m tall. **Branches** erect, old stems grey or brown or black (at least on herbarium specimens); branchlets green (sts tinged maroon) to brown, glabrous or minutely puberulent, hairs bifarious or rarely uniform; internodes (5–)12–22(–36) mm; leaf decurrencies us. somewhat evident (often with a slight ridge along medial line that can give branchlets an angular or flattened appearance). **Leaf bud** distinct; sinus small and rounded or square to oblong. **Leaves** erecto-patent or patent; *lamina* linear-lanceolate, coriaceous, slightly m-shaped in TS, 70–105(–145) × (7.2–)8–16(–19.3) mm; *apex* acute or subacute; brochidodromous *secondary veins* sts evident in fresh leaves; *margin* puberulent to ciliolate (sts sparsely), entire or distantly denticulate; *upper surface* dark green, glossy or dull, without evident stomata, hairy along midrib; *lower surface* green or light green; *petiole* 2.5–3.2 mm, hairy along margins and above. **Inflorescences** with 100–120 flowers, lateral, unbranched, 8–14.5 cm; peduncle 1.6–2.7 cm; rachis 6.6–12.1 cm. **Bracts** alternate, lanceolate, subacute or acute, sts sparsely hairy outside. **Flowers** [Jul–]Aug–Mar, ⚥. **Pedicels** 1.4–4 mm. **Calyx** 2.5–4 mm; *lobes* deltoid or ovate or oblong, acuminate to obtuse, glabrous outside (but often hairy inside). **Corolla** tube hairy inside, 3.5–5 × 1.9–2.4 mm, slightly expanded in lower half, longer than calyx; *lobes* white or tinged mauve at anthesis, ovate or elliptic, obtuse, suberect to patent, shorter than corolla tube, sts with a few hairs toward base on inner surface and sts ciliate (near base). **Stamen filaments** 4.5–5 mm; anthers pale mauve, (1.5–)1.9–2.3 mm. **Ovary** 0.8–1.1 mm; ovules *c.* 10–13 per locule; style (4.5–)6–9 mm. **Capsules** [Jun–]Apr, subacute, (3.6–)4.7–6(–7) × (2.3–)3–4 mm, loculicidal split extending ¼-way to base. **Seeds** strongly flattened, broad ellipsoid to discoid, not winged to only weakly winged, ± smooth, pale brown, 1.3–2.2 × 1.1–1.7 mm, MR 0.4–0.6 mm. 2*n* = 80.

DISTRIBUTION AND HABITAT Endemic to NI, from the Hunua Ranges in the north to the NW Ruahine Ra. in the south, and between the Raukumara Ra. in the east and the Pouakai Ra. (Taranaki) in the west. It grows in a range of situations from near-coastal lowland scrub to montane or subalpine forests.

NOTES Distinguished from other NI species by the combination of: an obvious leaf bud sinus; coriaceous, linear-lanceolate leaves; and corolla tubes longer than calyces and corolla lobes.

The species has similar leaves and fruit to *H. macrocarpa*. The two are probably closely related and may intergrade. They differ primarily in the presence/absence of a leaf bud sinus, and plants with both conditions co-occur in some localities near the geographic boundary between them (e.g. Kauaeranga Va., southern Coromandel Peninsula; CHR 221248, 221299, 221291, 179302, AK 132833). They are retained here as distinct species (cf. Druce 1993; Clarkson et al. 2002), primarily because they differ consistently in the sinus character over broad geographic areas. Revision of this classification might be appropriate if further studies more firmly establish a close relationship and/or substantial introgression between them.

ETYMOLOGY Honours Mr D. H. L. Corrigan, who discovered the species.

Hebe corriganii — Large-leaved APERTAE

FIG. 147   **A** plant on Kaweka Ra. **B** sprig. **C** leaf bud with small round sinus. **D** lower and upper leaf surfaces. **E** inflorescence. **F** frontal view of flower. **G** lateral view of flower. **H** septicidal view of capsule. **I** loculicidal view of capsule.

84. *Hebe adamsii* (Cheeseman) Cockayne et Allan

DESCRIPTION Spreading low or bushy shrub to 1 m tall. **Branches** ascending to erect, old stems brown to grey; branchlets yellowish-green and sts tinged purplish, puberulent or glabrous, hairs bifarious or uniform; internodes 5–22 mm; leaf decurrencies obscure, or evident (as a narrow ridge along medial line of each decurrency). **Leaf bud** distinct; sinus small and rounded or acute. **Leaves** erecto-patent to patent; *lamina* lanceolate or narrowly elliptic, coriaceous, flat to weakly m-shaped in TS, (30–)50–100 × (7–)10–18(–28) mm; *apex* acute to subacute; *base* abruptly cuneate; faint brochidodromous *secondary veins* us. evident in fresh leaves; *margin* narrowly cartilaginous, ciliolate or glabrous, sts tinged red; *upper surface* bronze-green to yellowish-green, glossy, with few stomata, hairy along midrib; *lower surface* bronze-green to yellowish-green, sparsely covered with minute gl. hairs (these often inconspicuous and sitting in shallow depressions) and often sparsely hairy along midrib; *petiole* 1–2 mm. **Inflorescences** with (20–)35–65 flowers, lateral, unbranched, 4.5–10(–15) cm; peduncle 0.9–3.5 cm; rachis 3.5–9.5 cm. **Bracts** alternate, linear, acute. **Flowers** Jan–Jun(–Jul), ⚥. **Pedicels** longer than or equal to bracts, 2.5–4(–6) mm. **Calyx** 3–3.5 mm; *lobes* ovate, acuminate. **Corolla** tube hairy inside, 2.3–4 × 2–3 mm, broadly funnelform, slightly shorter to slightly longer than calyx; *lobes* white (usually) or violet or mauve at anthesis, white or mauve with age, ovate or elliptic to deltoid, subacute, erect, longer than corolla tube, sts with a few hairs toward base on inner surface; corolla throat mauve or white. **Stamen filaments** coloured or white, 6–9.5 mm; anthers mauve to purple, 1.9–2.5 mm. **Ovary** 1–1.3 mm; ovules *c.* 8–10 per locule; style 4–8.5 mm. **Capsules** Apr–Jun, acute or subacute, (3.4–)4–5(–6) × 2.8–4 mm, loculicidal split extending ¼–½-way to base. **Seeds** strongly flattened, ellipsoid to discoid, pale brown or brown, 1.4–2.5 × 1.1–2 mm, MR *c.* 0.3 mm. $2n = 80$.

DISTRIBUTION AND HABITAT Far north of NI, known only from Unuwhao Ridge, The Pinnacle, Pinnacle Ridge and Tarure Hill. It grows on east-facing, conglomerate cliffs, usually among *Astelia banksii*, or in vertical joints, sometimes in scrub.

NOTES First collected and named by Cheeseman (1925), and not re-collected in the wild until 1985 (de Lange 1991). Moore (in Allan 1961) thought it was a hybrid, and upon rediscovery it was, at first, considered an unnamed species, known informally in New Zealand as *H.* "Unuwhao". Grounds for reinstatement at species level were presented by Garnock-Jones & Clarkson (1994).

All specimens seen in the course of this revision have a small sinus in the leaf bud, although Garnock-Jones & Clarkson (1994) note that this may sometimes be absent. This, among other features, distinguishes the species from similar large-leaved species of northern New Zealand – for example, *H. macrocarpa*, *H. ligustrifolia* and *H. perbella* (for differences from *H. breviracemosa* see notes under that species). In this instance the presence of a sinus may not be a reliable indicator of relationship, and the species is potentially closely related to otherwise similar species of "Occlusae". Among these, the recently described *H. perbella* (de Lange 1998) is particularly similar, to the extent that the two might potentially be considered conspecific. They share a similar habit, habitat, leaf shape, sparsely flowered inflorescences with pointed buds, and flowers with short corolla tubes and subacute corolla lobes. They differ primarily in the presence/absence of a leaf bud sinus and ovary indumentum (glabrous in *H. adamsii*, sparsely puberulent in *H. perbella*), and in chromosome number. They are retained at species level chiefly because of differences in the presence/absence of a leaf bud sinus, and differences (based on only a few samples) in flavonoid chemistry.

ETYMOLOGY Honours James Adams (1839–1906), former headmaster of Thames High School, who accompanied T. F. Cheeseman on his expedition to North Cape in 1896.

Hebe adamsii — Large-leaved APERTAE

FIG. 148 **A** plant on Tarure Hill. **B** sprig (of an uncommon form with strongly coloured flowers). **C** leaf bud with small sinus. **D** lower and upper leaf surfaces. **E** inflorescence with purple flowers. **F** inflorescence of common white-flowered form. **G** lateral view of flower. **H** capsule in septicidal (left) and loculicidal view.

85. *Hebe townsonii* (Cheeseman) Cockayne et Allan

DESCRIPTION Openly branched bushy shrub to 2.5 m tall. **Branches** erect, old stems (at least on herbarium specimens) brown or greenish-grey; branchlets green or brown or black, puberulent (sts with short, harsh antrorse hairs with swollen bases) or glabrous, hairs bifarious; internodes (2–)5–15(–18) mm; leaf decurrencies evident. **Leaf bud** distinct; sinus broad and acute. **Leaves** erecto-patent to recurved (with age); *lamina* lanceolate (often narrowly) or linear, coriaceous, flat or slightly m-shaped in TS or concave, (24–)29–80 × (4–)5–8(–9) mm; *apex* acute; *margin* glabrous (us.) or ciliolate; *upper surface* dark green, without evident stomata, glabrous or hairy along midrib; *lower surface* with a regular series of short oblique domatia just within margin, light green; *petiole* 2–5 mm. **Inflorescences** with 21–42 flowers, lateral, almost always unbranched or rarely tripartite (seen only on some inflorescences of WELT 13542), (5–)8–12 cm; peduncle 0.5–3 cm; rachis 5–10.5 cm. **Bracts** alternate or opposite and decussate below and becoming alternate above, lanceolate or ovate, acute (us.) to obtuse. **Flowers** Sep–Nov(–Jan), ♂ or ♀ (on different plants). **Pedicels** 2–8 mm. **Calyx** (2.5–)3.5–5.8 mm, 4–5-lobed (5th lobe small, posterior); *lobes* ovate or lanceolate or deltoid, subacute or acute or acuminate, with mixed gl. and egl. cilia (but often appearing almost exclusively gl., with egl. cilia very short and infrequent). **Corolla** tube glabrous; tube of ♂ flowers 1–2.5 mm (corolla rather unevenly divided such that tube is *c.* 1–1.5 mm on posterior side, and *c.* 2.5 mm on anterior side), cylindric, shorter than calyx (posterior side considerably shorter, anterior side only just shorter or about equal); *lobes* white or mauve at anthesis, ovate, acute to obtuse, suberect to recurved (with age), longer than corolla tube. **Stamen filaments** white or faintly mauve, 4.5–6 mm; anthers purple or mauve or cream, 1.7–2.1 mm. **Ovary** sts hairy (with a few fine, short hairs), *c.* 1 mm; ovules 6–9 per locule; style 4.5–7 mm, occasionally sparsely hairy. **Capsules** [Oct–]Nov–Feb(–Aug), acute or subacute, 3.5–5.5 × 3–4 mm, glabrous, loculicidal split extending ¼–½-way to base. **Seeds** flattened (sts strongly), broad ellipsoid-ovoid to discoid, not winged or weakly winged, brown, 1–1.4(–1.6) × 0.9–1.3 mm, MR 0.2–0.5 mm. $2n = 40$.

DISTRIBUTION AND HABITAT Northern and western Nelson, and north Westland, SI. It is known from only a few localities, between Mt Burnett in the north and Punakaiki in the south, mostly near the West Coast, but with one record (CHR 256548) from the Graham Va., Arthur Ra. It grows in scrub on and around calcium-rich rocks.

NOTES A distinctive species readily distinguished from all others by the presence of two rows of marginal domatia on the undersides of its leaves (Figs 149D and E). It is widely cultivated in New Zealand.

H. townsonii was recorded from Mt Messenger, NI (not shown on distribution map), by Simpson (1945) on the basis of (presumably cultivated) plants reportedly derived from wild collections made by R. O. Green. Although subsequent authors (e.g. Moore, in Allan 1961; Eagle 1982; Heads 1993) have also listed this locality for the species, there are no wild-collected specimens; only cultivated specimens suggest this provenance – for example, CHR 103064, 132134, 132135, 180127, WELT 82176 (and these could possibly be derived from the same stock as those examined by Simpson 1945). Recent and concerted searches of Mt Messenger (Druce 1980; Clarkson & Boase 1982; B. D. Clarkson pers. comm. 2003) have failed to find the species there, and the report by R. O. Green remains unsubstantiated.

ETYMOLOGY Honours William L. Townson (1855–1926), who provided T. F. Cheeseman with specimens upon which the original description was based.

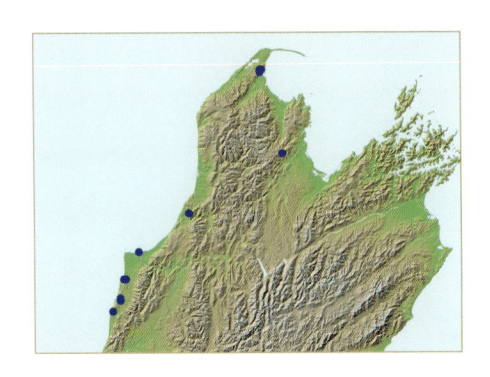

Hebe townsonii — Large-leaved APERTAE

FIG. 149 **A** plant on Mt Burnett. **B** sprig. **C** leaf bud with narrow acute sinus. **D** lower and upper leaf surfaces. Lower surface has two rows of domatia (arrowed). **E** close-up of domatia on underside of leaf. **F** capsule in loculicidal (left) and septicidal view. **G** inflorescence. **H** lateral view of flower.

86. *Hebe speciosa* (A.Cunn.) Andersen

DESCRIPTION Spreading low or bushy shrub to 1(–2) m tall. **Branches** spreading or ascending or pendent, old stems brown (often darkly so) or red-brown; branchlets green or red-brown, glabrous; internodes (5–)8–26 mm; leaf decurrencies evident, or obscure (with dried specimens us. having a narrow ridge decurrent with each leaf base and extending most of the length of the internode). **Leaf bud** distinct; sinus small and rounded. **Leaves** decussate, erecto-patent to patent; *lamina* obovate or broadly elliptic, coriaceous, ± flat or m-shaped in TS, 45–100 × (21–)25–51 mm; *apex* obtuse; very faint brochidodromous *secondary veins* sts evident in fresh leaves; *margin* us. puberulent (with gl. and egl. hairs prominent on the upper surface and sparse below) or glabrous, often tinged maroon-red; *upper surface* dark green (with midrib sts tinged maroon-red), glossy, without evident stomata, hairy along midrib and sts covered with minute gl. hairs; *lower surface* green (but petiole and midrib sts tinged maroon-red, esp. when young), glabrous or hairy along midrib or covered with minute gl. hairs; *petiole* (1.5–)2–6 mm. **Inflorescences** with 32–67(–116) flowers, lateral, unbranched, 4–14.5 cm; peduncle (0.8–)1.5–3.7 cm, hairy or glabrous; rachis 3.2–10.3 cm. **Bracts** alternate (basal pair sts appearing opposite), ovate or deltoid (sts narrowly), obtuse to acute. **Flowers** Jan–Nov, ⚥. **Pedicels** longer than or equal to bracts, 1–6.5 mm, sts recurved in fruit. **Calyx** 2.5–4 mm; *lobes* ovate, subacute or acute or obtuse, egl. ciliate or with mixed gl. and egl. cilia. **Corolla** tube hairy inside, 2.5–4 × 2.9–3.4 mm, broadly funnelform or contracted at base, longer than calyx; *lobes* magenta at anthesis, magenta with age, narrowly to broadly ovate, obtuse, suberect to recurved, longer than corolla tube (except anterior), ciliolate and often sparsely hairy inside; corolla throat magenta. **Stamen filaments** coloured, straight or incurved at apex in bud, (7–)8–13 mm; anthers magenta, (2.5–)2.9–3.2 mm. **Ovary** sts hairy (along septum only), 1.2–1.6 mm; ovules 40–45 per locule, in 2–3 layers; style (6.5–)9–15 mm. **Capsules** Jan–Jul(–Oct), obtuse or subacute, 3.5–7 × (3–)3.5–6.6 mm, sts hairy (with very fine hairs along septum), loculicidal split extending ¼–⅓-way to base. **Seeds** flattened (sts strongly), ± broad ellipsoid to discoid, not winged to winged, brown (sts pale), 1.1–1.8(–1.9) × (0.8–)1.1–1.6 mm, MR 0.2–0.5 mm. $2n = 40$.

DISTRIBUTION AND HABITAT Occurs in New Zealand at a range of widely separated localities on the east coast of the NI (southward from Hokianga Harbour), and in the Marlborough Sounds, SI. It grows in coastal sites on a range of substrates. It is apparently locally naturalised at a number of sites in New Zealand (de Lange & Cameron 1992) and in Ireland (Webb 1972), and it and/or a hybrid derived from it are reportedly locally naturalised in Hawai'i and on the Monterey Peninsula, California, USA (de Lange & Cameron 1992).

NOTES Distinguished from most other hebes by its magenta flowers and large, thick, dark green, glossy leaves. It is distinguished from *H. brevifolia*, the only other *Hebe* that can have similarly coloured flowers, by possession of a conspicuous leaf bud sinus (Fig. 150C) and by usually broader leaves.

H. speciosa is common in cultivation, and is a parent of many widely cultivated hybrids (Heenan 2001; Metcalf 2001). Detailed notes on the distribution of the species are provided by de Lange & Cameron (1992) and Ganley & Collins (1999). Armstrong & de Lange (2005) present a study of genetic variation showing that the three northern-most populations contain the majority of the species' genetic diversity, and suggesting that populations south of Auckland may be the product of historical cultivation and dispersal by Māori.

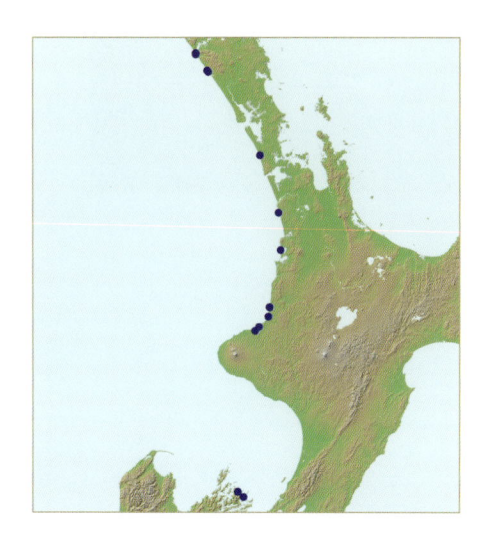

ETYMOLOGY L. (*speciosus* = showy, splendid).

Hebe speciosa — Large-leaved APERTAE

FIG. 150 **A** plant on Maunganui Bluff. **B** sprig. **C** leaf bud with rounded sinus. **D** lower and upper leaf surfaces. **E** inflorescence (left) and infructescence. **F** close-up of inflorescence. **G** seeds. **H** lateral view of flower. **I** capsule in septicidal (left) and loculicidal view.

TAXONOMIC TREATMENT 285

"Grandiflorae"

CRITICAL FEATURES spreading low shrub, decumbent, open, to 0.3(–0.5) m tall; **leaf buds indistinct** and surrounded by recently diverged leaves; leaf bases free; leaves not appressed or scale-like, **conspicuously toothed**, not glaucous, abscising at nodes; inflorescences lateral; **capsule angustiseptate**.

NUMBER OF SPECIES 1   DISTRIBUTION Mountains of SI

87. *Hebe macrantha*   (Hook.f.) Cockayne et Allan

DESCRIPTION Spreading low shrub to 0.3(–0.5) m tall. **Branches** erect or spreading or decumbent, old stems mottled grey; branchlets green or red-brown or brown, glabrous or puberulent (us. only on very youngest branchlets), hairs uniform or bifarious; internodes 1–14 mm; leaf decurrencies obscure to swollen. **Leaf bud** indistinct and tightly surrounded by recently diverged leaves. **Leaves** erecto-patent to patent; *lamina* obovate or elliptic (often broadly) or spathulate or sub-circular, coriaceous, slightly concave or flat, 5.5–30.2 × 2.5–13.5 mm; *apex* subacute or obtuse; *midrib* not thickened, or depressed to grooved above and thickened below, sts evident in fresh leaves; *margin* glabrous or sparsely ciliolate, sts tinged red, deeply toothed; *upper surface* green (sts tinged red near base), dull, with many stomata, glabrous or hairy along midrib; *lower surface* green, hairy along midrib (us.) or glabrous; *petiole* 0.5–5.4 mm. **Inflorescences** with 2–7 flowers, lateral, unbranched, 0.8–5.7 cm, with all flowers (including those near the apex) generally developing to maturity (but inflorescence us. terminated by a pair of empty bracts); peduncle (0.15–)0.3– 3.1 cm; rachis 0.2–2.7 cm. **Bracts** lowermost pair opposite, then subopposite or alternate above, narrowly deltoid or linear, obtuse (with a ± squarish tip, often with a sunken apical gland). **Flowers** (Nov–)Dec–Feb(–Apr), ⚥. **Pedicels** shorter than bracts, (0.5–)1–6(–14.7) mm. **Calyx** 4.7–10.2 mm; *lobes* ovate or lanceolate, acuminate (with a ± squarish tip, often with a sunken apical gland), glabrous outside (but hairy inside). **Corolla** tube glabrous, 4.5–5.3 × 2.5–4.5 mm (longer on anterior side), funnelform, shorter than calyx; *lobes* white at anthesis, ovate (often broadly), obtuse, patent to recurved, longer than corolla tube; corolla throat white or yellow. **Stamen filaments** thick and white, diverging slightly with age or remaining erect (and us. slightly incurved at the apex), 8–9.5 mm (varying from very shortly fused to the base of the corolla tube, to fused to the corolla tube for up to ⅔ its length); anthers creamy white or yellow, 2.6–3.5 mm. **Ovary** 1.6–2.5 mm; ovules 15–28 per locule, scattered on a hemispherical placenta; style 5.5–9.3 mm. **Capsules** [Dec–]Feb–May(–Nov), angustiseptate, acute, 6.5–12.4 mm long, 4.5–6.5 mm thick, loculicidal split extending ¼–all way to base. **Seeds** flattened (sts strongly), discoid, winged, pale brown, 1.5–2.7 × 1.2–2.2 mm, MR 0.6–1 mm. $2n = 42$.

Var. *macrantha* Leaves narrowly elliptic or oblanceolate, ratio of lamina length/width 1.2–3.3(–3.8), distance from leaf base (including petiole) to widest point (5.1–)10–20(–23.2) mm, number of teeth on one side of leaf (2–)3–7(–11). Peduncle (6–)7–30.7 mm long. Lowermost bracts on inflorescences (4–)5–9.1 mm long.

Var. *brachyphylla* (Cheeseman) Cockayne et Allan   Leaves broadly elliptic, ratio of lamina length/width (0.9–)1.1–2.2(–2.9), distance from leaf base (including petiole) to widest point (4.4–)6–11(–13.8) mm, number of teeth on one side of leaf (0–)1–4(–5). Peduncle (1.5–)3–13 mm long. Lowermost bracts on inflorescences 2–4(–8) mm long.

DISTRIBUTION AND HABITAT Var. *macrantha* occurs on mountains of SI, chiefly on or west of the Main Divide, from near Lk. Tennyson, to the Franklin Mts. Var. *brachyphylla* occurs from the Anatoki Ra., to the Hanmer Ra. The varieties overlap in distribution at Lk. Tennyson, southern Nelson, and possibly also (not verified by specimens) at localities between there and Lewis Pass. Both grow in penalpine grassland or low shrubland.

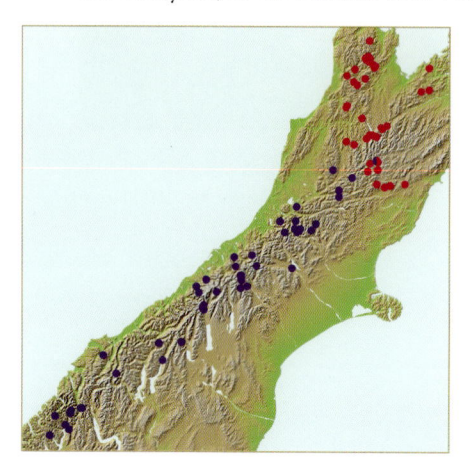

NOTES A distinctive species distinguished from other hebes by its large flowers, toothed leaves, indistinct leaf bud and laterally compressed capsules. Morphological variation, and the taxonomic status of the two varieties, are discussed by Bayly et al. (2004). Var. *macrantha* is more variable, particularly in leaf shape and size, than var. *brachyphylla* (Appendix 4).

ETYMOLOGY *Macrantha* (Gk *macros* = large, great; *anthos* = flower) refers to the characteristically large flowers; *brachyphylla* (Gk *brachys* = short, little; *phyllon* = leaf), means short-leaved.

- Var. *brachyphylla*
- Var. *macrantha*

Hebe macrantha   GRANDIFLORAE

FIG. 151 **A** var. *brachyphylla* on Mt Arthur. **B** sprigs of var. *macrantha* (left) and var. *brachyphylla*. **C** apex of shoot showing indistinct leaf bud (young leaves diverge early). **D** lower and upper leaf surfaces of var. *macrantha* (top) and var. *brachyphylla*. **E** frontal view of flower. **F** lateral view of flower. **G** inflorescence with some corollas removed (left) and infructescence. **H** capsule in septicidal (left) and loculicidal view.

TAXONOMIC TREATMENT   287

"Pauciflorae"

CRITICAL FEATURES spreading low shrub, decumbent to erect, to 0.2 m tall; **leaf buds surrounded by recently diverged leaves; leaf bases very minutely connate; leaves overlapping on stem but not appressed or scale-like**, entire, not glaucous, either abscising above nodes or not readily abscising and persistent along the stem for some distance; **inflorescences lateral; capsule angustiseptate**.

NUMBER OF SPECIES 1   DISTRIBUTION Mountains of Fiordland, SI

88. *Hebe pauciflora*   G.Simpson et J.S.Thomson

DESCRIPTION Spreading low shrub to 0.2 m tall. **Branches** decumbent to erect, old stems brown to grey; branchlets green or red-brown or brown, pubescent, hairs bifarious; internodes 1–2(–3) mm; leaf decurrencies evident and extended for length of internode; leaves either abscising above nodes or not readily abscising. **Leaf bud** tightly surrounded by recently diverged leaves; sinus broad and acute. **Leaves** very shortly connate, erecto-patent; *lamina* ovate to circular or deltoid (leaves spathulate when shape of petiole is considered), rigid, flat to concave, (2–)3–5(–6) × (2–)3–4(–6) mm; *apex* subacute to obtuse or subapiculate; *midrib* slightly thickened below (forming keel), but not strongly evident in fresh leaves; *margin* minutely papillate and ciliate (often with tangled, branching hairs); *upper surface* dark green to yellowish-green, glossy, with many stomata, glabrous; *lower surface* dark green to yellowish-green, glossy; *petiole* 1–1.5(–2) mm. **Juvenile and reversion leaves** entire, ciliate. **Inflorescences** with 2(–4) flowers, lateral, unbranched, 0.6–1.35 cm, longer than subtending leaves (if including the length of flowers); peduncle 0–0.2 cm; rachis 0–0.2 cm. **Bracts** opposite and decussate, connate, lanceolate to deltoid, subacute. **Flowers** Dec–Mar, ♀. **Pedicels** always shorter than bracts, *c.* 1 mm, glabrous or hairy. **Calyx** 4–6 mm; *lobes* oblanceolate, acute, minutely egl. ciliate (with branching hairs). **Corolla** tube glabrous, *c.* 3 × 1 mm, slightly contracted at mouth, equalling calyx; *lobes* white at anthesis, circular or elliptic (anterior only), obtuse, suberect to recurved, equalling corolla tube. **Stamen filaments** remaining erect, 2 mm; anthers pink to magenta or mauve, 1.5–2 mm. **Ovary** globose to ellipsoid, *c.* 1 mm; ovules 12–14 per locule; style 1.8–3 mm. **Capsules** Jan–Mar, angustiseptate, didymous, 4.5–5 mm long, 1.5–2 mm wide, *c.* 4 mm thick, septicidal split extending ¾-way to base, loculicidal split extending to base. **Seeds** weakly flattened, ellipsoid, pale brown, 1–1.5 × 0.8–1.2 mm. $2n = 42$.

DISTRIBUTION AND HABITAT Mountains of Fiordland, from Caswell Sound in the northwest to near Lk. Hauroko in the southeast. It grows in alpine grassland or low shrubland.

NOTES A distinctive species, recognised by having: entire, petiolate leaves with a fringe of fine, white, tangled and sometimes branched hairs on the margins (Fig. 152F); lateral inflorescences; large flowers with broad corolla lobes; and angustiseptate capsules. Its relationship to other members of *Hebe* is not clear. It resembles species of "Buxifoliatae" (with which it was included by Moore, in Allan 1961) and "Connatae" in vegetative characters, but differs in capsule and flower shape (and in these, presumably plesiomorphic, features resembles some members of *Parahebe*).

ETYMOLOGY L. (*pauci-* = few), means few-flowered.

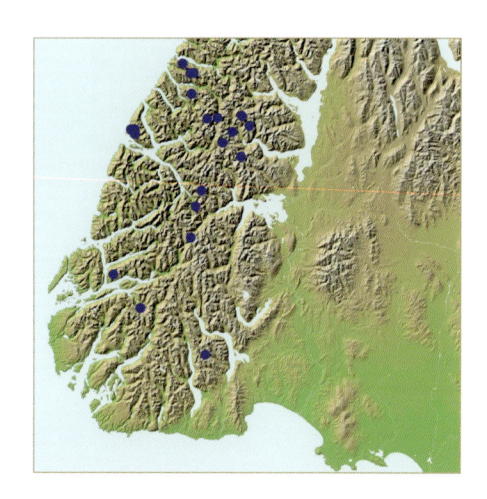

Hebe pauciflora PAUCIFLORAE

FIG. 152　A plant near Centre Pass, Southland. B sprig. C leaf bud that is not tightly closed, but has a sinus between the petioles of a leaf pair. D lower and upper leaf surfaces. E apical view of leaf bud. F close-up of leaf margin showing white tangled hairs. Also evident are stomata on the upper leaf surface. G frontal view of flower. H lateral view of flower. I septicidal view of capsule. J loculicidal view of capsule.

Leonohebe Heads

DESCRIPTION Subshrubs or bushy shrubs, 0.05–1.5(–2) m tall. **Branches** prostrate to erect; branchlets hairy on connate leaf bases or glabrous, 1–8.5 mm wide (including leaves); internodes 0.15–6.5 mm. **Leaf bud** tightly surrounded by recently diverged leaves. **Leaves** decussate, connate, appressed (us.) to erecto-patent, oblong or deltoid or semicircular, 0.8–4 × 0.4–2 mm; *lamina surfaces* glossy to glaucous; lower surface glabrous or covered with minute gl. hairs; *apex* obtuse to acute; *margin* hairy, entire; *venation* not evident in fresh leaves. **Juvenile leaves and reversion leaves** pinnatifid to entire, glabrous or hairy. **Inflorescences** with 2–22 flowers, lateral (sect. *Leonohebe*) or terminal (*L. cupressoides*), unbranched, 0.2–3.7 cm, longer than subtending leaves; peduncle 0.05–0.4 cm; rachis 0.2–3.25 cm; bracts opposite and decussate, connate or free. **Flowers** ♂ or ♀ (on different plants) or ⚥. **Pedicels** always shorter than bracts, 0–1 mm. **Calyx** 1.3–3.5 mm, (2–)4-lobed, anterior lobes free for most of their length or (in *L. cupressoides*) partly or wholly united, posterior lobes free for most of their length or (in *L. cupressoides*) united. **Corolla** tube sts hairy inside, 0.5–2.3 × 0.8–2 mm (♂ flowers 1–2.3 × 1.3–2 mm; ♀ flowers 0.5–1.6 × 1–1.6 mm), cylindric to funnelform, shorter than or equalling calyx; *lobes* white or coloured, obtuse, suberect to recurved, longer than or equalling corolla tube, glabrous or sts papillate (inside); corolla throat white or coloured. **Stamen filaments** 2, epipetalous and inserted on either side of posterior corolla lobe, white (us.) or coloured, straight at apex in bud, remaining erect or diverging with age, 0.4–3 mm; anthers pink or magenta or mauve or violet or purple, 0.9–1.7 mm; sterile anthers of ♀ flowers magenta or purple, 0.3–0.8 mm. **Nectarial disc** glabrous or ciliate. **Ovary** ovoid or globose or ellipsoid, glabrous, 0.4–1.3 mm, apex (in septicidal view) acute to didymous, bilocular; ovules 3–10 per locule, in 2 vertical rows or scattered on placenta; style 0.8–4 mm, glabrous; stigma capitate or bilobed or no wider than style. **Capsules** angustiseptate, emarginate or obtuse, 1.5–4.3 mm long, 0.9–3.3 mm thick (i.e. measurement at right angles to septum), glabrous, septicidal split extending ⅓ to all the way to base, loculicidal split extending ⅓ to all way to base. **Seeds** flattened (sts weakly), ellipsoid or ovoid or obovoid or discoid or oblong, not winged, smooth or finely papillate, brown (us. pale), 0.7–1.3 × 0.4–1.2 mm, MR 0.1–0.3 mm. $2n = 42$.

NUMBER OF SPECIES 5

DISTRIBUTION (Fig. 9) Mountains of SI, New Zealand

ETYMOLOGY Honours Leon C. M. Croizat (1894–1982) (Heads 1994c).

Sect. LEONOHEBE

Sect. *Leonohebe* (the Semiwhipcords)

CRITICAL FEATURES subshrubs, to *c.* 30 cm tall; leaf buds indistinct and tightly surrounded by recently diverged leaves; leaf bases connate; **leaves us. appressed and scale-like** (diverging most in *L. ciliolata*), entire, not glaucous, not readily abscising, persisting on the stem for some time; **inflorescences lateral**; capsule angustiseptate.

NUMBER OF SPECIES 4

DISTRIBUTION Mountains of SI

KEY TO SPECIES

1 Leafy branchlets almost square in cross section (e.g. Fig. 156D);
     leaves closely appressed .... 91. *Leonohebe cheesemanii*
   Leafy branchlets ± cruciform in cross section; leaves projecting out from stem 2

2 Leaves ± oblong above broad base (e.g. Fig. 154C); leaf apex square, usually with
     a small, sunken hydathode (e.g. Fig. 154E) .... 89. *Leonohebe ciliolata*
   Leaves deltoid or ovate; leaf apex not square, lacking hydathode 3

3 Leaves tumidly swollen, outer surface convex (e.g. Fig. 155C) .... 90. *Leonohebe tumida*
   Leaves thin, outer surface flat, or slightly concave in middle (e.g. Fig. 157C) .... 92. *Leonohebe tetrasticha*

FIG. 153 Male inflorescence of *Leonohebe cheesemanii*.

89. *Leonohebe ciliolata* (Hook.f.) Heads

DESCRIPTION Subshrub to 0.3 m tall, of semiwhipcord form. **Branches** decumbent; internodes 0.4–2(–2.5) mm; branchlets, including leaves, 3–8.5 mm wide, cruciform in TS; connate leaf bases glabrous; leaves not readily abscising, persistent along the stem for some distance. **Leaf bud** tightly surrounded by recently diverged leaves. **Leaves** connate, appressed to erecto-patent; *lamina* narrow oblong (above a broad base); venation not evident in fresh leaves; *margin* ciliate; *lower surface* dark green, glossy. **Inflorescences** with 2–6 flowers, lateral (obscuring vegetative tip when numerous), unbranched, 0.4–1 cm; peduncle 0.05–0.4 cm, hairy or glabrous; rachis glabrous or hairy. **Bracts** opposite and decussate, connate, deltoid or narrowly oblong, obtuse. **Flowers** (Oct–)Nov–Feb(–Jul), ♂ or ♀ (on different plants). **Pedicels** absent or if evident then always shorter than bracts, 0–1 mm, glabrous or hairy. **Calyx** 2.3–3.5 mm; *lobes* deltoid or oblong (often narrowly so), obtuse, with mixed gl. and egl. cilia, rarely hairy outside (esp. toward base). **Corolla** tube glabrous; tube of ♂ flowers 1.6–2.3 × 1.3–2 mm, cylindric (or sts slightly expanded around middle), shorter than or equalling calyx; tube of ♀ flowers 1.2–1.6 × 1.4–1.6 mm, funnelform and contracted at base (may also be expanded near middle), shorter than or equalling calyx; *lobes* white at anthesis, us. broadly ovate or rhomboid or obovate, obtuse, suberect to recurved, longer than corolla tube (often more markedly so in ♀ flowers). **Stamen filaments** remaining erect, 0.4–2.5 mm (♀ 0.4–1 mm; ♂ 2–2.5 mm); anthers purple or magenta, 1.3–1.7 mm; sterile anthers of ♀ flowers purple or magenta, 0.3–0.8 mm. **Ovary** ovoid or ellipsoid, 0.9–1.3 mm; ovules 5–10 per locule, in 2 vertical rows on placenta or scattered on a hemispherical placenta; style 1.5–2.7 mm (us. shorter in ♀ flowers than ♂ flowers); stigma larger in ♀ flowers, with long, multicellular papillae (papillae on stigmas of ♂ flowers are not prominent). **Capsules** Jan–Mar(–Aug), angustiseptate, obtuse, 3–4.3 mm long, 1.6–3.3 mm thick, septicidal split extending ½ to all way to base (us. to base), loculicidal split extending ⅓–¾-way to base. **Seeds** 0.8–1(–1.3) mm. $2n = 42$.

DISTRIBUTION AND HABITAT Mountains of SI, chiefly on or west of the Main Divide, from near Boulder Lk., northwest Nelson, to the Ben Ohau Ra., Canterbury, and possibly to Mt Alta, Otago (see below). It grows on alpine rock outcrops and boulder fields, often in exposed situations.

NOTES A distinctive species, distinguished from other semiwhipcords by having leaves ± oblong above a broad base (Fig. 154C), and a ± square leaf apex that usually has a small, sunken hydathode (Fig. 154E). It possibly intergrades, or hybridises, with *L. tumida* at some localities (see notes for that species). Wilson (1996) suggested that it may "merge with *H. tetrasticha*" [= *L. cheesemanii* as defined here], but herbarium specimens show no evidence of this.

The record from Mt Alta (southernmost point on the distribution map), is based on the account of Buchanan (1882; as *Mitrasacme hookeri*). However, the specimen illustrated in that article (Herb. Buchanan, WELT) gives no locality information. Buchanan's article also, and probably erroneously, recorded *L. cheesemanii* from Mt Alta.

ETYMOLOGY L. (*ciliolatus* = with small cilia, or fringing hairs), refers to the leaf margins.

Leonohebe ciliolata   Sect. **LEONOHEBE**

FIG. 154   **A** plant on Mt Haast. **B** sprig. **C** branchlet. **D** apical view of shoot apex, showing that branchlet is cruciform in cross section. **E** close-up of leaf apex with a small sunken hydathode (arrowed). **F** flowers of male plant. **G** flowers of female plant. **H** flowers of male plant, including partial lateral view of one opening flower. **I** branchlet with lateral infructescences.

TAXONOMIC TREATMENT   293

90. *Leonohebe tumida* (Kirk) Heads

DESCRIPTION Subshrub to 0.2 m tall, of semiwhipcord form. **Branches** decumbent; internodes (0.5–)0.8–2.6(–3.4) mm; branchlets, including leaves, 1.3–2.5(–3) mm wide, cruciform in TS; connate leaf bases glabrous; leaves not readily abscising, persistent along the stem for some distance. **Leaf bud** tightly surrounded by recently diverged leaves. **Leaves** connate, appressed; *lamina* deltoid (shortly and often bluntly so); venation not evident in fresh leaves; *margin* ciliate; *lower surface* dark green, glossy or dull. **Juvenile and reversion leaves** entire, ciliate and pubescent (with egl. hook-shaped hairs). **Inflorescences** with 2–8 flowers, lateral, unbranched, 0.2–0.9 cm; peduncle 0.15–0.3 cm. **Bracts** opposite and decussate, apparently free, deltoid or oblong, obtuse. **Flowers** Nov–Feb, us. ♂ or ♀ (on different plants) (but some specimens are possibly ⚥, appearing to have pollen and fruit). **Pedicels** absent or if evident then always shorter than bracts, 0–1 mm, glabrous or hairy. **Calyx** *c.* 2–2.5 mm; *lobes* oblong to slightly obovate, apex thickened and obtuse, with mixed gl. and egl. cilia (gl. cilia may be obscure). **Corolla** tube glabrous; tube of ♂ flowers 1.5–1.8 × *c.* 1.5 mm, cylindric or funnelform, shorter than or equalling calyx; tube of ♀ flowers shorter than calyx; *lobes* white at anthesis (pink to mauve in bud), ovate or circular or broadly elliptic, obtuse, suberect to patent, equalling corolla tube. **Stamen filaments** remaining erect, 1–1.2 mm; anthers purple, *c.* 1.1–1.3 mm; sterile anthers of ♀ flowers lighter purple. **Ovary** 0.5–0.7 mm; ovules 5–7 per locule, in 2 vertical rows on placenta; style 1–4 mm; stigma larger in ♀ flowers. **Capsules** Jan–Apr, angustiseptate, 1.5–3.2 mm long, 1.3–2.7 mm thick, septicidal split extending ⅓ to all way to base, loculicidal split extending to base. **Seeds** flattened, ellipsoid to discoid, pale brown, 0.8–1.1 × 0.7–1.2 mm, MR 0.1–0.3 mm. $2n = 42$.

DISTRIBUTION AND HABITAT Mountains of northern SI, including the Bryant, Richmond, Gordon, Travers, St Arnaud and Raglan ranges, as well as mountains between the Leatham and Waihopai rivers. It grows on alpine rock outcrops and sometimes on scree.

NOTES Distinguished from other semiwhipcords by tumidly swollen leaves that are convex on the outer surface (Fig. 155C).

In western and southern parts of its range it probably intergrades with *L. cheesemanii*. In the west, some specimens from Bounds (e.g. CHR 282641, 370042) and Pinnacle (CHR 366210) resemble *L. cheesemanii* in having leaves slightly more imbricate and less protruding/swollen than usual, making branchlets squarer in cross section. Specimens from nearby, on the Black Birch Ra. (CHR 174933, 470183; WELT 48113, 80937), are more extreme in these respects, and are assigned here to *L. cheesemanii*. These specimens, together with one from Crystal Peak, Crimea Ra. (CHR 282924; close to the southern geographic boundary between *L. tumida* and *L. cheesemanii*), have leaves more protruding/swollen than typically seen in *L. cheesemanii* (hence resembling *L. tumida*), but are included in that species on the basis of their branchlets, which are quite square in cross section.

L. tumida possibly also intergrades, or hybridises, with *L. ciliolata* at some localities. Specimens from Mt Robert, Travers Ra., are particularly variable; some (e.g. WELT 14044, 82477) most closely match *L. tumida*, some match *L. ciliolata* (e.g. CHR 379237), and others (e.g. CHR 122681, 270861, 184752, 379240) are intermediate, with leaves that are shorter and plumper than generally found in *L. ciliolata*, but both squarer at the apex and more oblong in their upper portion than usual in *L. tumida*. A sprig from a plant of this last kind is shown in Fig. 155B.

ETYMOLOGY L. (*tumidus* = swollen, protruding), refers to the shape of the leaves.

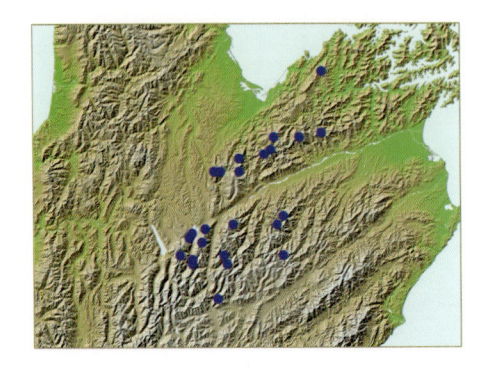

Leonohebe tumida Sect. LEONOHEBE

FIG. 155   A plant on Mt Richmond. B sprig of atypical plant from Mt Robert. C branchlet. D apical view of shoot apex. E inflorescences of male plant showing frontal view of flowers. F inflorescences of female plant showing frontal view of flowers. G inflorescence of male plant showing dorsal view of flower. H branchlet with lateral infructescence.

TAXONOMIC TREATMENT   295

91. *Leonohebe cheesemanii* (Buchanan) Heads

DESCRIPTION Subshrub to 0.3 m tall, of semiwhipcord form. **Branches** decumbent; internodes (0.15–)0.2–0.7(–0.8) mm; branchlets, including leaves, (1.3–)1.5–2.4(–2.7) mm wide, square in TS (or almost so, esp. older parts of branchlets, away from the apex); connate leaf bases glabrous; leaves not readily abscising, persistent along the stem for some distance. **Leaf bud** tightly surrounded by recently diverged leaves. **Leaves** connate, appressed; *lamina* deltoid (often broadly so) or semicircular (more rarely); *apex* obtuse to acute; venation not evident in fresh leaves; *margin* ciliate; *lower surface* light to dark green. **Inflorescences** with 2–6 flowers, lateral, unbranched, 0.3–0.5(–0.75) cm; peduncle 0.1–0.35 cm; rachis hairy or glabrous. **Bracts** opposite and decussate, free (us.) or connate, deltoid (and lowermost often keeled beneath), obtuse. **Flowers** Dec–Jan, ♂ or ♀ (on different plants). **Pedicels** absent or if evident then always shorter than bracts, 0–0.5 mm, glabrous or hairy. **Calyx** 1.3–2.2 mm; *lobes* deltoid or ovate or oblong, obtuse, with mixed gl. and egl. cilia (gl. cilia may not be apparent, stalk cell may be very short), rarely hairy outside. **Corolla** tube glabrous; tube of ♂ flowers *c.* 1–1.1 mm, shorter than calyx; tube of ♀ flowers 0.5–1.2 × *c.* 1 mm, contracted at base (and ± expanded around middle), shorter than or equalling calyx; *lobes* white or pink at anthesis, elliptic or ovate or rhomboid (and auriculate above contracted base), obtuse, suberect to recurved, longer than corolla tube, papillate inside. **Stamen filaments** remaining erect, 1.1–1.4 mm (♀ *c.* 1.1 mm; ♂ *c.* 1.4 mm); anthers magenta, 1–1.1 mm; sterile anthers of ♀ flowers magenta, *c.* 0.5 mm. **Nectarial disc** glabrous or ciliate. **Ovary** ovoid or ellipsoid, 0.4–0.6 mm; ovules 4–5 per locule, in 2 vertical rows on placenta; style 1.5–2.2 mm; stigma us. larger in ♀ flowers. **Capsules** Dec–Feb, angustiseptate, obtuse, 1.6–2.2 mm long, 1.3–2 mm thick, septicidal split extending ⅓ to all way to base, loculicidal split extending ½–¾-way to base. **Seeds** weakly flattened to ± trigonal, ellipsoid, brown, 1–1.2 × 0.5–0.7 mm, MR 0.2–0.3 mm. $2n = 42$.

DISTRIBUTION AND HABITAT Mountains of SI, chiefly east of the Main Divide, from Black Birch Ra., Marlborough, to Kirkliston Ra., south Canterbury. It grows on alpine rock outcrops and scree.

NOTES Possibly intergrades with *L. tumida*, and differences from *L. tetrasticha* are not always clear-cut (see notes under those species).

ETYMOLOGY Honours Thomas F. Cheeseman (1846–1923), curator of the Auckland Museum and author of *Manual of the New Zealand Flora*.

Leonohebe cheesemanii — Sect. LEONOHEBE

FIG. 156 **A** plant on Four Peaks Ra. **B** sprig. **C** branchlets from two different plants. **D** apical view of shoot apex showing that branchlet is square in cross section. **E** inflorescence of male plant showing frontal view of flowers. **F** inflorescences of female plant. **G** dorsal view of male flower. **H** branchlet with lateral infructescence.

92. *Leonohebe tetrasticha* (Hook.f.) Heads

DESCRIPTION Subshrub to 0.2 m tall, of semiwhipcord form. **Branches** decumbent; internodes (0.15–)0.25–0.5 mm; branchlets, including leaves, (1.5–)2–3.5(–4) mm wide, cruciform in TS; connate leaf bases glabrous; leaves not readily abscising, persistent along the stem for some distance. **Leaf bud** tightly surrounded by recently diverged leaves. **Leaves** connate, appressed; *lamina* deltoid; venation not evident in fresh leaves; *margin* ciliate; *lower surface* light to dark green. **Juvenile leaves** entire, pubescent (with egl. hook-shaped hairs). **Inflorescences** with 2–6 flowers, lateral (obscuring vegetative tip when numerous), unbranched, (0.2–)0.3–0.7 cm; peduncle 0.05–0.2 cm. **Bracts** opposite and decussate, connate or free (lowest us. free, but sts shortly connate; upper often shortly connate), deltoid, obtuse. **Flowers** (Nov–)Dec–Jan, ♂ or ♀ (on different plants). **Pedicels** absent or if evident then always shorter than bracts, 0–0.7 mm. **Calyx** 1.3–2 mm; *lobes* ovate or deltoid, obtuse, with mixed gl. and egl. cilia (but gl. hairs may be obscure). **Corolla** tube glabrous; tube of ♂ flowers *c.* 1.5 × *c.* 1.5 mm, contracted at base, equalling calyx; tube of ♀ flowers *c.* 1 × *c.* 1 mm, contracted at base, shorter than calyx; *lobes* white at anthesis, ovate or rhomboid (♂ flowers only), obtuse, suberect to patent, longer than or *c.* equalling corolla tube. **Stamen filaments** remaining erect, 0.7–1.5 mm (♂ *c.* 1.5 mm; ♀ 0.7–1 mm); anthers purple or violet to magenta, 1–1.2 mm; sterile anthers of ♀ flowers *c.* 0.5 mm. **Ovary** 0.5–0.6 mm; ovules 3–6 per locule, in 2 vertical rows on placenta; style 0.8–1.5 mm (often longer in ♂ flowers than ♀ flowers); stigma more prominent in ♀ flowers. **Capsules** Dec–Feb, angustiseptate, obtuse, 2–3 mm long, 1.5–2.5 mm thick. **Seeds** flattened, ellipsoid to oblong, smooth or finely papillate, pale brown (to orange), 0.8–1 × 0.5–0.7 mm, MR 0.1–0.3 mm. $2n = 42$.

DISTRIBUTION AND HABITAT Mountains of Canterbury (mostly) and Westland, from the Otira Va. in the northwest and Puketeraki Ra. in the northeast to Mt Somers in the south. It grows on alpine rocks and scree.

NOTES Differences from *L. cheesemanii* require clarification. The two species are distinguished primarily on differences in the profile of branchlets in TS (square, at least on older branchlets, in *L. cheesemanii*; cruciform in *L. tetrasticha*). However, branchlet profiles vary between the two extremes, and differences are not always clear-cut. Plants identified here as *L. cheesemanii*, but with features approaching *L. tetrasticha*, occur as far north as the Amuri District, as far west as the Leibig Ra. (Aoraki/Mt Cook NP), and as far south as the Kirkliston Ra. Some of these plants have branchlets that are prominently cruciform near the apex (where leaves are still expanding), but become ± square with age. Whether some of these plants are better placed under *L. tetrasticha* is debatable, and at least some have been treated as such by Wilson (1978, 1996), Macdonald (1980) and Heads (1994*a*).

From the distributions given here, *L. tetrasticha* effectively "replaces" *L. cheesemanii* in parts of mid-Canterbury. A possible explanation for such a pattern is that *L. tetrasticha* has differentiated in this area from an historically more widespread, *L. cheesemanii*-like ancestor. Of course, other scenarios are possible (different interpretations of the limits of the species might contradict this one), and whether *L. tetrasticha* and *L. cheesemanii* are most closely related is also not known. Analysis of ITS sequences (Wagstaff et al. 2002) does not support a sister relationship, but similarity of flavonoid profiles (Markham et al. 2005) might.

ETYMOLOGY L., means "in four rows" (*tetra* = preface meaning four of; *stichos* = a row or line of things), and refers to the leaves.

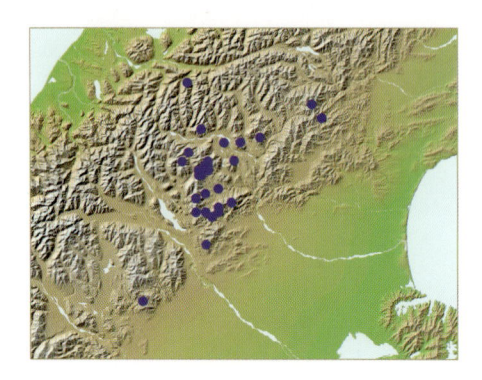

Leonohebe tetrasticha — Sect. LEONOHEBE

FIG. 157 **A** plant near Lk. Lyndon. **B** sprig. **C** branchlet. **D** apical view of shoot apex showing that branchlet is cruciform in cross section. **E** inflorescences of male plant showing frontal view of flowers. **F** inflorescences of female plant showing frontal view of flowers. **G** inflorescences of male plant showing lateral view of flowers. **H** branchlet with young, lateral infructescences.

TAXONOMIC TREATMENT 299

Sect. *Aromaticae* (*Leonohebe cupressoides*)

CRITICAL FEATURES Erect shrub, to 1.5(–2) m tall; leaf buds indistinct and tightly surrounded by recently diverged leaves; leaf bases connate; **leaves appressed, scale-like**, entire, **us. glaucous to glaucescent**, persistent on non-woody stems; **inflorescences terminal**; capsule angustiseptate.

NUMBER OF SPECIES 1   DISTRIBUTION Mountains of SI

93. *Leonohebe cupressoides*   (Hook.f.) Heads

DESCRIPTION Bushy shrub to 1.5(–2) m tall, of whipcord form. **Branches** erect; branchlets green or grey-green, ± glaucous; internodes 1.5–6.5 mm; branchlets, including leaves, 1–3.7 mm wide (0.5–1.5 mm excluding leaves); connate leaf bases hairy or glabrous; nodal joint distinct, exposed; leaves not readily abscising and persistent along the stem for some distance (although either obscure or eventually abscising on woodier branches). **Leaf bud** tightly surrounded by recently diverged leaves. **Leaves** connate, appressed (or slightly spreading when dry); *lamina* deltoid, 0.8–2 × 0.4–2 mm; *apex* acute to obtuse; *margin* ciliolate or gl.-ciliate; *lower surface* glaucous or glaucescent or yellowish-green, veins not visible, glabrous or covered with minute gl. hairs. **Juvenile leaves** pinnatifid, glabrous or puberulent. **Inflorescences** with 2–22 flowers, terminal, unbranched, 0.3–3.7 cm; rachis 0.2–3.3 cm, glabrous or hairy. **Bracts** opposite and decussate, shortly connate or free, ovate to deltoid, obtuse or subacute, hairy outside (hairs gl.). **Flowers** Dec–Jan[–Feb], ⚥. **Pedicels** absent or if evident then always shorter than bracts. **Calyx** 1.3–2 mm, 2–4-lobed, with anterior lobes united ⅔ (us.) to all way to apex, with posterior lobes united (us. all the way or almost so); *lobes* acuminate or emarginate, gl. ciliolate, often densely hairy outside (with twin-headed gl. hairs). **Corolla** tube hairy inside, 0.9–1.4 × 0.8–1.1 mm; *lobes* white or pale blue or pink or mauve at anthesis, white or pink or mauve with age, obtuse, suberect to recurved, longer than corolla tube, papillate inside; corolla throat pink or mauve or white. **Stamen filaments** coloured (when young), fading white, 2.1–3 mm; anthers reddish-pink to purplish-mauve, 0.9–1.2 mm. **Nectarial disc** glabrous or ciliolate. **Ovary** ovoid or globose, 0.8–1.1 mm, apex (in septum view) didymous; ovules scattered on a hemispherical placenta. **Capsules** Feb–May(–Dec), angustiseptate (but also ± flattened on narrow-obovate anterior and posterior faces), grooved along the septum, emarginate, 1.9–2.4 mm long, 0.9–1.4 mm thick, septicidal split extending ⅓-way to base, loculicidal split extending up to ⅓-way to base. **Seeds** weakly flattened, ovoid to ellipsoid-oblong or obovoid, pale brown, 0.7–1.1 × 0.4–0.6 mm, MR *c.* 0.1 mm. $2n = 42$.

DISTRIBUTION AND HABITAT Drier mountains of SI, east of the Main Divide, from Wairau Gorge, Nelson (historical records only), to The Remarkables, Otago. It grows in grey scrub (defined by Meurk et al. 1987 and Wardle 1991) on rock outcrops, bouldery moraine, and slips, usually near rivers and lakes.

NOTES A distinctive species, not readily confused with other leonohebes. It is distinguished from whipcord *Hebe* species by the combination of grey-green branchlets, widely spaced leaves, pink to mauve flowers, conspicuous nodal joints, fused anterior and posterior calyx lobes, and laterally compressed ovaries and fruit. The presence of terpenes (Perry & Foster 1994) probably gives rise to the often aromatic foliage. *L. cupressoides* is commonly cultivated in New Zealand, but does not usually flower in NI gardens. Detailed notes on distribution (present and past), ecology and conservation are provided by Widyatmoko & Norton (1997) and Norton (2000). Their distribution maps differ slightly from that shown here, because they include details from additional herbarium specimens, and some extant populations not represented by specimens.

ETYMOLOGY Resembling cypress, genus *Cupressus*, family Cupressaceae (the Gk suffix -*oides* = resembling, like).

Leonohebe cupressoides

Sect.
AROMATICAE

FIG. 158 A plant near Cave Stm, mid-Canterbury. **B** sprig. **C** branchlet. **D** frontal view of flower. **E** inflorescence showing lateral view of flowers. **F** young infructescence. **G** loculicidal view of ovary. **H** fused anterior calyx lobes of flower. **I** fused posterior calyx lobes of flower. **J** septicidal view of ovary.

TAXONOMIC TREATMENT 301

Nomenclature

The application of botanical names and designation of type specimens in this work follows the rules and recommendations of the ICBN (Greuter et al. 2000). The term isolectotype, which is not defined in the ICBN, is used here, as in other similar works, for any specimen that is considered a duplicate of the lectotype. The symbols "≡" and "=" indicate nomenclatural and taxonomic synonyms, respectively. An exclamation mark (!) after a specimen indicates that it has been examined by us. Abbreviations of author names follow Brummitt & Powell (1992).

Accepted here are a number of combinations in *Hebe* published by Andersen (1926). These combinations were not used in the treatment of Moore & Ashwin (in Allan 1961), who attributed combinations for most of the same taxa in *Hebe* to the later work of Cockayne & Allan (1926c). Since Andersen's names seem acceptable under the ICBN, names later published by Cockayne & Allan have no formal status, but are included here in the lists of synonyms to help allay confusion about correct author citation.

NAMES OF GENERA

Hebe Comm. ex Juss., *Genera Plantarum*: 105 (1789).
TYPE: *Hebe magellanica* J.F.Gmel., *Systema Naturae* 2: 27 (1791) [= *H. elliptica* (G.Forst.) Pennell].

= *Panoxis* Raf., *Medical Flora* 2: 109 (1830).
LECTOTYPE (here designated): *Veronica salicifolia* G.Forst.

Leonohebe Heads, *Botanical Society of Otago Newsletter* 5: 4 (1987).
TYPE: *Leonohebe ciliolata* (Hook.f.) Heads.

NAMES OF SUBGENERA

This list includes subgenus names that could be applied to taxa placed within the circumscriptions of *Hebe* and *Leonohebe* adopted here (even though these names currently exist only in the genus *Veronica*).

Veronica subg. *Koromika* J.B.Armstr., *Transactions of the New Zealand Institute* 13: 349 (1881).
LECTOTYPE (designated by Bayly & Kellow 2004b): *Veronica pubescens* Benth.

Veronica subg. *Pseudoveronica* J.B.Armstr., *Transactions of the New Zealand Institute* 13: 351 (1881).
LECTOTYPE (designated by Bayly & Kellow 2004b): *Veronica lycopodioides* Hook.f.

NAMES OF SECTIONS

This list includes section names that could be applied to taxa placed within the circumscriptions of *Hebe* and *Leonohebe* adopted here. Several of the names that exist only in *Leonohebe* are based on types that are here placed in *Hebe*.

Veronica sect. *Hebe* (Comm. ex Juss.) Benth. in DC., *Prodromus Systematis Naturalis Regni Vegetabilis. Vol. 10*: 459 (1846).
TYPE: *Hebe magellanica* J.F.Gmel. [= *H. elliptica* (G.Forst.) Pennell].

Leonohebe Heads sect. *Leonohebe*, an autonym established by the publication of other section names in *Leonohebe* by Heads (1987).
TYPE: *Leonohebe ciliolata* (Hook.f.) Heads.

Leonohebe sect. *Connatae* Heads, *Botanical Society of Otago Newsletter* 5: 6 (1987).
TYPE: *Leonohebe epacridea* (Hook.f.) Heads.

Leonohebe sect. *Apiti* Heads, *Botanical Society of Otago Newsletter* 5: 7 (1987).
TYPE: *Leonohebe benthamii* (Hook.f.) Heads.

Leonohebe sect. *Salicornioides* Heads, *Botanical Society of Otago Newsletter* 5: 7 (1987).
TYPE: *Leonohebe saliconioides* (Hook.f.) Heads.

Leonohebe sect. *Aromaticae* Heads, *Botanical Society of Otago Newsletter* 5: 8 (1987).
TYPE: *Leonohebe cupressoides* (Hook.f.) Heads.

Leonohebe sect. *Flagriformes* Heads, *Botanical Society of Otago Newsletter* 5: 8 (1987).
TYPE: *Leonohebe hectorii* (Hook.f.) Heads.

Leonohebe sect. *Buxifoliatae* Heads, *Botanical Society of Otago Newsletter* 5: 10 (1987).
TYPE: *Leonohebe odora* (Hook.f.) Heads.

Hebe sect. *Hebe*, an autonym established by the publication of other section names in *Hebe* by Heads (1987).
TYPE: *Hebe magellanica* J.F.Gmel. [= *H. elliptica* (G.Forst.) Pennell].

Hebe sect. *Subdistichae* Heads, *Botanical Society of Otago Newsletter* 5: 11 (1987).
TYPE: *Hebe diosmifolia* (A.Cunn.) Andersen.

Hebe sect. *Glaucae* Heads, *Botanical Society of Otago Newsletter* 5: 11 (1987).
TYPE: *Hebe pinguifolia* (Hook.f.) Cockayne et Allan.

NOTE: Hooker (1864) used the rank of section in his classification (evident from the notes presented prior to his key), but labelled these with numbers rather than names.

NAMES OF SERIES

Hebe sect. *Hebe* ser. *Hebe*, an autonym established by the publication of other series names in *Hebe* by Heads (1987).
TYPE: *Hebe magellanica* J.F.Gmel. [= *H. elliptica* (G.Forst.) Pennell].

Hebe sect. *Hebe* ser. *Occlusae* Heads, *Botanical Society of Otago Newsletter* 5: 11 (1987).
TYPE: *Hebe macrocarpa* (Vahl) Cockayne et Allan.

SUBSECTIONAL NAMES WITHOUT A CLEAR INDICATION OF RANK

Of the following names, those provided by Bentham (in DC., *Prodromus Systematis Naturalis Regni Vegetabilis. Vol. 10*: 1846) are derived from the names of included species. In accordance with the ICBN (Greuter et al. 2000; Art. 22.6), these names have the same type as the constituent species from which they are derived.

Veronica 1. *Speciosae* Benth. in DC., *Prodromus Systematis Naturalis Regni Vegetabilis. Vol. 10*: 459 (1846).
TYPE: *Veronica speciosa* R.Cunn. ex A.Cunn.

Veronica 2. *Decussatae* Benth. in DC., *Prodromus Systematis Naturalis Regni Vegetabilis. Vol. 10*: 460 (1846).
TYPE: *Veronica decussata* Aiton [= *H. elliptica* (G.Forst.) Pennell].

Veronica 1. *Integrae* Wettst., *Die Natürlichen Pflanzenfamilien* 4(3b): 86 (1891).
TYPE: None designated.

Veronica 2. *Serratae* Wettst., *Die Natürlichen Pflanzenfamilien* 4(3b): 86 (1891).
LECTOTYPE (here designated): *Veronica benthamii* Hook.f. [This is one of the two species listed as examples for the name, the other being *Veronica hulkeana* F.Muell. *V. benthamii* better matches the (very brief) description in that it more commonly has a five-partite corolla].

NAMES OF SPECIES AND INFRASPECIFIC TAXA IN *HEBE*

Hebe acutiflora Cockayne, *Transactions of the New Zealand Institute* 60: 468 (1929).
≡ *Veronica acutiflora* Benth. in DC., *Prodromus Systematis Naturalis Regni Vegetabilis. Vol. 10*: 460 (1846),

nom. illeg., non Roem. et Schult. *Systema Vegetabilium* 1: 112 (1817); ≡ *Veronica ligustrifolia* var. *acutiflora* Hook.f., *Flora Novae-Zelandiae* 1: 192 (1853).
HOLOTYPE: Northn Isld, New Zealand, *A. Cunningham 377*, 1838, Herb. Hookerianum, K!. ISOTYPES: K!, WELT 79332!.
NOTES: Bentham clearly stated that the type was in Herb. Hookerianum, so the specimen in that herbarium is the holotype. The isotype at K, not part of Hooker's herbarium, was not accessioned there until after publication of the protologue.

Hebe adamsii (Cheeseman) Cockayne et Allan, *Transactions of the New Zealand Institute* 57: 15 (1926).
≡ *Veronica adamsii* Cheeseman, *Manual of the New Zealand Flora*, 2nd edn: 786 (1925).
LECTOTYPE (designated by Garnock-Jones & Clarkson 1994): Kapowairua to Tom Bowline's Bay, North Cape, *T. F. C[heeseman]*, Jan 1896, 1545 to Kew, AK 7666!. ISOLECTOTYPES: AK 7665!, 203330!; WELT 79581! (although the wording of locality details on the label of this last sheet differs slightly from that on the lectotype).

Hebe albicans (Petrie) Cockayne, *Transactions of the New Zealand Institute* 60: 468 (1929) [Note: Moore (in Allan 1961) incorrectly cites p. 469].
≡ *Veronica albicans* Petrie, *Transactions of the New Zealand Institute* 49: 53 (1917).
LECTOTYPE (designated by Kellow et al. 2005): Mt Cobb, N.W. Nelson, *H. J. Matthews*, Feb 1909, WELT 16954!.

= *Hebe recurva* G.Simpson et J.S.Thomson, *Transactions of the Royal Society of New Zealand* 70: 32 (1940).
LECTOTYPE (designated by Moore, in Allan 1961): near Bainham, Aorere River, Nelson, rock platforms on river banks, *G. Simpson & J. S. Thomson*, CHR 33032!.

Hebe amplexicaulis (J.B.Armstr.) Cockayne et Allan, *Transactions of the New Zealand Institute* 56: 26 (1926).
≡ *Veronica amplexicaulis* J.B.Armstr., *New Zealand Country Journal* 3: 56 (1879).
HOLOTYPE: Upper Rangitata, *J. F. Armstrong*, 1869, CHR (Herb. Armstrong)!.

f. *amplexicaulis*
= *Hebe amplexicaulis* var. *erecta* Cockayne et Allan, *Transactions of the New Zealand Institute* 56: 26 (1926).
LECTOTYPE (designated by Moore, in Allan 1961): Eastern South Island, stream entering Rangitata from south above gorge, *H. H. A[llan]*, 2 Jan 1919, erect form with close strict branches, CHR 10805! [piece mounted on right of sheet only; piece on left is *H. amplexicaulis* f. *hirta*].
NOTES: The small sprig on CHR 89156! could be from the same branch as the lectotype, and should probably be considered part of the type collection. CHR 10805 is considered a lectotype, rather than a holotype, because at least one other specimen, in a different herbarium (e.g. on rocks by streamside, upper Rangitata river, *H. H. Allan*, 2 Jan 1919, K!), also matches details cited in the protologue. That K specimen could potentially be considered an isolectotype, but differences in wording of the locality information leave some doubt as to whether it should be considered part of the same "gathering" as the lectotype.

= *Hebe amplexicaulis* var. *suberecta* Cockayne et Allan, *Transactions of the New Zealand Institute* 56: 26 (1926).
LECTOTYPE (designated by Moore, in Moore & Edgar 1970): Eastern South Island, upper gorge of Lynn Stream, Mt Peel, *c.* 400 m, rock crevice, decumbent, abundant on rocks of northern and southern faces from 350–850 m, *H. H. A[llan]*, 3 Jan 1919, CHR 10820!. ISOLECTOTYPE: CHR 89155!.
NOTES: Moore (in Allan 1961) originally selected a different lectotype, but later amended this in the corrigenda to her treatment (in Moore & Edgar 1970).

f. *hirta* Garn.-Jones et Molloy, *New Zealand Journal of Botany* 20: 395 (1983).
≡ *Hebe allanii* Cockayne, *Transactions of the New Zealand Institute* 56: 25 (1926).
LECTOTYPE (designated by Moore, in Allan 1961): Eastern South Island, upper gorge of Lynn Stream near waterfall, Mt Peel, pubescent form, decumbent, *H. H. A[llan]*, 3 Jan 1919, CHR 10823!.
NOTES: Although a cover date of 1982 is shown on the volume of the *New Zealand Journal of Botany* in which the name f. *hirta* is proposed, this volume was not, as recorded in the subsequent volume, published until 11 Jan 1983.

Hebe angustissima (Cockayne) Bayly et Kellow, comb. nov.
≡ *Veronica salicifolia* var. *angustissima* Cockayne, *Transactions of the New Zealand Institute* 50: 184 (1918).
NEOTYPE (here designated): New Zealand, North Island, Wellington, Otaki Gorge, north of Pukehinau Stream, 100 m, rock face beside road, shrubs to *c*. 1.2 m tall, common, *M. J. Bayly 1177*, 13 Mar 1999, WELT 81521!.
NOTES: No original material has been found, hence the designation here of a neotype.

Hebe annulata (Petrie) Andersen, *Transactions of the New Zealand Institute* 56: 693 (1926).
≡ *Veronica armstrongii* J.B.Armstr. var. *annulata* Petrie, *Transactions of the New Zealand Institute* 45: 273 (1913); ≡ *Veronica annulata* (Petrie) Cheeseman, *Manual of the New Zealand Flora*, 2nd edn: 819 (1925) [Ashwin (in Allan 1961) incorrectly cited the page on which this combination was published]; ≡ *Hebe annulata* (Petrie) Cockayne et Allan, *Transactions of the New Zealand Institute* 57: 41 (1926), *nom. illeg.*, non Andersen; ≡ *Leonohebe annulata* (Petrie) Heads, *Botanical Society of Otago Newsletter* 5: 7 (1987).
LECTOTYPE (designated by Ashwin, in Allan 1961): on rock face of northern slope of Takitimu Mts, at about 900 m altitude, *L. Cockayne 7617*, 13 Mar 1912, WELT 5347!. ISOLECTOTYPES: K!, CHR 328337!, WELT 17493! [the Cockayne number 6717 on the last specimen is probably a transposition error].
NOTES: Cheeseman (1925) implied that the combination *Veronica annulata* was published by Cockayne (1919). This name was certainly used by Cockayne (pages 105 and 192), but since he made no reference to the basionym and provided no description, the name was not validly published in his book (Art. 33; Greuter et al. 2000). The authority could also be cited as *Veronica annulata* (Petrie) Cockayne ex Cheeseman, as suggested by Cockayne & Allan (1926c). WELT 17497! is probably also part of the type collection, but it is undated and has no Cockayne number.

Hebe arganthera Garn.-Jones, Bayly, W.G.Lee et Rance, *New Zealand Journal of Botany* 38: 380 (2000).
HOLOTYPE: Lake Monk, southwest Fiordland, limestone face within beech forest, 780 m, *B. D. Rance*, 4 Apr 1993, CHR 489369!.

Hebe armstrongii (J.B.Armstr.) Cockayne et Allan, *Transactions of the New Zealand Institute* 57: 40 (1926).
≡ *Veronica armstrongii* Johnson ex J.B.Armstr., *New Zealand Country Journal* 3: 59 (1879); ≡ *Veronica armstrongii* Kirk, *nom. illeg.* (superfluous because type of Armstrong's name included; see note by Ashwin, in Allan 1961), *Transactions of the New Zealand Institute* 11: 464 (1879); ≡ *Leonohebe armstrongii* (J.B.Armstr.) Heads, *Botanical Society of Otago Newsletter* 5: 7 (1987).
LECTOTYPE (designated by Ashwin, in Allan 1961): Rangitata Sources, 4–5000 ft., *J. F. A[rmstrong]*, 1869, CHR! (Herb. Armstrong). Possible isolectotypes: AK 8252! [this differs from the lectotype in the stated altitude, "4–6000 ft"], K! [this is a duplicate of AK 8252 (both have the number 1620, from T. F. Cheeseman), but gives the collector as J. B. Armstrong].

Hebe barkeri (Cockayne) Cockayne, *Transactions of the New Zealand Institute* 60: 469 (1929).
≡ *Veronica barkeri* Cockayne, *Transactions of the New Zealand Institute* 31: 421 (1899).
LECTOTYPE (designated by Moore, in Allan 1961): cultivated at Christchurch, plant originally from the Chatham Islands, *L. Cockayne*, AK 7663!.

Hebe benthamii (Hook.f.) Cockayne et Allan, *Transactions of the New Zealand Institute* 57: 43 (1926).
≡ *Veronica benthamii* Hook.f., *Flora Antarctica* 1: 60, Plates 39 & 40 (1844); ≡ *Leonohebe benthamii* (Hook.f.) Heads, *Botanical Society of Otago Newsletter* 5: 7 (1987).
LECTOTYPE (designated by Moore, in Allan 1961): Lord Auckland's Isd, *J. D. H[ooker]*, Nov. 1840, K!.
NOTES: Another sheet at K labelled "lord Auckland's group, J. D. H.", but with no date, could possibly be part of the same gathering as the lectotype. A further Hooker collection is dated 1845; since Hooker did not visit the Auckland Islands in 1845, the specimen must be incorrectly labelled. The name was originally published as "*benthami*", but this is treated as an error to be corrected (ICBN Art. 60.11, Rec. 60C.1).

= *Veronica finaustrina* Hombron et Jacq., *Voyage au Pôle Sud et dans l'Océanie – Botanique*: Plate 9, Fig. y (1845).
HOLOTYPE: *Voyage au Pôle Sud et dans l'Océanie – Botanique*: Plate 9, Fig. y (1845)!.
NOTES: Notes on the typification of *V. finaustrina* are provided by Kellow et al. (2003b).

Hebe biggarii (Cockayne) Cockayne, *Transactions of the New Zealand Institute* 60: 469 (1929).
≡ *Veronica biggarii* Cockayne, *Transactions of the New Zealand Institute* 48: 199 (1916).
LECTOTYPE (here designated): specimen from cultivated plant originally from the Eyre Mts, subalpine, *L. C[ockayne] ex Hal Poppelwell*, CHR 332289! (ex CANTY). Probable isolectotype: AK 107833!.

Hebe bishopiana (Petrie) D.Hatch, *Auckland Botanical Society Newsletter* 23: 1 (1966).
≡ *Veronica* ×*bishopiana* Petrie, *Transactions of the New Zealand Institute* 56: 15 (1926).
LECTOTYPE (designated by Moore, in Allan 1961): hill at Huia near Manukau Heads, *J. J. Bishop, H. Carse, E. Jenkins*, April 1924, WELT 5329!.
NOTES: Detailed notes on nomenclature and typification are provided by de Lange (1996).

Hebe bollonsii (Cockayne) Cockayne et Allan, *Transactions of the New Zealand Institute* 57: 15 (1926).
≡ *Veronica bollonsii* Cockayne, *Proceedings of the New Zealand Institute* 44: 50 (1912).
LECTOTYPE (designated by Moore, in Allan 1961): Poor Knights Islands, in coastal scrub, *L. Cockayne 9033*, WELT 5296!.

Hebe brachysiphon Summerh., *Kew Bulletin*: 397 (1927).
Based on *Veronica traversii* Hook.f. in *Botanical Magazine* 104: Plate 6390 (1878), non Hook.f. in *Handbook of the New Zealand Flora*: 208 (1864); ≡ *Veronica brachysiphon* (Summerh.) Bean, *Kew Bulletin*: 224 (1934).
LECTOTYPE (designated by Bayly & Kellow 2004*b*): from Sir J. D. Hooker's garden, 26 June 1893, K!, flowering pieces on top left and bottom right only (these are mounted on the same sheet as pieces collected in March 1893, and another specimen – comprising two pieces – collected at Edinburgh Botanical Gardens).

Hebe brevifolia (Cheeseman) de Lange, *New Zealand Journal of Botany* 35: 1 (1997).
≡ *Veronica speciosa* var. *brevifolia* Cheeseman, *Manual of the New Zealand Flora*: 500 (1906); ≡ *Hebe macrocarpa* var. *brevifolia* (Cheeseman) L.B.Moore, *Flora of New Zealand* 1: 908 (1961).
LECTOTYPE (designated by Moore, in Allan 1961): North Cape, *T. F. C[heeseman]*, Jan 1896, 1535 to Kew, AK 7653!. ISOLECTOTYPES: *T. F. C[heeseman] 1535*, K; WELT 16641! (undated), 16642!, 16643!.
NOTES: The type sheet chosen by Moore (in Allan 1961) includes five separate sprigs. de Lange (1997) considered that it was necessary to nominate a single one of these pieces as the lectotype (he chose that on the lower right, encircled in pencil). However, under the current rules of the ICBN (Art 8.2; and see Preface), this further choice is unnecessary and irrelevant.

Hebe breviracemosa (W.R.B.Oliv.) Andersen, *Transactions of the New Zealand Institute* 56: 693 (1926).
≡ *Veronica breviracemosa* W.R.B.Oliv., *Transactions of the New Zealand Institute* 42: 170 (1910); ≡ *Hebe breviracemosa* (W.R.B.Oliv.) Cockayne et Allan, *Transactions of the New Zealand Institute* 57: 17 (1926), *nom. illeg.*, non Andersen.
LECTOTYPE (designated by Moore, in Allan 1961): cliffs above Denham Bay, Sunday Island [Raoul Island], *W. R. B. Oliver*, 10 May 1908, WELT 5292!. ISOLECTOTYPES: CHR 291122!, OTA 18705!. Possible isolectotype (although collection details do not match exactly): K!.

Hebe buchananii (Hook.f.) Cockayne et Allan, *Transactions of the New Zealand Institute* 57: 36 (1926).
≡ *Veronica buchananii* Hook.f., *Handbook of the New Zealand Flora*: 211 (1864).
LECTOTYPE (designated by Bayly & Kellow 2004*b*): Otago, Lake district, *Hector & Buchanan no. 9*, Herb. Hookerianum, K!, two pieces on lower left of sheet.

= *Veronica buchananii* var. *exigua* Cheeseman, *Manual of the New Zealand Flora*: 527 (1906).
LECTOTYPE (designated by Moore, in Allan 1961): Hooker Glacier, 3500 ft, *T. F. C[heeseman]*, Jan 1898, 1604 to Kew, AK 8147!.

= *Veronica buchananii* var. *major* Cheeseman, *Manual of the New Zealand Flora*: 527 (1906); ≡ *Hebe buchananii* var. *major* (Cheeseman) A.Wall, *Transactions of the New Zealand Institute* 60: 384 (1929).
SYNTYPES: Herb. Cheeseman, AK [see notes by Moore, in Allan 1961; Herrick & Cameron 1994].

Hebe calcicola Bayly et Garn.-Jones, *New Zealand Journal of Botany* 39: 57 (2001).
HOLOTYPE: S of Salisbury Hut, Mt Arthur Tableland, NW Nelson, 3500 ft, limestone cliff beside pothole, *A. P. Druce*, Jan 1975, CHR 277568!.

Hebe canterburiensis (J.B.Armstr.) L.B.Moore, *Flora of New Zealand* 1: 899 (1961).
≡ *Veronica canterburiensis* J.B.Armstr. (as "*canterburiense*"), *New Zealand Country Journal* 3: 58 (1879); ≡ *Hebe vernicosa* var. *canterburiensis* (J.B.Armstr.) Cockayne et Allan, *Transactions of the New Zealand Institute* 57: 30 (1926).
LECTOTYPE (designated by Moore, in Allan 1961): Arthurs Pass, 3–4000 ft., *J. B. A*[*rmstrong*], CHR! (Herb. Armstrong).
NOTES: The type is treated as a lectotype, rather than a holotype, because there is another specimen in Herb. Armstrong (from Bealey Gorge, 1873) that could match the protologue citation "Arthur's Pass and other places".

Hebe carnosula (Hook.f.) Cockayne, *Transactions of the New Zealand Institute* 60: 469 (1929).
≡ *Veronica laevis* var. *carnosula* Hook.f., *Flora Novae-Zelandiae* 1: 194 (1853); ≡ *Veronica carnosula* (Hook.f.) Hook.f., *Handbook of the New Zealand Flora*: 210 (1864), *nom. illeg.*, non Lam., *Encyclopédique et Méthodique des Trois Règnes de la Nature. Botanique.* 1: 47 (1791).
HOLOTYPE: near Nelson, Morse's Mountain, 5000 ft., *Bidwill n. 10*, K!.
NOTES: All pieces on the type sheet are assumed to be part of a single gathering. The locality "Morse's Mountain", not found on any modern or historical maps, is probably Red Hills Ridge, Wairau Va. (as also assumed by Moore & Edgar 1970), and may have been named for Nathaniel Morse who, together with a Dr Cooper, was one of the first Europeans to settle and run sheep in the Tophouse area of Wairau Va.

Hebe chathamica (Buchanan) Cockayne et Allan, *Transactions of the New Zealand Institute* 57: 22 (1926).
≡ *Veronica chathamica* Buchanan, *Transactions of the New Zealand Institute* 7: 338 (1875).
LECTOTYPE (here designated): *Transactions of the New Zealand Institute* 7: 338, Plate 13, Fig. 1 (1875).
NOTES: No undisputable original specimens have been found.

= *Veronica coxiana* Kirk, *Transactions of the New Zealand Institute* 28: 529 (1896); ≡ *Veronica chathamica* var. *coxiana* (Kirk) Cheeseman, *Manual of the New Zealand Flora*, 2nd edn: 794 (1925); ≡ *Hebe coxiana* (Kirk) Cockayne, *Transactions of the New Zealand Institute* 60: 470 (1929).
LECTOTYPE (designated by Moore, in Allan 1961): Chatham Islands, *F. A. D. Cox*, WELT 5295!.

Hebe cockayneana (Cheeseman) Cockayne et Allan, *Transactions of the New Zealand Institute* 57: 32 (1926).
≡ *Veronica cockayneana* Cheeseman, *Manual of the New Zealand Flora*: 522 (1906).
LECTOTYPE (designated by Moore, in Allan 1961): Humboldt Mountains, Otago, *L. Cockayne 7949*, 19 Feb 1897, Herb. T. F. Cheeseman (1588 to Kew), AK 8054!.
NOTES: There is some confusion over the origin of the specimen chosen as the lectotype. Although Cheeseman (1906), and the specimen label, clearly indicate that it is from the Humboldt Mountains, Cockayne & Allan (1926c) assert that the "…species was based on specimens from one plant collected by L. Cockayne, not on the Humboldt Mountains, as given by Cheeseman, but in the cirque at the head of the Earnslaw Creek…". Specimens with the same Cockayne collecting number (7949) include specimens labelled both "Humboldt Mts" (CHR 331810!, WELT 47652!) and "Earnslaw Creek" (CHR 331811!, WELT 47679!).

The lectotype is labelled "1588 to Kew", but no matching specimen was located at K. The epithet was spelled "*cockayniana*" in the protologue, but this is considered an orthographic error requiring correction, as noted by Garnock-Jones (in Connor & Edgar 1987).

= *Veronica willcoxii* Petrie, *Transactions of the New Zealand Institute* 45: 272 (1913); ≡ *Hebe willcoxii* (Petrie) Cockayne et Allan, *Transactions of the New Zealand Institute* 57: 34 (1926).
LECTOTYPE (designated by Moore, in Allan 1961): top of Routeburn Valley, nr Lake Wakatipu, *D. Petrie*, Feb 1911, WELT 13453!.
NOTES: Moore (in Allan 1961) cited the number of the type specimen as WELT 5361. It appears that the number 5361 was not stamped on the type specimen when it was first registered, and that it was inadvertently re-registered as 13435. The sheet WELT 13435 otherwise exactly matches the details given by Moore, and has her annotation on a supplementary label (dated 13 Aug 1958), which says "This type specimen of *Veronica willcoxii* Petrie agrees well with Cockayne's specimen from Humboldt Mts. (Ak Mus. 8054) which is the type of *V. cockayniana* Cheesem. (1906)". The details for both WELT 5361 and 13435 in hand-written register at WELT are identical and match those of the presumed type (except for the locality, which has been abbreviated to "Routeburn Valley").

Hebe colensoi (Hook.f.) Cockayne, *Transactions of the New Zealand Institute* 60: 469 (1929).
≡ *Veronica colensoi* Hook.f., *Handbook of the New Zealand Flora*: 209 (1864).
LECTOTYPE (designated by Moore, in Allan 1961): high stony ridge above the River Taruarau, *Colenso 4062*, K!.
NOTES: As suggested by Moore (in Allan 1961) there are probably duplicates of the type, but without collecting numbers, at AK and WELT.

= *Veronica hillii* Colenso, *Transactions of the New Zealand Institute* 28: 606 (1896); ≡ *Hebe hillii* (Colenso) A.Wall, *Transactions of the New Zealand Institute* 60: 348 (1929); ≡ *Hebe colensoi* var. *hillii* (Colenso) L.B.Moore, *Flora of New Zealand* 1: 895 (1961).
LECTOTYPE (designated by Moore, in Allan 1961): *H. Hill*, 1894, K (n.v.).
NOTES: An additional name for the same entity, *Veronica hillii* Kirk, *Transactions of the New Zealand Institute* 28: 524 (1896), was published in the same volume as *V. hillii* Colenso. Subsequent combinations have been based on *V. hillii* Colenso, whereas *V. hillii* Kirk has generally been ignored. Kirk's name has not, however, been placed in direct synonymy under Colenso's name – for example, Moore (in Allan 1961) only mentions Kirk's name in her notes under *H. colensoi*, not in the list of synonyms (cf. the way other synonyms are treated in her account of *Hebe*). Whether or not Moore's treatment constitutes a choice of names in the sense of either ICBN Art. 11.5 or 53.6 (Greuter et al. 2000) is possibly debatable, and it might be argued that both names still have equal priority. Whichever name is considered to have priority, the other name is illegitimate. If Colenso's name has priority, then Kirk's name (currently untypified) would also have the same type (since the specimens cited in the protologue included the type of *V. hillii* Colenso).

Hebe corriganii Carse, *Transactions of the New Zealand Institute* 60: 573 (1930).
LECTOTYPE (designated by Bayly & Kellow 2004*b*): McLarens Falls, Wairoa River, Bay of Plenty, *B. Sladden*, Carse Herbarium 1237/6a, CHR 328473!.

Hebe crenulata Bayly, Kellow et de Lange, *New Zealand Journal of Botany* 40: 592 (2002).
HOLOTYPE: New Zealand, South Island, Nelson, Peel Range, Lake Peel near outlet, 1340 m, low shrubland over rock boulders, *M. J. Bayly 1316 & T. Galloway*, 16 Feb 2000, WELT 81743!. ISOTYPE: CHR.

Hebe cryptomorpha Bayly, Kellow, G.Harper et Garn.-Jones, *New Zealand Journal of Botany* 40: 596 (2002).
HOLOTYPE: New Zealand, South Island, Marlborough, Mt Richmond Forest Park, beside road to Mt Patriarch, *Hebe* shrubland, *M. J. Bayly 846 & R. Ansell*, 21 Dec 1997, WELT 80780!. ISOTYPES: AK, CHR.

Hebe decumbens (J.B.Armstr.) Cockayne et Allan, *Transactions of the New Zealand Institute* 57: 34 (1926).
≡ *Veronica decumbens* J.B.Armstr., *New Zealand Country Journal* 3: 57 (1879).
LECTOTYPE (designated by Moore, in Allan 1961): Rutherfords Bridge, Waiau, Nelson, *J. B. A[rmstrong]*, Dec 1869, CHR! (Herb. Armstrong).
NOTES: The type is treated as a lectotype, rather than a holotype, because there are other specimens in Herb. Armstrong that could match the protologue citation "from the Waiau. J. B. Armstrong".

Hebe dieffenbachii (Benth.) Cockayne et Allan, *Transactions of the New Zealand Institute* 57: 14 (1926).
≡ *Veronica dieffenbachii* Benth. in DC., *Prodromus Systematis Naturalis Regni Vegetabilis. Vol. 10*: 459 (1846).
HOLOTYPE: Chatham Islands, New Zealand, *Dieffenbach*, Herb. Hookerianum, K! (mounted on right of sheet that also includes collections by Enys and Travers).

= *Veronica dorrien-smithii* Cockayne, *Transactions of the New Zealand Institute* 44: 51 (1912); ≡ *Hebe dorrien-smithii* (Cockayne) Cockayne et Allan, *Transactions of the New Zealand Institute* 57: 14 (1926).
LECTOTYPE (designated by Bayly & Kellow 2004*b*): growing overhanging the water of L. Tekua Taupo [Lake Tuku a taupo], tobacco country, Chatham Island, *L. Cockayne 8003*, Feb 1901, WELT 5293!.
ISOLECTOTYPES: CHR 328354!, 328355!; AK 7660!.

Hebe dilatata G.Simpson et J.S.Thomson, *Transactions of the Royal Society of New Zealand* 73: 164 (1943).
HOLOTYPE: Blue Lake, Garvie Mountains, Otago, 1370 m above sea-level, *G. Simpson & J. S. Thomson*, CHR 63426! (a collection mounted on three sheets, labelled 63426A!, 63426B! and 63426C!).
NOTES: Moore (in Allan 1961) considered the type of *H. dilatata* to be CHR 63426A, and the other sheets to be isotypes. However, since the protologue mentions only one specimen (when describing new taxa

Simpson was usually particular about whether there was one or more than one "type" specimen), and all parts of CHR 63426 are likely to represent one "gathering" (sensu ICBN Art. 8.2), and the sheets are cross-labelled (see ICBN Art. 8.3), the whole collection is considered here to constitute the holotype.

= *Hebe crawii* Heads, *Botanical Society of Otago Newsletter* 5: 11 (1987).
HOLOTYPE: Excelsior Peak, Takitimu Mountains, 4800 ft, occasional spreading shrubs *c.* 40 cm tall in fellfield, *A. F. Mark*, 2 Feb 1971, OTA 31283!.

Hebe diosmifolia (A.Cunn.) Andersen, *Transactions of the New Zealand Institute* 56: 693 (1926).
≡ *Veronica diosmifolia* A.Cunn., *Botanical Magazine* 63: Sub-plate 3461 (1836); ≡ *Hebe diosmifolia* (A.Cunn.) Cockayne et Allan, *Transactions of the New Zealand Institute* 57: 25 (1926), *nom. illeg.*, non Andersen.
LECTOTYPE (designated by Bayly & Kellow 2004*b*): a slender twiggy shrub from 3–12 feet high found first at the head of the Wycaddy [Waikare] River and afterwards below the fall of the Keri Keri – also on the South Head of Hokianga, New Zealand, *R. Cunningham no. 301*, 1834, Allan Cunningham's New Zealand herbarium, K!, piece mounted on the lower left of a sheet (which also includes material collected by Hector).

= *Veronica menziesii* Benth. in DC., *Prodromus Systematis Naturalis Regni Vegetabilis. Vol. 10*: 461 (1846); ≡ *Hebe menziesii* (Benth.) Cockayne et Allan, *Transactions of the New Zealand Institute* 57: 25 (1926).
LECTOTYPE (designated by Bayly & Kellow 2004*b*): New Zealand, *Menzies*, Herb. Hookerianum, K!, piece on left of sheet only.

= *Veronica trisepala* Colenso, *Transactions of the New Zealand Institute* 15: 324 (1883); ≡ *Veronica diosmifolia* var. *trisepala* (Colenso) Kirk, *Transactions of the New Zealand Institute* 28: 525 (1896); ≡ *Hebe diosmifolia* var. *trisepala* (Colenso) A.Wall, *Transactions of the New Zealand Institute* 60: 384 (1929).
LECTOTYPE (designated by Moore, in Allan 1961): Kaweka Range, *A. Hamilton*, WELT 5352!, Herb. Petrie. ISOLECTOTYPES: WELT 79807!, K! (two pieces in upper right-hand corner of a sheet that also includes material from near Cape Reinga, *T. F. Cheeseman*).

= *Hebe diosmifolia* var. *vernalis* Carse, *Transactions of the New Zealand Institute* 60: 306 (1929).
LECTOTYPE (designated by Moore, in Allan 1961): on bank of Mangere Creek, Whangarei, *H. Carse*, 22 Oct [18]98, Carse Herbarium 1249, CHR 332300! (transferred from CANTY, May 1975). ISOLECTOTYPE: K!. Possible isolectotype: WELT 13213! (Mangere Falls, *H. Carse*, 22 Oct. 1898).

Hebe divaricata (Cheeseman) Cockayne et Allan, *Transactions of the New Zealand Institute* 56: 20 (1926).
≡ *Veronica menziesii* var. *divaricata* Cheeseman, *Manual of the New Zealand Flora*: 512 (1906).
LECTOTYPE (designated by Moore, in Allan 1961): Rai Va., *J. H. McMahon 49*, AK 7909!. ISOLECTOTYPE: WELT 47643!.

= *Hebe subfulvida* G.Simpson et J.S.Thomson, *Transactions of the Royal Society of New Zealand* 73: 163 (1943).
LECTOTYPE (designated by Moore, in Allan 1961): Pelorus Va., stream banks, *G. Simpson*, CHR 76003!. ISOLECTOTYPE: CHR 76012!.
NOTES: CHR 76009! has similar label details to the type specimens, except that Thomson is not listed as a collector. Since it has flowers, whereas the types have fruits, it might have been collected at a different time from them. AK 231192! is also similarly labelled and includes one sterile sprig and one sprig with young fruit.

= *Hebe corymbosa* G.Simpson, *Transactions of the Royal Society of New Zealand* 79: 428 (1952).
LECTOTYPE (designated by Bayly & Kellow 2004*b*): From plant in cultivation, collected by Mr N. Potts at Dun Mtn, Nelson, *G. Simpson*, Jan 1949, CHR 75693!, two flowering pieces at top of sheet. ISOLECTOTYPE: K!

Hebe elliptica (G.Forst.) Pennell, *Rhodora* 23: 39 (1921).
≡ *Veronica elliptica* G.Forst., *Florulae Insularum Australium Prodromus*: 3 (1786); ≡ *Veronica forsteri* F.Muell., *Vegetation of the Chatham Islands*: 45 (1864), *nom. illeg.* (superfluous: the name that ought to have been adopted was *Veronica elliptica* G.Forst.).
LECTOTYPE (designated by Moore, in Allan 1961): Forster Herbarium, K!.
NOTES: Nicolson & Fosberg (2004) present a list of potential syntypes.

= *Veronica decussata* Moench, *Vereichniss ausländischer Bäume und Stauden*: 137 (1785), *nom. utique rej. prop.*
TYPE: unknown.

NOTES: This little-used name, which pre-dates the publication of *Veronica elliptica* G.Forst., has been proposed for rejection (under ICBN Art. 56) by Bayly & Kellow (2004*a*).

= *Veronica decussata* Aiton, *Hortus Kewensis* Vol. 1: 20 (1789), *nom. illeg.*, non Moench, *Vereichniss ausländischer Bäume und Stauden*: 137 (1785).
TYPE: None designated.
NOTES: A potential type is a specimen at K! ("Veronica decussata sp. nova, 1782"). Descriptions in volume 1 of *Hortus Kewensis* are generally regarded as being written by Solander (who died in 1782) and edited and emended by Dryander, and, for this name, the author citation is often given as "Sol. in Aiton" (e.g. by Moore, in Allan 1961). However, according to ICBN Art. 4.6, and Ex. 24 (Greuter et al. 2000), this is inappropriate, given that the description is not specifically attributed to an author other than Aiton.

= *Hebe magellanica* J.F.Gmel., *Systema Naturae* 2: 27 (1791).
TYPE: unknown. The locality of J. F. Gmelin's herbarium and types is unknown; he was mainly a compiler and may not have had an herbarium of his own of any size (Stafleu & Cowan 1976).

= *Veronica marginata* Colenso, *Transactions of the New Zealand Institute* 28: 608 (1896).
LECTOTYPE (here designated): from a garden of Mr A. Wall, Porirua, near Wellington, 1895, K!.

= *Hebe elliptica* var. *crassifolia* Cockayne et Allan, *Transactions of the New Zealand Institute* 57: 27 (1926).
LECTOTYPE (designated by Moore, in Allan 1961): Kapiti Island, near Waterfall Rock, in rock crevices exposed to salt spray, *H. H. Allan*, Easter Monday, 1924, WELT 5297!.

Hebe epacridea (Hook.f.) Andersen, *Transactions of the New Zealand Institute* 56: 693 (1926).
≡ *Veronica epacridea* Hook.f., *Handbook of the New Zealand Flora*: 213 (1864); ≡ *Hebe epacridea* (Hook.f.) Cockayne et Allan, *Transactions of the New Zealand Institute* 57: 42 (1926), *nom. illeg.* non Andersen; ≡ *Leonohebe epacridea* (Hook.f.) Heads, *Botanical Society of Otago Newsletter* 5: 6 (1987).
LECTOTYPE (designated by Moore, in Allan 1961): New Zealand, Nelson, Tarndale, 3500 ft, *Sinclair*, 1861, K!.
NOTES: The lectotype consists of two sprigs mounted in the lower right-hand corner of a sheet containing two other collections, one from Discovery Peaks, Nelson, *Travers*, 1860 (one of the syntypes), and one from Canterbury, *Sinclair & von Haast*, 1860–1.

Hebe evenosa (Petrie) Cockayne et Allan, *Transactions of the New Zealand Institute* 57: 29 (1926).
≡ *Veronica evenosa* Petrie, *Transactions of the New Zealand Institute* 48: 189 (1916).
LECTOTYPE (designated by Moore, in Allan 1961): Mt Holdsworth, Tararuas, *c.* 3500 ft, at upper edge of forest, *D. P[etrie]*, 25 Jan 1908, WELT 5334!.

Hebe flavida Bayly, Kellow et de Lange sp. nov.
DIAGNOSTIC DESCRIPTION: erect shrub or small tree to *c.* 8 m tall, trunk diameter to *c.* 10 cm at breast height; leaf bud without a sinus; leaves lanceolate or oblanceolate or narrowly elliptic, (30–)50–100(–135) × (6–)10–20(–29) mm, acuminate or acute, petiole and base of midvein usually conspicuously yellow on upper side; corolla tubes shorter than or equalling calyces; corolla lobes usually subacute; inflorescences held erect, even in fruit.

Frutex erectus vel arbor parva usque ad *c.* 8 m alta, diametro trunci ad 10 cm d.b.h.; folii gemma sinu carens; folia lanceolata vel oblanceolata vel anguste elliptica, (30–)50–100(–135) × (6–)10–20(–29) mm, acuminata vel acuta, petiolo et nervi basi plerumque conspicue flava in pagina superiore; corollae tubi calyces aequantes vel eis breviores; corollae lobi plerumque subacuti; inflorescentiae erectae, etiam ubi fructiferae.
HOLOTYPE: New Zealand, North Island, North Auckland, "Waima Range", Hauturu State Forest, Frampton Block, 600 m, common on exposed outcrops and bluffs, growing with *Quintinia serrata*, *Weinmannia silvicola*, *Ackama rosifolia*, *A.* sp. nov. [= *A. nubicola*], *P. J. de Lange 5163*, 12 Feb. 2001, WELT 82916!.
ISOTYPE: AK 288628.

Hebe gibbsii (Kirk) Cockayne et Allan, *Transactions of the New Zealand Institute* 56: 20 (1926).
≡ *Veronica gibbsii* Kirk, *Transactions of the New Zealand Institute* 28: 524 (1896).
LECTOTYPE (designated by Moore, in Allan 1961): Mt Rintoul, ex Herb. T. Kirk, AK 8098!.
NOTES: The lectotype chosen by Moore (in Allan 1961) gives no details of collector, date or altitude. Other specimens from the private herbarium of T. Kirk at WELT, probably not available to Moore at the time of her revision, provide more of these details (linking them to the protologue), and might have made better

lectotypes. However, since Moore's type may simply lack information and is not necessarily "in serious conflict with the protologue", there are no strong grounds for it to be superseded.

Hebe glaucophylla (Cockayne) Cockayne, *Transactions of the New Zealand Institute* 60: 471 (1929).
≡ *Veronica glaucophylla* Cockayne, *Transactions of the New Zealand Institute* 31: 422 (1899).
NEOTYPE (first designated by Cockayne 1929, then more precisely by Moore, in Allan 1961): cultivated plant, originally from Craigieburn Mts, *L. Cockayne No. 8037*, 11 Jan 1902, AK 7970!. ISONEOTYPES: WELT 47659!, 47658!; CHR 331797!.
NOTES: No material matching original collection details, particularly with respect to the date (1890), has been found. Cockayne's (1929) type designation, and the refinements of Moore (in Allan 1961) are neotypifications because the chosen specimens are not part of the original material, being collected after publication of the protologue.

= *Veronica traversii* var. *fallax* Cheeseman, *Manual of the New Zealand Flora*: 519 (1906).
LECTOTYPE (here designated): St James Station, Clarence River, 3000 ft, *T. Kirk n. 775*, 1577 to Kew, AK 7978!.
NOTES: The protologue does not cite any specimens or localities, but the lectotype and one other sheet in Herb. Cheeseman (a specimen of *H. topiaria*, Mt Mantell, *W. Townson 613*, undated, AK7989!) are labelled with this name. The lectotype sheet includes two pieces, of which the upper, larger piece (labelled "A" in pencil) best matches the original description. Although both pieces are identified here as *H. glaucophylla*, the upper piece has corolla tubes slightly longer than usual, but still within the range, for that species.

Hebe haastii (Hook.f.) Cockayne et Allan, *Transactions of the New Zealand Institute* 57: 42 (1926).
≡ *Veronica haastii* Hook.f., *Handbook of the New Zealand Flora*: 213 (1864); ≡ *Leonohebe haastii* (Hook.f.) Heads var. *haastii*, *Botanical Society of Otago Newsletter* 5: 6 (1987).
LECTOTYPE (designated by Kellow et al. 2003b): [Mt Dobson], Canterbury, New Zealand, *J. Haast 625*, 1862, K! (two sprigs on upper left of sheet).

Hebe hectorii (Hook.f.) Cockayne et Allan, *Transactions of the New Zealand Institute* 57: 40 (1926).
≡ *Veronica hectorii* Hook.f., *Handbook of the New Zealand Flora*: 212 (1864); ≡ *Leonohebe hectorii* (Hook.f.) Heads, *Botanical Society of Otago Newsletter* 5: 8 (1987).
LECTOTYPE (designated by Ashwin, in Allan 1961): Otago, Mt Alta, *Hector no. 27*, 1863, Herb. Hookerianum, K! (small broken piece, mounted on lower right of sheet that also includes material collected by *Sinclair & Haast*, and by *Hector & Buchanan*).
NOTES: The epithet was spelled "*hectori*" in the protologue, but this is considered an orthographic error requiring correction, as noted by Garnock-Jones (in Connor & Edgar 1987).

subsp. *hectorii*
= *Veronica laingii* Cockayne, *Report on a Botanical Survey of Stewart Island*: 44 (1909); ≡ *Hebe laingii* (Cockayne) Andersen, *Transactions of the New Zealand Institute* 56: 693 (1926); ≡ *Hebe laingii* (Cockayne) Cockayne et Allan, *Transactions of the New Zealand Institute* 57: 40 (1926), *nom. illeg.*, non Andersen; ≡ *Hebe hectorii* subsp. *laingii* (Cockayne) Wagstaff et Wardle, *New Zealand Journal of Botany* 37: 33 (1999); ≡ *Leonohebe laingii* (Cockayne) Heads, *Botanical Society of Otago Newsletter* 5: 8 (1987).
LECTOTYPE (designated by Ashwin, in Allan 1961): near summit of Mt Anglem, Stewart Island, *L. Cockayne 9157*, CHR 333997!. ISOLECTOTYPES: AK 107837!, K!, WELT 5305!.

subsp. *coarctata* (Cheeseman) Wagstaff et Wardle, *New Zealand Journal of Botany* 37: 33 (1999).
≡ *Veronica coarctata* Cheeseman, *Manual of the New Zealand Flora*: 531 (1906), *pro parte*; ≡ *Hebe coarctata* (Cheeseman) Cockayne et Allan, *Transactions of the New Zealand Institute* 57: 40 (1926); ≡ *Leonohebe coarctata* (Cheeseman) Heads, *Botanical Society of Otago Newsletter* 5: 8 (1987).
LECTOTYPE (designated by Ashwin, in Allan 1961): Mt Arthur Plateau, Nelson, 4000 ft, *T. F. C[heeseman]*, Jan 1886, AK 8233!.

subsp. *demissa* (G.Simpson) Wagstaff et Wardle, *New Zealand Journal of Botany* 37: 33 (1999).
≡ *Hebe demissa* G.Simpson, *Transactions of the Royal Society of New Zealand* 75: 193 (1945); ≡ *Hebe hectorii* var. *demissa* (G.Simpson) Ashwin, *Flora of New Zealand* 1: 931 (1961); ≡ *Leonohebe hectorii* var. *demissa* (G.Simpson) Heads, *Botanical Society of Otago Newsletter* 5: 8 (1987).
LECTOTYPE (designated by Ashwin, in Allan 1961): ex Rock and Pillar Range, garden grown, Dunedin, *G. Simpson*, flowering early Jan, Dunedin, CHR 48080A!.

NOTES: CHR 48080B! is probably part of the same gathering as the lectotype, and the concept of the type could be expanded to include this sheet also, in which case a duplicate of CHR 48080B at K! would be an isolectotype. AK 22921!, with two flowering pieces, is possibly also an isolectotype (but the similarly labelled AK 22922! has mature fruit, and was probably collected at a different time).

= *Hebe subulata* G.Simpson, *Transactions of the Royal Society of New Zealand* 79: 427 (1952); ≡ *Leonohebe subulata* (G.Simpson) Heads, *Botanical Society of Otago Newsletter* 5: 9 (1987); ≡ *Hebe hectorii* subsp. *subulata* (G.Simpson) Wagstaff et Wardle, *New Zealand Journal of Botany* 37: 33 (1999).
LECTOTYPE (here designated): Old Man Range, Central Otago, *Owen Fletcher*, (cult), 10 Jan 1950, CHR 195571!.
NOTES: The protologue states that the type is "in the Herbarium, Plant Research Bureau, Wellington", which is now part of CHR. Ashwin (in Allan 1961) was unable to locate any type. The specimen here chosen as the lectotype is presumably part of Simpson's original material, but was not brought to CHR (from Simpson's private herbarium) until 1969.

Hebe imbricata Cockayne et Allan, *Transactions of the New Zealand Institute* 57: 42 (1926).
≡ *Veronica imbricata* Petrie, *Transactions of the New Zealand Institute* 48: 189 (1916), *nom. illeg.*, non Woerl., *Bericht des botanischen Vereines in Landshut* 8: 199 (1882); ≡ *Leonohebe imbricata* (Cockayne et Allan) Heads, *Botanical Society of Otago Newsletter* 5: 9 (1987).
LECTOTYPE (designated by Ashwin, in Allan 1961): Mt Cleughearn, *J. Crosby Smith*, WELT 5345!.
NOTES: WELT 5346! is possibly an isolectotype (note that this collection is dated Jan 1915, while annotations on the sheet by D. Petrie are dated 14 July 1914; one of these dates must be incorrect).

= *Veronica poppelwellii* Cockayne, *Transactions of the New Zealand Institute* 48: 200 (1916); ≡ *Hebe poppelwellii* (Cockayne) Cockayne et Allan, *Transactions of the New Zealand Institute* 57: 41 (1926); ≡ *Leonohebe poppelwellii* (Cockayne) Heads, *Botanical Society of Otago Newsletter* 5: 9 (1987); ≡ *Hebe imbricata* subsp. *poppelwellii* (Cockayne) Wagstaff et Wardle, *New Zealand Journal of Botany* 37: 33 (1999).
LECTOTYPE (designated by Ashwin, in Allan 1961): cultivated plant originally from Garvie Mts, *L. Cockayne No. 8116*, WELT 5306!. Possible isolectotype (but without Cockayne number): CHR 331861! (ex CANTY).

= *Veronica hectorii* var. *gracilior* Petrie ex Poppelw., *Transactions of the New Zealand Institute* 47: 140 (1915), *nom. nud.*
NOTES: In the absence of a description or type specimen, this name is assumed, given the comments of Cockayne (1916), to apply to *Veronica poppelwellii* Cockayne (although *H. hectorii* subsp. *demissa* also occurs in the Garvie Mts).

Hebe insularis (Cheeseman) Cockayne et Allan, *Transactions of the New Zealand Institute* 57: 25 (1926).
≡ *Veronica insularis* Cheeseman, *Transactions of the New Zealand Institute* 29: 392 (1897).
LECTOTYPE (designated by Moore, in Allan 1961): Three Kings Islands, *T. F. Cheeseman*, Nov 1889, AK 7888!. ISOLECTOTYPES: AK 7890!, 7889!; K!.

Hebe leiophylla (Cheeseman) Andersen, *Transactions of the New Zealand Institute* 56: 693 (1926).
≡ *Veronica leiophylla* Cheeseman, *Manual of the New Zealand Flora*: 509 (1906); ≡ *Hebe leiophylla* (Cheeseman) Cockayne et Allan, *Transactions of the New Zealand Institute* 57: 23 (1926), *nom. illeg.*, non Andersen; ≡ *Veronica parviflora* var. *phillyreaefolia* Hook.f., *Flora Novae-Zelandiae* 1: 192 (1853).
LECTOTYPE (designated by Bayly & Kellow 2004b): Nelson, New Zealand, *Bidwill no. 13*, K!.

= *Veronica ligustrifolia* var. *gracillima* Kirk, *Transactions of the New Zealand Institute* 28: 527 (1896); ≡ *Veronica gracillima* (Kirk) Cheeseman, *Manual of the New Zealand Flora*: 510 (1906); ≡ *Hebe gracillima* (Kirk) Cockayne et Allan, *Transactions of the New Zealand Institute* 57: 24 (1926).
HOLOTYPE: near Westport, *Dr Gaze*, Herb. T. Kirk, WELT 5337!.
NOTES: The combination *Hebe gracillima* (Kirk) Cockayne et Allan was originally published as × *Hebe gracillima*. Following ICBN Art. 50 the author citation is the same, regardless of whether or not the species is considered a hybrid.

Hebe ligustrifolia (A.Cunn.) Cockayne et Allan, *Transactions of the New Zealand Institute* 57: 16 (1926).
≡ *Veronica ligustrifolia* A.Cunn., *Botanical Magazine* 63: Sub-plate 3461 (1836).

HOLOTYPE: shady woods on the hills above the Kauakaua River, Bay of Islands, *R. C[unningham]*, 1833, K! (mounted on sheet that includes a mixture of other collections).

Hebe lycopodioides (Hook.f.) Andersen, *Transactions of the New Zealand Institute* 56: 693 (1926).
≡ *Veronica lycopodioides* Hook.f., *Handbook of the New Zealand Flora*: 211 (1864); ≡ *Hebe lycopodioides* (Hook.f.) Cockayne et Allan, *Transactions of the New Zealand Institute* 57: 40 (1926), *nom. illeg.*, non Andersen; ≡ *Leonohebe lycopodioides* (Hook.f.) Heads, *Botanical Society of Otago Newsletter* 5: 9 (1987).
LECTOTYPE (designated by Bayly & Kellow 2004*b*): Wairau Gorge, 4–5000 ft, *Travers 27*, Herb. Hookerianum, K!, three flowering pieces on upper left of sheet (which also includes material collected by *Hector* (Clarence Va., 4000 ft), and *Sinclair*).

= *Hebe lycopodioides* var. *patula* G.Simpson et J.S.Thomson, *Transactions of the Royal Society of New Zealand* 73: 164 (1943); ≡ *Leonohebe lycopodioides* var. *patula* (G.Simpson et J.S.Thomson) Heads, *Botanical Society of Otago Newsletter* 5: 9 (1987); ≡ *Hebe lycopodioides* subsp. *patula* (G.Simpson et J.S.Thomson) Wagstaff et Wardle, *New Zealand Journal of Botany* 37: 34 (1999).
HOLOTYPE: upper slopes of Mount Technical, Lewis Pass, grassland at 1200–1600 m, *G. Simpson & J. S. Thomson*, CHR 76005!.

Hebe macrantha (Hook.f.) Cockayne et Allan, *Transactions of the New Zealand Institute* 57: 43 (1926).
≡ *Veronica macrantha* Hook.f., *Handbook of the New Zealand Flora*: 213 (1864).
LECTOTYPE (designated by Moore, in Allan 1961): Canterbury, New Zealand, *Haast 562*, 1862, Herb. Hookerianum, K! (information from records of received specimens in the library at K indicates that this collection was made on a journey to the sources of the Waitaki R.).

var. *brachyphylla* (Cheeseman) Cockayne et Allan, *Transactions of the New Zealand Institute* 57: 43 (1926).
≡ *Veronica macrantha* var. *brachyphylla* Cheeseman, *Manual of the New Zealand Flora*: 537 (1906).
LECTOTYPE (designated by Moore, in Allan 1961): Mt Arthur, Nelson, 5000 ft, *T. F. C[heeseman]*, Jan 1886, 1663 to Kew, AK 58896!. ISOLECTOTYPES: K!; WELT 13108!, 13122!. Possible isolectotype: WELT 13114!.

Hebe macrocalyx (J.B.Armstr.) G.Simpson, *Transactions of the Royal Society of New Zealand* 79: 427 (1952).
≡ *Veronica macrocalyx* J.B.Armstr., *Transactions of the New Zealand Institute* 13: 353 (1881); ≡ *Veronica haastii* var. *macrocalyx* (J.B.Armstr.) Cheeseman, *Manual of the New Zealand Flora*: 534 (1906); ≡ *Hebe haastii* var. *macrocalyx* (J.B.Armstr.) Cockayne et Allan, *Transactions of the New Zealand Institute* 57: 42 (1926); ≡ *Leonohebe haastii* var. *macrocalyx* (J.B.Armstr.) Heads, *Botanical Society of Otago Newsletter* 5: 6 (1987).
LECTOTYPE (designated by Moore, in Allan 1961): Black Range and Mt Armstrong, 6000 ft., *J. B. A[rmstrong]*, 1867, CHR! (Herb. Armstrong).

var. *humilis* G.Simpson, *Transactions of the Royal Society of New Zealand* 79: 428 (1952).
≡ *Hebe haastii* var. *humilis* (G.Simpson) L.B.Moore, *Flora of New Zealand* 1: 940 (1961); ≡ *Leonohebe haastii* var. *humilis* (G.Simpson) Heads, *Botanical Society of Otago Newsletter* 5: 6 (1987).
LECTOTYPE (designated by Moore, in Allan 1961): from a plant in cultivation at Geo. Simpson's garden Dunedin, collected from slopes of Mt French at 1525 m altitude, *G. Simpson & J. S. Thomson*, Feb 1932, CHR 76135!.

Hebe macrocarpa (Vahl) Cockayne et Allan, *Transactions of the New Zealand Institute* 57: 20 (1926).
≡ *Veronica macrocarpa* Vahl, *Symbolae Botanicae* 3: 4 (1794); ≡ *Panoxis macrocarpa* (Vahl) Raf., *Medical Flora* 2: 109 (1830), *nom. illeg.* (combination not definitely indicated, ICBN Art. 33.1).
HOLOTYPE: nova zelandia [written on back of sheet], Hb. Vahlii [IDC microfiche foto Vahl. 78^{III}, 2–3], C!
NOTES: The single specimen in Vahl's herbarium at C (see also Hansen & Wagner 1998) is assumed to be the holotype. The protologue describes fruit, which the specimen lacks, but it is possible, rather than having seen other material, that Vahl borrowed the description of fruit from the earlier, unpublished description by Solander (in which all the terms of Vahl's description relating to fruit are found). It is possible that the type has a minute sinus in the leaf bud (only one on the specimen), a feature not usually found in *H. macrocarpa* as circumscribed here. Likewise, the corolla tubes are not as broad as those often seen in the species under our circumscription. Given variation in *H. macrocarpa*, we have not been able to interpret these features with certainty, but it is possible that correct application of the name *macrocarpa* is worthy of further study; more detailed comparison with *H. corriganii*, as defined here, would also be useful.

= *Veronica latisepala* Kirk, *Transactions of the New Zealand Institute* 28: 530 (1896); ≡ *Veronica macrocarpa* var. *latisepala* (Kirk) Cheeseman, *Manual of the New Zealand Flora*: 505 (1906); ≡ *Hebe macrocarpa* var. *latisepala* (Kirk) Cockayne et Allan, *Transactions of the New Zealand Institute* 57: 20 (1926).
LECTOTYPE (designated by Moore, in Allan 1961): cult^d plant from Port Fitzroy, *T. Kirk*, WELT 5320!. Possible isolectotype: K! (*Kirk no. 1428*).

Hebe masoniae (L.B.Moore) Garn.-Jones, *Australian Systematic Botany* 6: 478 (1993).
≡ *Hebe pauciramosa* var. *masoniae* L.B.Moore, *Flora of New Zealand* 1: 926 (1961); ≡ *Leonohebe masoniae* (L.B.Moore) Heads var. *masoniae*, *Botanical Society of Otago Newsletter* 5: 10 (1987).
HOLOTYPE: Head of Cobb Va., Nelson, *R. Mason*, 23 Feb 1946, CHR 54435!.
NOTES: The epithet was spelled "*masonae*" in the protologue, but this is considered an orthographic error requiring correction, as noted by Garnock-Jones (in Connor & Edgar 1987).

= *Leonohebe masoniae* var. *rotundata* Heads, *Botanical Society of Otago Newsletter* 5: 11 (1987).
HOLOTYPE: Mineral belt north of Cobb Reservoir, tussockland, 3500 ft, *A. P. Druce*, Nov 1980, CHR 389212.

Hebe mooreae (Heads) Garn.-Jones, *Australian Systematic Botany* 6: 479 (1993).
≡ *Leonohebe mooreae* Heads var. *mooreae*, *Botanical Society of Otago Newsletter* 5: 10 (1987).
HOLOTYPE: Douglas Range, South Westland, 4100', in low scrub on a steep gully side, *P. Wardle*, 16 Dec 1978, CHR 321236.

= *Leonohebe mooreae* var. *telmata* Heads, *Botanical Society of Otago Newsletter* 5: 10 (1987).
HOLOTYPE: Douglas Range, South Westland, 3500', abundant in *Chionochloa* grassland on rolling country, *P. Wardle*, 16 Dec 1978, CHR 231327.

Hebe murrellii G.Simpson et J.S.Thomson, *Transactions of the Royal Society of New Zealand* 73: 165 (1943).
≡ *Hebe petriei* var. *murrellii* (G.Simpson et J.S.Thomson) L.B.Moore, *Flora of New Zealand* 1: 938 (1961); ≡ *Leonohebe petriei* var. *murrellii* (G.Simpson et J.S.Thomson) Heads, *Botanical Society of Otago Newsletter* 5: 6 (1987).
HOLOTYPE: Kepler Range at sources of the Freeman River, moist shaded openings amongst rocks, *G. Simpson & J. S. Thomson*, March 1942 (note written on sheet in pencil, under label, states "Fowler Pass…leg. et ident. *G. Simpson*"), CHR 75695!. Probable isotype: AK 22904!.

Hebe obtusata (Cheeseman) Cockayne et Allan, *Transactions of the New Zealand Institute* 57: 15 (1926).
≡ *Veronica macroura* var. *dubia* Cheeseman, *Manual of the New Zealand Flora*: 501 (1906); ≡ *Veronica obtusata* Cheeseman, *Transactions of the New Zealand Institute* 48: 213 (1916).
LECTOTYPE (designated by Bayly & Kellow 2004b): Muriwai cliffs near Motutara, *T. F. Cheeseman*, March 1884, AK 7671!. ISOLECTOTYPE: AK 7672!.

Hebe ochracea Ashwin, *Flora of New Zealand* 1: 936 (1961).
≡ *Leonohebe ochracea* (Ashwin) Heads, *Botanical Society of Otago Newsletter* 5: 7 (1987).
HOLOTYPE: Cobb Va., N.W. Nelson, *F. G. Gibbs*, CHR 97077!. ISOTYPE: AK 8243! [according to Ashwin (in Allan 1961), although details on labels differ slightly].

Hebe odora (Hook.f.) Cockayne, *Transactions of the New Zealand Institute* 60: 472 (1929).
≡ *Veronica odora* Hook.f., *Flora Antarctica* 1: 62, Plate 41 (1844); ≡ *Veronica buxifolia* var. *odora* (Hook.f.) Kirk, *Transactions of the New Zealand Institute* 28: 523 (1896); ≡ *V. elliptica* var. *odora* (Hook.f.) Cheeseman, *Manual of the New Zealand Flora*: 516 (1906); ≡ *Hebe buxifolia* var. *odora* (Hook.f.) Andersen, *Transactions of the New Zealand Institute* 56: 693 (1926); ≡ *Leonohebe odora* (Hook.f.) Heads, *Botanical Society of Otago Newsletter* 5: 10 (1987).
HOLOTYPE: among the woods in Ld Auckland's Islands, [*J. D. Hooker*] *1460*, Nov. 1840, Herb. Hookerianum, K!.
NOTES: The same number, 1460, is shown on another Hooker specimen from the Auckland Ids, but this is dated 1845. Since Hooker did not visit the Auckland Ids in 1845, the specimen must be incorrectly labelled.

= *Veronica buxifolia* Benth. in DC., *Prodromus Systematis Naturalis Regni Vegetabilis. Vol. 10*: 462 (1846); ≡ *Hebe buxifolia* (Benth.) Andersen, *Transactions of the New Zealand Institute* 56: 693 (1926); ≡ *Hebe buxifolia* (Benth.) Cockayne et Allan, *Transactions of the New Zealand Institute* 57: 32 (1926), nom. illeg., non Andersen.

HOLOTYPE: Mounts of Intr., N. Island, N. Zealand, *Dieffenbach*, Herb. Hookerianum, K! (upper pieces on a sheet that also includes material collected by *Colenso*).

= *Veronica anomala* Armstr., *Transactions of the New Zealand Institute* 4: 291 (1872); ≡ *Hebe anomala* (Armstr.) Cockayne, *Transactions of the New Zealand Institute* 60: 468 (1929).
LECTOTYPE (designated by Moore, in Allan 1961): Upper Rakaia, *J. F. Armstrong*, 1865, CHR! (Herb. Armstrong).

= *Veronica haustrata* J.B.Armstr., *New Zealand Country Journal* 3: 58 (1879); ≡ *Hebe haustrata* (J.B.Armstr.) Andersen, *Transactions of the New Zealand Institute* 56: 693 (1926).
LECTOTYPE (designated by Moore, in Allan 1961): Upper Rangitata, 4000 ft, *J. F. A[rmstrong]*, 1869, CHR! (Herb. Armstrong).
NOTES: Although no other potential types have been seen, the type of *V. haustrata* is listed as a lectotype, rather than a holotype, because the protologue (which gives a vague locality and the names of two collectors) suggests that more than one specimen was used.

= *Veronica buxifolia* var. *patens* Cheeseman, *Manual of the New Zealand Flora*: 523 (1906).
LECTOTYPE (designated in part by Moore, in Allan 1961; designated more precisely by Bayly & Kellow 2004b): Mt Arthur Plateau, Nelson, alt. 4000 ft, *T. F. Cheeseman*, AK 8076!, two uppermost pieces only. ISOLECTOTYPE: WELT 5359!.

= *Veronica buxifolia* var. *prostrata* Cockayne, *Report on a Botanical Survey of Stewart Island*: 44 (1909); ≡ *Hebe buxifolia* var. *prostrata* (Cockayne) Andersen, *Transactions of the New Zealand Institute* 56: 693 (1926).
TYPE: None designated.
NOTES: No Cockayne specimens labelled var. *prostrata* have been found. Prostrate plants of *H. odora* are common on parts of Stewart Id.

Hebe paludosa (Cockayne) D.A.Norton et de Lange, *New Zealand Journal of Botany* 36: 532 (1998).
≡ *Veronica salicifolia* var. *paludosa* Cockayne, *Transactions of the New Zealand Institute* 48: 202 (1916); ≡ *Hebe salicifolia* var. *paludosa* (Cockayne) Cockayne et Allan, *Transactions of the New Zealand Institute* 57: 18 (1926).
LECTOTYPE (designated by Moore, in Allan 1961): Swamp, Lake Ianthe, Westland, *L. Cockayne 8118*, AK 7776!. ISOLECTOTYPE: WELT 16439!.
NOTES: Norton & de Lange (1998) provide detailed notes on typification.

Hebe pareora Garn.-Jones et Molloy, *New Zealand Journal of Botany* 20: 398 (1983).
HOLOTYPE: Upper Pareora Gorge, South Canterbury, on cliffs overhanging river, *P. J. Garnock-Jones 1512, Molloy & Anderson*, 4 Feb 1981, CHR 363050!. ISOTYPES: AK 179708!, WELT 78753!.
NOTES: Although the cover date on the volume of *New Zealand Journal of Botany* containing the protologue is 1982, it was not published until 11 Jan 1983, as recorded in the subsequent volume

Hebe parviflora (Vahl) Andersen, *Transactions of the New Zealand Institute* 56: 693 (1926).
≡ *Veronica parviflora* Vahl, *Symbolae Botanicae* 3: 4 (1794); ≡ *Hebe parviflora* (Vahl) Cockayne et Allan, *Transactions of the New Zealand Institute* 57: 23 (1926), *nom. illeg.*, non Andersen.
LECTOTYPE (designated by Bayly et al. 2000): nova Zeland, mis: Dr Montin, Hb. Vahlii, C!.

= *Veronica arborea* Buchanan, *Transactions of the New Zealand Institute* 6: 242 (1874); ≡ *Veronica parviflora* var. *arborea* (Buchanan) Kirk, *Transactions of the New Zealand Institute* 28: 527 (1896); ≡ *Hebe parviflora* var. *arborea* (Buchanan) L.B.Moore, *Flora of New Zealand* 1: 913 (1961).
HOLOTYPE: "*Veronica arborea* Buch:", Herb. Buchanan Vol. VII, WELT!.
NOTES: Moore (in Allan 1961) stated that the type was at OTM, which is where the Buchanan herbarium resided at that time. Specimens in the Buchanan volumes have generally been considered to include the types of his published names (Adams 2002). In the case of *V. arborea*, they include only one possible type, which is considered a holotype (rather than a lectotype designated by L. B. Moore).

Hebe pauciflora G.Simpson et J.S.Thomson, *Transactions of the Royal Society of New Zealand* 73: 166 (1943).
≡ *Leonohebe pauciflora* (G.Simpson et J.S.Thomson) Heads, *Botanical Society of Otago Newsletter* 5: 10 (1987).
HOLOTYPE: Kepler Range, near Fowler Pass, grassland in open situations, *G. Simpson*, Mar. 1942, CHR 75689!. Probable isotypes: CHR 97607! (perhaps deposited sometime after 75689?), K!, AK 22903!.

Hebe pauciramosa (Cockayne et Allan) L.B.Moore, *Flora of New Zealand* 1: 925 (1961).
≡ *Hebe buxifolia* var. *pauciramosa* Cockayne et Allan, *Transactions of the New Zealand Institute* 56: 27 (1926); ≡ *Leonohebe pauciramosa* (Cockayne et Allan) Heads, *Botanical Society of Otago Newsletter* 5: 10 (1987).
LECTOTYPE (designated by Moore, in Allan 1961): wet ground up to Lake Harris, 4000 ft. or less, *L. Cockayne No. 8129*, 7 May 1921, WELT 5354!. ISOLECTOTYPES: AK 107674!, K!.

Hebe perbella de Lange, *New Zealand Journal of Botany* 36: 399 (1998).
HOLOTYPE: New Zealand, Northland, Tutamoe Ecological District, Waima Forest, Hauturu High Point Track, *P. J. de Lange 3169 & I. McFadden*, 7 Nov 1996, AK 230119!. ISOTYPES: CHR 487605!, K!, WELT 79994!.

Hebe petriei (Buchanan) Cockayne et Allan, *Transactions of the New Zealand Institute* 57: 42 (1926).
≡ *Mitrasacme petriei* Buchanan, *Transactions of the New Zealand Institute* 14: 349 (1882); ≡ *Veronica petriei* (Buchanan) Kirk, *Transactions of the New Zealand Institute* 28: 517 (1896); ≡ *Leonohebe petriei* (Buchanan) Heads, *Botanical Society of Otago Newsletter* 5: 6 (1987).
LECTOTYPE (designated by Moore, in Allan 1961): WELT! in Herb. Buchanan, *Petrie*, 1881 [cited by Moore as being in OTM, where Herb. Buchanan resided at the time]. Probable isolectotypes: WELT 5118!, 5119!; AK 8283!
NOTE: The probable isolectotypes give locality information, but the lectotype does not.

Hebe pimeleoides (Hook.f.) Cockayne et Allan, *Transactions of the New Zealand Institute* 57: 38 (1926).
≡ *Veronica pimeleoides* Hook.f., *Flora Novae-Zelandiae* 1: 195 (1853); ≡ *Hebe pimeleoides* var. *rupestris* Cockayne et Allan, *Transactions of the New Zealand Institute* 57: 39 (1926), *nom. illeg.* (superfluous).
HOLOTYPE: Port Cooper, New Zealand, *Lyall*, K! [sprigs mounted in the upper right-hand corner of a sheet that includes several other collections].
NOTES: As noted by Cockayne & Allan (1926c), and by Moore (in Allan 1961), *H. pimeleoides* is not known from the type locality (Port Cooper, a name formerly used for Lyttleton), as given on the Lyall specimen (the holotype) cited by Hooker (1853). Cockayne & Allan (1926c) noted that Lyall and others made an excursion inland to near what is now Culverden, and suggested that the specimen may have been collected at that time.

subsp. *pimeleoides*
= *Veronica pimeleoides* var. *minor* Hook.f., *Handbook of the New Zealand Flora*: 738 (1867); ≡ *Hebe pimeleoides* var. *minor* (Hook.f.) Cockayne et Allan, *Transactions of the New Zealand Institute* 57: 38 (1926).
HOLOTYPE: Shingle beds of River Cameron, near Lake Heron, *von Haast*, 27 Oct. 1864, K! [sprigs mounted in the lower right corner of a sheet that also includes material, collected after the species was described, from Clarence Va., *T. F. Cheeseman*]. Probable isotype: CHR 22697! [probably a piece removed from holotype].

subsp. *faucicola* Kellow et Bayly, *New Zealand Journal of Botany* 41: 242 (2003).
HOLOTYPE: Otago, Clyde Dam lookout point, *c.* 300 m NE of dam wall, *M. J. Bayly 1492 & A. V. Kellow*, 21 Jan 2001, WELT 82445!.

Hebe pinguifolia (Hook.f.) Cockayne et Allan, *Transactions of the New Zealand Institute* 57: 36 (1926).
≡ *Veronica pinguifolia* Hook.f., *Handbook of the New Zealand Flora*: 210 (1864).
LECTOTYPE (designated by Bayly & Kellow 2004b): Canterbury, New Zealand, *Haast 574*, Herb. Hookerianum, K! (piece mounted on bottom left of a sheet that also includes two other collections).
NOTES: This species was figured by Hook.f., *Botanical Magazine* Plate 6587 (1881), as *V. carnosula*.

Hebe propinqua (Cheeseman) Cockayne et Allan, *Transactions of the New Zealand Institute* 57: 41 (1926).
≡ *Veronica propinqua* Cheeseman, *Manual of the New Zealand Flora*: 533 (1906); ≡ *Leonohebe propinqua* (Cheeseman) Heads, *Botanical Society of Otago Newsletter* 5: 9 (1987).
LECTOTYPE (designated by Ashwin, in Allan 1961): Mt Maungatua, near Dunedin, 2900 ft, *D. Petrie*, AK 8258!.

= *Veronica cupressoides* var. *variabilis* N.E.Br., *Gardeners' Chronicle* 1: 20, Fig. 5 (1888).
LECTOTYPE (designated by Bayly & Kellow 2004b): Edinborough [sic] Botanic Gardens, Sept 1887, K!, single piece with mature leaves, in upper left corner of sheet (which also includes material from Kew Gardens and Hay Lodge, as well as additional Edinburgh specimens collected in 1893).

= *Veronica propinqua* var. *major* Cockayne ex Cheeseman, *Manual of the New Zealand Flora*, 2nd edn: 820 (1925); ≡ *Hebe propinqua* var. *major* (Cockayne ex Cheeseman) Cockayne et Allan, *Transactions of the New Zealand Institute* 57: 41 (1926).
HOLOTYPE: Mt Dick, *L. C[ockayne] 8133*, Herb. Cheeseman, AK 50973! [this same sheet was incorrectly cited by Ashwin, in Allan 1961, as AK 50978]. ISOTYPE: AK 107836!.
NOTES: AK 50973 is the only sheet in Herb. Cheeseman that bears the name *Veronica propinqua* var. *major*. In line with the spirit of ICBN Recommendation 9A.4, this sheet is considered the holotype whereas that in the main collection at AK (labelled only "*Hebe propinqua*", but having the same Cockayne number) is considered an isotype.

Hebe pubescens (Benth.) Cockayne et Allan, *Transactions of the New Zealand Institute* 57: 17 (1926).
≡ *Veronica pubescens* Benth. in DC., *Prodromus Systematis Naturalis Regni Vegetabilis. Vol. 10*: 460 (1846).
HOLOTYPE: "Prope Opuragi" [Mercury Bay], Nova Zealandia, *J. Banks & D. Solander*, 1769, BM 603447!.

subsp. *rehuarum* Bayly et de Lange, *New Zealand Journal of Botany* 41: 40 (2003).
HOLOTYPE: New Zealand, North Auckland, Great Barrier (Aotea) Island, Port Fitzroy, Rarohara Bay, Old Lady Track, Lookout Rock, *P. J. de Lange 5192*, 29 Mar 2001, WELT 82548!. ISOTYPES: AK, CHR, K, MEL, WAIK, BM.

subsp. *sejuncta* Bayly et de Lange, *New Zealand Journal of Botany* 41: 42 (2003).
HOLOTYPE: New Zealand, North Auckland, Little Barrier (Hauturu) Island, Tirikakawa Stream, near trackside, *P. J. de Lange 5187*, 17 Mar 2001, WELT 82549!. ISOTYPES: AK, CHR, K, BM, MEL, WAIK, OTA.

Hebe rakaiensis (J.B.Armstr.) Cockayne, *Transactions of the New Zealand Institute* 60: 472 (1929).
≡ *Veronica rakaiensis* J.B.Armstr., *Transactions of the New Zealand Institute* 13: 356 (1881)
LECTOTYPE (designated by Moore, in Allan 1961): Rakaia Va., CHR! (Herb. Armstrong).
NOTES: There are several specimens in Herb. Armstrong that are potential types, but only one is labelled "Rakaia Valley" (others say "Rakaia" or "Upper Rakaia"), i.e. matches Moore's statement of the type locality, and is interpreted here as being the lectotype.
= *Hebe scott-thomsonii* Allan, *Transactions of the Royal Society of New Zealand* 69: 274 (1939).
LECTOTYPE (designated by Moore, in Allan 1961): Deep Stream, rocks by river at bridge, *H. H. Allan*, 15 Aug 1937, CHR 18230!.

Hebe ramosissima G.Simpson et J.S.Thomson, *Transactions of the Royal Society of New Zealand* 72: 29 (1943).
≡ *Leonohebe ramosissima* (G.Simpson et J.S.Thomson) Heads, *Botanical Society of Otago Newsletter* 5: 7 (1987).
LECTOTYPE (designated by Moore, in Allan 1961): Mount Tapuaenuku, moist debris at 2150 m alt., *G. Simpson*, CHR 75691!. ISOLECTOTYPE: AK 107861!.

Hebe rapensis (F.Br.) Garn.-Jones, *New Zealand Journal of Botany* 14: 79 (1976).
≡ *Veronica rapensis* F.Br., *Bernice P. Bishop Museum Bulletin* 130: 266 (1935).
SYNTYPES: Rapa, Mitiperu, steep slope, 1150 feet, *J. F. G. S[tokes] 372*, 26 Oct. 1921, BISH 581153, BISH 581154!.
NOTES: The protologue clearly states that *Stokes 372* is the type. Garnock-Jones (1976*a*) considered BISH 581154 (labelled *Stokes 372*) to be the holotype, but was unaware that additional material from the same collection was also mounted on another sheet. Subsequent annotations on specimens at BISH (fide D. Gowing Nov. 1990) suggest that BISH 581153 is the holotype, and BISH 581154 an isotype. The reasons for this assertion are unknown. If the two sheets are not cross-labelled (see Greuter et al. 2000, Art. 8, Ex. 4) they should be regarded as separate specimens; one should be the lectotype, the other an isolectotype.

Hebe rigidula (Cheeseman) Cockayne et Allan, *Transactions of the New Zealand Institute* 56: 20 (1926).
≡ *Veronica rigidula* Cheeseman, *Manual of the New Zealand Flora*: 514 (1906).
LECTOTYPE (designated by Moore, in Allan 1961): Pelorus River, Marlborough, *J. H. McMahon*, AK 7919!.
NOTES: Possible isolectotypes include WELT 13284! and 13286!, and a sheet at K!, but given the limited information on these specimens it is difficult to be certain if any are duplicates of the lectotype. Moore (in

Allan 1961) and Bayly et al. (2002) incorrectly attributed publication of the combination *Hebe rigidula* (Cheeseman) Cockayne et Allan to Cockayne & Allan (1926*b*); it was actually published by Cockayne & Allan (1926*a*) eight months earlier.

var. *sulcata* Bayly et Kellow, *New Zealand Journal of Botany* 40: 585 (2002).
HOLOTYPE: New Zealand, D'Urville Island (Rangitoto ke te Tonga), eastern side of Attempt Hill near road, *P. J. de Lange 5043 & G. M. Crowcroft*, 19 Jan 2001, AK 252335. ISOTYPES: WELT 82582!, CHR.

Hebe rupicola (Cheeseman) Cockayne et Allan, *Transactions of the New Zealand Institute* 57: 26 (1926).
≡ *Veronica rupicola* Cheeseman, *Manual of the New Zealand Flora*: 514 (1906).
LECTOTYPE (designated by Moore, in Allan 1961): gorge of the Conway River, Marlborough, *L. Cockayne 8000*, Herb. T. F. Cheeseman (1570 to Kew), AK 7926!. Probable isolectotype (but lacking Cockayne number): WELT 78119! [ex Herb. Cheeseman].
NOTES: Although the type is labelled "1570 to Kew", no Cockayne specimens were found at K. Further specimens at WELT and CHR (333985!, ex CANTY), from Herb. Cockayne, are possibly from the type gathering, but at least one of these, WELT 13271, has a different Cockayne number.

= *Hebe lapidosa* G.Simpson et J.S.Thomson, *Transactions of the Royal Society of New Zealand* 70: 31 (1940).
HOLOTYPE: Dee River Gorge, Clarence Basin, Marlborough, rock debris on rock benches and flood-beds, *G. Simpson*, CHR 56636! (mounted on two sheets, labelled 56636A and 56636B). ISOTYPE: AK 22163!.
NOTES: Moore (in Allan 1961) considered the type of *H. lapidosa* to be the sheet CHR 56636B (in accordance with annotations by H. H. Allan, 1952, who was of the same opinion and who considered sheet 56636A to be an isotype). Since both sheets of CHR 56636 are presumably part of the same "gathering" (sensu ICBN Art. 8.2) and are cross-labelled (ICBN Art. 8.3), the whole collection is here considered to constitute the holotype.

Hebe salicifolia (G.Forst.) Pennell, *Rhodora* 23: 39 (1921).
≡ *Veronica salicifolia* G.Forst., *Florulae Insularum Australium Prodromus*: 3 (1786); ≡ *Panoxis salicifolia* (G.Forst.) Raf., *Medical Flora* 2: 109 (1830), *nom. illeg.* (combination not definitely indicated, ICBN, Art. 33.1).
LECTOTYPE (designated by Bayly & Kellow 2004*b*): habitat in Nova Zeelandia, [*Forster*], the Forster Herbarium, presented by the corporation of Liverpool, August 1885, K!.
NOTES: Nicolson & Fosberg (2004) present a list of potential syntypes.

= *Veronica fonkii* Phil., *Linnaea* 29: 110 (1857–8); ≡ *Hebe fonkii* (Phil.) Cockayne et Allan, *Transactions of the New Zealand Institute* 57: 21 (1926).
TYPE(s): En las playas y barrancas de Chonos [on the beaches and slopes of Chonos], *Dr. Fonk*, SGO 56269, 43153 (n.v.).
NOTES: Without seeing the type of *V. fonkii* its identity is assumed from details given in the protologue, and from the opinions of Pennell presented by Cockayne & Allan (1926*c*). The two sheets listed here are those cited by Muñoz Pizarro (1960).

= *Veronica salicifolia* var. *communis* Cockayne, *Transactions of the New Zealand Institute* 48: 201 (1916); ≡ *Hebe salicifolia* var. *communis* (Cockayne) Cockayne et Allan, *Transactions of the New Zealand Institute* 57: 17 (1926).
LECTOTYPE (designated by Moore, in Allan 1961): scrub on bank of R. Kowai, Canterbury, *L. Cockayne 8041*, 16 Feb 1902, CHR 328786! (ex CANTY). ISOLECTOTYPES: AK 7739!, CHR 328785!.

Hebe salicornioides (Hook.f.) Cockayne et Allan, *Transactions of the New Zealand Institute* 57: 40 (1926).
≡ *Veronica salicornioides* Hook.f., *Handbook of the New Zealand Flora*: 212 (1864); ≡ *Leonohebe salicornioides* (Hook.f.) Heads, *Botanical Society of Otago Newsletter* 5: 7 (1987).
LECTOTYPE (designated by Ashwin, in Allan 1961): Nelson Mts, *Rough*, 1859, Herb. Hookerianum, K! (piece on right side of a sheet that also includes two pieces of *H. armstrongii*, *Sinclair & Haast*).

Hebe scopulorum Bayly, de Lange et Garn.-Jones, *New Zealand Journal of Botany* 40: 586 (2002).
HOLOTYPE: New Zealand, North Island, South Auckland, Kawhia region, Rock Peak, 520 m, limestone bluffs around summit, *M. J. Bayly 1444, P. J. Garnock-Jones & P. J. de Lange*, 10 Oct 2000, WELT 82488/A!. ISOTYPES: AK!, CHR!.

Hebe societatis Bayly et Kellow, *New Zealand Journal of Botany* 40: 576 (2002).
HOLOTYPE: New Zealand, South Island, Nelson, Braeburn Range, Mount Murchison, 1450 m, steep northeast facing slopes in *Chionochloa australis* grassland, *M. J. Bayly 1471 & A. V. Kellow*, 12 Jan 2001, WELT 82424!.

Hebe speciosa (A.Cunn.) Andersen, *Transactions of the New Zealand Institute* 56: 693 (1926).
≡ *Veronica speciosa* R.Cunn. ex A.Cunn., *Botanical Magazine* 63: Sub-plate 3461 (1836); ≡ *Hebe speciosa* (A.Cunn.) Cockayne et Allan, *Transactions of the New Zealand Institute* 57: 14 (1926), *nom. illeg.*, non Andersen.
LECTOTYPE (designated by Moore, in Allan 1961): *R. Cunningham No. 373*, 1834, K!. ISOLECTOTYPE: WELT 79342!
NOTES: The lectotype designated by Moore (in Allan 1961) is dated 1834, but the protologue gives the date as 1833.

Hebe stenophylla (Steudel) Bayly et Garn.-Jones, *New Zealand Journal of Botany* 38: 173 (2000).
≡ *Veronica stenophylla* Steudel, *Nomenclator Botanicus*, 2nd edn: 760 (1841); ≡ *Veronica angustifolia* A.Rich., *Essai d'une Flore de la Nouvelle Zélande*: 187 (1832), *nom. illeg.*, non Fisch. ex Link, *Enumeratio Plantarum Horti Regi Berolinensis Altera* 1: 19 (1821); nec S.F. Gray, *A Natural Arrangement of British Plants* 2: 306, Plate 2, (1872), nec Bernh., nec Steud.; ≡ *Hebe angustifolia* Cockayne et Allan, *Transactions of the New Zealand Institute* 57: 23 (1926).
LECTOTYPE (designated by Bayly et al. 2000): *Veronica angustifolia* Nob., Nlle Zelande, P!, Herb. Richard.

var. *stenophylla*
= *Veronica parviflora* var. *angustifolia* Hook.f., *Botanical Magazine*: t. 5965 (1872); ≡ *Hebe parviflora* var. *angustifolia* (Hook.f.) L.B.Moore, *Flora of New Zealand* 1: 912 (1961).
LECTOTYPE (designated by Bayly et al. 2000): *Botanical Magazine*: t. 5965 (1872).

= *Veronica squalida* Kirk, *Transactions of the New Zealand Institute* 28: 528 (1896).
LECTOTYPE (designated by Moore, in Allan 1961): Matori, Nelson Province, *T. Kirk*, 9 Feb 1877, WELT 5339!.

= *Veronica angustifolia* var. *abbreviata* Petrie, *Transactions of the New Zealand Institute* 53: 371 (1921).
HOLOTYPE: Valley of the Ure R., Marlbro [Marlborough], *B. C. Aston*, early Apr 1915, WELT 5340!.

var. *hesperia* Bayly et Garn.-Jones, *New Zealand Journal of Botany* 38: 180 (2000).
HOLOTYPE: Nelson, c. 2 km southwest of Kaihoka Lakes, on bluffs beside Limestone Road, *M. J. Bayly 1149*, 28 Jan 1999, WELT 81486!. ISOTYPES: AK, CHR.

var. *oliveri* Bayly et Garn.-Jones, *New Zealand Journal of Botany* 38: 182 (2000).
HOLOTYPE: Stephens Island, *W. R. B. Oliver*, Jan 1922, WELT 64970!. ISOTYPE: WELT 15028.

Hebe stricta (Benth.) L.B.Moore, *Flora of New Zealand* 1: 904 (1961).
≡ *Veronica stricta* Benth. in DC., *Prodromus Systematis Naturalis Regni Vegetabilis. Vol. 10*: 459 (1846); ≡ *Veronica salicifolia* var. *stricta* (Benth.) Hook.f. *Flora Novae-Zelandiae* 1: 191 (1853); ≡ *Hebe salicifolia* var. *stricta* (Benth.) Cockayne et Allan, *Transactions of the New Zealand Institute* 57: 17 (1926).
LECTOTYPE (designated by Moore, in Allan 1961; designated more precisely by Bayly & Kellow 2004*b*): Auckland, N.Z., *Sinclair*, Herb. Hookerianum K!, four uppermost pieces on sheet only (sheet also includes another Sinclair specimen, of two pieces, from Thames).
NOTES: The name *Veronica stricta* listed in *Index Kewensis* as "Lodd. Cat. (1823); ex Schult. Mant. i. Add. II. 228…" is a *nom. nud.* (i.e. with no description provided by either Loddiges or Schultes) and has no status under the ICBN.

var. *stricta*
= *Veronica parkinsoniana* Colenso, *Transactions of the New Zealand Institute* 21: 97 (1889); ≡ *Hebe parkinsoniana* (Colenso) Cockayne, *Transactions of the New Zealand Institute* 60: 472 (1929).
TYPE: None designated.
NOTES: No potential types have been found in New Zealand herbaria or at K. The description suggests, as also gathered by Moore (in Allan 1961), that the name applies to *H. stricta*.

= *Hebe salicifolia* var. *longiracemosa* (Cockayne) Cockayne et Allan, *Transactions of the New Zealand Institute*

57: 18 (1926); ≡ *Veronica salicifolia* var. *longiracemosa* Cockayne, *Transactions of the New Zealand Institute* 49: 61 (1917).

LECTOTYPE (designated by Moore, in Allan 1961): on steep bank at outskirts of forest, Moumahaki (Egmont-Wanganui bot. distr. of N.Z.), lowland belt, *L. Cockayne 8163*, March 1916, K!.

NOTES: AK 8459!, labelled by Cockayne as "*Veronica salicifolia* Forst.f. var. *longiracemosa* Cockayne" and "Type specimen", has a different collection number (8119; cited by Moore as "AK 8119") from the lectotype, and for this reason is not considered part of the type collection. CHR 328778!, probably the specimen cited by L. B. Moore as from Canterbury Museum, is potentially an isotype, but it has no collection number and the locality information resembles that of AK 8459, rather than the lectotype.

var. *atkinsonii* (Cockayne) L.B.Moore, *Flora of New Zealand* 1: 906 (1961).

≡ *Veronica salicifolia* var. *atkinsonii* Cockayne, *Transactions of the New Zealand Institute* 48: 200 (1916); ≡ *Hebe salicifolia* var. *atkinsonii* (Cockayne) Andersen, *Transactions of the New Zealand Institute* 56: 693 (1926); ≡ *Hebe salicifolia* var. *atkinsonii* (Cockayne) Cockayne et Allan, *Transactions of the New Zealand Institute* 57: 18 (1926), *nom. illeg.*, non Andersen.

TYPE: None designated.

NOTES: See notes by Moore (in Allan 1961).

var. *macroura* (Benth.) L.B.Moore, *Flora of New Zealand* 1: 906 (1961).

≡ *Veronica macroura* Benth. in DC., *Prodromus Systematis Naturalis Regni Vegetabilis. Vol. 10*: 459 (1846); ≡ *Hebe macroura* (Benth.) Cockayne et Allan, *Transactions of the New Zealand Institute* 57: 15 (1926).

HOLOTYPE: "*Veronica macroura* Hook fil", N. Zealand, Herb. Benthamianum, K!.

= *Veronica cookiana* Colenso, *Transactions of the New Zealand Institute* 20: 201 (1888); ≡ *Veronica macroura* var. *cookiana* (Colenso) Cheeseman, *Manual of the New Zealand Flora*: 501 (1906); ≡ *Hebe cookiana* (Colenso) Cockayne et Allan, *Transactions of the New Zealand Institute* 57: 16 (1926).

LECTOTYPE (designated by Moore, in Allan 1961): Table Cape plant, *H. Hill*, 1887, Herb. Colenso, WELT 5315 [details taken mostly from the handwritten register at WELT].

NOTES: The lectotype specimen has not been seen by us. WELT 5287 (Table Cape Plant, *H. Hill*, Herb. Petrie, undated) is possibly an isolectotype.

var. *egmontiana* L.B.Moore, *Flora of New Zealand* 1: 907 (1961).

HOLOTYPE: Mt Egmont, subalpine scrub, *L. Cockayne No. 8162*, 18 March 1916, WELT 5365!. ISOTYPES: AK 7778!; CHR 328767! (large piece on left only; small piece on right is probably *H. paludosa* – in other words, not part of the same gathering as the holotype), 328766!; K!; WELT 79016! (mounted on two sheets).

NOTES: The isotypes lack Cockayne's specimen number (8162), but otherwise match the details of the holotype. Both of the CHR isotypes are ex CANTY.

var. *lata* L.B.Moore, *Flora of New Zealand* 1: 907 (1961).

HOLOTYPE: Kaweka Mountains, west of Kuripapango Hill, *c.* 4400 ft, on rocky outcrop, *A. P. Druce (457)*, 5 May 1959, CHR 76144!. ISOTYPE: K!.

Hebe strictissima (Kirk) L.B.Moore, *Flora of New Zealand* 1: 916 (1961).

≡ *Veronica parviflora* var. *strictissima* Kirk, *Transactions of the New Zealand Institute* 28: 527 (1896); ≡ *Veronica leiophylla* var. *strictissima* (Kirk) Cockayne, *Cawthron Lecture* 3: 11 (1920); ≡ *Hebe leiophylla* var. *strictissima* (Kirk) Cockayne et Allan, *Transactions of the New Zealand Institute* 57: 24 (1926).

LECTOTYPE (designated by Moore, in Allan 1961): Akaroa, *T. Kirk*, Jan 1876, WELT 5341!. ISOLECTOTYPE: AK 7879!.

Hebe subalpina (Cockayne) Andersen, *Transactions of the New Zealand Institute* 56: 693 (1926).

≡ *Veronica subalpina* Cockayne, *Transactions of the New Zealand Institute* 31: 420 (1899); ≡ *Hebe subalpina* (Cockayne) Cockayne et Allan, *Transactions of the New Zealand Institute* 57: 29 (1926), *nom. illeg.*, non Andersen.

LECTOTYPE (designated by Moore, in Allan 1961): Mt Rangi Taipo, Westland, *L. Cockayne 8030*, Jan 1896, AK 8012!.

= *Veronica montana* J.B.Armstr., *New Zealand Country Journal* 3: 58 (1879), *nom. illeg.*, non L., *Centuria Plantarum* 1: 3 (1755), nec Pallas, *Reise* 2: 522 (1776); ≡ *Veronica monticola* J.B.Armstr., *Transactions of the*

New Zealand Institute 13: 354 (1881), *nom. illeg.*, non Trautv., *Bulletin de l'Académie Impériale des Sciences de Saint-Pétersbourg* 10: 398 (1866); ≡ *Hebe monticola* Andersen, *Transactions of the New Zealand Institute* 56: 693 (1926); ≡ *Hebe montana* Cockayne et Allan, *Transactions of the New Zealand Institute* 57: 31 (1926), *nom illeg.*, based on the same type as *H. monticola* Andersen; ≡ *Hebe monticola* A.Wall, *Transactions of the New Zealand Institute* 60: 385 (1929), *nom. illeg.*, non Andersen.

LECTOTYPE (designated by Moore, in Allan 1961): Rangitata Va., *J. F. Armstrong*, 1869, CHR! (Herb. Armstrong).

NOTES: Moore (in Allan 1961) designated a single piece (the larger) on the type sheet as the lectotype. Under the current rules of the ICBN (Art 8.2; and see Preface), it seems more appropriate to regard the whole sheet as comprising the lectotype. An Armstrong specimen in Herb. T. Kirk at WELT, which lacks collecting data, might match the lectotype (Moore, in Allan 1961).

= *Hebe fruticeti* G.Simpson et J.S.Thomson, *Transactions of the Royal Society of New Zealand* 70: 30 (1940).
LECTOTYPE (designated by Moore, in Allan 1961): Head of Estuary Burn, Lake Wanaka, Otago, sub-alpine scrub, *G. Simpson & J. S. Thomson*, CHR 33029!.

Hebe tairawhiti B.D.Clarkson et Garn.-Jones, *New Zealand Journal of Botany* 34: 51 (1996).
HOLOTYPE: New Zealand, Gisborne Land District, Makorori Beach, on mudstone banks and cliffs, *B. D. Clarkson*, 10 May 1994, CHR 454678!. ISOTYPES: AK 229879!, NZFRI, WAIK.

Hebe tetragona (Hook.) Andersen, *Transactions of the New Zealand Institute* 56: 693 (1926).
≡ *Veronica tetragona* Hook., *Icones Plantarum* 6: Plate 580 (1843); ≡ *Hebe tetragona* (Hook.) Cockayne et Allan, *Transactions of the New Zealand Institute* 57: 39 (1926), *nom. illeg.*, non Andersen; ≡ *Leonohebe tetragona* (Hook.) Heads, *Botanical Society of Otago Newsletter* 5: 9 (1987).
LECTOTYPE (designated by Ashwin, in Allan 1961): *Bidwill 50*, Herb. Hookerianum, K!.
NOTES: The heavily flowering middle piece of the lectotype probably matches the piece figured in the protologue. The lectotype is mounted on the same sheet as the other syntype (*Colenso 63*). When publishing his new combination, Heads (1987) incorrectly cited the basionym as "*Veronica tetragona* Hook. f.". This error does not invalidate his combination (ICBN Art. 33.4).

subsp. *tetragona*
= *Podocarpus ? dieffenbachii* Hook., *Icones Plantarum* 6: Plate 547 (1843).
TYPE: "Queen Charlotte Sound, New Zealand. *Dr. Dieffenbach*", K? (n.v.).

subsp. *subsimilis* (Colenso) Bayly et Kellow, comb. nov.
≡ *Veronica subsimilis* Colenso, *Transactions of the New Zealand Institute* 31: 278 (1899); ≡ *Leonohebe subsimilis* (Colenso) Heads, *Botanical Society of Otago Newsletter* 5: 9 (1987); ≡ *Hebe subsimilis* (Colenso) Ashwin, *Flora of New Zealand* 1: 929 (1961); ≡ *Hebe hectorii* subsp. *subsimilis* (Colenso) Wagstaff et Wardle, *New Zealand Journal of Botany* 37: 33 (1999).
TYPE: Ruahine Range, *H. Hill*, Herb. Colenso, WELT 5342.
NOTES: We have not been able to locate the type specimen. It has certainly been registered at WELT, and was apparently seen by Moore or Ashwin (in Allan 1961).

= *Veronica astonii* Petrie, *Transactions of the New Zealand Institute* 40: 288 (1908); ≡ *Hebe astonii* (Petrie) Cockayne et Allan, *Transactions of the New Zealand Institute* 57: 39 (1926); ≡ *Hebe subsimilis* var. *astonii* (Petrie) Ashwin, *Flora of New Zealand* 1: 930 (1961); ≡ *Leonohebe subsimilis* var. *astonii* (Petrie) Heads, *Botanical Society of Otago Newsletter* 5: 9 (1987).
LECTOTYPE (designated by Ashwin, in Allan 1961): Tararuas, Mt Hector, *D. Petrie*, 29 Jan 1907, WELT 5307! [Ashwin, apparently incorrectly, cited the date on this specimen as "25.1.07".] ISOLECTOTYPE: AK 8191!.

Hebe topiaria L.B.Moore, *Flora of New Zealand* 1: 917 (1961).
HOLOTYPE: Mt Arthur Tableland, Nelson, common at Cundy's Creek, etc., *F. G. Gibbs* (no. 576 to T. F. Cheeseman), (July 1910 to T. F. C.), CHR 76137!. ISOTYPES: AK 8051!.

Hebe townsonii (Cheeseman) Cockayne et Allan, *Transactions of the New Zealand Institute* 57: 20 (1926).
≡ *Veronica townsonii* Cheeseman, *Transactions of the New Zealand Institute* 45: 95 (1913); ≡ *Veronica macrocarpa* var. *crassifolia* Cheeseman, *Manual of the New Zealand Flora*: 505 (1906).
LECTOTYPE (designated by Moore, in Allan 1961): Karamea Hill, N.W. Nelson, *W. Townson 21B*, AK 7799!.

NOTES: The name *V. townsonii* is an avowed substitute for the earlier name, *V. macrocarpa* var. *crassifolia* (the epithet *crassifolia* not being available at species rank), and it therefore has the same type (ICBN Art. 7.3), even though this specimen is not explicitly listed among those Cheeseman cited in the description of *V. townsonii*. There are other Townson specimens that are potential isolectotypes, but these either give slightly different locality information, have different collecting numbers, or lack a collecting number (e.g. CHR 331763!).

Hebe traversii (Hook.f.) Andersen, *Transactions of the New Zealand Institute* 56: 694 (1926).
≡ *Veronica traversii* Hook.f., *Handbook of the New Zealand Flora*: 208 (1864); ≡ *Hebe traversii* (Hook.f.) Cockayne et Allan, *Transactions of the New Zealand Institute* 57: 29 (1926), *nom. illeg.*, non Andersen.
LECTOTYPE (designated by V. S. Summerhayes in *Kew Bulletin*: 397 (1927)): Hurunui, 3–4000 ft, *Travers*, Herb. Hookerianum, K! (mounted on upper left of sheet that includes several other collections).

= *Veronica traversii* var. *elegans* Cheeseman, *Manual of the New Zealand Flora*: 519 (1906).
LECTOTYPE (designated by Moore, in Allan 1961): Craigieburn, upper Waimakariri, Canterbury, *L. Cockayne 8018*, Herb. T. F. Cheeseman (1574 to Kew), AK 8004!. ISOLECTOTYPE: WELT 79056!

Hebe treadwellii Cockayne et Allan, *Transactions of the New Zealand Institute* 56: 27 (1926).
LECTOTYPE (here designated): in open places in scrub near grassline of Mt Ollivier, Sealey [Sealy] Range, *L. Cockayne*, 17 Feb 1919, CHR 332342!.
NOTES: Two specimens (both the same species) have been found that match the locality/specimen information given in the protologue – the lectotype and WELT 47641!, Sealey [Sealy] Range by Mueller Glacier, 4000 ft, Herb. L. Cockayne [no date]. Although the WELT specimen is labelled "*Hebe Treadwellii*" and "Type" in Cockayne's hand (see ICBN Rec. 9A.3), it was not chosen as the lectotype. This is primarily because it is a fruiting specimen, and the protologue clearly states "Capsula non visa", suggesting that the specimen was probably not used in drafting the original description. The lectotype, on the other hand, includes flowers (described in detail in the protologue but missing from WELT 47641) and no fruit. It was also clearly considered by Cockayne to represent an undescribed species (although he labelled it "*Veronica Wrightii* Cockayne sp. nov. ined.") and was collected before publication of the protologue.

A further specimen at CHR (6336!) includes a note by L. B. Moore suggesting that it should be regarded as the type. That specimen is, however, from Mt Sebastopol, a locality not mentioned in the protologue, and is dated "1928", two years after publication of the name.

= *Hebe brockiei* G.Simpson et J.S.Thomson, *Transactions of the Royal Society of New Zealand* 72: 28 (1942).
LECTOTYPE (designated by Moore, in Allan 1961): hills above Amuri Pass, north Canterbury, grassland, *G. Simpson & J. S. Thomson*, Dec 1940, CHR 56679!.

Hebe truncatula (Colenso) L.B.Moore, *Flora of New Zealand* 1: 912 (1961).
≡ *Veronica truncatula* Colenso, *Transactions of the New Zealand Institute* 31: 276 (1899).
LECTOTYPE (designated by Moore, in Allan 1961): Ruahine Range, *H. Hill*, ex Herb. Colenso, AK 8436!.

Hebe urvilleana W.R.B.Oliv., *Records of the Dominion Museum, Wellington* 1: 212 (1944).
LECTOTYPE (designated by Moore, in Allan 1961): Bald Spur, D'Urville Island, low manuka scrub on serpentine, *W. R. B. O[liver]*, 9 Feb 1943, WELT 5335!. ISOLECTOTYPES: CHR 89147!, WELT 47651!.

Hebe venustula (Colenso) L.B.Moore, *Flora of New Zealand* 1: 897 (1961).
≡ *Veronica venustula* Colenso, *Transactions of the New Zealand Institute* 27: 393 (1895).
LECTOTYPE (designated by Moore, in Allan 1961): east side of Ruahine Range, *A. Olsen*, Dec. 1893, AK 7891. ISOLECTOTYPE: K!.

= *Veronica azurea* [azunea] Colenso, *Transactions of the New Zealand Institute* 31: 277 (1899), *nom. illeg.*, non Link, *Enumeratio Plantarum Horti Regi Berolinensis Altera* 1: 22 (1821).
LECTOTYPE (designated by Moore, in Allan 1961): Ruahine Range, *H. Hill*, WELT 5316.

= *Hebe laevis* Cockayne et Allan, *Transactions of the New Zealand Institute* 57: 26 (1926); ≡ *Veronica laevis* Benth. in DC., *Prodromus Systematis Naturalis Regni Vegetabilis. Vol. 10*: 461 (1846), *nom. illeg.*, non Lam., *Flore Françoise* 2: 444 (1778).
LECTOTYPE (designated by Moore, in Allan 1961): *Colenso 4060*, K!.
NOTES: According to Moore (in Allan 1961) there is an isolectotype at WELT (n.v.).

Hebe vernicosa (Hook.f.) Cockayne et Allan, *Transactions of the New Zealand Institute* 57: 30 (1926).
≡ *Veronica vernicosa* Hook.f., *Handbook of the New Zealand Flora*: 208 (1864).
LECTOTYPE (designated by Moore, in Allan 1961): upper Wairau, *Dr Munro no. 12*, Jan 1854, K! (pieces on left side of sheet that also includes material from Canterbury Hills, *Travers*).

= *Veronica greyi* J.B.Armstr., *New Zealand Country Journal* 3: 57 (1879) [note that Armstrong spelled this *grayi* in *Transactions of the New Zealand Institute* 13: 354 (1881)]; ≡ *Hebe greyi* (J.B.Armstr.) Cockayne, *Transactions of the New Zealand Institute* 60: 471 (1929).
HOLOTYPE: Waiau Va., 3000 ft, *J. B. Armstrong*, CHR! (Herb. Armstrong).

= *Veronica vernicosa* var. *gracilis* Cheeseman, *Manual of the New Zealand Flora*: 520 (1906).
LECTOTYPE (designated by Moore, in Allan 1961): Mt Arthur Plateau, Nelson, 3500 ft, *T. F. Cheeseman*, 1562 to Kew, AK 8032!.
NOTES: The lectotype is labelled "1562 to Kew". At K there is a specimen from Herb. Cheeseman with the number 1563. This is labelled "Mount Arthur, 4500 ft, Jan. 1886"; in other words, it gives a different altitude to the lectotype, and is therefore not from the same locality and is not an isotype.

= *Veronica vernicosa* var. *multiflora* Cheeseman, *Manual of the New Zealand Flora*: 520 (1906).
LECTOTYPE (designated by Moore, in Allan 1961): cultivated in Mr Matthews' garden, Dunedin, *H. J. Matthews*, AK 8029!.

NAMES OF SPECIES IN *LEONOHEBE*

Leonohebe cheesemanii (Buchanan) Heads, *Botanical Society of Otago Newsletter* 5: 5 (1987)
≡ *Mitrasacme cheesemanii* Buchanan, *Transactions of the New Zealand Institute* 14: 348 (1882); ≡ *Veronica quadrifaria* Kirk, *Transactions of the New Zealand Institute* 28: 521 (1896), non *V. cheesemanii* Benth. in Hook., *Icones Plantarum* t. 1336A (1881); ≡ *Hebe cheesemanii* (Buchanan) Cockayne et Allan, *Transactions of the New Zealand Institute* 57: 39 (1926).
LECTOTYPE (designated by Moore, in Allan 1961): WELT in Herb. Buchanan! [the bottom sprig on this page matches the illustration in the protologue, and is presumably the specimen (formerly held by the Otago Museum) to which Moore (in Allan 1961) referred, even though it is not furnished with collecting information]. ISOLECTOTYPE: AK 8174!.

Leonohebe ciliolata (Hook.f.) Heads, *Botanical Society of Otago Newsletter* 5: 5 (1987).
≡ *Logania ciliolata* Hook.f., *Handbook of the New Zealand Flora*: 737 (1867); ≡ *Veronica gilliesiana* Kirk, *Transactions of the New Zealand Institute* 28: 519 (1896); ≡ *Hebe ciliolata* (Hook.f.) Cockayne et Allan, *Transactions of the New Zealand Institute* 57: 39 (1926).
LECTOTYPE (here designated): slopes above Browning's Pass, 4000–6000 ft, *Haast 95*, 1 April 1866 [assuming the date at the bottom of the label, 21 Nov. 1866, is the date the specimen was received at K], K!.
NOTES: Although the label of only one specimen (the lectotype) exactly matches the locality and altitude specified in the protologue, another specimen (*Haast 82*, amongst rocks on Browning's Pass, 1 April 1866, K!) has details basically consistent with those specified, and could possibly compete for type status. Given this, it is appropriate to designate a lectotype.

When Kirk published the name *V. gilliesiana*, he possibly thought that the epithet "*ciliolata*" was already occupied in *Veronica*, by the name *V. ciliolata* (Hook.f.) Benth. et Hook.f., *Genera Plantarum* 2: 964 (1876). That latter name, thought to be based on *Logania ciliolata* Hook.f. [≡ *Chionohebe ciliolata* (Hook.f.) B.G.Briggs et Ehrend.], was also cited by Ashwin (in Allan 1961) but, under ICBN Art. 33.1 (see Ex. 2), it is not validly published. If both *Leonohebe ciliolata* and *Chionohebe ciliolata* were transferred to *Veronica* (as proposed by Albach et al. 2004a), the epithet *ciliolata* could be used for either (the choice being up to the transferring author).

= *Mitrasacme hookeri* Buchanan, *Transactions of the New Zealand Institute* 14: 348 (1882).
HOLOTYPE: Herb. Buchanan, WELT! [formerly housed at Otago Museum]. ISOTYPE: Mt Alta, 5,000 ft, "given to me by Mr Buchanan as part of his type specimen of 'Mitrasacme Hookeri'", Herb. T. Kirk, WELT 13044!.
NOTES: Although the holotype shows no collecting information that might link it to the protologue, the far

right branch matches the original illustration exactly. Specimens in the Buchanan volumes have generally been considered to include the types of his published names (Adams 2002), and treating this specimen as a holotype (rather than selecting a lectotype) is consistent with the spirit of ICBN Recommendation 9A.4.

Leonohebe cupressoides (Hook.f.) Heads, *Botanical Society of Otago Newsletter* 5: 8 (1987).
≡ *Veronica cupressoides* Hook.f., *Handbook of the New Zealand Flora*: 212 (1864); ≡ *Hebe cupressoides* (Hook.f.) Andersen, *Transactions of the New Zealand Institute* 56: 693 (1926); ≡ *Hebe cupressoides* (Hook.f.) Cockayne et Allan, *Transactions of the New Zealand Institute* 57: 42 (1926), *nom. illeg.*, non Andersen.
LECTOTYPE (first designated by Moore, in Allan 1961; designated here more precisely): Lindis Pass & Lake Dis[trict], on river flats, *Hector & Buchanan no. 7*, K! (sprig in lower left corner of sheet only).
NOTES: The lectotype designated by Moore (in Allan 1961) potentially includes material from more than one location collected at different times (i.e. potentially more than one "gathering" sensu ICBN Art. 8.2). Given this, it is appropriate to further refine that lectotypification.

Leonohebe tetrasticha (Hook.f.) Heads, *Botanical Society of Otago Newsletter* 5: 5 (1987).
≡ *Veronica tetrasticha* Hook.f., *Handbook of the New Zealand Flora*: 212 (1864); ≡ *Hebe tetrasticha* (Hook.f.) Andersen, *Transactions of the New Zealand Institute* 56: 694 (1926); ≡ *Hebe tetrasticha* (Hook.f.) Cockayne et Allan, *Transactions of the New Zealand Institute* 57: 39 (1926), *nom. illeg.*, non Andersen.
LECTOTYPE (designated by Moore, in Allan 1961): Canterbury, New Zealand, *Haast 763*, 1862, Herb. Hookerianum, K! [information from records of received specimens – in the library at K – indicates that this collection is from "River Hopkins forming Lake Ohau"].

Leonohebe tumida (Kirk) Heads, *Botanical Society of Otago Newsletter* 5: 5 (1987).
≡ *Veronica tumida* Kirk, *Transactions of the New Zealand Institute* 28: 521 (1896); ≡ *Hebe tumida* (Kirk) Cockayne et Allan, *Transactions of the New Zealand Institute* 57: 39 (1926).
LECTOTYPE (designated by Bayly & Kellow 2004*b*): Ben Nevis, *F. G. Gibbs*, 15 Jan. 1896, private herbarium of T. Kirk., WELT 43493/A!.

INCERTAE SEDIS

The following is a list of names of uncertain application. Some may represent hybrids, and as such could be placed in the next section.

Hebe darwiniana (Colenso) Cockayne, *Transactions of the New Zealand Institute* 60: 470 (1929); ≡ *Veronica darwiniana* Colenso, *Transactions of the New Zealand Institute* 25: 332 (1895).
TYPE: None designated; there are potential syntypes at AK (Herrick & Cameron 1994) and at K!.

Hebe gigantea (Cockayne) Cockayne et Allan, *Transactions of the New Zealand Institute* 57: 19 (1926); ≡ *Veronica gigantea* Cockayne, *Transactions of the New Zealand Institute* 34: 319 (1902); ≡ *Veronica salicifolia* var. *gigantea* (Cockayne) Cheeseman, *Manual of the New Zealand Flora*: 504 (1906).
TYPE: not designated.
NOTES: Details of the original description suggest that this name probably applies, as indicated by Moore (in Allan 1961), to *Hebe barkeri* (Cockayne) Cockayne. As also noted by Moore (in Allan 1961), only one Cockayne specimen has been found (juvenile leaves from a plant growing in the moist Te Awatapu Forest, *L. Cockayne*, Jan 1901, WELT 5294!). The juvenile leaves on this sheet cannot be identified by us with confidence, and could potentially be either *H. barkeri* or *H. dieffenbachii*. Given this, we refrain from lectotypifying the name and hope that either some clear criteria for identifying this specimen, or further original specimens, are found.

Hebe glauca-caerulea (J.B.Armstr.) Cockayne, *Transactions of the New Zealand Institute* 60: 471 (1929); ≡ *Veronica glauca-caerulea* J.B.Armstr., *New Zealand Country Journal* 3: 57 (1879); ≡ *Veronica pimeleoides* var. *glauca-caerulea* (J.B.Armstr.) Cheeseman, *Manual of the New Zealand Flora*: 527 (1906); ≡ *Hebe pimeleoides* var. *glauco-caerulea* (J.B.Armstr.) Cockayne et Allan, *Transactions of the New Zealand Institute* 57: 38 (1926).
LECTOTYPE (designated by Kellow et al. 2003*a*): Clyde and Rangitata Vallies [*sic*], *J. F. Armstrong*, 1869, CHR! (Herb. Armstrong).
NOTES: See Kellow et al. (2003*a*) for further details.

Hebe longiracemosa Cockayne, *Transactions of the New Zealand Institute* 60: 471 (1929); ≡ *Veronica longiracemosa* Petrie, *Transactions of the New Zealand Institute* 49: 52 (1917), *nom. illeg.*, non Colenso, *Transactions of the New Zealand Institute* 20: 203 (1888).
HOLO(?)TYPE: Awatere Va., *H. J. Matthews*, cult., Mar 1909, WELT 5312.

Hebe matthewsii (Cheeseman) Cockayne, *Transactions of the New Zealand Institute* 60: 471 (1929); ≡ *Veronica matthewsii* Cheeseman, *Manual of the New Zealand Flora*: 517 (1906).
LECTOTYPE (designated by Moore, in Allan 1961): Humboldt Mountains, Otago, *H. J. Matthews*, Herb. T. F. Cheeseman (1581 to Kew), AK 7955!. ISOLECTOTYPE: K!.

Hebe myrtifolia (Benth.) Cockayne, *Transactions of the New Zealand Institute* 60: 472 (1929); ≡ *Veronica myrtifolia* Benth. in DC., *Prodromus Systematis Naturalis Regni Vegetabilis. Vol.* 10: 460 (1846); ≡ *Veronica macrocarpa* var. *myrtifolia* (Benth.) Hook.f., *Flora Novae-Zelandiae* 1: 192 (1853).
SYNTYPES: collected by *Edgerly & Logan* (not found).

Veronica obovata Kirk, *Transactions of the New Zealand Institute* 9: 502 (1877); ≡ *Hebe obovata* (Kirk) Cockayne, *Transactions of the New Zealand Institute* 60: 472 (1929).
LECTOTYPE (designated by Bayly & Kellow 2004*b*): Broken River, *T. Kirk 685*, Herb. T. Kirk., WELT 47647!.
ISOLECTOTYPE: K!.
NOTES: Similar to *H. rakaiensis*, particularly in length of corolla tube, hairy ovaries and leaf margins. It differs from that species in its comparatively broad, thick leaves, which, on some specimens, appear ± glaucous. Kirk's original collections (represented in many herbaria) are not well matched by any other specimens.

Veronica parviflora var. *obtusa* Kirk, *Transactions of the New Zealand Institute* 28: 527 (1896).
HOLOTYPE: Hawke's Bay, *A. Hamilton*, WELT 16734!.
NOTES: Morphology and confusion over the WELT number of the type specimen are discussed by Bayly et al. (2000). The type closely resembles some specimens of *H. parviflora*, but could plausibly be the hybrid *H. stenophylla* × *H. stricta*, as indirectly implied by Moore (in Allan 1961).

NAMES OF POSSIBLE WILD HYBRIDS

The following list is based on that of Moore (in Allan 1961); for most names it does not provide any more detail than is presented in that work. It lists only published names for suspected hybrids – that is, it is not a list of known (but un-named) hybrid combinations. Names are arranged in alphabetical order by epithet. Identification of parentage, particularly from herbarium specimens, is difficult and we have generally refrained from making strong assertions in this area. No attempt is made to identify which names might be synonymous.

Hebe ×*affinis* (Cheeseman) Cockayne et Allan, *Transactions of the New Zealand Institute* 57: 20 (1926); ≡ *Veronica macropcarpa* var. *affinis* Cheeseman, *Manual of the New Zealand Flora*: 505 (1906); ≡ *Hebe macrosala* Cockayne et Allan, *Transactions of the New Zealand Institute* 57: 20 (1926), *nom. nud.*
HOLOTYPE: cliffs at Northcote, Auckland Harbour, *T. F. C[heeseman]*, 1875, AK 7730.
NOTES: Said to be *Hebe macrocarpa* × *H. salicifolia* [probably *H. stricta* var. *stricta*].

Hebe ×*amabilis* (Cheeseman) Cockayne et Allan, *Transactions of the New Zealand Institute* 57: 21 (1926); ≡ *Veronica amabilis* Cheeseman, *Manual of the New Zealand Flora*: 506 (1906); ≡ *Veronica salicifolia* var. *gracilis* Kirk, *The Forest Flora of New Zealand*: Plate 120 (1889).
LECTOTYPE (designated by Moore, in Allan 1961): The Bluff, *T. Kirk*, WELT 5321!.
NOTES: Probably *H. elliptica* × *H. salicifolia*.

Hebe ×*angustisala* Cockayne et Allan, *Transactions of the New Zealand Institute* 57: 46 (1926).
TYPE: None designated.
NOTES: Described as *Hebe angustifolia* [*H. stenophylla*] × *H. salicifolia* [possibly *H. salicifolia* or *H. stricta* var. *atkinsonii*, depending on locality of type].

Hebe blanda (Cheeseman) Pennell, *Rhodora* 23: 39 (1921); ≡ *Veronica amabilis* var. *blanda* Cheeseman, *Manual of the New Zealand Flora*: 506 (1906); ≡ *Hebe amabilis* var. *blanda* (Cheeseman) Andersen, *Transactions of the New Zealand Institute* 56: 693 (1926).

LECTOTYPE (designated by Moore, in Allan 1961): Otago Harbour, *D. Petrie*, Feb 1893, AK 7804.
NOTES: Probably *H. elliptica* × *H. salicifolia*.

Hebe carsei (Petrie) Cockayne, *Transactions of the New Zealand Institute* 60: 469 (1929); ≡ *Veronica carsei* Petrie, *Transactions of the New Zealand Institute* 55: 96 (1924).
TYPE: None designated. Syntypes are at WELT and AK (the latter are listed by Herrick & Cameron 1994).
NOTES: Thought to be *Hebe stricta* var. *stricta* × *H. venustula*.

Hebe cassinioides (Petrie) Cockayne, *Transactions of the New Zealand Institute* 60: 469 (1929); ≡ *Veronica cassinioides* Petrie, *Transactions of the New Zealand Institute* 47: 52 (1915).
LECTOTYPE (designated by Moore, in Allan 1961): Takitimu Mts, [*H. J. Matthews*], 1886, Herb. Petrie, WELT 5324!.
NOTES: Probably a hybrid between *H. odora* and a whipcord species.

Hebe dartonii (Petrie) Cockayne, *Transactions of the New Zealand Institute* 60: 470 (1929); ≡ *Veronica dartonii* Petrie, *Transactions of the New Zealand Institute* 55: 98 (1924).
LECTOTYPE (designated by Moore, in Allan 1961): Firewood Creek near Cromwell, in bottom of valley on rocky slope, *D. Petrie*, Jan 1911, WELT 5325!.
NOTES: The type is consistent with the parentage *H. pimeleoides* [probably subsp. *faucicola*] × *H. salicifolia* suggested by Cockayne (1929).

Hebe divergens (Cheeseman) Cockayne, *Transactions of the New Zealand Institute* 60: 470 (1929); ≡ *Veronica divergens* Cheeseman, *Manual of the New Zealand Flora*: 502 (1906).
LECTOTYPE (designated by Moore, in Allan 1961): Brighton, south of Westport, *W. Townson*, AK 7691!.
ISOLECTOTYPE: K!.
NOTES: Possibly *H. elliptica* × *H. leiophylla*. Additional specimens at AK and CHR are potentially original material.

Hebe ×*ellipsala* Cockayne et Allan, *Transactions of the New Zealand Institute* 57: 46 (1926).
TYPE: None designated.
NOTES: Described as *H. elliptica* × *H. salicifolia*.

Hebe ×*kirkii* (J.B.Armstr.) Cockayne et Allan, *Transactions of the New Zealand Institute* 57: 19 (1926); ≡ *Veronica kirkii* J.B.Armstr., *New Zealand Country Journal* 3: 58 (1879); ≡ *Veronica salicifolia* var. *kirkii* (J.B.Armstr.) Cheeseman, *Manual of the New Zealand Flora*: 504 (1906).
LECTOTYPE (here designated): Upper Rangitata, *J. F. A*[*rmstrong*], 1869, CHR! (Herb. Armstrong).
NOTES: Probably a hybrid between *H. salicifolia* and a small-leaved species. Moore (in Allan 1961) suggests *H. rakaiensis* as the other parent; this notion is consistent with the presence of hairs on ovaries of the lectotype.

Hebe ×*laevastoni* Cockayne et Allan, *Transactions of the New Zealand Institute* 57: 47 (1926).
TYPE: None designated.
NOTES: Described as *H. laevis* [*H. venustula*] × *H. astonii* [*H. subsimilis*].

Hebe ×*laevisala* Cockayne et Allan, *Transactions of the New Zealand Institute* 57: 46 (1926).
TYPE: None designated.
NOTES: Described as *H. laevis* [*H. venustula*] × *H. salicifolia* [probably *H. stricta* var. *stricta*].

Hebe ×*leiosala* Cockayne et Allan, *Transactions of the New Zealand Institute* 57: 47 (1926).
TYPE: None designated.
NOTES: Described as *H. leiophylla* × *H. salicifolia*.

Hebe lewisii (J.B.Armstr.) A.Wall, *Transactions of the New Zealand Institute* 60: 384 (1929); ≡ *Veronica lewisii* J.B.Armstr., *Transactions of the New Zealand Institute* 13: 357 (1881).
LECTOTYPE (designated by Bayly & Kellow 2004b): Timaru Downs, *J. F. A*[*rmstrong*], CHR! (Herb. Armstrong).
NOTES: Could be *H. elliptica* × *H. salicifolia*, as suggested by Moore (in Allan 1961).

Hebe loganioides (J.B.Armstr.) A.Wall, *Transactions of the New Zealand Institute* 60: 384 (1929); ≡ *Veronica loganioides* J.B.Armstr., *New Zealand Country Journal* 3: 59 (1879).
LECTOTYPE (here designated): Upper Rangitata, *J. F. A*[*rmstrong*], 1869, CHR! (Herb. Armstrong).
ISOLECTOTYPE: CHR! (Herb. Armstrong).

NOTES: It is not clear what the parents of this hybrid are (but see Moore, in Allan 1961, for a discussion of some characters).

Hebe ×*macrosala* Cockayne et Allan, *Transactions of the New Zealand Institute* 57: 46 (1926).
TYPE: None designated.
NOTES: Described as *Hebe macrocarpa* × *H. salicifolia* [probably *H. stricta* var. *stricta*].

Hebe ×*simmonsii* (Cockayne) Cockayne et Allan, *Transactions of the New Zealand Institute* 56: 20 (1926); ≡ *Veronica* ×*simmonsii* Cockayne, *Transactions of the New Zealand Institute* 48: 202 (1916).
LECTOTYPE (designated by Moore, in Allan 1961): French Pass, shore scrub, *L. Cockayne 8117*, WELT 5323!. ISOLECTOTYPES: CHR 328385!; K!; WELT 84018A, B.
NOTES: Specimens are consistent with the parentage (*H. stenophylla* × *H. stricta* var. *atkinsonii*) suggested in the protologue.

NAMES OF HORTICULTURAL FORMS

This list includes names relating to plants assumed to be of garden origin. It includes only names covered by the ICBN, not cultivar names or *nomina nuda*. Some names included in the similar list of Moore (in Allan 1961), and in *Index Kewensis*, are not treated here as validly published. These are: *Veronica jasminoides*, attributed to Bean, *Trees and Shrubs Hardy in the British Isles, Vol. 3* (1933), but listed therein (and in subsequent editions) only as a synonym of *V. diosmifolia* (ICBN Art 34.1); *V. hendersonii*, supposedly published by Sieb. & Voss, in *Vilmorin's Blumengärtnerei*, 3rd edn, i: 784 (1894), but listed there only under *H. speciosa*; and *V.* ×*myrtifolia* R.Linds, which is a later homonym of *Veronica myrtifolia* Benth.

Hebe ×*andersonii* (Lindl. et Paxton) Cockayne, *Transactions of the New Zealand Institute* 60: 468 (1929); ≡ *Veronica andersonii* Lindley et Paxton, *Paxton's Flower Garden* 2: t. 38 (1851).
LECTOTYPE (Heenan 1994b): Illustration in *Paxton's Flower Garden* 2: t. 38 (1851).
NOTES: Very widely cultivated. Heenan (1994b) concluded it is *H. stricta* var. *stricta* × *H. speciosa*.

Hebe balfouriana (Hook.f.) Cockayne, *Transactions of the New Zealand Institute* 60: 468 (1929); ≡ *Veronica balfouriana* Hook.f., *Botanical Magazine* 123: t. 7556 (1897).
TYPE: From Sir J. D. Hooker's garden, 28 Jun 1894, K.
NOTES: Cockayne & Allan (1926c) suggest this is a hybrid involving *H. vernicosa*, and Moore (in Allan 1961) suggested that *H. pimeleoides* was possibly the other parent.

Hebe carnea A.Wall, *Transactions of the New Zealand Institute* 60: 384 (1929); ≡ *Veronica carnea* J.B.Armstr., *Transactions of the New Zealand Institute* 13: 357 (1881), *nom. illeg.*, non Vitm. 1789–92 nec DC. nec Schultes 1822.
TYPE: None designated, but there is a specimen with this name in CHR! (Herb. Armstrong).

Hebe erecta (Kirk) Cockayne, *Transactions of the New Zealand Institute* 60: 470 (1929); ≡ *Veronica erecta* Kirk, *Transactions of the New Zealand Institute* 28: 517 (1896).
LECTOTYPE (designated by Moore, in Allan 1961): Mt Bonpland, as Mr Martin believes, cultivated in Mr Martin's nursery, WELT 5332!.
NOTES: Thought by Cockayne (1929) and Cheeseman (1925) to be similar to members of *Heliohebe*, but the type is consistent with the parentage *H. pimeleoides* × *H. salicifolia*, suggested by Moore (in Allan 1961).

Hebe ×*franciscana* (Eastw.) Souster, *Journal of the Royal Horticultural Society* 81: 498 (1956); ≡ *Veronica franciscana* Eastw., *Leaflets of Western Botany* 3: 221 (1943).
HOLOTYPE: Cultivated Golden Gate Park, *A. Eastwood*, Oct. 1913, CAS 20685 (Heenan 1994a).
NOTES: This is *H. elliptica* × *H. speciosa* and is commonly cultivated. History, morphology and various cultivars are discussed by Heenan (1994a).

Hebe godefroyana (Carrière) Cockayne, *Transactions of the New Zealand Institute* 60: 471 (1929); ≡ *Veronica godefroyana* Carrière, *Revue Horticole* 60: 455 (1888).
TYPE: None designated.

Hebe imperialis (Bouch.) Cockayne, *Transactions of the New Zealand Institute* 60: 471 (1929); ≡ *Veronica speciosa* var. *imperialis* Bouch., *Flore des Serres* 22: 97, t. 2317 (1878).
TYPE: Unknown.

Hebe rotundata (Kirk) Cockayne, *Transactions of the New Zealand Institute* 60: 472 (1929); ≡ *Veronica rotundata* Kirk, *Transactions of the New Zealand Institute* 28: 530 (1896).
TYPE: None designated.

Veronica ×*edinensis* R.Linds., *Gardeners' Chronicle* 48: 103 (1910).
TYPE: None designated.
NOTES: Stated in the protologue to be *V. hectorii* (female parent) × *V. pimeleoides* (ex Royal Botanic Gardens, Edinburgh), and to resemble *V. epacridea*.

Veronica kermesina Loudon, *Encyclopedia of Plants*: 1546 (1855); ≡ *Hebe speciosa* var. *kermesina* "A Devonian" [a pseudonym used by John Luscombe], *Gardeners' Chronicle* 32: 500 (1850).
TYPE: None designated.
NOTES: From the original description this is probably either *Veronica speciosa* R.Cunn. ex A.Cunn or, as suggested by Metcalf (2001), a hybrid derived from it.

Veronica lindleyana Paxton, *Paxton's Magazine of Botany* 12: 247 (1846), non Wall., *A Numerical List of Dried Specimens*: 404 (1847), *nom. nud.*; ≡ *Veronica stricta* var. *lindleyana* (Paxton) J.B.Armstr., *Transactions of the New Zealand Institute* 13: 356 (1881).
TYPE: None designated.

EXCLUDED NAMES

Excluded names are listed first for *Hebe* and then for *Leonohebe*. The list does not include names excluded from *Hebe* that are synonymous with names in *Leonohebe*, or vice versa (these are listed in the preceding sections). Accepted names are shown in bold.

Names Excluded from Hebe

Hebe albiflora Pennell, *Journal of the Arnold Arboretum* 24: 257 (1943). ≡ **Parahebe albiflora** (Pennell) P.Royen et Ehrend., *Taxon* 19: 483 (1970).

Hebe bidwillii (Hook.) A.Wall, *Transactions of the New Zealand Institute* 60: 384 (1929). ≡ *Hebe bidwillii* (Hook.) Allan, *Transactions of the New Zealand Institute* 69: 276 (1939) *nom. illeg.*, non A.Wall. ≡ **Parahebe ×bidwillii** (Hook.) W.R.B.Oliv. (see Ashwin, in Allan 1961).

Hebe brassii Pennell, *Brittonia* 2: 185 (1936). = **Parahebe giulianettii** (Schltr.) P.Royen, *Taxon* 19: 483 (1970).

Hebe canescens A.Wall, *Transactions of the New Zealand Institute* 60: 384 (1929). ≡ **Parahebe canescens** (A.Wall) W.R.B.Oliv., *Records of the Dominion Museum, Wellington* 1: 229 (1944).

Hebe carstensensis (Wernham) Diels, *Botanische Jahrbücher für Systematik, Pflanzengeshichte und Pflanzengeographie* 62: 491 (1929). ≡ **Parahebe carstensensis** (Wernham) P.Royen, *Taxon* 19: 483 (1970).

Hebe catarractae (G.Forst.) A.Wall, *Transactions of the New Zealand Institute* 60: 384 (1929). ≡ **Parahebe catarractae** (G.Forst.) W.R.B.Oliv., *Records of the Dominion Museum, Wellington* 1: 230 (1944).

Hebe ciliata Pennell, *Journal of the Arnold Arboretum* 24: 258 (1943). ≡ **Parahebe ciliata** (Pennell) P.Royen et Ehrend, *Taxon* 19: 483 (1970).

Hebe dasyphylla (Kirk) Cockayne et Allan, *Transactions of the New Zealand Institute* 57: 42 (1926). = **Chionohebe densifolia** (F.Muell.) B.G.Briggs et Ehrend., *Contributions from Herbarium Australiense* 25: 2 (1976), although the generic placement of this species requires attention, as part of a general revision of the limits of *Chionohebe* and *Parahebe* (cf. Heads 2003; Garnock-Jones & Lloyd 2004).

Hebe diosmoides (Schltr.) Pennell, *Brittonia* 2: 184 (1936). ≡ **Parahebe diosmoides** (Schltr.) P.Royen et Ehrend., *Botanische Jahrbücher für Systematik, Pflanzengeschichte und Pflanzengeographie* 91: 402 (1972).

Hebe ×fairfieldii (Hook.f.) A.Wall, *Transactions of the New Zealand Institute* 60: 384 (1929). ≡ **Heliohebe ×fairfieldii** (Hook.f.) Garn.-Jones, *New Zealand Journal of Botany* 31: 337 (1993).

Hebe formosa (R.Br.) Cockayne, *Transactions of the New Zealand Institute* 6: 470 (1929). ≡ **Derwentia formosa** (R.Br.) Bayly comb. nov., based on *Veronica formosa* R.Br., *Prodromus Florae Novae Hollandiae*: 434 (1810).

Hebe hookeriana (Walp.) Allan, *Transactions of the New Zealand Institute* 69: 276 (1939). ≡ **Parahebe hookeriana** (Walp.) W.R.B.Oliv., *Records of the Dominion Museum, Wellington* 1: 230 (1944).

Hebe hulkeana (F.Muell.) Andersen, *Transactions of the New Zealand Institute* 56: 693 (1926). ≡ **Heliohebe** *hulkeana* (F.Muell.) Garn.-Jones **subsp.** *hulkeana*, *New Zealand Journal of Botany* 31: 328 (1993).

Hebe lavaudiana (Raoul) Andersen, *Transactions of the New Zealand Institute* 56: 693 (1926). ≡ ***Heliohebe lavaudiana*** (Raoul) Garn.-Jones, *New Zealand Journal of Botany* 31: 331 (1993).

Hebe lendenfeldii (F.Muell.) Pennell, *Brittonia* 2: 184 (1936). ≡ ***Parahebe lendenfeldii*** (F.Muell.) P.Royen, *Taxon* 19: 483 (1970).

Hebe linifolia (Hook.f.) Andersen, *Transactions of the New Zealand Institute* 56: 693 (1926). ≡ ***Parahebe linifolia*** (Hook.f.) W.R.B.Oliv., *Records of the Dominion Museum, Wellington* 1: 230 (1944).

Hebe lyallii (Hook.f.) A.Wall, *Transactions of the New Zealand Institute* 60: 385 (1929). ≡ ***Parahebe lyallii*** (Hook.f.) W.R.B.Oliv., *Records of the Dominion Museum, Wellington* 1: 230 (1944).

Hebe olsenii (Colenso) A.Wall, *Transactions of the New Zealand Institute* 60: 385 (1929). = ***Parahebe hookeriana*** (Walp.) W.R.B.Oliv., *Records of the Dominion Museum, Wellington* 1: 230 (1944).

Hebe polyphylla Pennell, *Journal of the Arnold Arboretum* 24: 257 (1943). ≡ ***Parahebe polyphylla*** (Pennell) P.Royen et Ehrend., *Taxon* 19: 483 (1970).

Hebe raoulii (Hook.f.) Cockayne et Allan, *Transactions of the New Zealand Institute* 57: 44 (1926). ≡ ***Heliohebe raoulii*** (Hook.f.) Garn.-Jones **subsp.** *raoulii*, *New Zealand Journal of Botany* 31: 333 (1993).

Hebe raoulii var. *maccaskillii* Allan, *Transactions of the Royal Society of New Zealand* 69: 273 (1940). ≡ ***Heliohebe raoulii*** **subsp.** ***maccaskillii*** (Allan) Garn.-Jones, *New Zealand Journal of Botany* 31: 335 (1993).

Hebe raoulii var. *pentasepala* L.B.Moore, *Flora of New Zealand* 1: 942 (1961). ≡ ***Heliohebe pentasepala*** (L.B.Moore) Garn.-Jones, *New Zealand Journal of Botany* 31: 332 (1993).

Hebe rigida Pennell, *Journal of the Arnold Arboretum* 24: 259 (1943). ≡ ***Parahebe rigida*** (Pennell) P.Royen et Ehrend., *Taxon* 19: 483 (1970).

Hebe rubra Pennell, *Brittonia* 2: 184 (1936). ≡ ***Parahebe rubra*** (Pennell) P.Royen et Ehrend., *Taxon* 19: 483 (1970).

Hebe tenuis Pennell, *Journal of the Arnold Arboretum* 24: 259 (1943). ≡ ***Parahebe tenuis*** (Pennell) P.Royen et Ehrend., *Taxon* 19: 483 (1970).

Hebe thymelaeoides Pennell, *Brittonia* 2: 186 (1936). = ***Parahebe diosmoides*** (Schltr.) P.Royen et Ehrend., *Botanische Jahrbücher für Systematik, Pflanzengeschichte und Pflanzengeographie* 91: 402 (1972).

Hebe uniflora (Kirk) Cockayne et Allan, *Transactions of the New Zealand Institute* 57: 43 (1926). Possibly a hybrid (Wagstaff & Garnock-Jones 2000) between *Chionohebe ciliolata* (Hook.f.) B.G.Briggs et Ehrend. and either *C. densifolia* (F.Muell.) B.G.Briggs et Ehrend. or *Parahebe planopetiolata* (Simpson et Thomson) W.R.B.Oliv. Even if it is not a hybrid, as suggested by Heads (1994*a*), it is still better placed in either *Chionohebe* B.G.Briggs et Ehrend. or *Parahebe* W.R.B.Oliv.

Hebe vanderwateri (Wernham) Van Steenis, *Bulletin du Jardin Botanique de Buitenzorg III* 13: 252 (1934). ≡ ***Parahebe vanderwateri*** (Wernham) P.Royen, *Taxon* 19: 483 (1970).

Names Excluded from Leonohebe

Leonohebe sect. *Densifoliae* Heads, *Botanical Society of Otago Newsletter* 5: 4 (1987). ≡ ***Chionohebe*** B.G.Briggs et Ehrend. ***p.p.***, although the limits of that genus and *Parahebe* require revision (cf. Heads 2003; Garnock-Jones & Lloyd 2004).

Leonohebe densifolia (F.Muell.) Heads, *Botanical Society of Otago Newsletter* 5: 4 (1987). ≡ ***Chionohebe densifolia*** (F.Muell.) B.G.Briggs et Ehrend., *Contributions from Herbarium Australiense* 25: 2 (1976), although the generic placement of this species requires attention, as part of a general revision of the limits of *Chionohebe* and *Parahebe* (cf. Heads 2003; Garnock-Jones & Lloyd 2004).

Leonohebe uniflora (Kirk) Heads, *Botanical Society of Otago Newsletter* 5: 5 (1987). Possibly a hybrid (Wagstaff & Garnock-Jones 2000) between *Chionohebe ciliolata* (Hook.f.) B.G.Briggs et Ehrend. and either *C. densifolia* (F.Muell.) B.G.Briggs et Ehrend. or *Parahebe planopetiolata* (Simpson et Thomson) W.R.B.Oliv. Even if it is not a hybrid, as suggested by Heads (1994*a*), it is still better placed in either *Chionohebe* B.G.Briggs et Ehrend. or *Parahebe* W.R.B.Oliv.

Common and Māori Names

ENGLISH COMMON NAMES

Very few *Hebe* species are widely known by common (informal) names in English. Moore (in Allan 1961) noted that the name "boxwood" refers to some small-leaved hebes of mountain areas, but this name is not used as commonly now as it may have been in the past. The common names in most general use today are the Māori names koromiko (used for *Hebe* in general, but in particular for willow-leaved species similar to *H. stricta* and *H. salicifolia*), tītīrangi and napuka (both used for *H. speciosa*).

The lack of English common names has driven some recent authors to invent them. Salmon (1968, 1991, 1992), in particular, coined many new "common" names for hebes of mountain areas, and more recent publications by DoC staff (e.g. Crisp et al. 2000; Brandon et al. 2004) use apparently new names for some geographically restricted and/or threatened species.

In many cases these newly suggested names are neither simpler to remember, nor more descriptive or informative than the scientific names. For example, the names Colenso's hebe, mountain koromiko, varnished koromiko, and black-barked mountain hebe are given for *H. colensoi*, *H. subalpina*, *H. vernicosa* and *H. decumbens*, respectively (Salmon 1968). Such names are otherwise not in "common" use, and their introduction seems to achieve no other purpose than to create more names for people to remember.

The following list of English common names is taken from a survey of recent New Zealand publications (examples of which are given in brackets after the names). It is not comprehensive, and most of the names are not widely used outside of these publications.

Awaroa koromiko	*Hebe scopulorum* (Brandon et al. 2004)
Barker's koromiko	*Hebe barkeri* (Crisp et al. 2000)
Bartlett's hebe	*Hebe perbella* (de Lange 1998; Forester & Townsend 2004)
Beech forest hebe	*Hebe vernicosa* (Smith-Dodsworth 1991)
Black-barked mountain hebe	*Hebe decumbens* (Salmon 1968, 1991, 1992)
Blue-flowered mountain hebe	*Hebe pimeleoides* (Salmon 1968, 1991, 1992)
Boulder Lake hebe	unknown (Salmon 1968)
Canterbury hebe	*Hebe canterburiensis* (Salmon 1968, 1991)
Canterbury whipcords	*Leonohebe cheesemanii, L. tetrasticha* (Salmon 1991, 1992)
Chatham Island koromiko	*Hebe chathamica* (Crisp et al. 2000)
Clubmoss hebe	*Hebe lycopodioides* (Smith-Dodsworth 1991)
Clubmoss whipcord	*Hebe lycopodioides* (Salmon 1968, 1991, 1992)
Colenso's hebe	*Hebe colensoi* (Salmon 1968, 1991, 1992)
Cypress koromiko	*Leonohebe cupressoides* (Nicol 1997)
Cypress whipcord	*Hebe propinqua* (Salmon 1968, 1992)
Cypress-like hebe	*Leonohebe cupressoides* (Salmon 1991, 1992)
Dieffenbach's koromiko	*Hebe dieffenbachii* (Crisp et al. 2000)
Dish-leaved hebe	*Hebe treadwellii* (Salmon 1991, 1992), *H. pinguifolia* (Salmon 1968, 1991, 1992)
Dwarf whipcord	*Hebe hectorii* subsp. *demissa* (Salmon 1968, 1992)
Hollow-leaved hebe	*Hebe buchananii* (Salmon 1968)
Kermadec koromiko	*Hebe breviracemosa* (de Lange & Stanley 1999)
Large-flowered hebe	*Hebe macrantha* (Salmon 1968, 1992; Smith-Dodsworth 1991; Wilson 1996)

Leathery-leaved mountain hebe	*Hebe odora, H. mooreae* (Salmon 1968, 1991, 1992)
Mt Arthur hebe	*Hebe albicans* (Salmon 1968, 1991, 1992)
Mountain koromiko	*Hebe subalpina* (Salmon 1968, 1991, 1992)
Nelson mountain hebe	*Hebe gibbsii* (Salmon 1968, 1991, 1992)
North Cape hebe	*Hebe brevifolia* (Salmon 1991)
North Island whipcord	*Hebe tetragona* (Smith-Dodsworth 1991)
Northern forest hebe	*Hebe diosmifolia* (Salmon 1991)
Northern koromiko	*Hebe obtusata* (Salmon 1991)
North-west Nelson hebe	*Hebe hectorii* subsp. *coarctata* (Salmon 1991, 1992)
Ochreous whipcord	*Hebe ochracea* (Salmon 1968, 1991, 1992)
Pumice whipcord	*Hebe tetragona* (Salmon 1968, 1992)
Purple hebe	*Hebe speciosa* (Nicol 1997)
Purple koromiko	*Hebe speciosa* (Smith-Dodsworth 1991)
Ruapehu hebe	*Hebe venustula* (Salmon 1968, 1991, 1992)
Scree hebe	*Hebe epacridea* (Salmon 1991)
Shore hebe	*Hebe elliptica* (Nicol 1997)
Shore koromiko	*Hebe elliptica* (Nicol 1997)
Spiny whipcord	*Leonohebe ciliolata* (Salmon 1968, 1991, 1992)
Swamp hebe	*Hebe pauciramosa, H. masoniae* (Salmon 1992)
Takaka hebe	*Hebe divaricata* (Salmon 1968, 1992)
Tararua hebe	*Hebe evenosa* (Salmon 1968, 1991, 1992)
Thick leaved hebe	*Hebe buchananii* (Salmon 1991)
Trailing whipcord	*Hebe epacridea* (Salmon 1968, in error); *H. haastii, H. macrocalyx* (Salmon 1991, 1992)
Traver's hebe	*Hebe traversii* (Salmon 1991)
Travers' hebe	*Hebe traversii* (Salmon 1968)
Varnished koromiko	*Hebe vernicosa* (Salmon 1968, 1991, 1992)
Waitakere rock koromiko	*Hebe bishopiana* (de Lange 1999)
Western mountain koromiko	*Hebe leiophylla* (Salmon 1991)
Whipcord hebe	*Hebe lycopodioides, H. salicornioides, Leonohebe cupressoides, H. propinqua, H. tetragona, H. armstrongii* (Wilson & Galloway 1993)

MĀORI NAMES

The following list of Māori names for hebes is based on that provided by Beever (1991). Where Beever indicated a long vowel sound by the use of double vowels, they are indicated here by use of a macron (¯), which is the common convention in written Te Reo.

Aute	*Hebe diosmifolia*
Kōkōmuka	*Hebe macrocarpa, H. stricta, H. salicifolia* and probably similar species.
Koromiko, kōkoromiko, kōkoromuka, korohiko, korokio, koromuka	*Hebe stricta* and/or *H. salicifolia*, and probably similar species also.
Koromiko tāranga, kōkōmuka tāranga	Beever (1991) records these names for *Hebe parviflora*, the classification of which has since been revised (Bayly et al. 2000). Without further information it is unclear whether these names apply to *H. parviflora, H. stenophylla* or both.
Tītīrangi, napuka	*Hebe speciosa*

NAMES FROM FRENCH POLYNESIA

According to Brown (1935), *H. rapensis* is known locally on Rapa by the names painaka, tutaipainaka or tutai painaka.

PART C

Indices

Appendix 1

Informal Names used by Druce (1980, 1993) and Eagle (1982)

Some of these names have been widely used by New Zealand botanists, and the purpose of this list is to indicate which species these informal taxa are assigned to in this book.

DRUCE (1980)	EAGLE (1982)	DRUCE (1993)	THIS BOOK
H. sp. (a)	*H. diosmifolia* p.p.	*H. diosmifolia* or "*H.* aff. *diosmifolia*"[1]	*H. diosmifolia* p.p.
H. sp. (b)	*H. diosmifolia* p.p.	*H. diosmifolia* or "*H.* aff. *diosmifolia*"[1]	*H. diosmifolia* p.p.
H. sp. (c)	*H. parviflora* var. *arborea*	"*H. arborea*"	*H. parviflora*
H. sp. (d) var. (i)	*H. parviflora* var. *angustifolia* p.p.	"*H. squalida*" p.p.	*H. stenophylla* var. *stenophylla* p.p.
H. sp. (d) var. (ii)	*H. parviflora* var. *angustifolia* p.p.	"*H. squalida*" p.p.	*H. stenophylla* var. *stenophylla* p.p.
H. sp. (d) var. (iii)	*H. parviflora* var. *angustifolia* p.p.	"*H. squalida*" p.p.	*H. stenophylla* var. *stenophylla* p.p.
H. sp. (e)	*H. macrocarpa* var. *brevifolia*	"*H. brevifolia*"	*H. brevifolia*
H. sp. (f)	*H. macrocarpa* var. *latisepala*	"*H. latisepala*"	*H. macrocarpa* p.p.
H. sp. (g)	*H. pauciramosa* var. *masonae*	"*H. masoniae*"	*H. masoniae*
H. sp. (h)	*H.* sp. (h)	"*H. bishopiana*"	*H. bishopiana*
H. sp. (i)	*H.* sp. (i)	"*H. mooreae*"	*H. mooreae*
H. sp. (l) var. (i)	*H. stricta* var. *egmontiana*	"*H. egmontiana*" p.p.	*H. stricta* var. *egmontiana*
H. sp. (l) var. (ii)	*H. stricta* var. *lata*	"*H. egmontiana*" p.p.	*H. stricta* var. *lata*
H. sp. (m)	*H.* sp. (m)	"*H.* Whangarei"	*H. ligustrifolia* p.p.
H. sp. (n)	*H.* sp. (n)	"*H.* Wairoa"	*H. tairawhiti*
H. sp. (o)	*H.* sp. (o)	"*H.* marble"	*H. calcicola*
H. sp. (q)	*H.* sp. (q)	"*H.* aff. *rigidula*"	*H. cryptomorpha*
H. sp. (t)	*H.* sp. (t)	*H. pareora*	*H. pareora*
H. sp. (u) var. (i)	*H.* sp. (u) form (i)	*H. anomala* p.p.	*H. odora* p.p.
H. sp. (u) var. (ii)	*H.* sp. (u) form (ii)	*H. anomala* p.p.	*H. odora* p.p.
H. sp. (v)	*H.* sp. (v)	"*H.* Mokohinau"	*H. pubescens* subsp. *sejuncta* p.p.
H. sp. (w)	—	"*H.* Great Barrier"	*H. macrocarpa* p.p.
H. sp. (x)	*H.* sp. (x)	"*H.* Bartlett"	*H. perbella*
H. glaucophylla var. (i)	*H. glaucophylla* p.p.	"*H. glaucophylla* NW Nelson" p.p.	*H. albicans* p.p.
H. glaucophylla var. (ii)	*H. glaucophylla* p.p.	"*H. glaucophylla* NW Nelson" p.p.	*H. albicans* p.p.
H. pimeleoides agg. p.p.	*H. pimeleoides* var. *rupestris*	"*H. rupestris*"	*H. pimeleoides* subsp. *faucicola*
H. pimeleoides agg. p.p.	*H. pimeleoides* p.p.	"*H.* aff. *pimeleoides*"	*H. pimeleoides* subsp. *pimeleoides* p.p.
H. pubescens var.	*H. pubescens* form (i)	"*H.* Little Barrier"	*H. pubescens* subsp. *rehuarum* and subsp. *sejuncta* p.p.
H. rigidula var. (i)	*H. rigidula* form (i)	"*H.* Lady"	*H. scopulorum*
H. rigidula var. (ii)	*H. rigidula* form (ii)	*H. rigidula* p.p.	*H. rigidula* var. *sulcata*
H. stricta var. (*H. salicifolia* var. *angustissima*)	*H. stricta* form (i)	"*H. angustissima*"	*H. angustissima*
—	—	"*H.* Bald Knob Ridge"	*H. treadwellii* p.p.
—	—	"*H. takahe*"	*H. arganthera*
—	—	"*H.* aff. *pinguifolia*"	*H.* cf. *pinguifolia*

1 These informal taxa are based largely on differences in chromosome number. Without knowing the chromosome number of typical *H. diosmifolia*, Druce (1993) was not able to specify which of his "species" should be given that name.

Appendix 2

Variation in *Hebe hectorii*

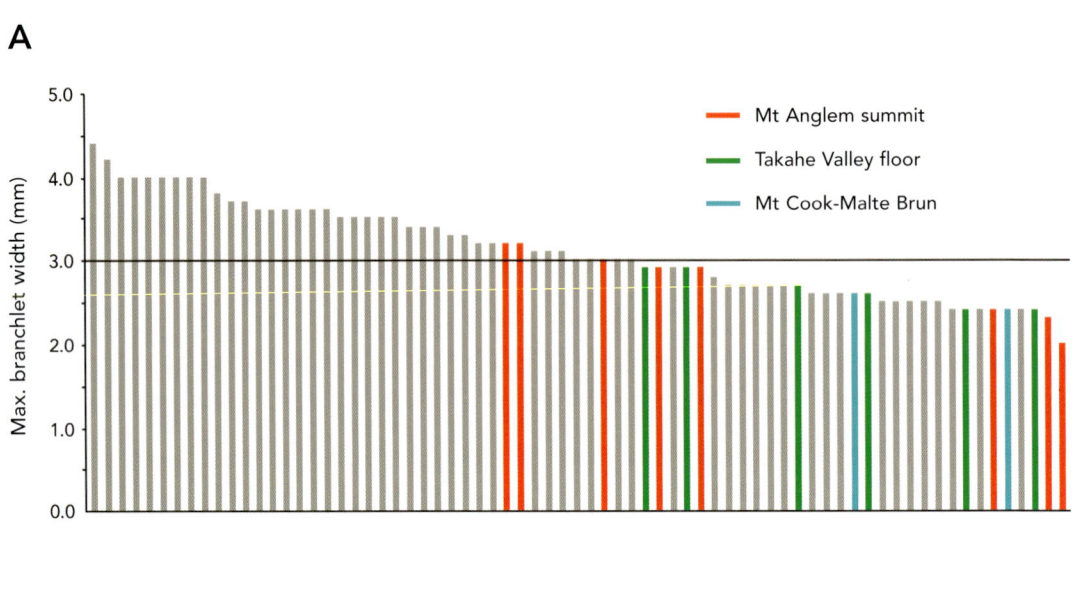

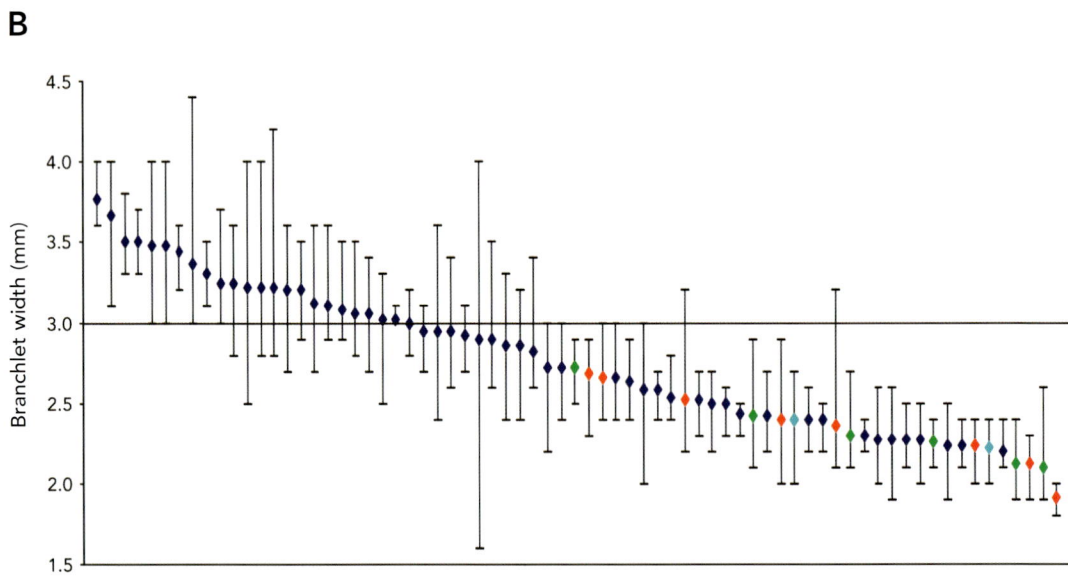

FIG. 159 Variation in subsp. *hectorii*. **A** maximum branchlet width. **B** branchlet width. Measurements are from herbarium specimens, for which the geographic spread is shown in Fig. 161 and herbarium numbers in Appendix 5. Points on B are arranged by mean values (from five measurements per specimen) and error bars show maximum and minimum values. Horizontal lines indicate the branchlet diameter (3 mm) used by Wagstaff & Wardle (1999) to distinguish subsp. *hectorii* from subsp. *laingii*. Colour coding highlights data for specimens from selected areas: Mt Anglem is the type locality of subsp. *laingii*; the floor of Takahe Va. (Murchison Mts) includes specimens placed in subsp. *laingii* by Wagstaff & Wardle; Mt Aspiring and the Aoraki/Mt Cook-Malte Brun areas include at least some specimens placed in subsp. *hectorii* by Wagstaff & Wardle.

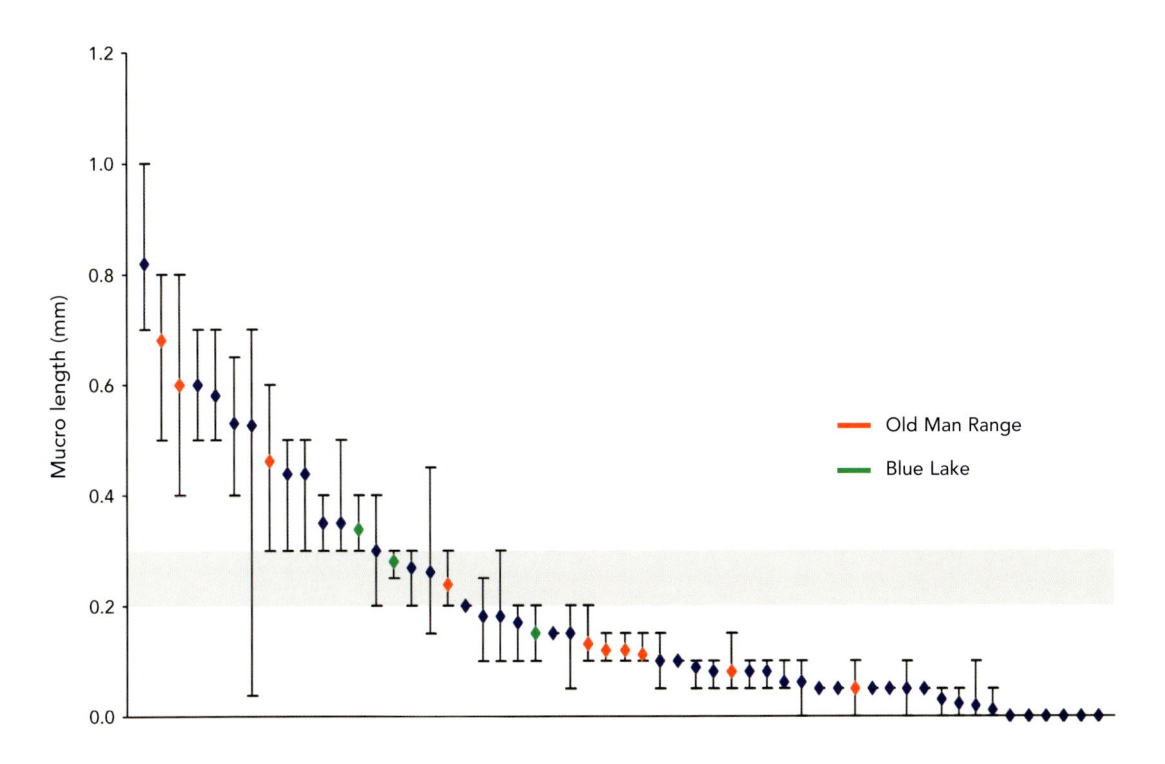

FIG. 160 Variation in leaf mucro length in subsp. *demissa*. Measurements are from herbarium specimens, for which the geographic spread is shown in Fig. 161 and herbarium numbers in Appendix 5. Points are arranged by mean values (from five measurements per specimen) and error bars show maximum and minimum values. The key of Wagstaff & Wardle (1999) assigned specimens with mucros greater than 0.3 mm (i.e. above the grey bar) to subsp. *subulata*, and those with mucros of *c.* 0.2 mm to subsp. *demissa*. Colour coding highlights data for specimens from selected areas: Old Man Ra. is the type locality of subsp. *subulata*; Blue Lk. (Garvie Mts) is a locality also recorded for subsp. *subulata* by Wagstaff & Wardle. Mucro length on the Old Man Ra. might be generally greater in plants from the southern areas, but herbarium specimens do not all have sufficient locality information to ascertain this.

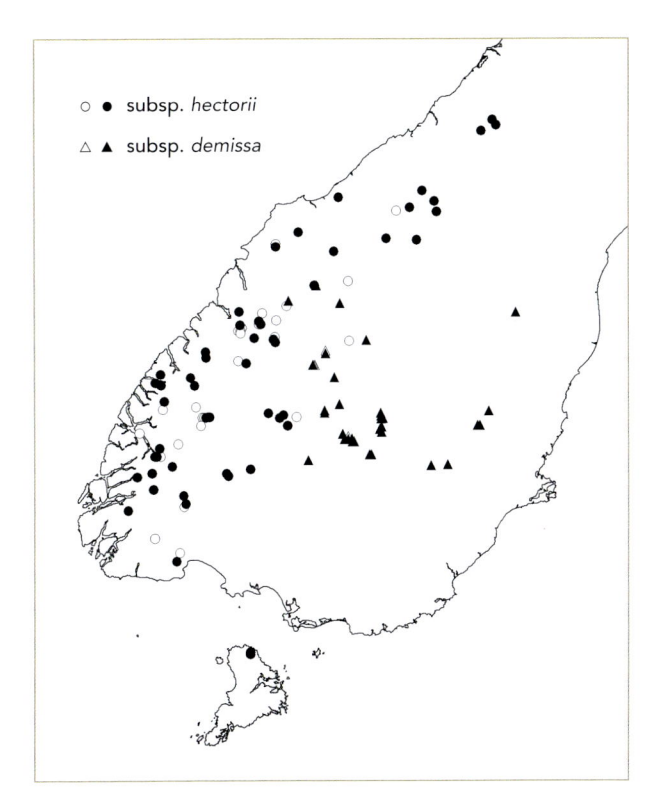

FIG. 161 Distributions of *H. hectorii* subsp. *hectorii* and subsp. *demissa*. Filled symbols indicate localities represented by measurements in Fig. 159 and Fig. 160.

Appendix 3

Variation in *Hebe lycopodioides*

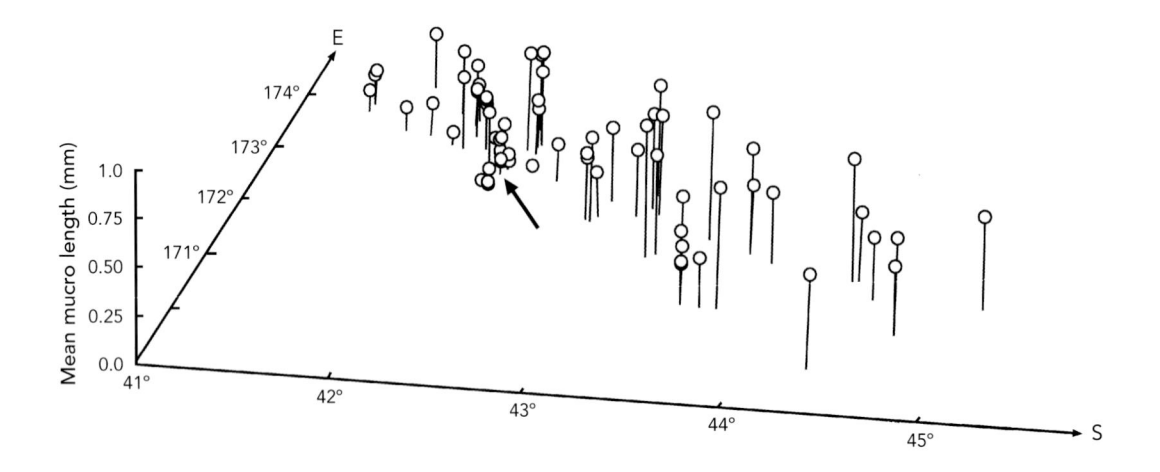

FIG. 162 Three-dimensional plot showing geographic variation in leaf mucro length in *H. lycopodioides*. The geographic arrangement of points is the same as that shown by filled symbols on Fig. 163. Measurements are from herbarium specimens (as outlined in Fig. 164). There is an obvious cluster of specimens with short mucros in the area near Lewis Pass (indicated by arrow).

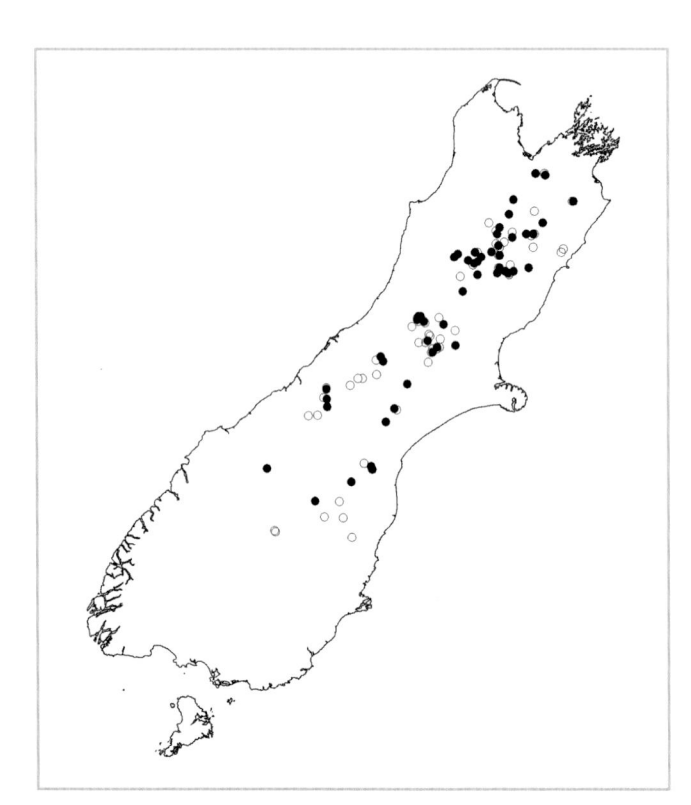

FIG. 163 Distribution of *H. lycopodioides*. Filled symbols indicate localities represented by measurements in Fig. 162 and Fig. 164.

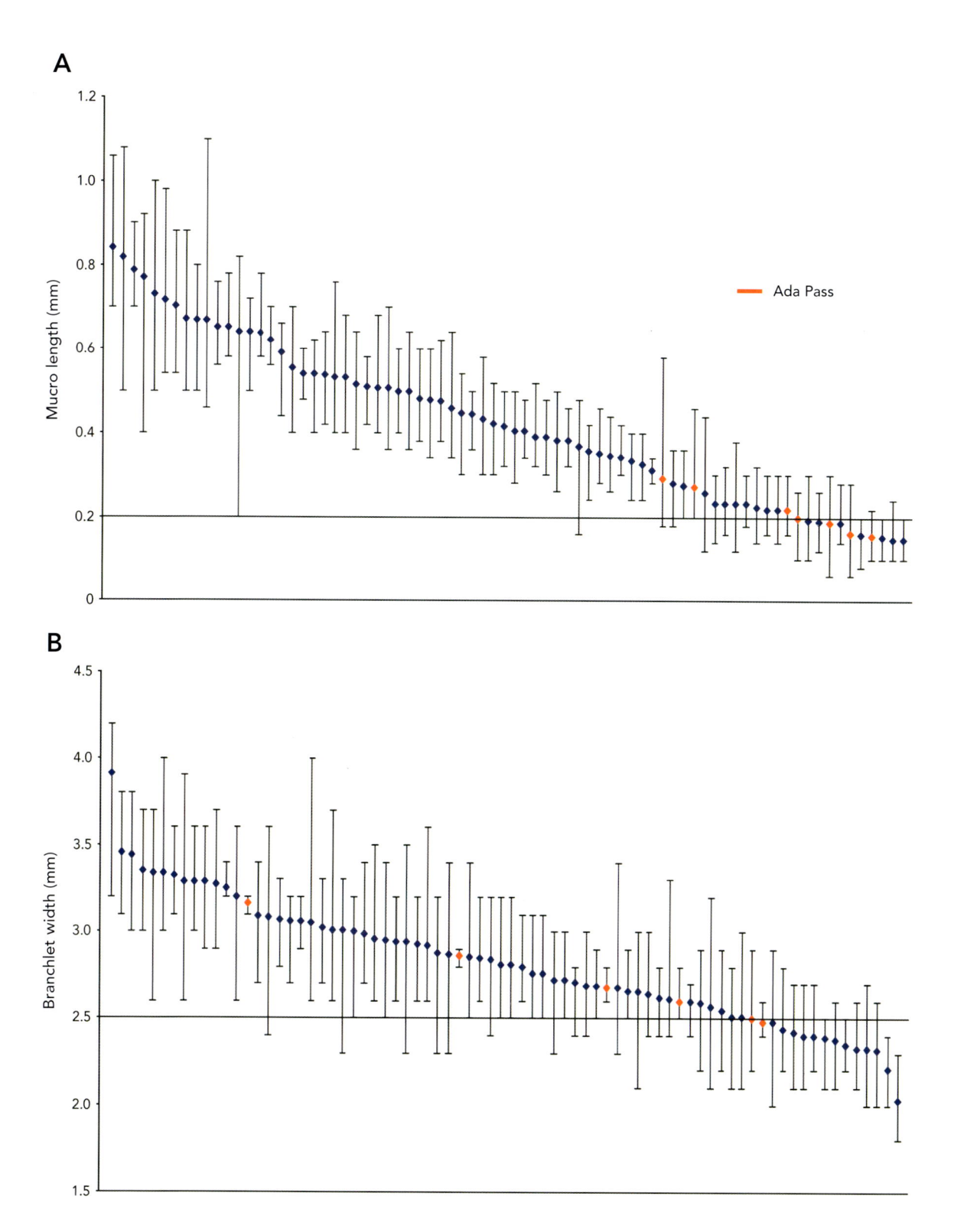

FIG. 164 Graphs showing variation in *H. lycopodioides* in: **A** leaf mucro length; **B** branchlet width. Measurements are from herbarium specimens, for which the geographic spread is shown in Fig. 163 and herbarium numbers in Appendix 5. Points each represent a single specimen, and those on each graph are arranged by mean values (from ten measurements, where possible, per specimen); error bars show maximum and minimum values. The horizontal line on A shows the mucro length (0.2 mm) used, in part, by Wagstaff & Wardle (1999) to distinguish subsp. *lycopodioides* from subsp. *patula*. The horizontal line on B shows the branchlet width (2.5 mm) also used, in part, by Wagstaff & Wardle to distinguish subsp. *lycopodioides* from subsp. *patula*; they ascribed branchlet widths of < 2.5 mm to subsp. *patula*, but noted that subsp. *lycopodioides* was variable in this feature. Colour coding highlights data for specimens from a single population (on a ridge south of Ada Pass, Spenser Mts) close to the type locality of subsp. *patula*.

Appendix 4

Examples of Leaf Outlines for Selected Species

In a range of *Hebe* species the shape and size of the leaves can show considerable variation between and/or within populations, and sometimes on individual plants. Illustrated below are examples of variation in leaf shape and size for selected species. These are included to supplement information on variation provided in the notes that follow species' descriptions, and as an aid to identification, so that specimens can be checked against the known range of variation for a species. The leaf outlines were drawn from mature leaves on herbarium specimens (as indicated), and are reproduced at life size. They are shown in alphabetical order by species name.

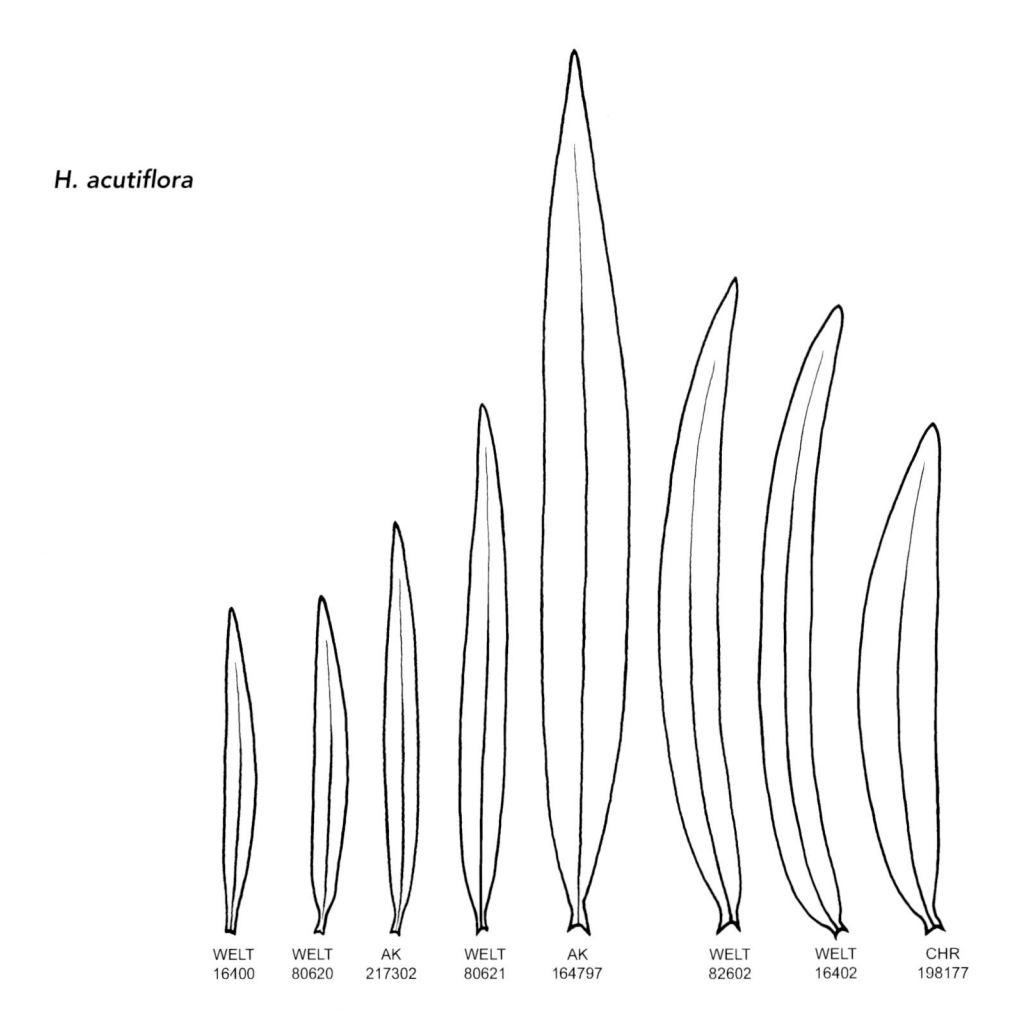

H. acutiflora

H. albicans

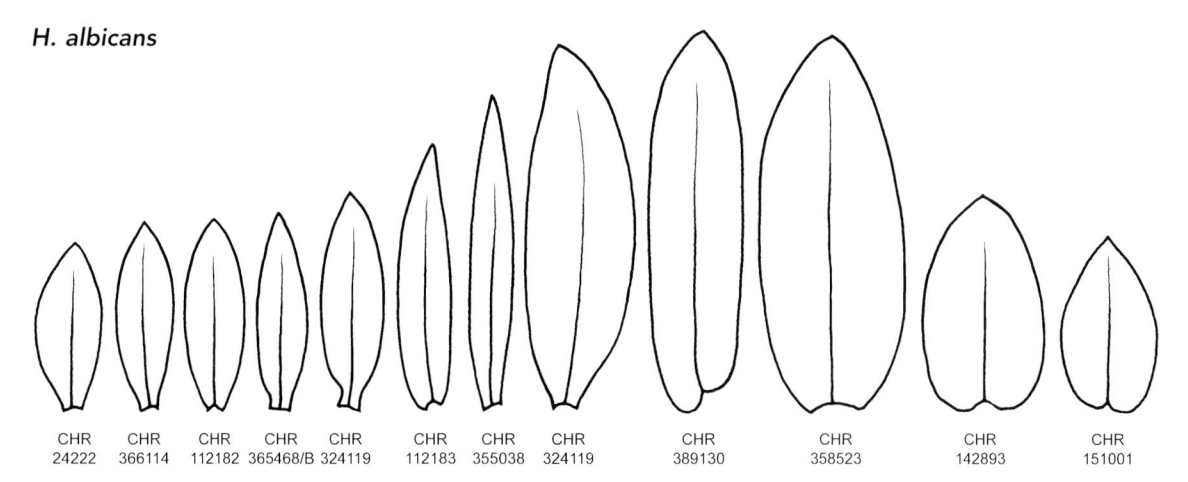

| CHR 24222 | CHR 366114 | CHR 112182 | CHR 365468/B | CHR 324119 | CHR 112183 | CHR 355038 | CHR 324119 | CHR 389130 | CHR 358523 | CHR 142893 | CHR 151001 |

H. angustissima

Otaki Gorge

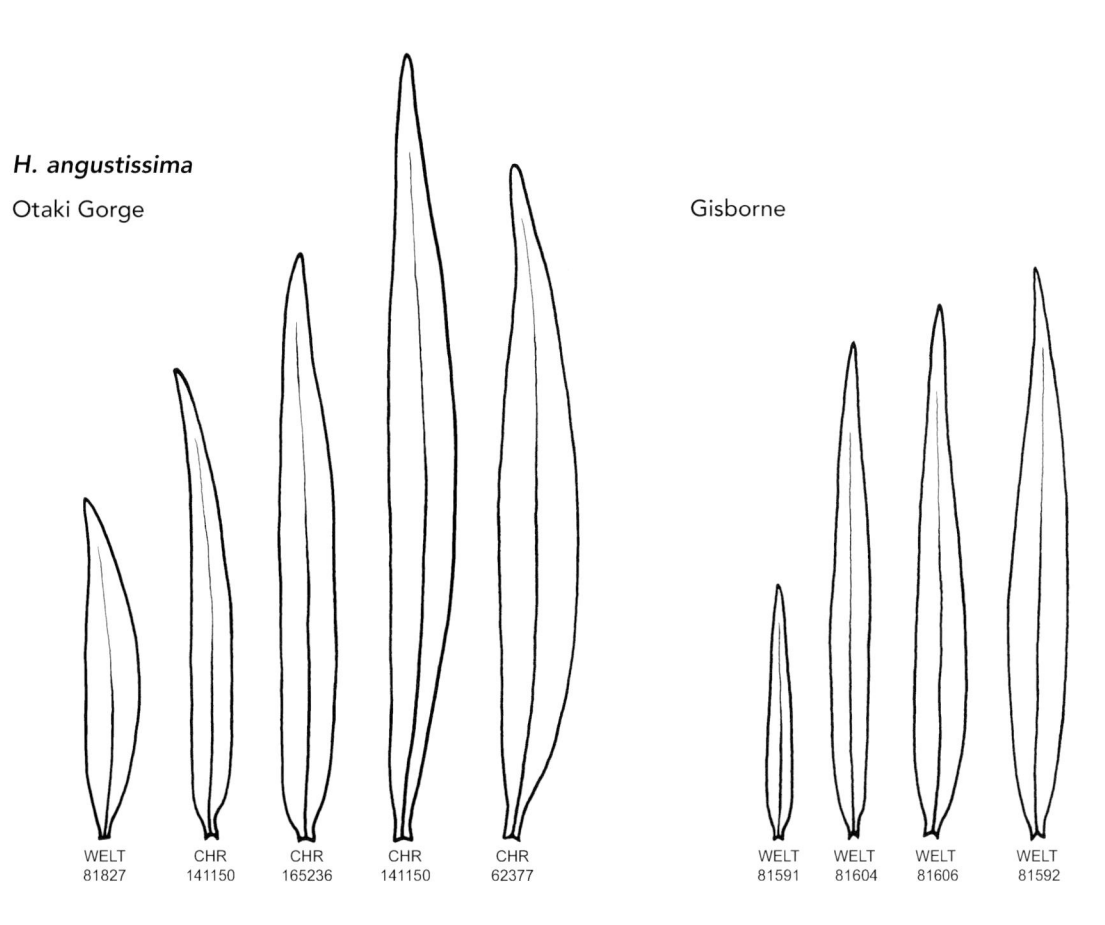

| WELT 81827 | CHR 141150 | CHR 165236 | CHR 141150 | CHR 62377 |

Gisborne

| WELT 81591 | WELT 81604 | WELT 81606 | WELT 81592 |

H. brachysiphon

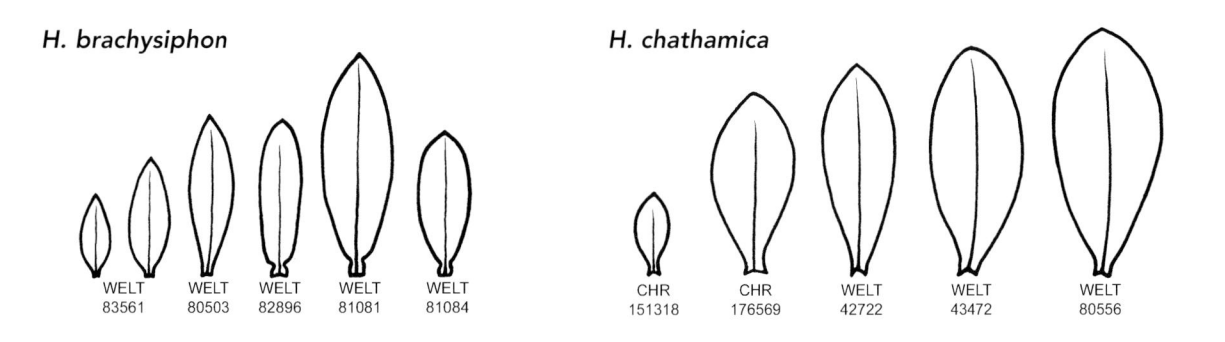

| WELT 83561 | WELT 80503 | WELT 82896 | WELT 81081 | WELT 81084 |

H. chathamica

| CHR 151318 | CHR 176569 | WELT 42722 | WELT 43472 | WELT 80556 |

APPENDIX 4 341

H. cockayneana

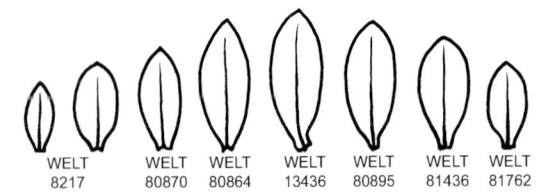

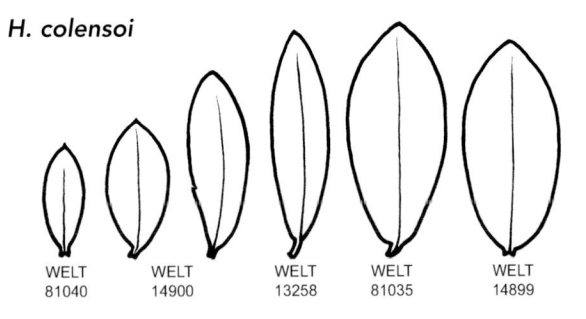

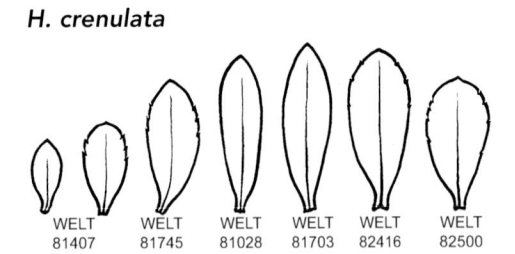

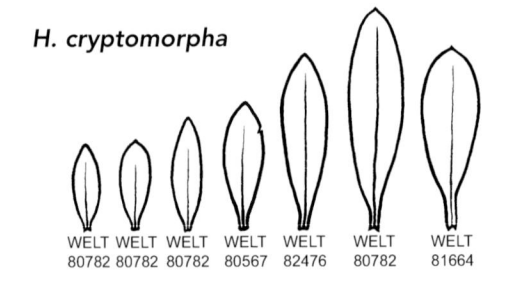

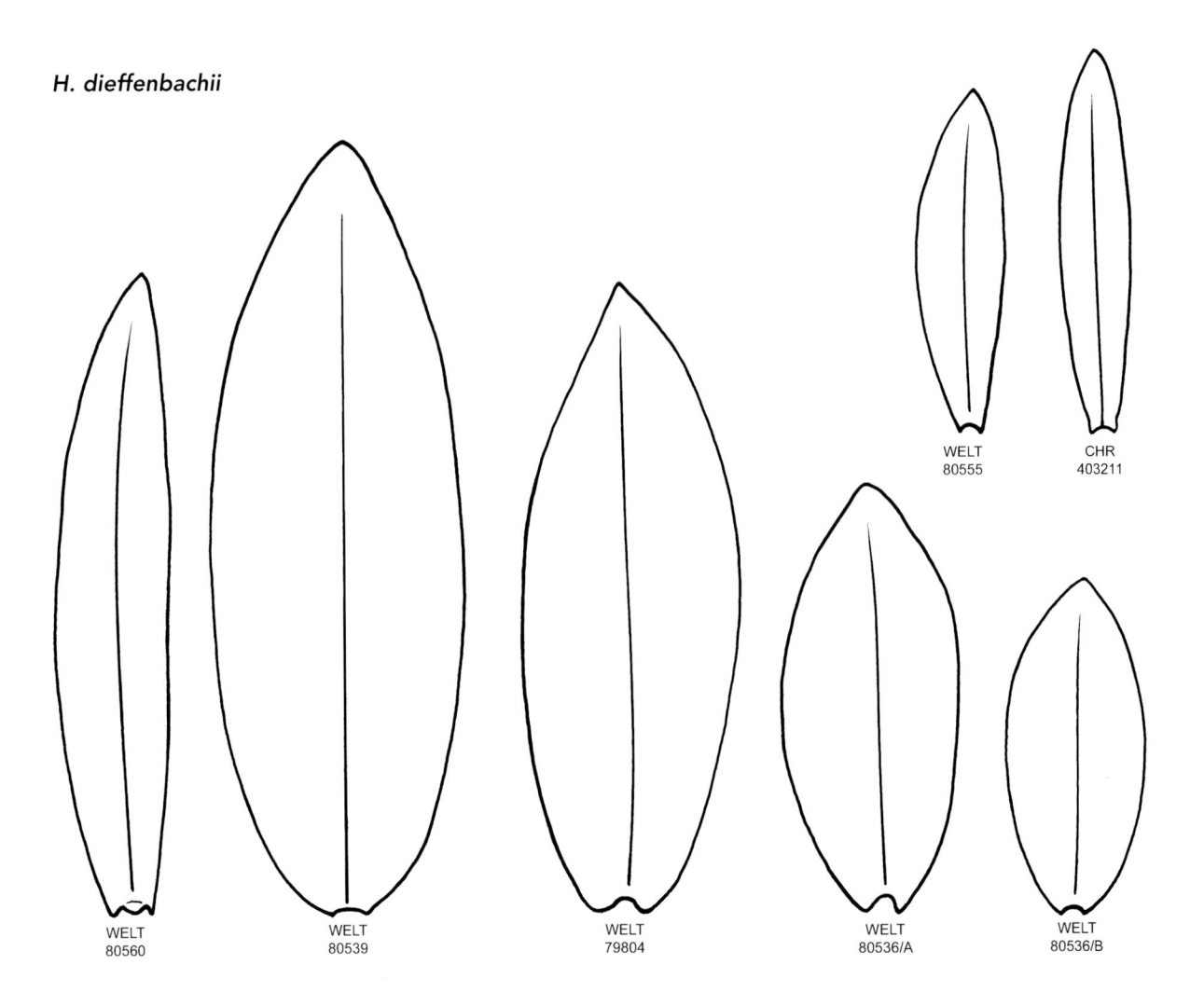

| WELT 8217 | WELT 80870 | WELT 80864 | WELT 13436 | WELT 80895 | WELT 81436 | WELT 81762 |

H. colensoi

| WELT 81040 | WELT 14900 | WELT 13258 | WELT 81035 | WELT 14899 |

H. crenulata

| WELT 81407 | WELT 81745 | WELT 81028 | WELT 81703 | WELT 82416 | WELT 82500 |

H. cryptomorpha

| WELT 80782 | WELT 80782 | WELT 80782 | WELT 80567 | WELT 82476 | WELT 80782 | WELT 81664 |

H. dieffenbachii

| WELT 80560 | WELT 80539 | WELT 79804 | WELT 80555 | CHR 403211 | WELT 80536/A | WELT 80536/B |

H. elliptica

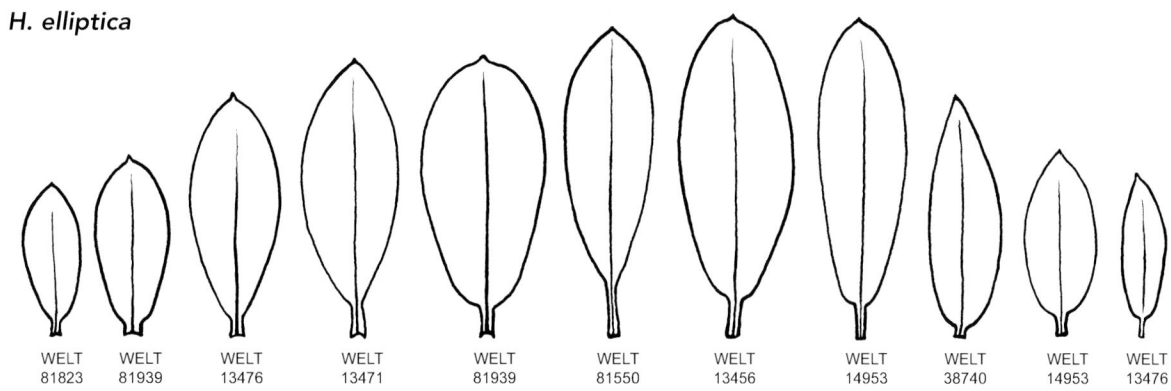

| WELT 81823 | WELT 81939 | WELT 13476 | WELT 13471 | WELT 81939 | WELT 81550 | WELT 13456 | WELT 14953 | WELT 38740 | WELT 14953 | WELT 13476 |

H. evenosa

| WELT 82501 | WELT 80630 | WELT 83565 |

H. flavida

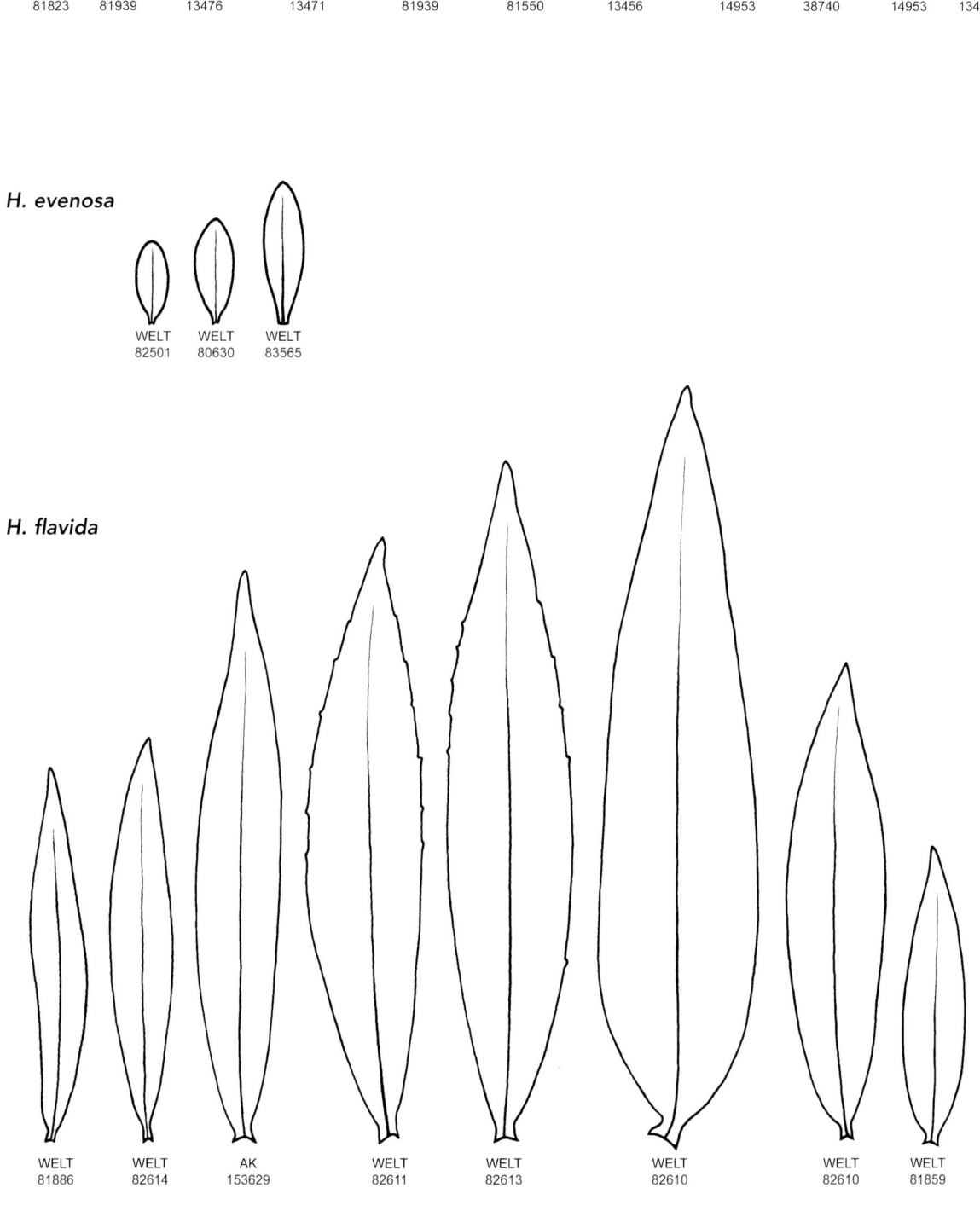

| WELT 81886 | WELT 82614 | AK 153629 | WELT 82611 | WELT 82613 | WELT 82610 | WELT 82610 | WELT 81859 |

APPENDIX 4   343

H. leiophylla

| WELT 82657 | WELT 80695 | WELT 15026 | WELT 14988 | WELT 80680 | WELT 80713 | WELT 80679 | WELT 81632 |

H. ligustrifolia

| WELT 82605 | WELT 81902 | WELT 81904 | WELT 81901 | WELT 16411 | WELT 14937 | AK 212531 |

H. macrantha var. *brachyphylla*

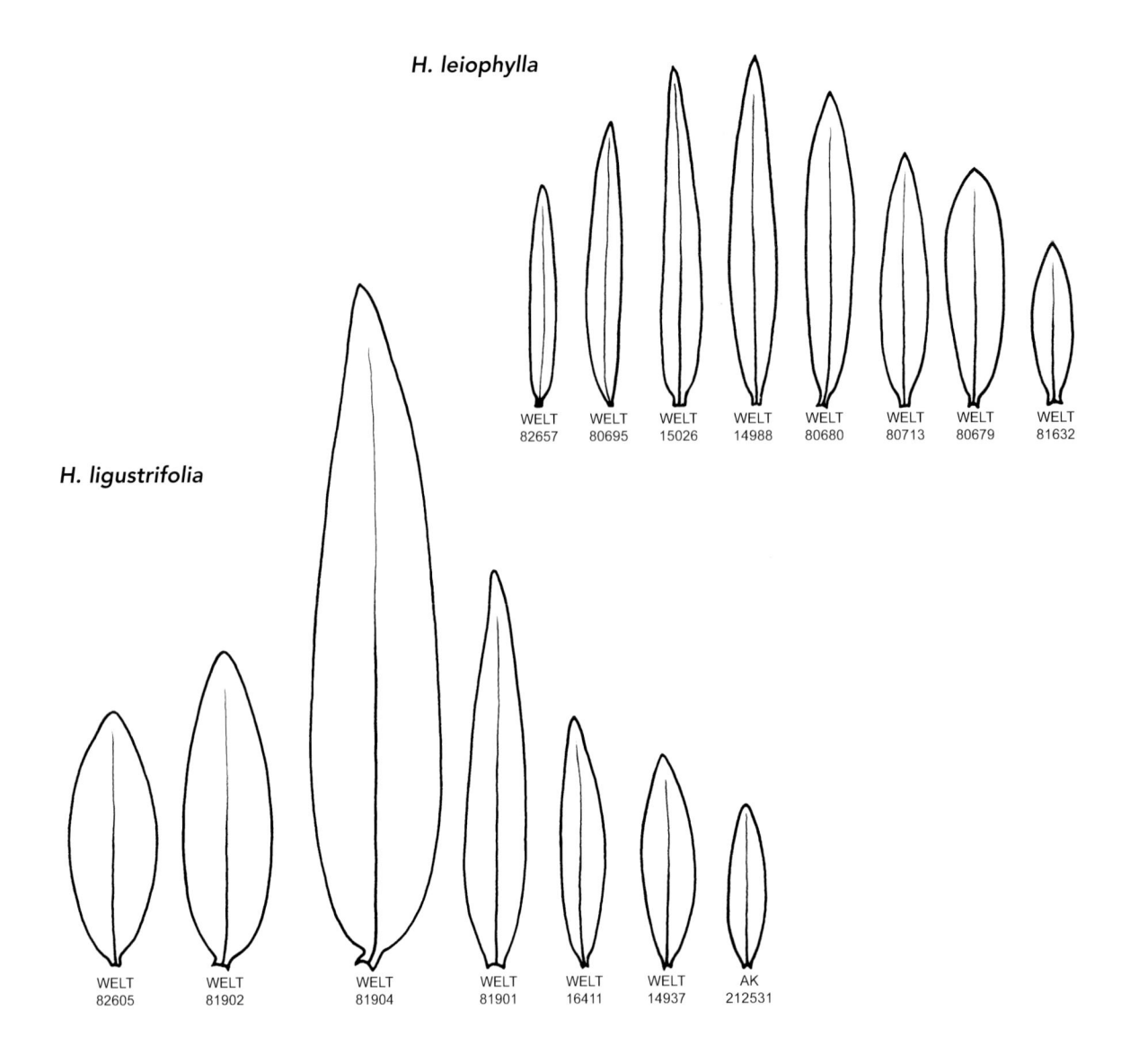

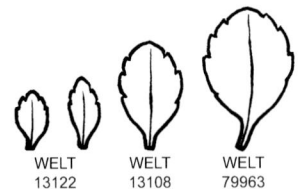

| WELT 13122 | WELT 13108 | WELT 79963 |

H. macrantha var. *macrantha*

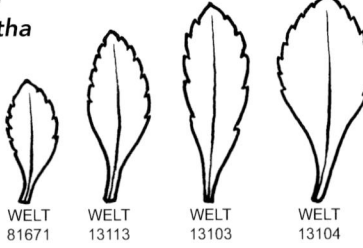

| WELT 81671 | WELT 13113 | WELT 13103 | WELT 13104 |

H. macrocalyx var. *humilis*

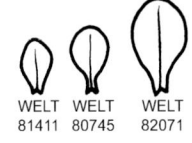

| WELT 81411 | WELT 80745 | WELT 82071 |

H. macrocalyx var. *macrocalyx*

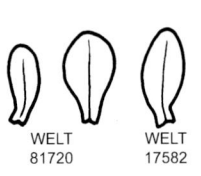

| WELT 81720 | WELT 17582 |

H. macrocarpa
 including var. *latisepala*

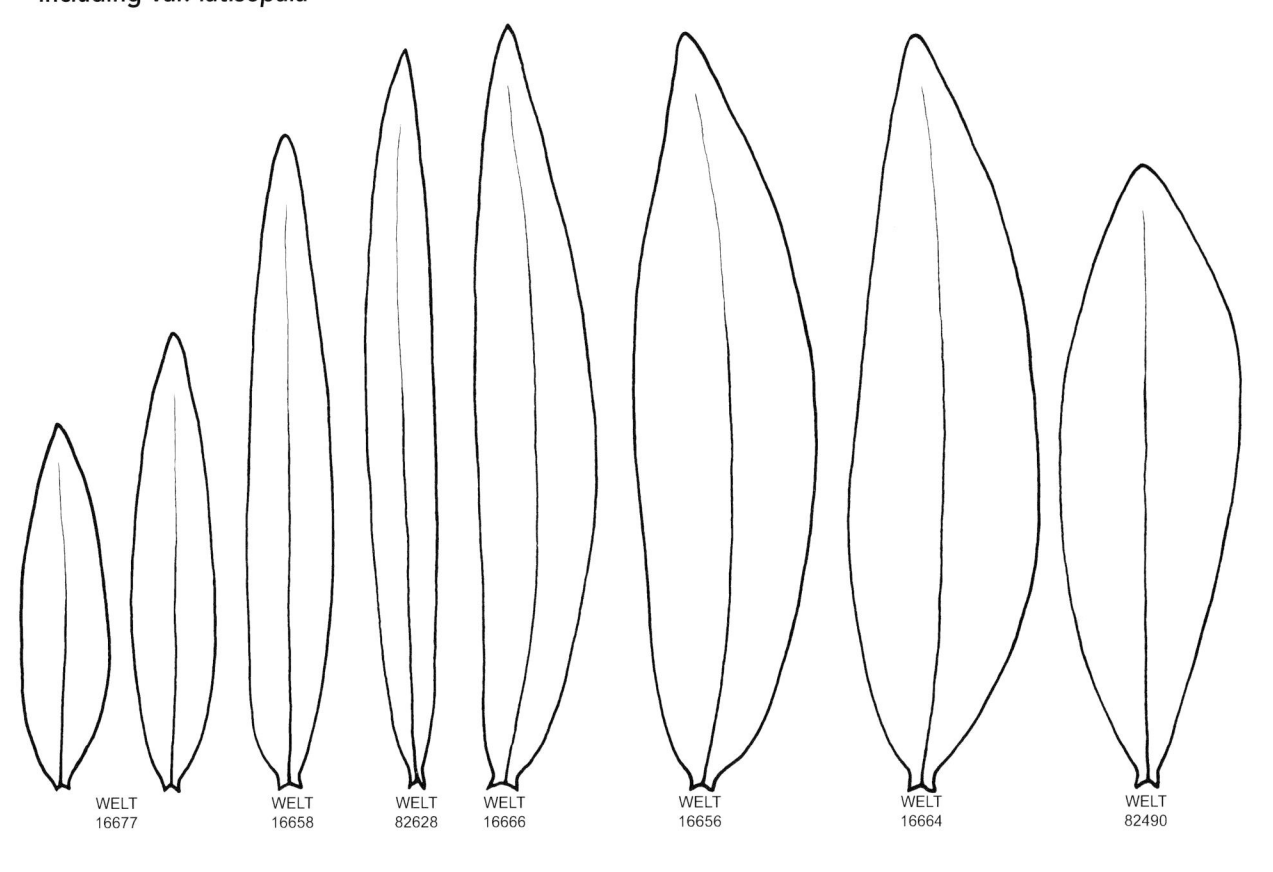

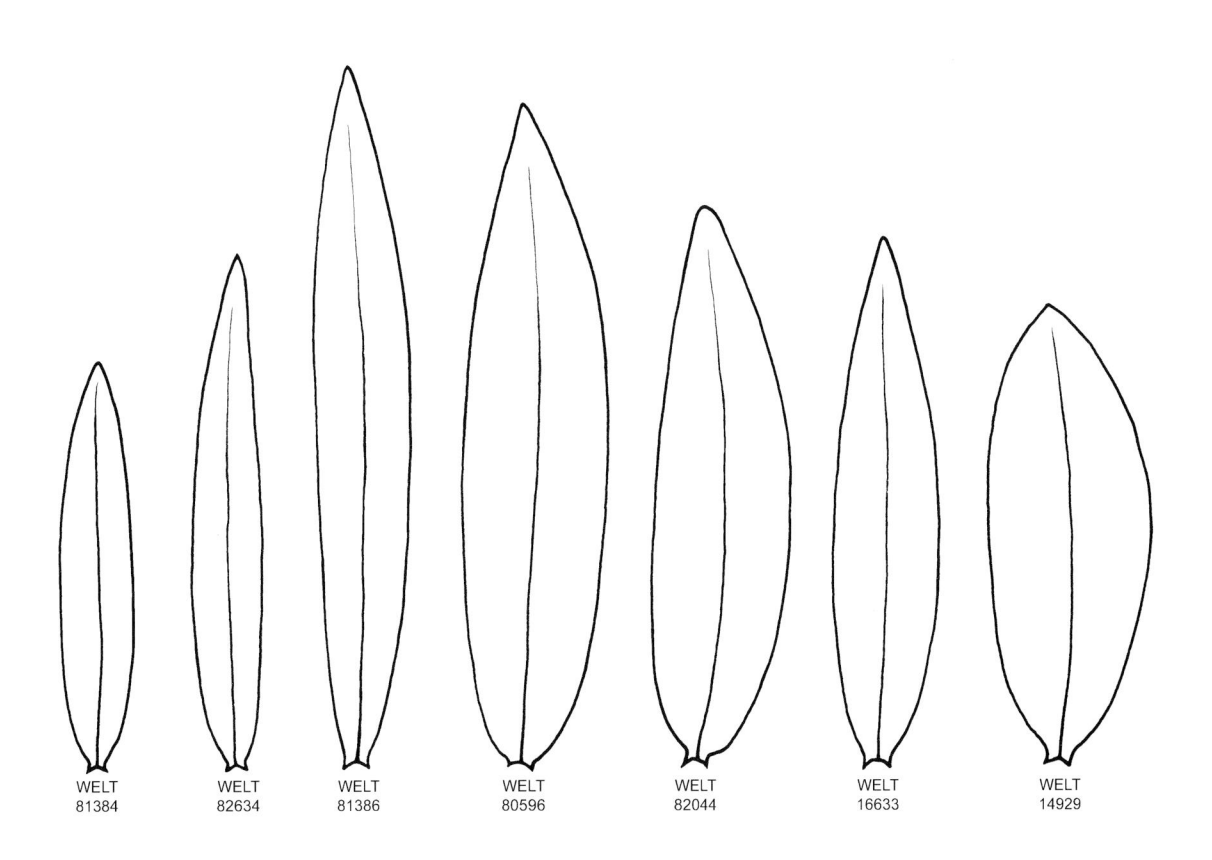

WELT 16677
WELT 16658
WELT 82628
WELT 16666
WELT 16656
WELT 16664
WELT 82490

WELT 81384
WELT 82634
WELT 81386
WELT 80596
WELT 82044
WELT 16633
WELT 14929

APPENDIX 4

H. odora

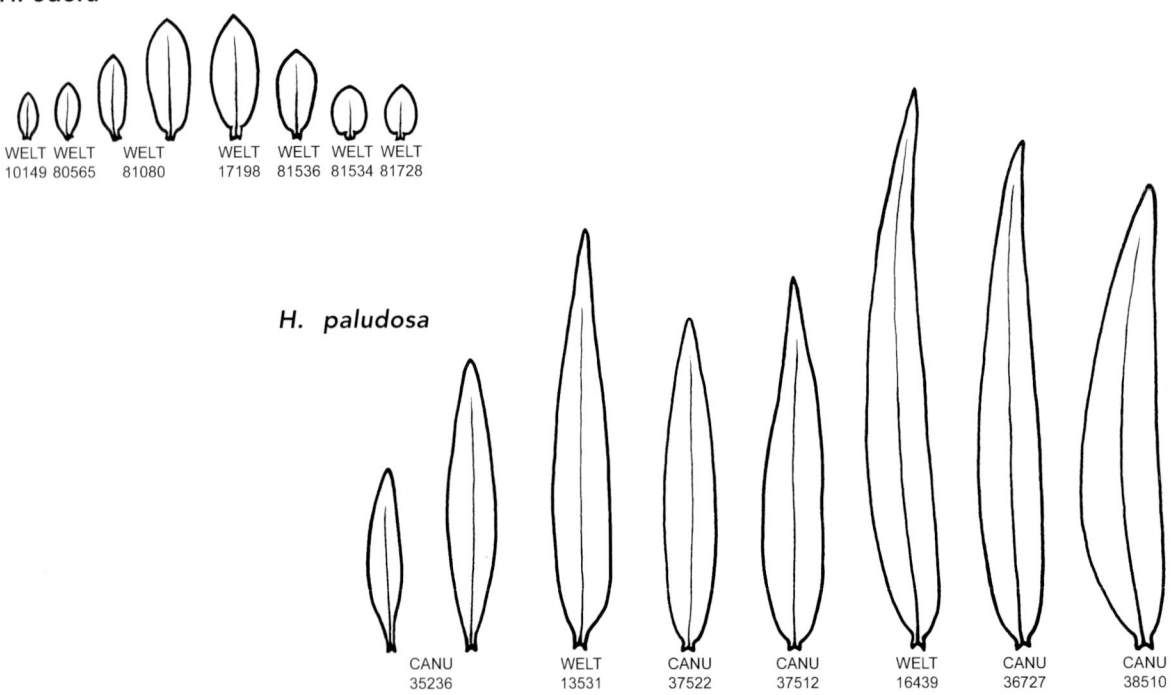

| WELT 10149 | WELT 80565 | WELT 81080 | WELT 17198 | WELT 81536 | WELT 81534 | WELT 81728 |

H. paludosa

| CANU 35236 | WELT 13531 | CANU 37522 | CANU 37512 | WELT 16439 | CANU 36727 | CANU 38510 |

H. pimeleoides subsp. *faucicola*

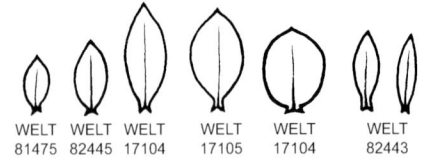

| WELT 81475 | WELT 82445 | WELT 17104 | WELT 17105 | WELT 17104 | WELT 82443 |

H. pimeleoides subsp. *pimeleoides*

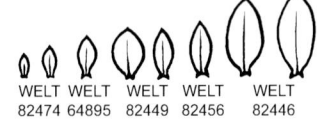

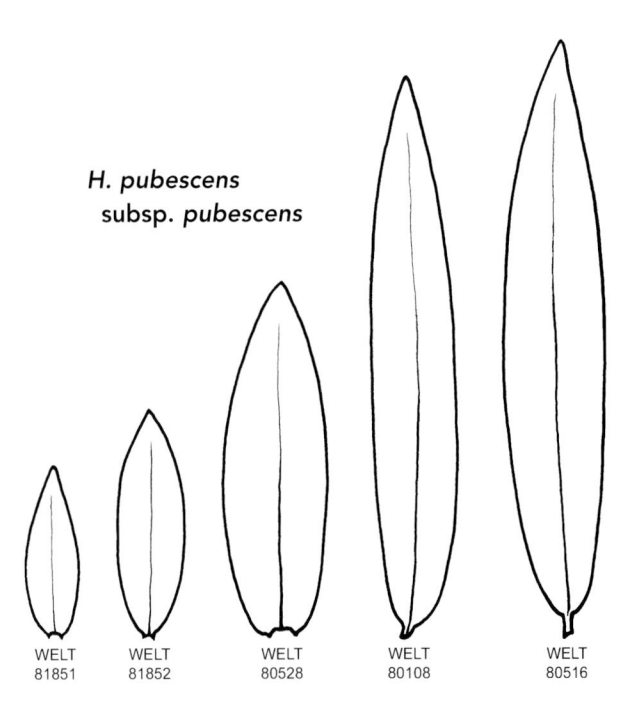

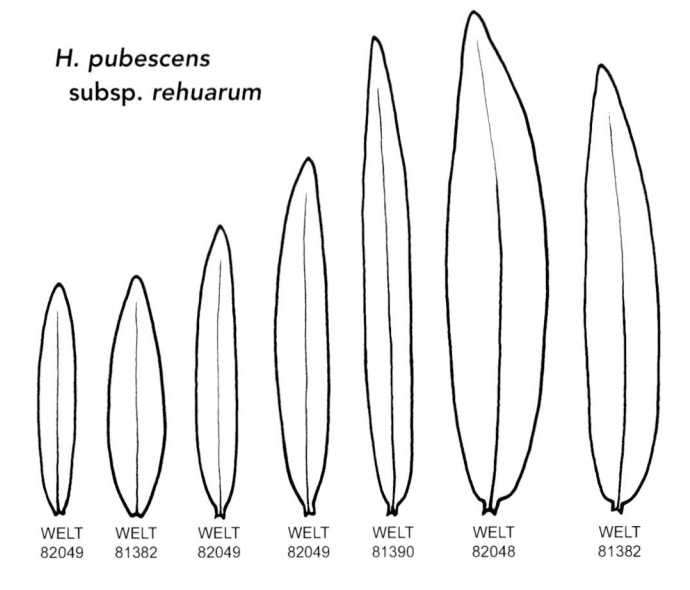

| WELT 82474 | WELT 64895 | WELT 82449 | WELT 82456 | WELT 82446 |

H. pubescens subsp. *pubescens*

| WELT 81851 | WELT 81852 | WELT 80528 | WELT 80108 | WELT 80516 |

H. pubescens subsp. *rehuarum*

| WELT 82049 | WELT 81382 | WELT 82049 | WELT 82049 | WELT 81390 | WELT 82048 | WELT 81382 |

H. pubescens
subsp. *sejuncta*

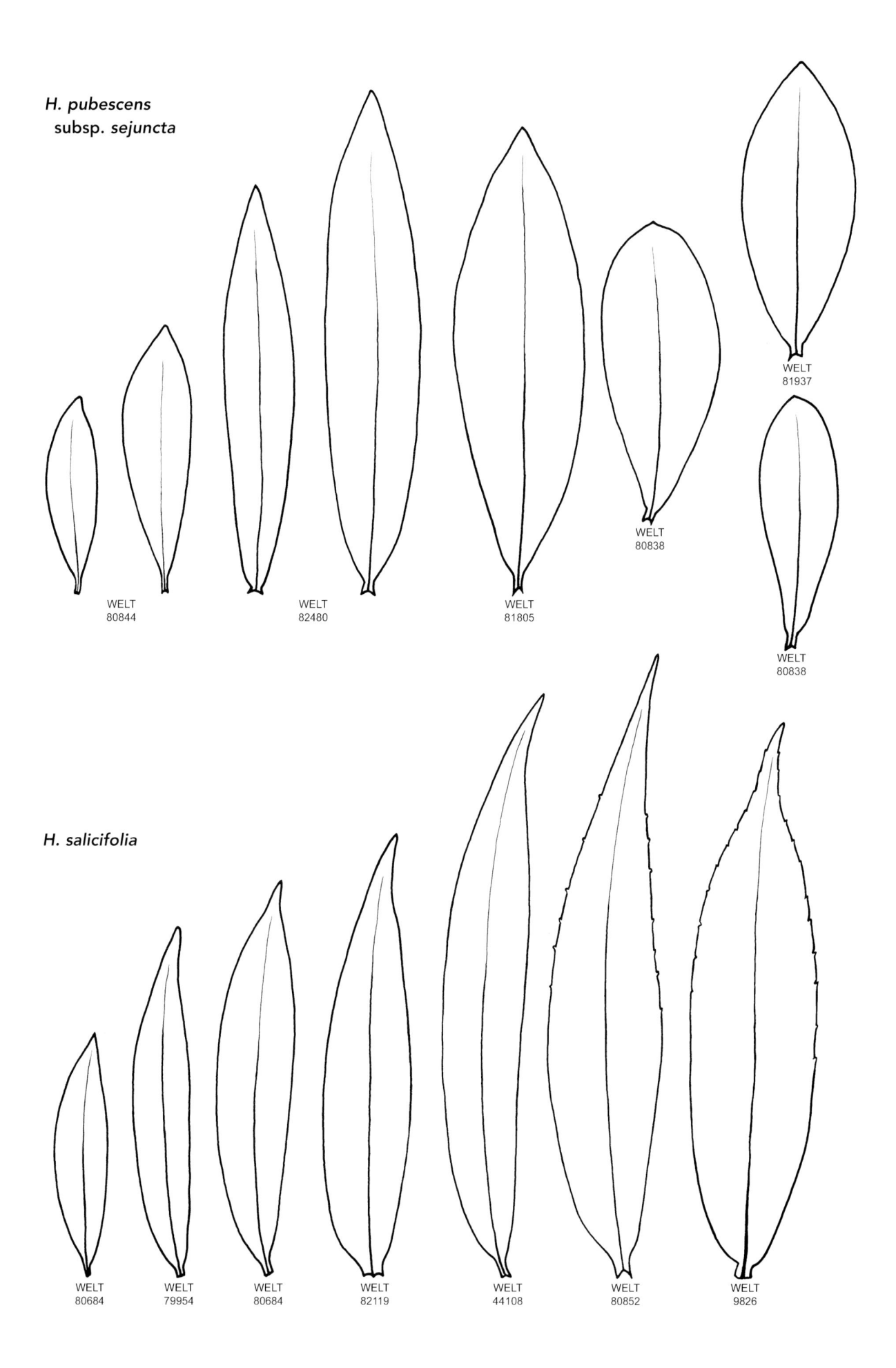

WELT 80844
WELT 82480
WELT 81805
WELT 80838
WELT 81937
WELT 80838

H. salicifolia

WELT 80684
WELT 79954
WELT 80684
WELT 82119
WELT 44108
WELT 80852
WELT 9826

APPENDIX 4 347

H. stricta
var. *atkinsonii*

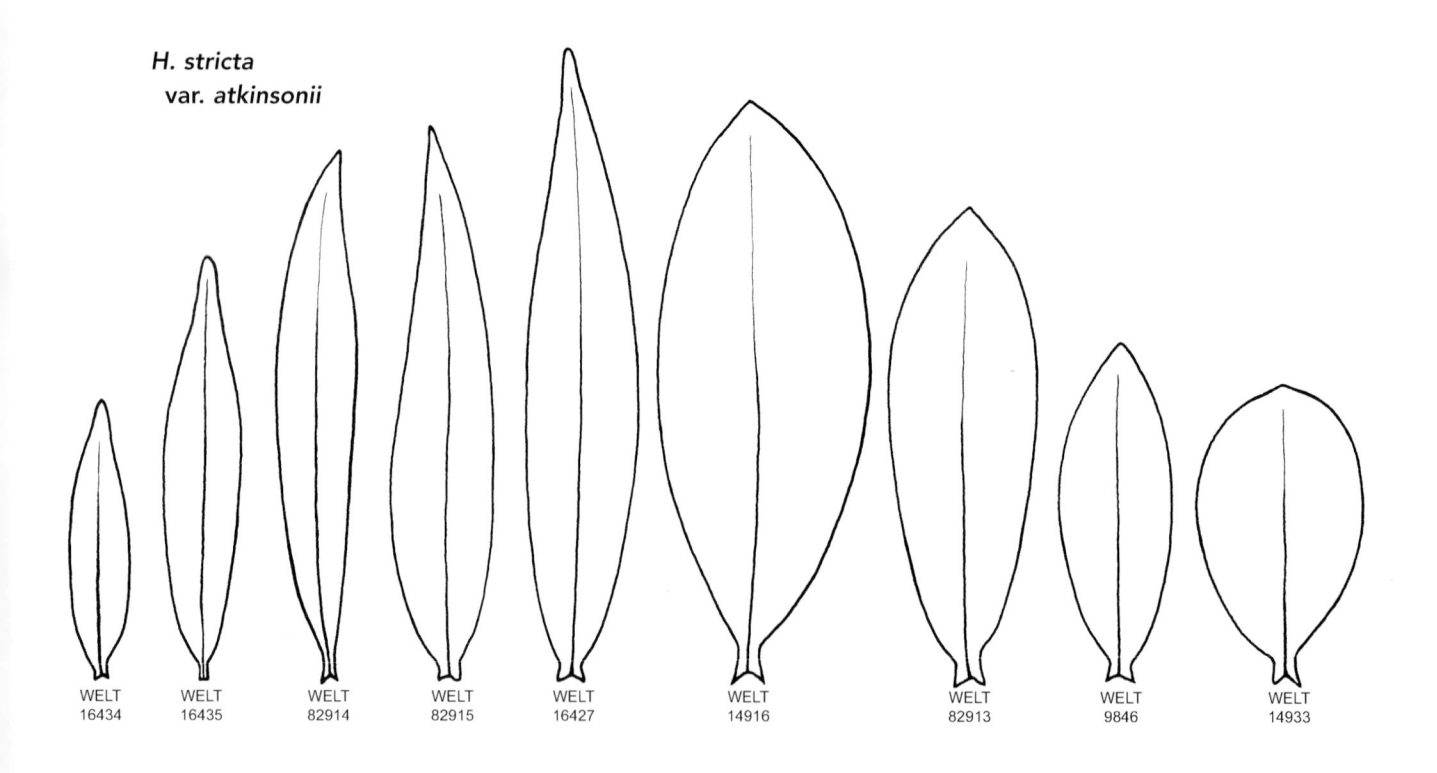

| WELT 16434 | WELT 16435 | WELT 82914 | WELT 82915 | WELT 16427 | WELT 14916 | WELT 82913 | WELT 9846 | WELT 14933 |

H. stricta
var. *egmontiana*

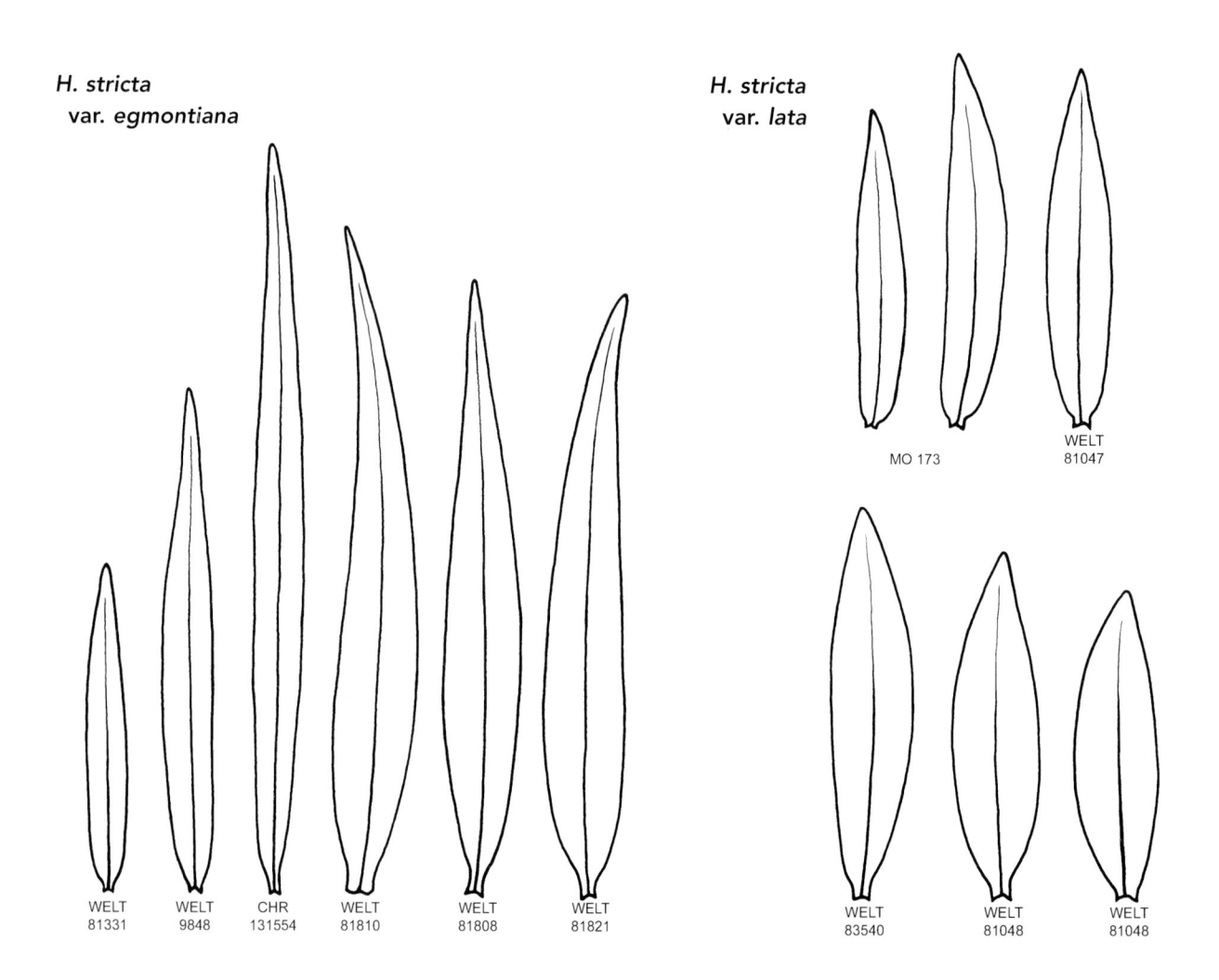

| WELT 81331 | WELT 9848 | CHR 131554 | WELT 81810 | WELT 81808 | WELT 81821 |

H. stricta
var. *lata*

MO 173   WELT 81047

WELT 83540   WELT 81048   WELT 81048

348   AN ILLUSTRATED GUIDE TO NEW ZEALAND HEBES

H. stricta
var. *macroura*
East Cape

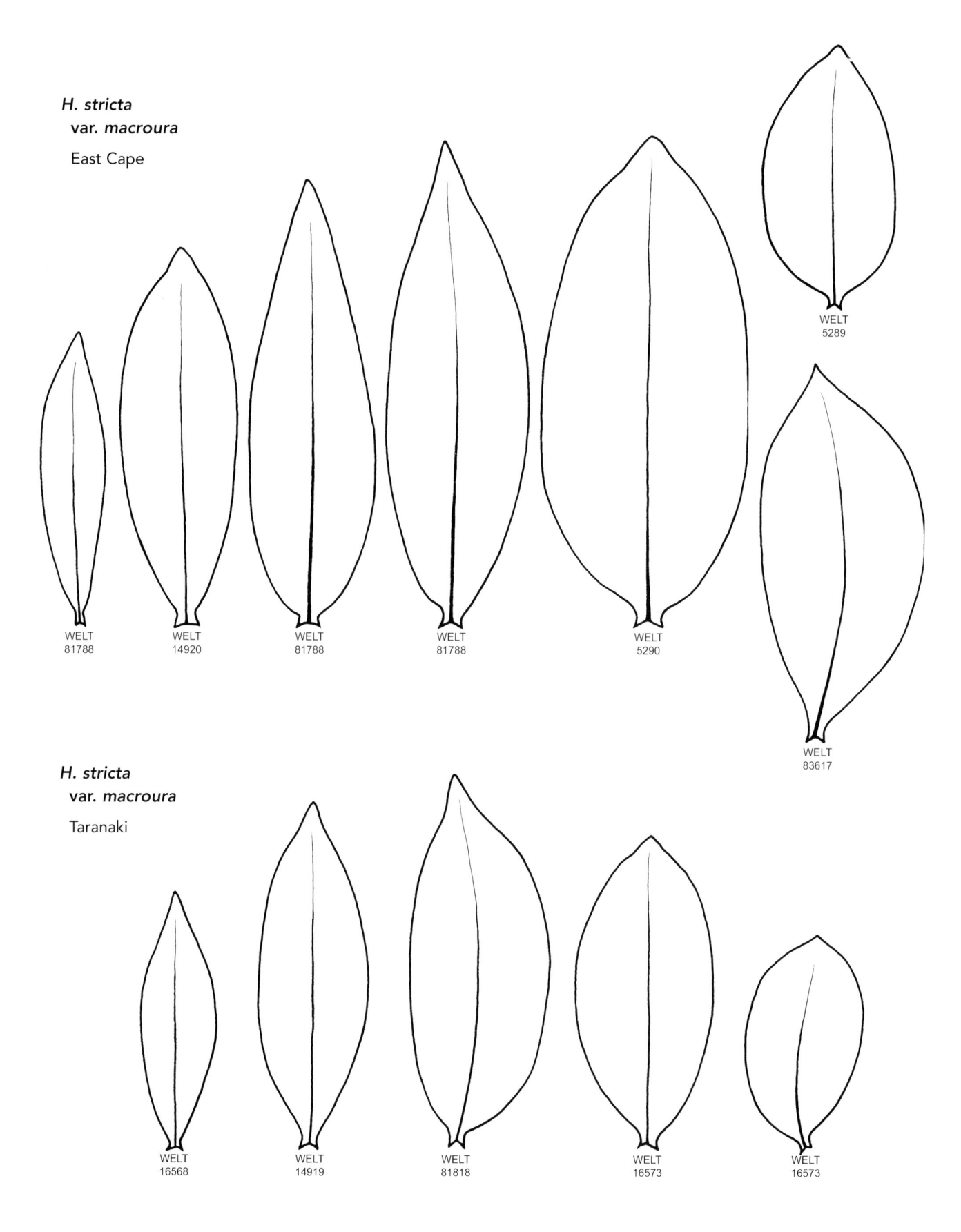

WELT 81788
WELT 14920
WELT 81788
WELT 81788
WELT 5290
WELT 5289
WELT 83617

H. stricta
var. *macroura*
Taranaki

WELT 16568
WELT 14919
WELT 81818
WELT 16573
WELT 16573

APPENDIX 4   349

H. stricta
   var. *stricta*

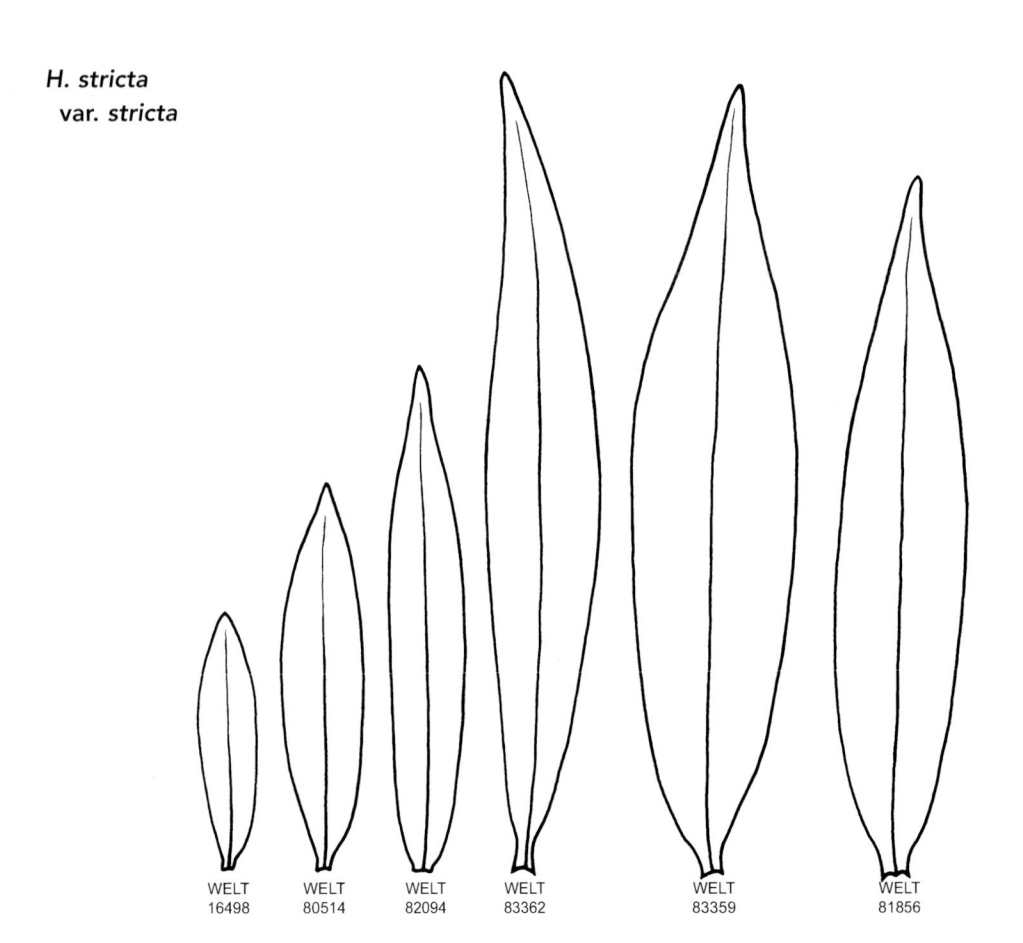

| WELT 16498 | WELT 80514 | WELT 82094 | WELT 83362 | WELT 83359 | WELT 81856 |

H. strictissima

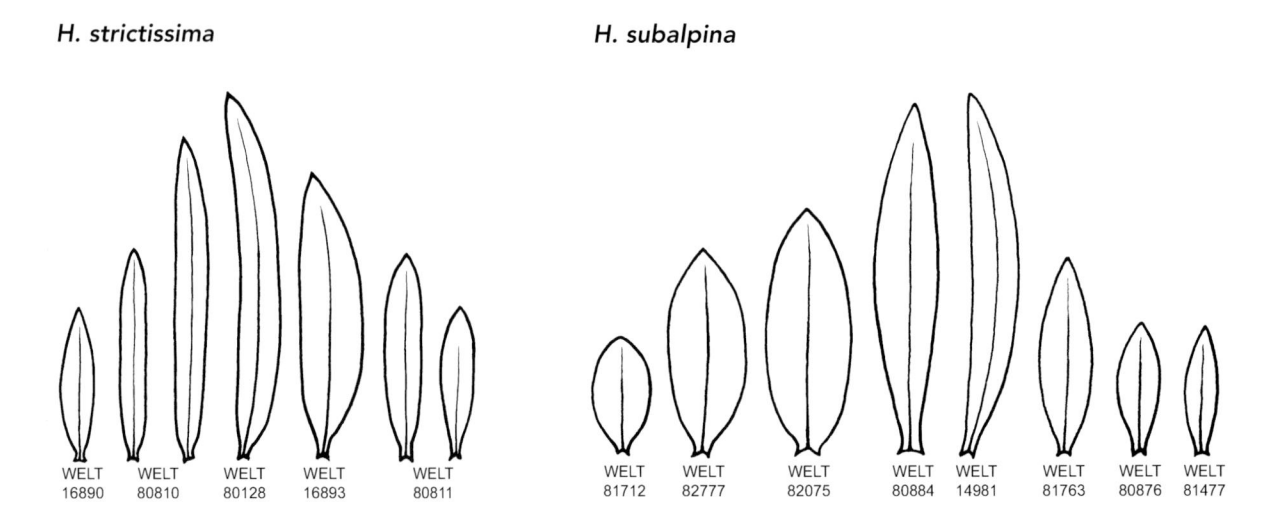

| WELT 16890 | WELT 80810 | WELT 80128 | WELT 16893 | WELT 80811 |

H. subalpina

| WELT 81712 | WELT 82777 | WELT 82075 | WELT 80884 | WELT 14981 | WELT 81763 | WELT 80876 | WELT 81477 |

350   AN ILLUSTRATED GUIDE TO NEW ZEALAND HEBES

H. topiaria

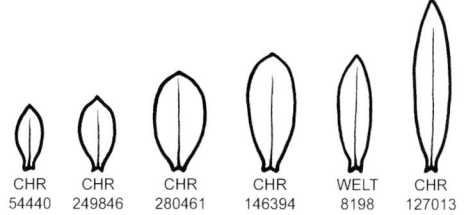

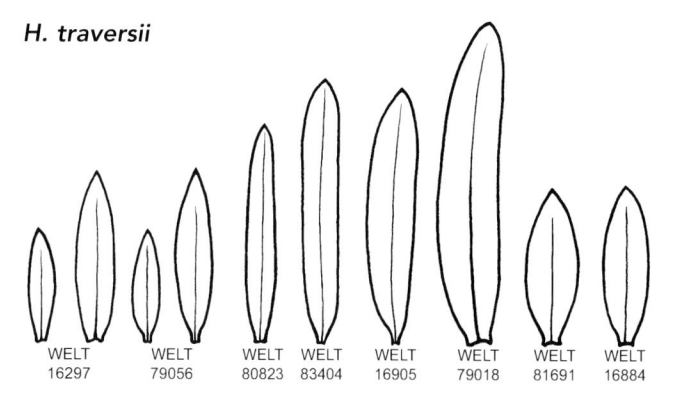

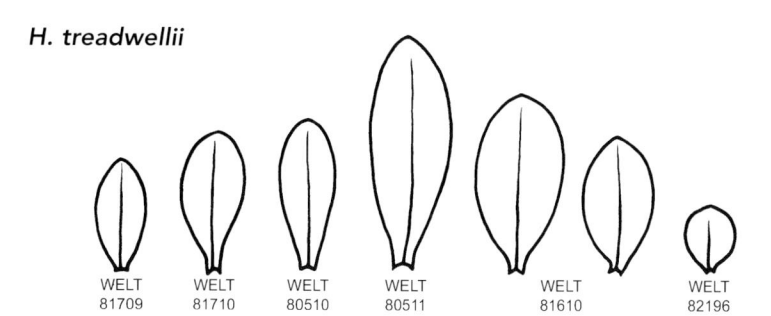

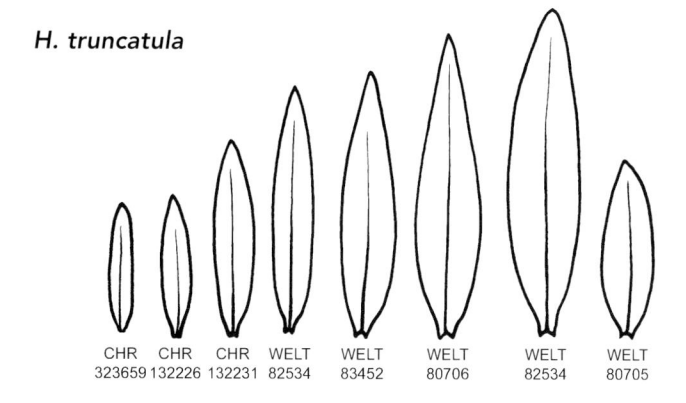

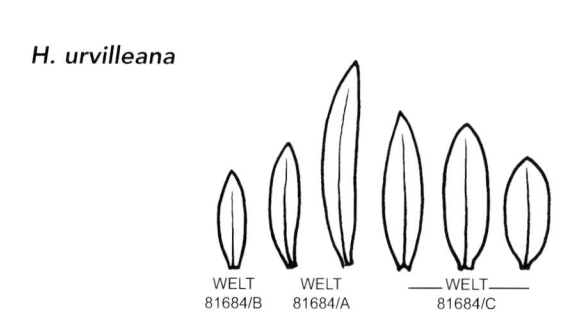

| CHR 54440 | CHR 249846 | CHR 280461 | CHR 146394 | WELT 8198 | CHR 127013 |

H. traversii

| WELT 16297 | WELT 79056 | WELT 80823 | WELT 83404 | WELT 16905 | WELT 79018 | WELT 81691 | WELT 16884 |

H. treadwellii

| WELT 81709 | WELT 81710 | WELT 80510 | WELT 80511 | WELT 81610 | WELT 82196 |

H. truncatula

| CHR 323659 | CHR 132226 | CHR 132231 | WELT 82534 | WELT 83452 | WELT 80706 | WELT 82534 | WELT 80705 |

H. urvilleana

| WELT 81684/B | WELT 81684/A | WELT 81684/C |

APPENDIX 4   351

Appendix 5

Details of Voucher Specimens for Photographs and Illustrations

An asterisk (*) after a voucher number indicates that the specimen is not necessarily the same individual as that shown in the photograph, but one from the same population, collected at the same time as the photograph was taken. The abbreviation MO is used for Micro-Optics voucher specimens, provided by Bill Malcolm; these are housed at WELT as a single collection (and individual sheets do not have separate WELT numbers). The abbreviations MJB and PGJ indicate collections made by M. J. Bayly and P. J. Garnock-Jones, respectively.

Fig. 32   *Hebe divaricata* MO 131.
Fig. 33   *Hebe subalpina* WELT 83553.
Fig. 48   Whipcord leaf apices **A**, WELT 82683; **B**, WELT 6608.
Fig. 49   *Hebe tetragona* **A**, WELT 80951*; **B, C1, D1**, WELT 83596; **C2**, unvouchered (collected by MJB and A. V. Kellow, Whanahuia Ra., 18 May 2003); **D2, F**, MO 512; **E**, MO 123.
Fig. 50   *Hebe tetragona* × *H. odora* **A**, WELT 82793; **B**, WELT 82794; **C**, WELT 82792.
Fig. 51   *Hebe hectorii* subsp. *hectorii* **A**, WELT 82433*; **B**, WELT 81433; **C1, E**, WELT 82683; **C2, D2**, WELT 82693; **D1**, MO 592; **F**, MO 259.
Fig. 52   *Hebe hectorii* subsp. *demissa* and subsp. *coarctata* **A1, E**, MO 107; **A2**, WELT 81445*; **B (upper)**, WELT 83592; **B (lower), C1**, MO 354; **C2**, MO 364; **C3, D2**, MO 593; **D1**, MO 544; **F**, WELT 82705.
Fig. 53   *Hebe propinqua* **A**, WELT 81462; **B**, WELT 81427*; **C**, WELT 83566; **D**, MO 586; **E, F**, MO 253.
Fig. 54   *Hebe lycopodioides* **A**, WELT 82465; **B, C1**, WELT 82460; **C2, D2**, MO 590; **D1**, MO 509; **E, F**, MO 184.
Fig. 55   *Hebe imbricata* **A**, WELT 81471; **B**, WELT 82682; **C1, D**, MO 589; **C2**, MO 376; **E**, MO 272 (MJB 1095); **F**, MO 258 (MJB 1132).
Fig. 56   *Hebe ochracea* **A, E, F**, MO 111; **B**, WELT 82795; **C**, MO 372; **D**, MO 587.
Fig. 57   *Hebe salicornioides* **A**, WELT 81009*; **B**, WELT 82473; **C, E**, MO 594; **D**, MO 542; **F**, MO 208.
Fig. 58   *Hebe armstrongii* **A**, unvouchered (Enys Reserve); **B**, WELT 83586; **C, D2**, WELT 83604; **D1**, MO 581; **E**, MO 159; **F**, MO 533.
Fig. 59   *Hebe annulata* **A**, WELT 81463; **B**, WELT 83587; **C, D**, MO 585; **E, F**, MO 331.
Fig. 61   *Hebe epacridea* **A**, WELT 81012; **B**, WELT 77357; **C**, WELT 82524.
Fig. 62   *Hebe epacridea* **A**, WELT 81010*; **B, E, M**, MO 358; **C, D**, MO 599; **F**, MO 395; **G, I**, MO 559; **H, J, K**, MO 558; **L**, WELT 82425.
Fig. 63   *Hebe macrocalyx* **A**, WELT 81721*; **B**, WELT 83614 (MO 536 is a duplicate); **C** (including inset), **D**, MO 326; **E** (including inset), MO 538; **F, H**, WELT 83439; **G, I**, WELT 83438; **J, K**, MO 536.
Fig. 64   *Hebe ramosissima* **A**, WELT 81688; **B, E**, WELT 83615; **C, D, I**, MO 323 (duplicate of WELT 81687); **F–H**, MO 378.
Fig. 65   *Hebe haastii* **A**, WELT 81718*; **B**, WELT 82451; **C–H, K, L**, MO 328 (duplicate of WELT 81718); **I**, MO 362 (duplicate of WELT 82466); **J**, WELT 82466.
Fig. 66   *Hebe petriei* **A**, WELT 81452; **B**, WELT 83580; **C**, MO 579; **D, E, H**, MO 267 (PGJ 2372); **F**, MO 264 (PGJ 2372); **G**, unvouchered, cult. Landcare Research, Lincoln, gardens accession number G18797B; **I**, WELT 83580.
Fig. 67   *Hebe murrellii* **A**, WELT 82686*; **B**, MO 518 (duplicate of WELT 82686); **C–E, H**, MO 270 (PGJ 2373); **F**, MO 269 (PGJ 2374); **G, I–K**, WELT 82686.
Fig. 68   *Hebe benthamii* **A, B**, WELT 83712; **C, D, G, H**, WELT 83542; **E, F**, WELT 83713; **I**, unvouchered (Vivienne Nichols, subantarctic islands).
Fig. 69   *Hebe buchananii* WELT 83919*.

Fig. 70   *Hebe amplexicaulis* **A**, WELT 82458*; **B, H**, WELT 82548; **C–E, G**, MO 345; **F**, unvouchered (plant cultivated by A. Dench, Wellington).

Fig. 71   *Hebe pareora* **A**, WELT 81766*; **B, H (left)**, same plant as MO 548, cult. Victoria University of Wellington (but specimens collected at different times); **C**, MO 275; **D–G**, WELT 81766; **H (right)**, MO 548.

Fig. 72   *Hebe gibbsii* **A**, WELT 83550*; **B, H**, WELT 83550; **B (inset), C–E**, MO 104; **F, G**, MO 360.

Fig. 73   *Hebe pinguifolia* **A**, WELT 83560*; **B, I**, WELT 83560; **C, D**, MO 540; **E, G**, MO 339 (duplicate of WELT 81677: MJB 1238); **F, H**, MO 336 (duplicate of WELT 81677: MJB 1237); **J, K**, MO 251.

Fig. 74   *Hebe buchananii* **A**, WELT 81444; **B–D**, MO 351; **E, H**, MO 163; **F, G**, MO 507; **I, J**, MO 252.

Fig. 75   *Hebe pimeleoides* **A**, WELT 81005*; **B**, WELT 82444*; **C (left)**, WELT 82456; **C (right), I**, WELT 82641; **D (upper), E–G**, MO 539; **D (lower), L**, MO 197; **H**, MO 522; **J**, MO 261; **K**, MO 232.

Fig. 76   *Hebe biggarii* **A**, WELT 81459*; **B, H**, WELT 83558; **C, D, J, K**, MO 161; **E, F**, MO 341; **G**, MO 348; **I**, MO 266.

Fig. 77   *Hebe parviflora* **A, C**, WELT 80709; *Hebe stenophylla* **B, D**, WELT 80979.

Fig. 78   *Hebe subalpina* **A**, WELT 81003*; **B**, WELT 82435; **C, D**, MO 219; **E**, WELT 83553; **F, H**, MO 217; **G**, MO 218; **I**, MO 303.

Fig. 79   *Hebe urvilleana* **A, F, H**, WELT 81684; **B**, WELT 83549; **C–E, G**, WELT 81683.

Fig. 80   *Hebe evenosa* **A**, WELT 80630*; **B**, WELT 83565; **C, D**, MO 276; **E–G**, MO 534; **H, I**, WELT 16844.

Fig. 81   *Hebe truncatula* **A**, WELT 82534*; **B, F–I**, WELT 82534; **C–E**, MO 274.

Fig. 83   *Hebe treadwellii* **A**, WELT 81709*; **B (upper), C–J** WELT 82450; **B (lower)**, WELT 81709.

Fig. 84   *Hebe decumbens* **A**, WELT 83563*; **B (including inset)**, WELT 83563; **C–H**, MO 170.

Fig. 85   *Hebe rakaiensis* **A**, unvouchered (same population as WELT 81467, but photographed one year later); **B**, WELT 82462; **C, D (inset)**, MO 528; **D**, MO 284; **E, F**, MO 112; **G**, WELT 17016.

Fig. 86   *Hebe calcicola* **A, G**, WELT 80728; **B**, WELT 83552; **C, D, H**, MO 189; **E**, MO 115; **F**, MO 124.

Fig. 87   *Hebe glaucophylla* **A**, WELT 81707; **B**, WELT 83562; **C–H**, MO 176.

Fig. 88   *Hebe topiaria* **A, C, D, G, I, J**, MO 108; **B**, WELT 83606; **D (inset), E, F, H**, MO 543.

Fig. 89   *Hebe albicans* **A**, WELT 83603; **B**, WELT 81756*; **C**, WELT 81751*; **D (left)**, WELT 83486; **D (right)**, WELT 83485; **E, G (left), H, J**, MO 324; **F, G (right), I, L**, MO 321; **H (centre)**, MO 320; **K**, WELT 82555.

Fig. 90   *Hebe stenophylla* **A**, WELT 81107*; **B**, WELT 83564; **C, D, H, I**, MO 113; **E**, MO 277; **F**, WELT 81517*; **G (upper)**, WELT 81486; **G (lower)**, MO 249.

Fig. 91   *Hebe traversii* **A**, WELT 81675*; **B**, WELT 83610; **C, D (including inset)**, MO 402 (duplicate of WELT 83610); **E, F**, MO 223; **G, H**, WELT 79018.

Fig. 92   *Hebe parviflora* **A**, WELTU (PGJ 2258); **B**, WELT 83573; **C, D, H**, MO 140; **D (inset)**, MO 596; **E–G**, MO 119 (PGJ 2257).

Fig. 93   *Hebe strictissima* **A**, unvouchered (same population as WELT 80811, but photographed eighteen months later); **B**, WELT 83574; **C, D (inset)**, MO 399; **D**, MO 279; **E–G**, MO 278; **H**, WELT 15007.

Fig. 94   *Hebe stricta* var. *stricta* and var. *macroura* **A**, unvouchered (near Rangiwahia); **B**, WELT 81783*; **C (left)**, WELT 83597; **C (right), E, L, M**, WELT 83617; **D, F**, MO 212; **G, I, K**, WELT 83605; **H, J**, MO 213.

Fig. 95   *Hebe stricta* var. *atkinsonii* **A, C–I**, MO 156; **B**, WELT 83735.

Fig. 96   *Hebe stricta* var. *egmontiana* and var. *lata* **A**, WELT 81811; **B**, unvouchered (Makahu Spur); **C (left)**, WELT 83569; **C (right), E, G, I, K–M**, WELT 83540; **D, F, H, J**, WELT 83595.

Fig. 97   *Hebe bishopiana* **A**, WELT 81398*; **B, E**, WELT 81398; **C**, WELT 80970; **D, H**, MO 241; **F, G**, MO 396.

Fig. 98   *Hebe obtusata* **A**, unvouchered (Takatu Head); **B–D (inset), G**, MO 403; **D–F, H, I**, MO 190.

Fig. 99   *Hebe tairawhiti* **A**, WELT 81789*; **B, E**, WELT 83568; **C, D, F–I**, MO 245.

Fig. 100 *Hebe angustissima* **A**, WELT 81521*; **B–I**, MO 514.

Fig. 101 *Hebe acutiflora* **A**, WELT 80623*; **B, E**, WELT 83570; **C, D (inset)**, MO 552; **D**, MO 158; **F, G**, MO 120; **H**, MO 145.

Fig. 102 *Hebe flavida* **A**, WELT 82611; **B, G**, WELT 83593; **C–F**, MO 244; **H, I**, WELT 82612.

Fig. 104 *Hebe ligustrifolia* **A**, unvouchered (Unuwhao); **B**, WELT 83591; **C, E–G**, MO 114; **D**, WELT 83616; **H**, WELT 80969.

Fig. 105   *Hebe bollonsii* **A**, unvouchered (Aorangi Island); **B**, WELT 83579; **C, E, G**, MO 349; **D, H, I**, MO 155; **F**, unvouchered (voucher details given on photographic slide are probably incorrect).

Fig. 106   *Hebe perbella* **A**, WELT 79994*; **B, E**, WELT 83575; **C, D, F–H**, MO 147.

Fig. 107   *Hebe macrocarpa* **A**, unvouchered (Shakespeare Cliff); **B**, unvouchered (Windy Canyon, Great Barrier Island); **C (left), E**, WELT 83607; **C (right)**, WELT 83577; **D (left)**, WELT 80972; **D (right), F, I**, MO 186 (although voucher lacks flowers); **G, H**, WELT 83541.

Fig. 108   *Hebe brevifolia* **A**, unvouchered (Surville cliffs); **B–F**, MO 404; **G–I**, MO 150.

Fig. 109   *Hebe barkeri* **A**, WELT 80559; **B**, WELT 83950; **C (leaves), D, F, G**, MO 160; **C (insets)**, WELT 83441; **E**, MO 239.

Fig. 110   *Hebe dieffenbachii* **A**, WELT 80550*; **B**, WELT 83584; **C, D (leaves), I**, MO 172; **D (insets)**, WELT 83611; **E, F, H**, MO 505; **G**, MO 171.

Fig. 111   *Hebe chathamica* **A**, WELT 80550*; **B, E**, WELT 83588; **C, D, H, I**, MO 167; **F, G**, MO 166.

Fig. 112   *Hebe rapensis* **A**, unvouchered (Jean-Yves Meyer, Rapa).

Fig. 114   *Hebe odora* WELT 82784*.

Fig. 115   *Hebe odora* **A**, WELT 81447; **B, H**, WELT 83555; **B (inset)**, WELT 81021; **C–E, J**, MO 535; **F, G**, MO 373; **I**, MO 191.

Fig. 116   *Hebe mooreae* **A**, WELT 81727; **B, D, E, J**, MO 571; **B (inset)**, WELT 81726; **C, F**, MO 179; **G, H**, MO 180; **I**, WELT 83557.

Fig. 117   *Hebe masoniae* **A**, WELT 82415*; **B (including inset), C**, MO 368; **D (including inset), E, H**, MO 110; **F, I, J**, MO 577; **G**, unvouchered (Mt Arthur).

Fig. 118   *Hebe pauciramosa* **A**, WELT 80938*; **B**, WELT 82431; **B (inset)**, WELT 17266; **C, D**, MO 374; **E, G, H**, MO 598; **F**, MO 332; **I**, MO 394.

Fig. 119   *Hebe vernicosa* **A, C–E**, unvouchered (Mt Arthur); **B**, WELT 83547; **F, G**, WELT 82637; **H, I**, MO 305

Fig. 120   *Hebe canterburiensis* **A**, MO 101; **B**, MO 532; **C, D**, MO 352; **E, G, H**, MO 101; **F**, MO 164.

Fig. 121   *Hebe dilatata* **A, C–F, H (left)**, WELT 81443; **B**, unvouchered; **G**, WELT 81450; **H (right)**, WELT 81438.

Fig. 122   *Hebe societatis* **A**, WELT 82423*; **B (left)**, WELT 82426 (which is a spirit collection only); **B (right), E**, WELT 82423; **C, D, F–H**, MO 369.

Fig. 123   *Hebe carnosula* **A**, unvouchered (Red Hills Ridge); **B**, WELT 83572; **C, D**, MO 306; **E**, WELT 83341; **F**, WELT 83343; **G, I**, WELT 83342; **H**, WELTU 14842.

Fig. 124   *Hebe rupicola* **A, B**, WELT 83567; **C, D, H**, MO 233; **E, G**, MO 311; **F**, MO 337.

Fig. 125   *Hebe rigidula* **A**, unvouchered (Pelorus Bridge); **B (left), D1**, WELT 82535; **B (right), D2**, WELT 82536; **C (upper)**, MO 576; **C (lower), E**, MO 380; **F**, WELT 83571; **G, H**, WELT 82574.

Fig. 126   *Hebe colensoi* **A**, unvouchered (Taruarau River); **B**, WELT 83582; **C**, MO 293; **D, E**, WELT 82565; **F**, MO 357; **G**, WELT 83583; **H, I**, WELT 81035.

Fig. 127   *Hebe scopulorum* **A**, WELT 82488*; **B**, WELT 82576; **C, F**, WELT 82576; **D, E, G**, WELT 82563; **H**, WELT 82485.

Fig. 128   *Hebe crenulata* **A**, WELT 81743; **B, E**, WELT 82416; **C, D**, MO 309; **F, G**, MO 322 (duplicate of WELT 81743).

Fig. 129   *Hebe cryptomorpha* **A**, WELT 83551*; **B, E**, WELT 83551; **C, D**, WELT 82573; **F–H**, WELT 82575.

Fig. 131   *Hebe cockayneana* **A**, WELT 80869*; **B**, WELT 82427; **C, E, F**, WELT 82571; **D**, MO 273; **G, H**, MO 572; **I**, WELT 83556; **J**, MO 573.

Fig. 132   Leaf margins **A**, WELT 81498; **B**, WELT 80869.

Fig. 133   *Hebe arganthera* **A**, WELT 82697*; **B, E–H**, MO 516; **C, D**, MO 220; **D (inset)**, MO 503.

Fig. 134   *Hebe diosmifolia* **A**, unvouchered (Cape Reinga); **B, F**, WELT 83578; **C, D**, MO 152; **E**, unvouchered (Waipoua River); **G**, MO 565; **H**, MO 285; **I**, MO 297.

Fig. 135   *Hebe venustula* **A**, unvouchered (Makahu Spur); **B**, WELT 83598; **C–F, I**, MO 388; **G, H**, MO 389.

Fig. 136   *Hebe brachysiphon* **A**, MO 350*; **B–I** MO 350 (MJB S-66).

Fig. 137   *Hebe divaricata* **A**, WELT 81668; **B**, WELT 82414; **C, D**, MO 118; **E, H**, WELT 83554; **F**, WELT 81665*; **G, J**, MO 531; **I**, MO 131.

Fig. 138   *Hebe insularis* **A**, WELT 81599*; **B, I**, WELT 83581; **C**, MO 237; **D–H**, WELT 81598.

Fig. 139   *Hebe elliptica* **A**, unvouchered (Pancake Rocks); **B**, WELT 83589; **C, E (upper), F–I**, MO 597; **D, E (lower)**, MO 175; **J, K**, MO 316.

Fig. 140 *Hebe leiophylla* **A**, unvouchered (near bridge at start of Buller River, Lake Rotoiti); **B**, WELT 83613; **B (inset)**, **C**, **D (inset)**, MO 566; **D**, **H**, **I**, MO 177; **E–G**, MO 562.

Fig. 142 *Hebe salicifolia* **A**, unvouchered (track to Red Hills Ridge); **B**, **E**; WELT 83585; **B (inset)**, **D (inset)**, MO 547; **C**, **I**, MO 206; **D**, MO 532; **F–H**, MO 207.

Fig. 143 *Hebe paludosa* **A**, WELT 81479*; **B**, **E**, WELT 83559; **B (inset)**, MO 574; **C**, **D**, **G**, **H**, WELT 83608 (= MO 256); **F**, MO 128.

Fig. 144 *Hebe pubescens* subsp. *pubescens* and subsp. *rehuarum* **A**, unvouchered (Cooks Beach); **B**, WELT 81854; **C**, **E**, WELT 81851; **D**, WELT 82049; **F**, MO 377; **G**, **I–L**, MO 344; **H**, WELT 82566.

Fig. 145 *Hebe pubescens* subsp. *sejuncta* **A**, unvouchered (Hokoromea Island); **B**, **C (upper)**, WELT 82480; **C (lower)**, **G**, **H**, WELT 82569; **D**, **K**, MO 281; **E**, **F**, WELT 82567; **I**, **J**, MO 238.

Fig. 146 *Hebe breviracemosa* **A**, unvouchered (Hutchies Track, Raoul Island); **B**, **D**, **G**, WELT 82627; **C**, **E**, **F**, **H**, MO 295.

Fig. 147 *Hebe corriganii* **A**, WELT 83599*; **B**, WELT 83599; **C**, **H**, **I**, MO 149; **D**, MO 185; **E–G**, WELT 83594.

Fig. 148 *Hebe adamsii* **A**, unvouchered (Tarure Hill); **B**, WELT 84062; **C**, unvouchered (cultivated at Percy Reserve, ex The Pinnacle, Unuwhao, Northland); **D**, **F**, MO 122; **E**, **G**, **H**, WELT 80962.

Fig. 149 *Hebe townsonii* **A**, WELT 81755*; **B**, **C**, MO 397; **D–F**, MO 133; **G**, **H**, MO 222.

Fig. 150 *Hebe speciosa* **A**, unvouchered (Maunganui Bluff, Aranga end, Sept 2001); **B**, **E**, WELT 83576; **C**, **D**, MO 209; **F**, **H**, MO 210; **G**, **I**, MO 302.

Fig. 151 *Hebe macrantha* **A**, WELT 80740*; **B (left)**, WELT 82429; **B (right)**, **E–G**, WELT 83544; **C**, **D (lower)**, WELT 82554; **D (upper)**, WELT 82429; **H**, WELT 83545.

Fig. 152 *Hebe pauciflora* **A**, WELT 83609*; **B**, **D**, **F–H**, WELT 83609; **C**, **E**, **I**, **J**, WELT 81759.

Fig. 153 *Leonohebe cheesemanii* MO 169 (WELT 80937*).

Fig. 154 *Leonohebe ciliolata* **A**, WELT 81673*; **B**, WELT 83548; **C**, **E–G**, **I**, WELT 82556; **D**, MO 526 (duplicate of WELT 82778); **H**, MO 280.

Fig. 155 *Leonohebe tumida* **A**, WELT 81661*; **B**, WELT 82477; **C–H**, MO 226.

Fig. 156 *Leonohebe cheesemanii* **A**, WELT 82457*; **B**, WELT 82457; **C1** WELT 82455; **C2**, **D**, **E**, **G** MO 169 (WELT 80937*); **F**, **H**, MO 168 (WELT 80936*).

Fig. 157 *Leonohebe tetrasticha* **A**, MO 384*; **B**, **C**, **H**, MO 384 (MJB S-65); **D**, MO 525; **E**, **G**, MO 555; **F**, MO 554.

Fig. 158 *Leonohebe cupressoides* **A**, unvouchered (Cave Stm); **B**, **G**, **J**, WELT 82816 (MJB S-147); **C**, MO 132; **D**, **E**, **H**, **I**, MO 595; **F**, MO 546.

Fig. 159 Variation in *H. hectorii* subsp. *hectorii* (in left to right order on graphs): **A**, CHR 517256, CHR 517261, CHR 10797, CHR 331753, CHR 331755, CHR 319115, CHR 252717, CHR 517262, CHR 517316, CHR 176652, CHR 20414, CHR 465659, CHR 517248, CHR 470101, CHR 182432, CHR 218347, CHR 125677, CHR 517253, CHR 465670, CHR 20492, CHR 63348, CHR 122640, CHR 309263, CHR 63349, CHR 465627, CHR 465621, CHR 175114, CHR 28263, CHR 119650, CHR 78149, WELT 5305, WELT 6595, CHR 10826, CHR 333998, CHR 119606, CHR 167050, CHR 520412, WELT 17431, CHR 338356, CHR 517249, CHR 517250, WELT 82692, CHR 480976, CHR 483453, WELT 17430, CHR 520414, CHR 520413, CHR 512588, CHR 371846, CHR 481023, CHR 80925, WELT 82706/4, CHR 487196, CHR 212606, CHR 189831, CHR 78151, WELT 82706/2, CHR 193023, CHR 310995, CHR 77393, CHR 309525, CHR 261907, CHR 227472, CHR 516150, CHR 311267, CHR 145121, CHR 252719, CHR 183433, WELT 82706/5, CHR 355507, CHR 201753; **B**, CHR 10797, CHR 331753, CHR 176652, CHR 20414, CHR 331755, CHR 319115, CHR 517248, CHR 517256, CHR 465670, CHR 465659, CHR 470101, CHR 252717, CHR 517262, CHR 517261, CHR 182432, CHR 20492, CHR 218347, CHR 125677, CHR 63348, CHR 122640, CHR 63349, CHR 175114, CHR 10826, CHR 119650, CHR 333998, CHR 517253, CHR 465627, CHR 119606, CHR 517316, CHR 309263, CHR 28263, CHR 78149, CHR 465621, CHR 167050, CHR 520412, CHR 517250, WELT 82692, WELT 17431, CHR 338356, CHR 480976, CHR 517249, CHR 520413, CHR 520414, WELT 5305, CHR 512588, CHR 371846, CHR 487196, CHR 193023, CHR 483453, CHR 481023, WELT 17430, CHR 80925, CHR 212606, CHR 310995, WELT 6595, WELT 82706/4, CHR 227472, CHR 189831, CHR 78151, CHR 77393, CHR 309525, CHR 516150, CHR 261907, CHR 311267, CHR 145121, CHR 252719, CHR 183433, WELT 82706/5, CHR 355507, WELT 82706/2, CHR 201753.

Fig. 160 Variation in *H. hectorii* subsp. *demissa* (in left to right order on graph): WELT 80910, CHR 517254, CHR 517247, WELT 81445, CHR 205020, CHR 195499, CHR 195571, CHR 471961, WELT 17468, CHR 517251, CHR 517269, CHR 517263, CHR 517244, WELT 80911, CHR 184558, CHR 517239, WELT 6613, WELT 17473, WELT 17423, CHR 471821, WELT 6612, WELT 81528, WELT 17471, CHR 517241, CHR 17190, CHR 517243, WELT 6011, CHR 517240, WELT 82705, CHR 395426, WELT 81524, WELT 80889, CHR 199643, WELT 17427, CHR 517246, WELT 81440, CHR 470386, CHR 395574, WELT 80891, WELT 81541, CHR 122649, CHR 145148, WELT 80888, WELT 81441, CHR 72729, WELT 17428, CHR 56678, WELT 80890, CHR 439582, WELT 17429, WELT 14047, WELT 41359, WELT 17425, CHR 283900, CHR 103057.

Fig. 164 Variation in *H. lycopodioides* (in left to right order on graphs): **A**, WELT 17414, CHR 402090 (left), WELT 82464, CHR 207223, CHR 517290, CHR 517278 (C), WELT 82012, CHR 517281, WELT 79827, WELT 81007, WELT 14025, CHR 536540 (lower left), CHR 517273 (upper right), CHR 517273 (lower left), CHR 455479, WELT 17418, WELT 17388, CHR 512480, WELT 17396, WELT 17416, WELT 17419, WELT 82551, CHR 203997, WELT 82460, WELT 17421, WELT 82465, WELT 17407, WELT 17392, CHR 536540 (upper right), WELT 14024, WELT 17417, WELT 17391, CHR 517278 (D), CHR 195532, WELT 80930, WELT 14023, WELT 80989, WELT 80981, CHR 252714, CHR 215937, WELT 80986, WELT 80983, WELT 17389, CHR 517278 (B), CHR 517278 (A), CHR 180496, CHR 257140, CHR 369064, WELT 79962, CHR 58405, WELT 80569, WELT 81657, WELT 81700/1, CHR 56603, CHR 270840, WELT 81700/5, CHR 223066 (top left), CHR 420845, CHR 517283 (A), WELT 81670, CHR 517283 (C), CHR 479179, CHR 127671, CHR 190336, WELT 81700/4, WELT 81700/3, CHR 189837, CHR 517283 (B), CHR 517285 (upper right), WELT 81700/6, WELT 81700/2, CHR 517284, WELT 81699, CHR 517315 (upper right), CHR 517315A, WELT 81669; **B**, CHR 512480, CHR 203997, WELT 80983, CHR 369064, WELT 80930, WELT 82551, WELT 80986, WELT 14025, WELT 79827, WELT 17396, CHR 517290, WELT 17417, WELT 80981, WELT 81700/5, CHR 58405, WELT 17392, WELT 17419, CHR 215937, CHR 195532, WELT 17418, WELT 81657, WELT 81007, CHR 517273 (lower left), CHR 223066 (top left), WELT 79962, WELT 82464, WELT 80989, CHR 56603, CHR 257140, WELT 17414, WELT 14023, CHR 402090 (left), WELT 82460, WELT 81700-3, CHR 517281, CHR 207223, CHR 536540 (lower left), CHR 420845, WELT 80569, CHR 455479, WELT 81700/1, WELT 82465, CHR 517283 (A), CHR 517278 (B), WELT 17416, WELT 17407, CHR 252714, WELT 81700/2, CHR 189837, CHR 127671, WELT 17391, WELT 17388, WELT 17389, CHR 517284, WELT 81700/4, CHR 517283 (C), CHR 536540 (upper right), CHR 517273 (upper right), WELT 14024, CHR 180496, WELT 82012, WELT 81699, WELT 81700/6, WELT 17421, CHR 517285 (upper right), CHR 517278 (C), CHR 517315A, CHR 190336, CHR 479179, CHR 270840, CHR 517278 (D), WELT 81670, CHR 517283 (B), CHR 517278 (A), CHR 517315 (upper right), WELT 81669.

Glossary

Definitions in this glossary have been extracted or adapted from a range of sources, including *Flora of New Zealand* (e.g. Allan 1961), *Flora of Australia* (McCusker 1981), *Flora of New South Wales* (e.g. Harden 2002), *Flora of Victoria* (e.g. Walsh & Entwisle 1999), *The Penguin Dictionary of Botany* (Tootill 1984), *The Language of Botany* (Debenham [1971]), *Anatomy of Seed Plants* (Esau 1977), *Botanical Latin* (Stearn 1973), and glossaries and discussion pages on the Internet, including those of Jim Croft (www.anbg.gov.au/glossary/croft.html), Mike Crisp (www.science.uts.edu.au/sasb/glossary.html), Curtis Clark and Nancy Charest (www.csupomona.edu/~jcclark/classes/bio406/glossary.html), and Peter Hoen (www.bio.uu.nl/~palaeo/glossary).

A

Abaxial Facing away from the axis or stem – for example, the lower surface of a leaf.

Abscission The normal shedding from a plant of an organ that is mature or aged, such as a ripe fruit or an old leaf.

Acuminate Tapering to a fine point, with sides ± concave (**Fig. 167**).

Acute Sharply pointed; converging edges forming an angle less than 90 degrees (**Fig. 167**). Cf. obtuse.

Alternate Arranged singly at different levels along a stem or inflorescence. Cf. opposite.

Amplexicaul Clasping the stem; describes leaf bases that are broad and that partly surround the stem.

Androecium Male component of a flower; the stamens of one flower collectively.

Aneuploidy An evolutionary change in chromosome number that is not in multiples of the haploid number, such as addition or loss of one or two whole chromosomes.

Angustiseptate with a narrow septum. For *Hebe*, this term is used in reference to the capsules, describing the condition in which the septum is narrow, and at right angles to the plane of compression of the fruit (**Fig. 29**). Cf. latiseptate.

Anterior On the side away from the axis – for floral parts this means toward the subtending bract, thus appearing in front. Cf. posterior.

Anther The part of the stamen that contains pollen, and which is usually borne on a stalk (filament) (**Fig. 25**).

Anthesis The stage at which a flower opens.

Anticlinal At right angles to the surface – for example, anticlinal walls of a leaf cell are those perpendicular to the leaf surface.

Antrorse Turned toward the apex – for example, relating to curved hairs on leaves or stems.

Apiculate Terminating in a short point (**Fig. 167**). Cf. mucronate (where the point is more abruptly narrowed and sharper).

Appressed Closely and flatly pressed against a surface – for example, of leaves (**Fig. 168**).

Arbuscular Of mycorrhizae, having bush-like projections of fungal hyphae (arbuscules) that penetrate cells of the root cortex of host plants.

Ascending Directed upwards gradually (at a narrow angle), often from a ± prostrate base – for example, of stems (**Fig. 165**).

Autecology The ecology of a single species, including the study of factors affecting growth, reproduction and environmental interactions through all phases in the life cycle.

Axil The upper angle formed by a leaf or bract and the branch or inflorescence on which it is borne.

Axillary Borne in, or arising from, an axil.

FIG. 165 Plant habits and orientation of branches (after Hutchins 1997).

Prostrate

Spreading

Decumbent

Ascending

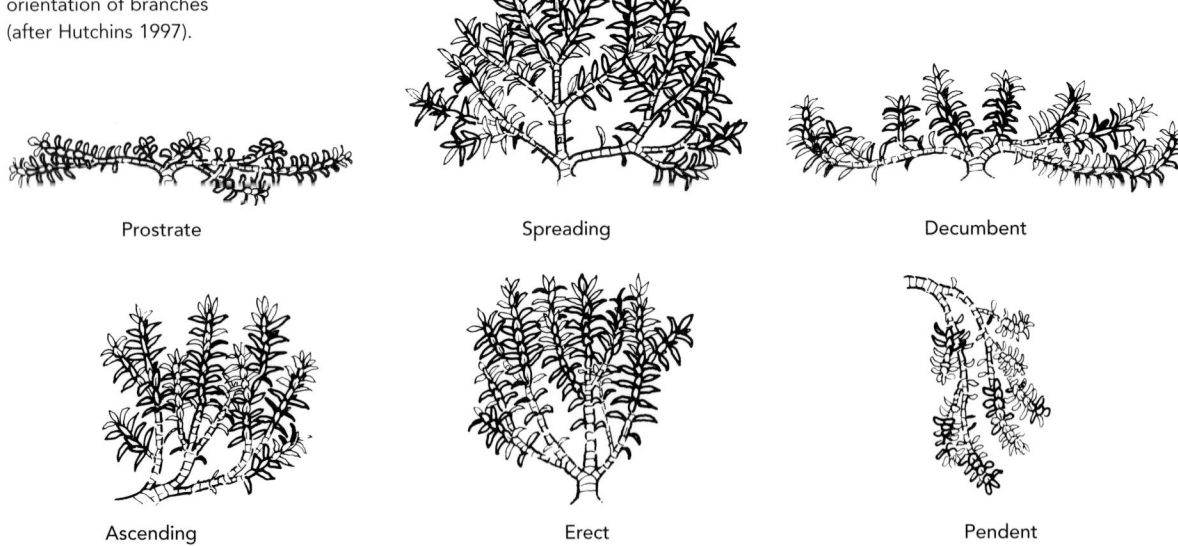

Erect

Pendent

FIG. 166 Lamina shapes.

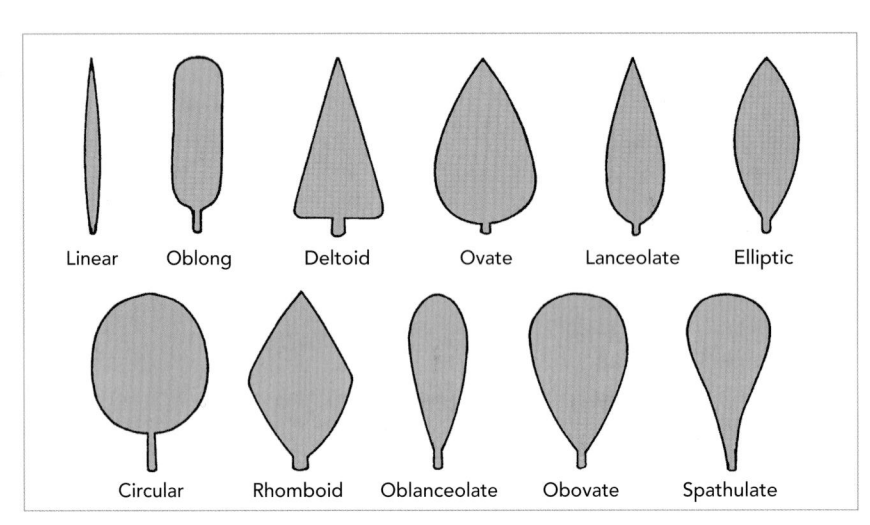

Linear　Oblong　Deltoid　Ovate　Lanceolate　Elliptic

Circular　Rhomboid　Oblanceolate　Obovate　Spathulate

FIG. 167 Apex shapes.

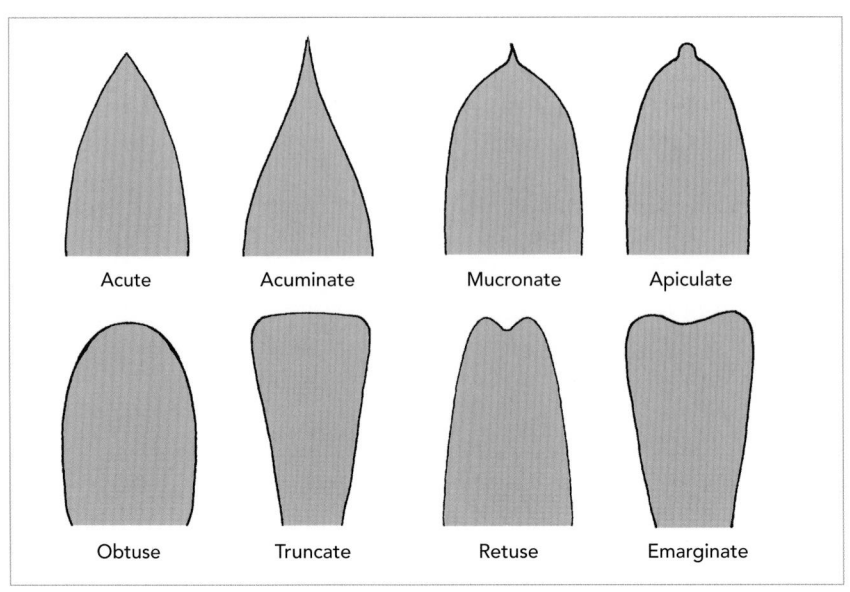

Acute　Acuminate　Mucronate　Apiculate

Obtuse　Truncate　Retuse　Emarginate

B

Bifarious Arranged in two opposite rows; used here to describe the arrangement of hairs on the stem (**Fig. 18**). Cf. uniform.

Bract A modified leaf, occurring in inflorescences, usually subtending a flower or inflorescence branch (**Fig. 25**).

Bracteate Subtended by a bract or bracts.

Branchlet A small branch.

Brochidodromous A form of leaf venation in which there is a single primary vein, with secondary veins arising from it that curve upwards toward the leaf margin and join together in a series of marginal loops.

C

Calyx the outermost, usually green, part of a flower (**Fig. 25**). In *Hebe* and *Leonohebe*, the calyx is partly divided into several lobes.

Capitate Head-like.

Capsule The dry type of fruit produced in *Hebe* and *Leonohebe*. This develops from the ovary and bears the seeds.

Carpel The structure that bears and encloses the ovules in flowering plants, usually comprising the ovary, style and stigma.

Cartilaginous Like cartilage. In *Hebe* leaf margins this describes a distinct narrow border that is colourless and/or translucent.

Cell The fundamental unit of a living organism. Plant cells, at least when young, consist of a protoplast (the living portion of the cell, which includes the cell membrane, nucleus, cytoplasm and organelles) surrounded by a cell wall.

Chromosomes The parts of a cell, found in the nucleus, that carry the genes. In plants they are visible as rod-like structures at certain stages of cell division (when stained for viewing under a microscope).

Cilia Hairs that are ± confined to the margins of an organ.

Ciliate Having hairs along the margin.

Ciliolate Diminutive form of ciliate.

Clade A monophyletic group; a non-terminal branch on a phylogenetic tree (**Fig. 169**).

Cladistic Pertaining to cladistics.

Cladistics Methods for assessing relationships and classifications of organisms developed largely from the work of Hennig (1966). These methods place emphasis on shared derived features, rather than overall similarity, when assessing the relationships of organisms.

Cline Gradual and continuous character variation throughout the range of a species, or sometimes between two species, usually along a geographic or environmental gradient, in which individuals at the two extremes may differ markedly.

Colpus (pl. colpi) An oblong to elliptic aperture in a pollen grain, at least twice as long as it is broad. Pollen with such apertures is termed colpate.

Connate Fused to another organ of the same kind, as in a pair of opposite leaves (**Figs 62E and F**) or bracts that are basally fused.

Conspecific Belonging to the same species.

Cordate Heart-shaped, with the notch at the base; a leaf base with a notch between two rounded basal lobes.

Coriaceous With a leathery texture.

Corolla The inner whorl of non-fertile parts of a flower. In *Hebe* and *Leonohebe*, this consists of 4(–6) corolla lobes (petals) that are basally united into a corolla tube (**Figs 25 and 26**). The region on the inner surface of the corolla, near the base of the lobes, is referred to here as the corolla throat.

Crenate With blunt or rounded teeth; scalloped.

Crenulate Minutely scalloped.

Cruciform In the form of a cross.

Cultivar Cultivated variety. An assemblage of cultivated individuals distinguished by any characters significant for the purposes of agriculture, forestry or horticulture, and which, when reproduced, retains its distinguishing features.

Cuneate Wedge-shaped, as of a leaf or leaf base.
Cytology The study of the structure and function of cells.

D

Decumbent Spreading horizontally, but with the ends growing upward (**Fig. 165**).
Decurrency A region of distinctive tissue extending downwards from a point of insertion. In *Hebe*, there is often a conspicuous decurrency below the point of insertion of each leaf on a stem.
Decussate Arranged in opposite pairs on the stem, with each pair at right angles to the preceding pair; the typical arrangement of leaves in *Hebe* and *Leonohebe*.
Dehiscence The opening and shedding of contents, as when an anther opens to release pollen, or a capsule splits to release seeds.
Deltoid Triangular (**Fig. 166**).
Denticulate With small teeth.
Dichotomous Forking into two equal branches.
Didymous Borne in pairs; or, when referring to ovaries or capsules, having two lobes (**Fig. 44**).
Dioecious Having male and female flowers on different plants.
Diploid Having two sets of chromosomes, twice the haploid number, in the nucleus of each somatic cell (cells other than gametes and their precursors).
Disc (disk) A plate or ring of structures derived from the receptacle, and occurring between whorls of floral parts. In *Hebe* and *Leonohebe*, a nectar-secreting disc surrounds the base of the ovary in each flower (**Fig. 28**).
Discoid Disc-shaped, flat and circular.
Distally Toward the free end or apex, away from the point of attachment.
Distichous Arranged in two rows on opposite sides of a stem and thus in the same plane.
Domatia Pits or depressions on the undersides of leaves; in *Hebe*, seen only in *H. townsonii* (**Fig. 149E**).
Dorsifixed Attached at or by the back, as with anthers on a filament.
Duplex Twofold; double.
Duplicate An herbarium specimen that represents the same gathering as another specimen (collected at the same place and time, by the same collector – and with the same collector's number, where this is provided).

E

Eglandular hair A hair without swollen head cell(s).
Ellipsoid The three-dimensional equivalent of elliptic.
Elliptic Two-dimensional in shape, oval in outline and broadest about the middle (**Fig. 166**).
Emarginate Having a broad, shallow notch at the apex (**Figs 44** and **167**).
Endemic Having a natural distribution confined to a particular geographic area.
Endotrophic mycorrhiza A form of mycorrhiza in which the fungus lives between and within the cells of the root cortex of a host plant, with growth on the outside of the root being limited.
Entire Having a smooth margin; not lobed, divided or toothed.
Epidermis The outermost layer of cells on a primary plant organ (e.g. on a leaf).
Epipetalous Borne on the petals.
Erect Upright. For floral parts (e.g. stamens and corolla lobes) or leaves (**Fig. 168**), this means ± straight in line from the point of attachment; for a shrub, this means ± perpendicular to the ground (**Fig. 165**).
Erecto-patent Orientation between erect and patent, so that the smallest angle to the point of attachment is *c.* 45 degrees (**Fig. 168**).
Exine The outer wall of a pollen grain.

F

Family A category of plant classification, comprising one or more genera.
Filament The stalk-like part of the stamen that supports the anther (**Fig. 25**).
Flavonoids A group of plant compounds (including flavones, anthocyanins, flavanones, chalcones, aurones and flavonols) that are based on a 2-phenylbenzopyran nucleus. Some of their properties, and their use as markers in plant systematics, are discussed in the introductory chapter "Flavonoid Biochemistry".

FIG. 168 Leaf postures (after Hutchins 1997).

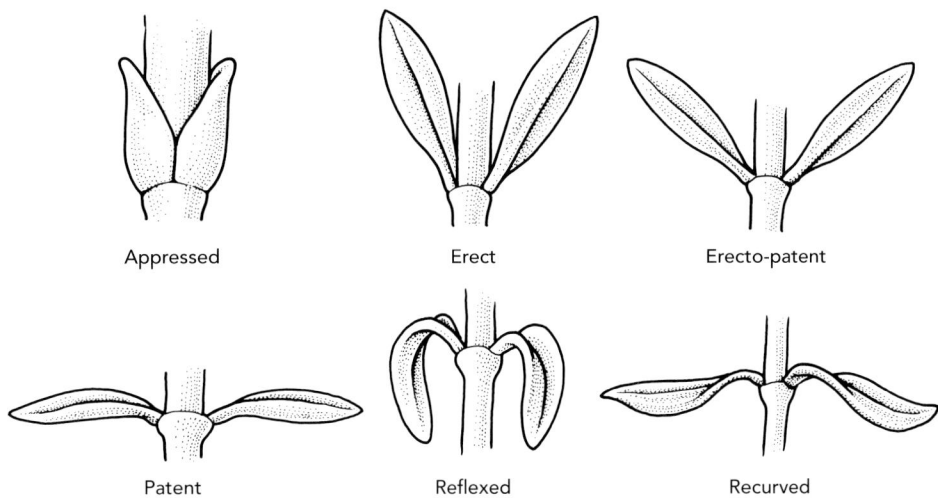

Appressed   Erect   Erecto-patent
Patent   Reflexed   Recurved

Forma A Latin term, often replaced with "form" in common usage: the lowest category used in plant classification. It is a subdivision of a species, ranked below subspecies and variety.

Funnelform Funnel-shaped, as with a flower whose corolla tube widens gradually from the base.

G

Gene The fundamental unit of inheritance, consisting of a segment of DNA on a chromosome that, through its composition of pairs of bases, codes for a specific protein or RNA molecule.

Genus (pl. genera) A category of plant classification, comprising one or more species. The genus name forms the first part of the two-word name (binomial) given to each species (e.g. *Hebe*).

Glabrous Without hairs.

Gland A secretory structure within or on the surface of a plant (often a smooth, usually shining, bead-like outgrowth).

Glandular hair A hair tipped with a gland. In *Hebe* and *Leonohebe*, glandular hairs either have one or, more commonly, two swollen head cells (**Fig. 17**).

Glaucescent Somewhat glaucous; becoming glaucous.

Glaucous Covered with a waxy bloom (which can usually be rubbed off) that gives a dull blue-green colour, as in the leaf surfaces of *H. amplexicaulis* (**Fig. 70D**).

Globose Globular or spherical in shape.

Gynodioecious Having some plants with hermaphrodite flowers and others with female flowers.

Gynoecium The female component of a flower, comprising one or more carpels (usually two fused carpels in *Hebe* and *Leonohebe*).

H

Haploid Having a single set of chromosomes (half the full set of genetic material) in the nucleus, a condition found in the embryo sac (containing the egg cell) and pollen (which produce the sperm cells) of flowering plants.

Heteroblasty A condition where juvenile and adult vegetative shoots differ morphologically (in characters of habit and/or of leaves).

Hexaploid Having six sets of chromosomes in the nucleus of each somatic cell (cells other than gametes and their precursors).

Holotype The one specimen or illustration used, or designated as the nomenclatural type by the author of a new species or infraspecific name.

Homoplasy (adj. homoplasious, homoplastic) Similarity due to independent evolutionary change (e.g. parallelisms or reversals), rather than shared inheritance from a common ancestor.

Hyaline Thin and translucent.

Hybrid An offspring of genetically different parents (the parents usually belonging to different taxa).

Hydathode A structure (e.g. a water pore) that exudes water on the surface or margin of a leaf.
Hypodermis A layer or layers of cells beneath the epidermis distinct from the underlying ground tissue (in leaves, this is a distinct layer/layers between the epidermis and mesophyll).

I

Imbricate Overlapping, like roofing tiles.
Incertae sedis A Latin term meaning of uncertain position or affinity.
Incised Cut deeply and (usually) unevenly.
Indumentum Any surface covering, such as hairs (in *Hebe* and *Leonohebe*) or scales; a collective term for such coverings.
Inflorescence The flower-bearing structure of a plant.
Infraspecific Within a species.
Infraspecific taxa Taxa below the rank of species – in other words, subspecies, variety and form.
Infructescence An inflorescence in the fruiting stage.
Internode The portion of a stem between two nodes.
Intramarginal Situated inside the margin, but close to it, as with the veins of some leaves.
Introgression (introgressive hybridisation) Incorporation of the genes of one lineage or taxon into another through successful hybridisation and subsequent backcrossing with one of the parental groups.
Intron A DNA sequence that interrupts the protein-coding sequence of a gene; introns are transcribed into messenger RNA but the sequences are eliminated from the RNA before it is used to make protein.
Introrse Facing inward or toward the axis – for example, introrse anthers open toward the centre of a flower.
Isolectotype A duplicate of a lectotype.
Isotype A duplicate of a holotype; it is always a specimen.

J

Jordanon A true-breeding group of similar individuals plainly distinct from any other such group.
Juvenile leaves Leaves formed on a young plant and different in form from the adult leaves.

L

Lamina A thin, flat organ or part, such as the blade of a leaf.
Lanceolate Lance-shaped; three to six times as long as broad, broadest below the middle and tapering to the apex (**Fig. 166**).
Lateral Attached to the side of an organ, as in the case of leaves or inflorescences (**Fig. 40**) on a stem.
Latiseptate With a broad septum. For *Hebe*, this term is used in reference to the capsules, describing the condition in which the septum is broad and parallel to the plane of compression of the fruit (**Fig. 29**). Cf. angustiseptate.
Leaf bud In *Hebe*, the cluster of young leaves that have not yet diverged from each other (the leaves of each pair generally adhering along their margins), and that surround the growing point at the apex of a shoot.
Lectotype A specimen or illustration designated from the original material (as defined in the ICBN) if no holotype was designated at the time of publication, or if it is missing, or if it is found to belong to more than one taxon.
Linear Very narrow; the length measuring more than twelve times the width, and with the sides mostly parallel (**Fig. 166**).
Lira (pl. lirae) A ridge that is part of the ornamentation of a pollen grain.
Locule A cavity or chamber. Used here to describe the chambers of an ovary.
Loculicidal Of a capsule, dehiscing by splitting along the medial line of the locules. Cf. septicidal.

M

Mammillate With nipple- or teat-shaped projections.
Mesophyll Photosynthetic cells of a leaf blade located between the two epidermal layers. This is commonly divided into distinct bands of palisade mesophyll (with tightly arranged elongated cells, whose long axis is perpendicular to the leaf surface) and spongy mesophyll (with cells of various shapes and conspicuous intercellular spaces).

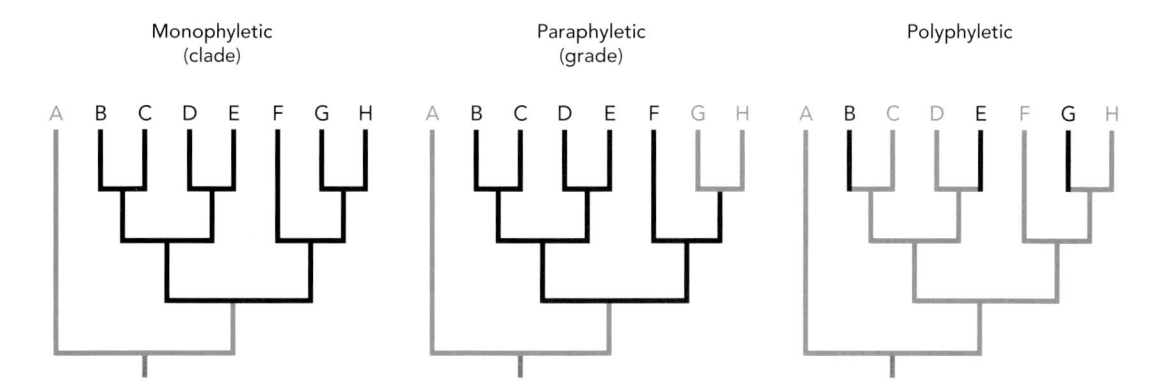

FIG. 169 Illustration of some cladistic terms. A–H are taxa, whose relationships are summarised by the phylogenetic trees. Branches shown in bold denote the taxa comprising a group.

Mesotrophic Of a habitat moderately nutrient-rich and moderately productive; one of four trophic state categories (between oligotrophic and eutrophic) used to describe production in wetlands, lakes and reservoirs.

Micropyle A minute opening in the integuments of an ovule through which the pollen tube enters; recognisable on mature seeds as a minute pore in the seed coat.

Midrib The central and usually most prominent vein of a leaf or leaf-like organ.

Monophyletic Of a group of organisms with a unique evolutionary origin – one that includes a single common ancestor and all of its descendants (**Fig. 169**).

Monotypic Containing only one taxon of a subordinate rank – for example, a family of one genus, or a genus of one species.

Morphological Pertaining to morphology.

Morphology The form and structure of an organism.

Mucro A sharp, usually abruptly constricted, apical point.

Mucronate Having a mucro (**Fig. 167**).

Mycorrhiza (pl. mycorrhizae) A symbiotic association between a fungus and a plant root.

N

Naturalised Originating elsewhere but now established and reproducing in a new area.

Nectar A usually sweet fluid secreted from a specialised gland or nectary. In *Hebe* and *Leonohebe*, nectar is secreted in flowers by a nectarial disc.

Neotype A specimen or illustration selected to serve as a nomenclatural type as long as all of the material on which the name of the taxon was based is missing.

Nodal joint In whipcord hebes, this is a distinct line or demarcation at the point where a leaf attaches to a stem, distinguishing tissue of the leaf from that of the internode below (**Fig. 24**). The nodal joint is not obvious in all whipcord species and, even when marked, may be hidden from view by overlapping leaves.

Node The position on a stem from which (one or more) leaves arise. In *Hebe* and *Leonohebe*, a pair of opposite leaves is produced at each node.

Nomenclatural synonyms In botanical nomenclature, synonyms based on the same type.

Nomen nudum A Latin term meaning a name published without an accompanying description or diagnosis (or without reference to a previously published description or diagnosis). Such names are not validly published, and have no status under the rules of the ICBN.

O

Ob- A prefix meaning the other way around, as in obovate.

Oblanceolate Two-dimensional in shape, three to six times as long as broad, broadest above the middle and tapering to the base (**Fig. 166**).

Oblate Two-dimensional in shape, almost circular, but slightly broader than long.

Oblong Two-dimensional in shape, sides being ± parallel and longer than wide (**Fig. 166**). Used here for length:width ratios between 3:2 and 3:1.

Obovate Two-dimensional in shape, with length one to three times breadth, and broadest above the middle (**Fig. 166**).

Obtuse Blunt or rounded at the apex; converging edges forming an angle greater than 90 degrees (**Fig. 167**). Cf. acute.

Opposite Describes leaves, bracts or flowers borne at the same level but on opposite sides of a stem or inflorescence. Cf. alternate.

Ovary The swollen, basal portion of the female part of a flower, enclosing the ovules (**Fig. 28**).

Ovate Two-dimensional in shape, with length one to three times breadth, and broadest below the middle (**Fig. 166**).

Ovule The structure that produces an egg cell, and which after fertilisation develops into a seed. In *Hebe* and *Leonohebe*, several to many ovules are produced in each locule of an ovary.

P

Papilla (pl. papillae; adj. papillate) A small, elongated protuberance on the surface of an organ, usually an extension of an epidermal cell.

Paraphyletic Of a group of organisms, consisting of a single ancestor and some, but not all, of its descendents (**Fig. 169**).

Parenchyma Relatively unspecialised tissue composed of cells with thin, non-lignified walls and living protoplasts.

Patent Spreading ± at right angles to the axis – for example, relating to leaves (**Fig. 168**) or corolla lobes.

Pedicel The stalk of a flower (**Fig. 25**).

Pedicellate Borne on a pedicel.

Peduncle The stalk of an inflorescence; that is, the portion of an inflorescence below the lowermost flowers or bracts.

Peltate With the stalk or point of attachment on the lower surface, away from the margin; ± shield-like or umbrella-like.

Penalpine The altitudinal vegetation belt between the subalpine and alpine belts, characterised in New Zealand by *Chionochloa* tussock grasses, robust species of *Astelia*, *Celmisia*, *Aciphylla* and *Ranunculus*, and shrubs < 1 m tall. This term was coined by Wardle (e.g. 1991).

Pendent Drooping, hanging downwards, as of a shrub (**Fig. 165**).

Petiole The stalk of a leaf.

Phyllotaxis The arrangement of leaves or other organs on a stem or axis.

Phylogenetic Pertaining to phylogeny.

Phylogeny The historical, evolutionary relationships of organisms.

Pinnatifid Of simple leaves and leaflets, with the lamina deeply cut into lobes on both sides of the midrib.

Placenta (pl. placentae) Tissue within the ovary to which the ovules are attached.

Plesiomorphic A relatively primitive or ancestral character state.

Plicate Longitudinally folded.

Pollen A collective term for the material, usually composed of many individual pollen grains, that is shed from anthers of flowering plants (and microsporangia of gymnosperms) and contains the male generative cells.

Polyphyletic Of a group of organisms having multiple evolutionary origins, and not including a unique common ancestor (**Fig. 169**). (Note that some authors suggest a more limited definition of this term, restricting it only to groups of hybrid origin, because distinction from the term paraphyletic can be somewhat arbitrary, depending only on the inclusion/exclusion of a common ancestor.)

Polyploid Having three or more sets of chromosomes in the nucleus of each somatic cell (cells other than gametes and their precursors).

Polyploidy The condition of being polyploid.

Posterior On the side toward the axis. Cf. anterior.

Prostrate Growing flat along the ground (**Fig. 165**).

Protologue Content of the publication establishing a new botanical name (Gk *protos* = first; *logos* =

discourse), including any text and illustrations pertinent to the new name.

Pseudodichotomous Apparently dichotomous with an undeveloped terminal bud and two equal lateral branches.

Puberulent Minutely pubescent; stubbly (**Fig. 18**).

Pubescent Covered with short, soft, erect hairs.

R

Raceme An indeterminate inflorescence (i.e. without a flower bud at the very apex) with pedicellate flowers.

Rachis The central axis of an inflorescence – the portion from the apex to the lowermost flowers or bracts.

Receptacle In flowering plants, the often ± expanded top of the stalk on which a flower or flower head arises.

Recurved Bent or curved backwards or downwards – for example, relating to leaves (**Fig. 168**) or corolla lobes.

Reflexed Bent sharply backwards or downwards, as of leaves (**Fig. 168**).

Retuse With an obtuse and slightly notched apex (**Fig. 167**).

Reversion leaves Leaves of juvenile form borne on an adult shoot. In *Hebe* these are most commonly seen in some whipcord species, particularly in cultivation, but also on wild-collected specimens.

Revolute Rolled under (downwards or backwards).

Rhomboid A four-sided figure whose opposite sides are parallel, but whose adjacent sides are of unequal length (i.e. like an oblique rectangle; **Fig. 166**).

Rupestral Growing on rocks.

S

Scarious Dry and more or less membranous.

Secondary chemistry The chemistry of secondary metabolites, which are those not essential for the normal growth, development or reproduction of a plant (not involved in respiration, photosynthesis, production of proteins and so on). Secondary metabolites include alkaloids, terpenoids, phenolics (which include flavonoids), irridoids, steroids, volatile oils and saponins.

Section A category of plant classification that is a subdivision of a genus, with a rank between subgenus and series.

Seed The reproductive body formed from a fertilised ovule.

Semiwhipcord A member of *Leonohebe* sect. *Leonohebe*; low-growing plants with scale-like appressed leaves, differing from "true" whipcords in having lateral inflorescences and angustiseptate capsules.

Septicidal Of a capsule, dehiscing by splitting along or through the septum. Cf. loculicidal.

Septum A dividing wall. Used here for the wall(s) that divide an ovary into two or more chambers (called locules).

Series A category of plant classification that is a subdivision of a genus, lower in rank than subgenus and section.

Serrate Toothed with forward-pointing asymmetrical teeth; like the cutting edge of a saw.

Shrub A woody perennial plant, smaller than a tree. In descriptions in this book, a "spreading low shrub" is generally less than 50 cm tall and usually broader than it is tall; and a "bushy shrub" is *c.* 20 cm to 3 m tall, usually taller than it is broad (if not, then it is more than 50 cm tall) and often with more than one main stem.

Simplex Simple; of a single series or kind.

Sinus The gap or recess between two lobes or segments; in *Hebe* leaf buds, it refers to the gap between the bases of the two leaves of a pair (**Fig. 21**)

Sister (or sister groups) In cladistic terminology, two taxa or groups of taxa that are most closely related. For example, in **Fig. 169**, taxa G and H are sisters, and taxon F is sister to the group G+H.

Spathulate Spoon-shaped; broad at the apex, and abruptly narrowed toward the base (**Fig. 166**).

Species (pl. species) A category of plant classification, between the rank of series and subspecies. In practice, it is the fundamental unit of classification, commonly being the lowest rank used to classify groups of plants. Species names comprise two words: a genus name (e.g. *Hebe*) and a specific epithet (e.g. *speciosa*). Different botanists hold different opinions on the criteria that should be used to define species. In practice, a species usually comprises a group of individuals or populations that share common features and/or ancestry, and is the smallest group that can be readily or consistently recognised (usually on the basis of morphological characters).

Spike An unbranched, indeterminate inflorescence (i.e. without a flower bud at the very apex) in which the flowers are not pedicellate.

Spreading Extending horizontally.

Stamen The male reproductive structure of a flower, composed of (**Fig. 25**) a filament (or stalk) and an anther (which produces pollen). *Hebe* and *Leonohebe* flowers each have two stamens.

Staminode A sterile stamen – in other words, one not producing pollen.

Sterile Of a herbarium specimen, lacking flowers or fruits.

Stigma The apical portion of the female part of flower, receptive to pollen (**Fig. 25**).

Stoma (pl. stomata) A pore on the surface of a leaf or other aerial parts that allows exchange of gases between plant tissue and the surrounding atmosphere.

Style The elongated middle portion of the female part of flower, between the ovary and stigma (**Fig. 25**).

Sub- A prefix meaning somewhat or almost.

Subacute Almost acute.

Subdistichous Somewhat or almost distichous. The leaves of some small-leaved "Apertae" (e.g. **Fig. 119B**) appear to be almost distichous because of the way their petioles are twisted.

Subgenus (pl. subgenera) A category of plant classification that is a subdivision of a genus, higher in rank than section and series.

Subshrub A small plant with partly herbaceous stems. The term is used here only for semiwhipcord members of *Leonohebe* and some "Connatae".

Subspecies A category of plant classification that is a subdivision of a species. It is higher in rank than variety or forma.

Subtending A term describing a leaf or bract whose axil gives rise to a branch, inflorescence, inflorescence branch or flower.

Synonym (adj. synonymous) One of two or more names for the same taxon.

Syntype Any specimen cited in the protologue when no holotype was designated, or any one of two or more specimens simultaneously designated as types.

T

Tannin A phenolic compound capable of precipitating proteins. Tannins are a diverse group of compounds and there are differing opinions on the precise definition of the word. Traditionally, a tannin is any substance capable of precipitating the gelatine of animal hides to an insoluble compound – in other words, changing hide into leather (the process of tanning).

Taxon (pl. taxa) A taxonomic group, of any rank (e.g. family, genus, species).

Taxonomic synonyms In botanical nomenclature, synonyms based on different types.

Tectum A layer of the outer part of the exine (the sexine) that forms a roof over underlying layers (columellae, granules and so on).

Terete Circular in cross section; ± cylindrical.

Terminal At the end of a stem or axis, as of an inflorescence at the apex of a shoot (Fig. 40).

Testa Seed coat.

Tetragonous Having four angles in cross section.

Tetraploid Having four sets of chromosomes in the nucleus of each somatic cell (cells other than gametes and their precursors).

Toothed Having teeth. Used here in a very broad sense to describe leaves with notches on the margins.

Tree Woody plant, taller (as used in this book) than 3 m, and generally with a single main stem.

Tripartite With three parts. Used here to describe inflorescences with a main axis and two lateral branches.

Truncate Flattened abruptly as though cut off (**Fig. 167**).

Tumid Swollen, protruding.

Turgid Swollen, firm; used here for capsules that are plump, and not obviously flattened (**Fig. 29**).

Type In botanical nomenclature, the single element (specimen, illustration and so on) to which a name is permanently attached. For higher plants, the type of a taxon at species rank or below is a specimen (usually) or an illustration, while the type of a genus is that of a designated species, and the type of a family is the same as that of the generic name on which it is based. The type method allows botanists to determine the correct application of names by reference to permanently preserved specimens or illustrations. Different sorts of types (e.g. holotype, lectotype, neotype and so on) are recognised, depending on the content of the original description of a taxon and on how the type was designated.

U

Uniform Evenly distributed; used here to describe the arrangement of hairs on the stem (**Fig. 18**). Cf. bifarious.

Uniseriate Composed of a single row of cells.

V

Vacuole A fluid-filled cavity, bounded by a membrane, in the cytoplasm of a cell.

Variety (*varietas* in Latin) A category of plant classification that is a subdivision of a species. It is the rank between subspecies and forma.

Vernicose Shining, as though varnished.

Voucher A specimen preserved (with documentation) to substantiate recorded observations. In this book, we list voucher specimens from plants shown in photographs or used to obtain chromosome counts; subsequent researchers can study these specimens to verify their identities.

W

Whipcord A shrub in which the leaves are scale-like, close-set and pressed against the stem (so that branches resemble plaited rawhide whips); a member of *Hebe* "Flagriformes".

Whorl A ring of organs (e.g. leaves, bracts or floral parts) borne at the same level on an axis.

References

Adams, C. J. D. (1979). Age and origin of the Southern Alps. In: Walcott, R. I. and Creswell, M. M., eds., *The Origin of the Southern Alps*. Royal Society of New Zealand Bulletin 18: 73–8. Wellington: Royal Society of New Zealand.

Adams, N. M. (2002). John Buchanan F.L.S. botanist and artist (1819–1898). *Tuhinga, Records of the Museum of New Zealand Te Papa Tongarewa* 13: 71–115.

Adamson, R. S. (1912). On the comparative anatomy of the leaves of certain species of *Veronica*. *Journal of the Linnean Society, Botany* 40: 247–74.

Aiton, W. (1789). *Hortus Kewensis*. London: George Nicol.

Albach, D. C. and Chase, M. W. (2001). Paraphyly of *Veronica* (Veroniceae; Scrophulariaceae): evidence from the internal transcribed spacer (ITS) sequences of nuclear ribosomal DNA. *Journal of Plant Research* 114: 9–18.

Albach, D. C. and Chase, M. W. (2004). Incongruence in Veroniceae (Plantaginaceae): evidence from two plastid and a nuclear ribosomal DNA region. *Molecular Phylogenetics and Evolution* 32: 183–97.

Albach, D. C., Martínez-Ortega, M. M., Fischer, M. A. and Chase, M. W. (2004*a*). A new classification of the tribe Veroniceae – problems and a possible solution. *Taxon* 53: 429–52.

Albach, D. C., Martínez-Ortega, M. M., Fischer, M. A. and Chase, M. W. (2004*b*). Evolution of Veroniceae: a phylogenetic perspective. *Annals of the Missouri Botanical Garden* 91: 275–302.

Albach, D. C., Martínez-Ortega, M. M. and Chase, M. W. (2004*c*). *Veronica*: parallel evolution and phylogeography in the Mediterranean. *Plant Systematics and Evolution* 246: 177–94.

Allan, H. H. (1939). Notes on New Zealand floristic botany, including descriptions of new species etc. – no. 7. *Transactions of the Royal Society of New Zealand* 69: 270–81.

Allan, H. H. (1961). *Flora of New Zealand. Vol. 1*. Wellington: Government Printer. [Note: L. B. Moore prepared the section on non-whipcord *Hebe* species, M. B. Ashwin prepared the section on whipcord *Hebe* species.]

Alston, R. E. and Turner, B. L. (1962). New techniques in the analysis of complex natural hybridisation. *Proceedings of the National Academy of Science (US)* 48: 130–37.

Alston, R. E. and Turner, B. L. (1963). Natural hybridisation among four species of *Baptisia* (Leguminosae). *American Journal of Botany* 50: 159–73.

Alston, R. E., Rosler, H., Naifeh, K. and Mabry, T. J. (1965). Hybrid compounds in natural interspecific hybrids. *Proceedings of the National Academy of Science* 54: 1458–65.

Andersen, J. C. (1926). Popular names of New Zealand plants. *Transactions and Proceedings of the New Zealand Institute* 56: 659–714.

APG (1998). An ordinal classification for the families of flowering plants. *Annals of the Missouri Botanical Garden* 85: 531–53.

APG (2003). An update of the Angiosperm Phylogeny Group classification for the orders and families of flowering plants: APG II. *Botanical Journal of the Linnean Society* 141: 399–436.

Armstrong, J. B. (1879). Descriptions of some new native plants. *New Zealand Country Journal* 3: 56–7.

Armstrong, J. B. (1881). A synopsis of the New Zealand species of *Veronica*, Linn., with notes on new species. *Transactions and Proceedings of the New Zealand Institute* 13: 344–59.

Armstrong, J. F. (1872). On some new species of New Zealand plants. *Transactions and Proceedings of the New Zealand Institute* 4: 290–1.

Armstrong, T. T. J. and de Lange, P. J. (2005). Conservation genetics of *Hebe speciosa* (Plantaginaceae) an endangered New Zealand shrub. *Botanical Journal of the Linnean Society*, 149: 229–239.

Bayly, M. J. and Kellow, A. V. (2004a). Proposal to reject the name *Veronica decussata* (Plantaginaceae). *Taxon* 53: 571–2.

Bayly, M. J. and Kellow, A. V. (2004b). Lectotypification of names of New Zealand members of *Veronica* and *Hebe* (Plantaginaceae). *Tuhinga, Records of the Museum of New Zealand Te Papa Tongarewa* 15: 43–52.

Bayly, M. J., Garnock-Jones, P. J., Mitchell, K. A., Markham, K. R. and Brownsey, P. J. (2000). A taxonomic revision of the *Hebe parviflora* complex (Scrophulariaceae), based on morphology and flavonoid chemistry. *New Zealand Journal of Botany* 38: 165–90.

Bayly, M. J., Garnock-Jones, P. J., Mitchell, K. A., Markham, K. R. and Brownsey, P. J. (2001). Description of *Hebe calcicola* (Scrophulariaceae), a new species from north-west Nelson, New Zealand, including details of flavonoid chemistry. *New Zealand Journal of Botany* 39: 55–67.

Bayly, M. J., Kellow, A. V., Mitchell, K., Markham, K. R., de Lange, P. J., Harper, G. E., Garnock-Jones, P. J. and Brownsey, P. J. (2002). Descriptions and flavonoid chemistry of new taxa in *Hebe* sect. *Subdistichae* (Scrophulariaceae). *New Zealand Journal of Botany* 40: 571–602.

Bayly, M. J., de Lange, P. J., Mitchell, K., Markham, K. R., Garnock-Jones, P. J., Kellow, A. V. and Brownsey, P. J. (2003). Geographic variation in morphology and flavonoid chemistry in *Hebe pubescens* and *H. bollonsii* (Scrophulariaceae), including a new infraspecific classification for *H. pubescens*. *New Zealand Journal of Botany* 41: 23–53.

Bayly, M. J., Kellow, A. V., Ansell, R., Mitchell, K. and Markham, K. R. (2004). Geographic variation in *Hebe macrantha* (Plantaginaceae): morphology and flavonoid chemistry. *Tuhinga, Records of the Museum of New Zealand Te Papa Tongarewa* 15: 27–41.

Beever, J. (1991). *A Dictionary of Maori Plant Names*. 2nd edn. Auckland: Auckland Botanical Society.

Bentham, G. (1846). Scrophulariaceae. In: de Candolle, A., ed., *Prodromus Systematis Naturalis Regni Vegetabilis. Vol. 10.* Paris: Masson. pp. 180–586.

Beuzenberg, E. J. and Hair, J. B. (1983). Contributions to a chromosome atlas of the New Zealand flora – 25. Miscellaneous families. *New Zealand Journal of Botany* 21: 13–20.

Blackman, C. J., Jordan, G. J. and Wiltshire, R. J. E. (2005). Leaf gigantism in coastal areas: morphological and physiological variation in four species on the Tasman Peninsula, Tasmania. *Australian Journal of Botany* 53: 91–100.

Bohm, B. A. (1998). *Introduction to Flavonoids*. Amsterdam: Harwood Academic Publishers.

Brandon, A. (1995). Species limits in the *Hebe subalpina* complex (Scrophulariaceae). Unpublished BSc. (Hons) thesis, Victoria University of Wellington, Wellington, New Zealand. [Copy held in the library of the Museum of New Zealand Te Papa Tongarewa, Wellington.]

Brandon, A., de Lange, P. and Townsend, A. (2004). *Threatened Plants of Waikato Conservancy*. Wellington: Department of Conservation.

Breitwieser, I. (1993). Comparative leaf anatomy of New Zealand and Tasmanian Inuleae (Compositae). *Botanical Journal of the Linnean Society* 111: 183–209.

Breitwieser, I. and Ward, J. M. (1998). Leaf anatomy of *Raoulia* Hook.f. (Compositae, Gnaphalieae). *Botanical Journal of the Linnean Society* 126: 217–35.

Briggs, B. G. and Ehrendorfer, F. (1976). *Chionohebe*, a new name for *Pygmea* Hook.f. (Scrophulariaceae). *Contributions from Herbarium Australiense* 25: 1–4.

Briggs, B. G. and Ehrendorfer, F. (1992). A revision of the Australian species of *Parahebe* and *Derwentia* (Scrophulariaceae). *Telopea* 5: 241–87.

Briggs, B. G. and Makinson, R. O. (1992). *Parahebe*. In: Harden, G. J., ed., *Flora of New South Wales. Vol. 3.* Kensington: New South Wales University Press. p. 574.

Briggs, B. G., Wiecek, B. and Whalen, A. J. (1992). *Veronica*. In: Harden, G. J., ed., *Flora of New South Wales. Vol. 3.* Kensington: New South Wales University Press. pp. 578–82.

Brown, F. (1935). Flora of Southeastern Polynesia III, Dicotyledons. *Bernice P. Bishop Museum Bulletin* 130: 1–386.

Brownsey, P. J. (2001). New Zealand's pteridophyte flora – plants of ancient lineage but recent arrival? *Brittonia* 53: 284–303.

Brummitt, R. K. (2002). How to chop up a tree. *Taxon* 51: 31–41.

Brummitt, R. K. and Powell, C. E. (1992). *Authors of Plant Names.* Kew: Royal Botanic Gardens.

Buchanan, J. (1882). On the alpine flora of New Zealand. *Transactions and Proceedings of the New Zealand Institute* 14: 342–56.

Burrows, C. J. (1965). Some discontinuous distributions of plants within New Zealand and their ecological significance. Part II: disjunctions between Otago-Southland and Nelson-Marlborough and related distribution patterns. *Tuatara* 13: 9–29.

Cameron, E. K., de Lange, P. J., Given, D. R., Johnson, P. N. and Ogle, C. C. (1993). New Zealand Botanical Society threatened and local plant lists (1993 revision). *New Zealand Botanical Society Newsletter* 32: 14–28.

Carlquist, S. (1988). *Comparative Wood Anatomy*. New York: Springer-Verlag.

Carse, H. (1929). Botanical notes and new varieties. *Transactions and Proceedings of the New Zealand Institute* 60: 305–7.

Chalk, D. (1988). *Hebes and Parahebes*. London: Christopher Helm.

Cheeseman, T. F. (1906). *Manual of the New Zealand Flora*. 1st edn. Wellington: Government Printer.

Cheeseman, T. F. (1925). *Manual of the New Zealand Flora*. 2nd edn. Wellington: Government Printer.

Clarkson, B. R. and Boase M. R. (1982). *Scenic Reserves of West Taranaki*. Biological Survey of Reserves Series No. 10. Wellington: Department of Lands and Survey.

Clarkson, B. R. and Garnock-Jones, P. J. (1996). *Hebe tairawhiti* (Scrophulariaceae): a new shrub species from New Zealand. *New Zealand Journal of Botany* 34: 51–6.

Clarkson, B., Merrett, M. and Downs, T. (2002). *Botany of the Waikato*. Hamilton: Waikato Botanical Society Inc.

Cockayne, L. (1898). An inquiry into the seedling forms of New Zealand phanerogams and their development. *Transactions and Proceedings of the New Zealand Institute* 31: 354–98.

Cockayne, L. (1902). A short account of the plant-covering of Chatham Island. *Transactions and Proceedings of the New Zealand Institute* 34: 243–325.

Cockayne, L. (1909). *Report on a Botanical Survey of Stewart Island*. Wellington: Government Printer.

Cockayne, L. (1916). Notes on New Zealand floristic botany, including descriptions of new species, &c. (No. 1). *Transactions and Proceedings of the New Zealand Institute* 48: 193–202.

Cockayne, L. (1918). Notes on New Zealand floristic botany, including descriptions of new species, &c. (No. 3). *Transactions and Proceedings of the New Zealand Institute* 50: 161–91.

Cockayne, L. (1919). *New Zealand Plants and Their Story*. 2nd edn. Wellington: Government Printer.

Cockayne, L. (1924). *The Cultivation of New Zealand Plants*. Wellington: Whitcombe & Tombs.

Cockayne, L. (1926). *Monograph on the New Zealand Beech Forests. Part 1. The ecology of the forests and the taxonomy of the beeches*. New Zealand Forest Service Bulletin 4. Wellington: New Zealand Forest Service.

Cockayne, L. (1929). New combinations in the genus *Hebe*. *Transactions and Proceedings of the New Zealand Institute* 60: 465–72.

Cockayne, L. and Allan H. H. (1926*a*). A proposed new Botanical District for the New Zealand Region. *Transactions and Proceedings of the New Zealand Institute* 56: 19–20.

Cockayne, L. and Allan H. H. (1926*b*). Notes on New Zealand floristic botany, including descriptions of new species, &c. (No. 4). *Transactions and Proceedings of the New Zealand Institute* 56: 21–33.

Cockayne, L. and Allan H. H. (1926*c*). The present taxonomic status of the New Zealand species of *Hebe*. *Transactions and Proceedings of the New Zealand Institute* 57: 1–47.

Cockayne, L. and Allan, H. H. (1934). An annotated list of groups of wild hybrids in the New Zealand flora. *Annals of Botany* 48: 1–55.

Colenso, W. (1883). Descriptions of a few new indigenous plants. *Transactions and Proceedings of the New Zealand Institute* 15: 320–39.

Connor, H. E. and Edgar, E. (1987). Name changes in the indigenous New Zealand flora, 1960–1980 and nomina nova IV, 1983–1986. *New Zealand Journal of Botany* 25: 115–70.

Crisp, P., Miskelly, C. and Sawyer, J. (2000). *Endemic Plants of the Chatham Islands*. Wellington: Department of Conservation.

Dallwitz, M. J., Paine, T. A. and Zurcher, E. J. (1993). *User's Guide to the DELTA System: a general system for processing taxonomic descriptions*. 4th edn. Canberra: CSIRO Division of Entomology.

Darlington, C. D. and Wylie, A. P. (1955). *Chromosome Atlas of Flowering Plants*. London: George Allen & Unwin.

Dawson, M. I. (2000). Index of chromosome numbers of indigenous New Zealand spermatophytes. *New Zealand Journal of Botany* 38: 47–150.

Dawson, M. I. and Beuzenberg, E. J. (2000). Contributions to a chromosome atlas of the New Zealand flora – 36. Miscellaneous families. *New Zealand Journal of Botany* 38: 1–23.

de Lange, P. J. (1986). Studies into an allopatric race of *Hebe rigidula* (Cheeseman) Cockayne et Allan. Unpublished undergraduate special topics dissertation in botany, University of Waikato Herbarium. [Copy held in the library of the Museum of New Zealand Te Papa Tongarewa, Wellington.]

de Lange, P. J. (1991). *Hebe* 'Unuwhao': notes on its distribution, ecology and conservation status. *New Zealand Botanical Society Newsletter* 24: 7–8.

de Lange, P. J. (1996). *Hebe bishopiana* (Scrophulariaceae) – an endemic species of the Waitakere Ranges, west Auckland, New Zealand. *New Zealand Journal of Botany* 34: 187–94.

de Lange, P. J. (1997). *Hebe brevifolia* (Scrophulariaceae) – an ultramafic endemic of the Surville Cliffs, North Cape, New Zealand. *New Zealand Journal of Botany* 35: 1–8.

de Lange, P. J. (1998). *Hebe perbella* (Scrophulariaceae) – a new and threatened species from western Northland, North Island, New Zealand. *New Zealand Journal of Botany* 36: 399–406.

de Lange, P. J. (1999). *Waitakere Rock Koromiko (*Hebe bishopiana*) Recovery Plan*. Waitakere City Council Miscellaneous Publications. Waitakere City Council.

de Lange, P. J. and Cameron E. K. (1992). Conservation status of titirangi (*Hebe speciosa*). *New Zealand Botanical Society Newsletter* 29: 11–15.

de Lange, P. J. and Murray, B. G. (2002). Contributions to a chromosome atlas of the New Zealand flora – 37. Miscellaneous families. *New Zealand Journal of Botany* 40: 1–23.

de Lange, P. J. and Norton, D. A. (1998). Revisiting rarity: a botanical perspective on the meanings of rarity and the classification of New Zealand's uncommon plants. In: *Ecosystems, Entomology and Plants*. Royal Society of New Zealand Miscellaneous Series 48. Wellington: Royal Society of New Zealand. pp. 145–60.

de Lange P. J. and Stanley, R. (1999). Kermadec koromiko (*Hebe breviracemosa*) comes back from the brink of extinction. *New Zealand Botanical Society Newsletter* 55: 9–12.

de Lange, P. J., Heenan, P. B., Given, D. R., Norton, D. A., Ogle, C. C., Johnson, P. N. and Cameron, E. K. (1999). Threatened and uncommon plants of New Zealand. *New Zealand Journal of Botany* 37: 603–28.

de Lange, P. J., Norton, D. A., Heenan, P. B., Courtney, S. P., Molloy, B. P. J., Ogle, C. C., Rance, B. D., Johnson, P. N. and Hitchmough, R. (2004*a*). Threatened and uncommon plants of New Zealand. *New Zealand Journal of Botany* 42: 45–76.

de Lange, P. J., Norton, D. A., Heenan, P. B., Courtney, S. P., Molloy, B. P. J., Ogle, C. C., Rance, B. D., Johnson, P. N. and Hitchmough, R. (2004*b*). Errata to threatened and uncommon plants of New Zealand. *New Zealand Journal of Botany* 42: 715.

de Lange, P. J., Murray, B. G. and Datson, P. M. (2004*c*). Contributions to a chromosome atlas of the New Zealand flora – 38. Counts for 50 families. *New Zealand Journal of Botany* 42: 873–904.

Debenham, C. [1971]. *The Language of Botany*. Chipping Norton, NSW: Society for Growing Australian Plants.

Delph, L. F. (1988). The evolution and maintenance of gender dimorphism in New Zealand *Hebe* (Scrophulariaceae). Unpublished Ph.D. thesis, University of Canterbury, Christchurch, New Zealand.

Delph, L. F. (1990*a*). The evolution of gender dimorphism in New Zealand *Hebe* (Scrophulariaceae) species. *Evolutionary Trends in Plants* 4: 85–97.

Delph, L. F. (1990*b*). Sex ratio variation in the gynodioecious shrub *Hebe strictissima* (Scrophulariaceae). *Evolution* 44: 134–42.

Delph, L. F. (1990*c*). Sex-differential resource allocation patterns in the subdioecious shrub *Hebe subalpina*. *Ecology* 71: 1342–51.

Delph, L. F. and Lively, C. M. (1992). Pollinator visitation, floral display, and nectar production of sexual morphs of a gynodioecious shrub. *Oikos* 63: 161–70.

Delph, L. F. and Lloyd, D. G. (1991). Environmental and genetic control of gender in the dimorphic shrub *Hebe subalpina*. *Evolution* 45: 1957–64.

Delph, L. F. and Lloyd, D. G. (1996). Inbreeding depression in the gynodioecious shrub *Hebe subalpina* (Scrophulariaceae). *New Zealand Journal of Botany* 34: 241–7.

Diels, F. L. E. (1929). Beiträge zur Flora des Saruwaged-Gebirges. *Botanische Jahrbücher für Systematik, Pflanzengeshichte und Pflanzengeographie* 62: 452–501.

Druce, A. P. (1968). Vascular plants of Mt Holdsworth, Tararua Ra. (including Pig Flat) 2500–4835 ft. Unpublished checklist held at Landcare Research, Lincoln, New Zealand.

Druce, A. P. (1980). Trees, shrubs, and lianes of New Zealand (including wild hybrids). Unpublished checklist held at Landcare Research, Lincoln, New Zealand. [Copy also held in the library of the Museum of New Zealand Te Papa Tongarewa, Wellington.]

Druce, A. P. (1993). Indigenous vascular plants of New Zealand. 9th revision. Unpublished checklist held at Landcare Research, Lincoln, New Zealand. [Copy also held in the library of the Museum of New Zealand Te Papa Tongarewa, Wellington.]

Druce, A. P. and Courtney, S. (1989). *Hebe matthewsii* rediscovered. *Wellington Botanical Society Bulletin* 45: 83–5.

Dugdale, J. S. (1975). The insects in relation to plants. In: Kuschel, G., ed., *Biogeography and Ecology in New Zealand*. The Hague: Junk. pp. 561–89.

Eagle, A. (1982). *Eagle's Trees and Shrubs of New Zealand*. 2nd series. Auckland: Collins.

Elder, N. L. (1939). The glaucous *Hebe* of the Inland Patea. *Veronica colensoi*, *V. hillii* and *V. darwiniana*. *Transactions of the Royal Society of New Zealand* 69: 373–7.

Elder, N. L. (1971). The glaucous hebe of the Inland Patea: a footnote. *Wellington Botanical Society Bulletin* 37: 64.

Esau, K., (1977). *Anatomy of Seed Plants*. 2nd edn. New York: John Wiley.

Fleming, C. A. (1979). *The Geological History of New Zealand and its Life*. Auckland: Auckland University Press.

Forester, L. and Townsend, A. (2004). *Threatened Plants of Northland Conservancy*. Wellington: Department of Conservation.

Forster, G. (1786). *Florulae Insularum Australium Prodromus*. Göttingen: Dieterich.

Frankel, O. H. (1940). Studies in *Hebe* II. The significance of male sterility in the genetic system. *Journal of Genetics* 40: 171–84.

Frankel, O. H. (1941). Cytology and taxonomy of *Hebe*, *Veronica* and *Pygmea*. *Nature* 147: 117.

Frankel, O. H. and Hair, J. B. (1937). Studies on the cytology, genetics and taxonomy of New Zealand *Hebe* and *Veronica* (part 1). *New Zealand Journal of Science and Technology* 18: 669–87.

Fuller, S. A. (1985). *Kapiti Island Vegetation. Report on a vegetation survey of Kapiti Island 1984/85*. Wellington: Department of Lands and Survey.

Ganley, E. and Collins, L. (1999). A new population of *Hebe speciosa* (titirangi) on the Waikato coast. *New Zealand Botanical Society Newsletter* 57: 16–18.

Garnock-Jones, P. J. (1976a). *Hebe rapensis* (F. Brown) Garnock-Jones comb. nov. and its relationships. *New Zealand Journal of Botany* 14: 79–83.

Garnock-Jones, P. J. (1976b). Breeding systems and pollination in New Zealand *Parahebe* (Scrophulariaceae). *New Zealand Journal of Botany* 14: 291–8.

Garnock-Jones, P. J. (1992). Scented hebes. *Hebe News* 7: 7–8.

Garnock-Jones, P. J. (1993a). Phylogeny of the *Hebe* complex (Scrophulariaceae: Veroniceae). *Australian Systematic Botany* 6: 457–79.

Garnock-Jones, P. J. (1993b). *Heliohebe* (Scrophulariaceae – Veroniceae), a new genus segregated from *Hebe*. *New Zealand Journal of Botany* 31: 323–39.

Garnock-Jones, P. J. and Clarkson, B. D. (1994). *Hebe adamsii* and *H. murrellii* (Scrophulariaceae) reinstated. *New Zealand Journal of Botany* 32: 11–15.

Garnock-Jones, P. J. and Lloyd, D. G. (2004). A taxonomic revision of *Parahebe* (Plantaginaceae) in New Zealand. *New Zealand Journal of Botany* 42: 181–232.

Garnock-Jones P. J. and Molloy, B. P. J. (1983a). Polymorphism and the taxonomic status of the *Hebe amplexicaulis* complex (Scrophulariaceae). *New Zealand Journal of Botany* 20: 391–9. [Note: although the cover date on this publication is 1982, it was not published until 11 January 1983, as recorded in the subsequent volume.]

Garnock-Jones P. J. and Molloy, B. P. J. (1983b). Protandry and inbreeding depression in *Hebe amplexicaulis* (Scrophulariaceae). *New Zealand Journal of Botany* 20: 401–4. [Note: although the cover date on this publication is 1982, it was not published until 11 January 1983, as recorded in the subsequent volume.]

Garnock-Jones, P. J., Bayly, M. J., Lee, W. G. and Rance, B. D. (2000). *Hebe arganthera* (Scrophulariaceae), a new species from calcareous outcrops in Fiordland, New Zealand. *New Zealand Journal of Botany* 38: 379–88.

Given, D. R. (1981). *Rare and Endangered Plants of New Zealand*. Wellington: A. H. & A. W. Reed.

Gmelin, J. F. (1791). *Systema Naturae. Tomus II. Pars I*. Leipzig: Georg Emanuel Beer.

Godley, E. J. (1967). Widely distributed species, land bridges and continental drift. *Nature* 214: 74–5.

Grayer-Barkmeijer, R. J. (1979). Chemosystematic investigations in *Veronica* L. (Scrophulariaceae) and related genera. Unpublished Ph.D. thesis, Rijksuniversiteit of Leiden, Leiden, Netherlands.

Greuter, W., McNeill, J., Barrie, F. R., Burdet, H. M., Demoulin, V., Filgueiras, T. S., Nicholson, D. H., Silva, P. C., Skog, J. E., Trehane, P., Turland, N. J. and Hawksworth, D. L. (2000). *International Code of Botanical Nomenclature (Saint Louis Code)*. Regnum Vegetabile 138. Königstein: Koeltz Scientific Books (for International Association for Plant Taxonomy).

Grindley, G. W. (1961). *Sheet 13. Golden Bay. Geological Map of New Zealand 1:250,000*. 1st edn. Wellington: DSIR.

Hair, J. B. (1966). Biosystematics of the New Zealand flora, 1945–1964. *New Zealand Journal of Botany* 4: 559–95.

Hair, J. B. (1967). Contributions to a chromosome atlas of the New Zealand flora – 10 *Hebe* (Scrophulariaceae). *New Zealand Journal of Botany* 5: 322–52.

Hansen, B. and Wagner, P. (1998). A catalogue of the herbarium specimens from Captain Cook's first and second expeditions housed in the Copenhagen herbarium (C). *Allertonia* 7: 307–61.

Harborne, J. B. (1967). *Comparative Biochemistry of the Flavonoids*. London: Academic Press.

Harden, G. J. (2002). *Flora of New South Wales. Vol. 2*. Revised edn. Sydney: University of New South Wales.

Hatch, E. D. (1966). *Hebe* × *bishopiana*. *Auckland Botanical Society Newsletter* 23: 1.

Heads, M. J. (1987). New names in New Zealand Scrophulariaceae. *Otago Botanical Society Newsletter* 5: 4–11.

Heads, M. (1989). Integrating earth and life sciences in New Zealand natural history: the parallel arcs model. *New Zealand Journal of Zoology* 16: 549–85.

Heads, M. J. (1992). Taxonomic notes on the *Hebe* complex (Scrophulariaceae) in the New Zealand mountains. *Candollea* 47: 583–95.

Heads, M. J. (1993). Biogeography and biodiversity in *Hebe*, a South Pacific genus of Scrophulariaceae. *Candollea* 48: 19–60.

Heads, M. J. (1994a). Biogeography and evolution in the *Hebe* complex (Scrophulariaceae): *Leonohebe* and *Chionohebe*. *Candollea* 49: 81–119.

Heads, M. J. (1994b). A biogeographic review of *Parahebe* (Scrophulariaceae). *Botanical Journal of the Linnean Society* 115: 65–89.

Heads, M. J. (1994c). Morphology, architecture and taxonomy in the *Hebe* complex (Scrophulariaceae). *Bulletin du Muséum National d'Histoire Naturelle, Section B, Adansonia* 16: 163–91.

Heads, M. (1998). Biogeographic disjunction along the Alpine fault, New Zealand. *Biological Journal of the Linnean Society* 63: 161–76.

Heads, M. J. (2003). *Hebejeebie* (Plantaginaceae), a new genus from the South Island, New Zealand, and Mt Kosciusko, SE Australia. *Botanical Society of Otago Newsletter* 36: 10–12.

Heads, M. J. and Craw, R. (2004). The Alpine Fault biogeographic hypothesis revisited. *Cladistics* 20: 184–90.

Heenan, P. B. (1994a). The origin and identification of *Hebe* ×*franciscana* and its cultivars (Scrophulariaceae). *Horticulture in New Zealand* 5: 15–20.

Heenan, P. B. (1994b). The origin and identification of *Hebe* ×*andersonii* and its cultivars (Scrophulariaceae). *Horticulture in New Zealand* 5: 21–5.

Heenan, P. B. (2001). A history of *Hebe* as a garden plant. In: Metcalf, L. J., ed., *International Register of Hebe Cultivars*. Lincoln: Royal New Zealand Institute of Horticulture (Inc.). pp. 16–60.

Heine, E. M. (1937). Observations on the pollination of New Zealand flowering plants. *Transactions of the Royal Society of New Zealand* 67: 133–48.

Heller, W. and Forkmann, G. (1994). Biosynthesis of flavonoids. In: Harborne, J. B., ed., *The Flavonoids – advances in research since 1986*. London: Chapman and Hall. pp. 499–535.

Hennig, W. (1966). *Phylogenetic Systematics*. Urbana: University of Illinois Press.

Herrick, J. F. and Cameron, E. K. (1994). Annotated checklist of type specimens of New Zealand plants in the Auckland Institute and Museum herbarium (AK). Part 5. Dicotyledons. *Records of the Auckland Institute and Museum* 31: 89–173.

Holmgren, P. K., Holmgren, N. H. and Barnett L. C. (1990). *Index herbariorum. Part I: The herbaria of the world.* 8th edn. Regnum Vegetabile 120. Bronx: New York Botanical Gardens (for International Association for Plant Taxonomy).

Hombron, J. B. and Jacquinot, H. (1845). *Voyage au Pôle Sud et dans l'Océanie sur les Corvettes* l'Astrolabe *et la Zelée – Botanique.* Paris: Gide et Cie. Atlas, Plate 9.

Hong, D. (1984). Taxonomy and evolution of the Veroniceae (Scrophulariaceae) with special reference to palynology. *Opera Botanica* 75: 5–60.

Hooker, J. D. (1844). *The Botany of the Antarctic Voyage of H. M. Ships* Erebus *and* Terror, *in the Years 1839–1843. Flora Antarctica. Part 1. Botany of Lord Auckland's Group and Campbell's Island.* London: Reeve Brothers. [Part 1 comprises pages up to p. 208 in combined volumes. Although publication of this part was not complete until May 1845, the sections covering *Veronica* species were published by October 1844.]

Hooker, J. D. (1853). *The Botany of the Antarctic Voyage of H. M. Ships* Erebus *and* Terror, *in the Years 1839–1843. II. Flora Novae-Zelandiae. Part 1. Flowering plants.* London: Lovell Reeve.

Hooker, J. D. (1864). *Handbook of the New Zealand Flora: a systematic description of the native plants of New Zealand and the Chatham, Kermadec's, Lord Auckland's, Campbell's, and Macquarrie's Islands. Part 1.* London: Reeve and Co. [Part 1 comprises pages up to p. 392 in combined volumes.]

Huber, A. (1927). Beiträge zu Klärung verwandtschaftlicher Beziehungen in der Gattung *Veronica. Jahrbücher für Wissenschaftliche Botanik* 66: 359–80.

Hutchins, G. (1997). *Hebes Here and There.* Caversham: Hutchins & Davies.

IUCN (1994). *IUCN Red List Categories.* Gland, Switzerland: International Union for the Conservation of Nature, Species Survival Commission.

IUCN (2000). *IUCN Red List Categories.* Gland, Switzerland: International Union for the Conservation of Nature, Species Survival Commission.

Johnson, P. N. and Campbell, D. J. (1975). Vascular plants of the Auckland Islands. *New Zealand Journal of Botany* 13: 665–720.

Judd, W. S., Campbell, C. S., Kellogg, E. A. and Stevens, P. F. (1999). *Plant Systematics, a Phylogenetic Approach.* Sunderland: Sinauer Associates.

Judd, W. S., Campbell, C. S., Kellogg, E. A., Stevens, P. F. and Donoghue, M. J. (2002). *Plant Systematics, a Phylogenetic Approach.* 2nd edn. Sunderland: Sinauer Associates.

Jussieu, A. L. de (1789). *Genera Plantarum Secundum Ordines Naturales Disposita Juxta Methodum in Horte Regio Parisiensi Excaratum, Anno 1774.* Paris: Herissant et Barrios.

Kampny, C. M. (1995). Pollination and flower diversity in Scrophulariaceae. *Botanical Review* 61: 350–66.

Kampny, C. M. and Dengler, N. G. (1997). Evolution of flower shape in Veroniceae (Scrophulariaceae). *Plant Systematics and Evolution* 205:1–25.

Kampny, C. M., Dickinson, T. A. and Dengler, N. G. (1993). Quantitative comparison of floral development in *Veronica chamaedrys* and *Veronicastrum virginicum* (Scrophulariaceae). *American Journal of Botany* 80: 449–60.

Kear, D. and Schofield, J. C. (1959). The Te Kuiti Group. *New Zealand Journal of Geology and Geophysics* 2: 685–717.

Kellow, A. V., Bayly, M. J., Mitchell, K. A., Markham, K. R. and Garnock-Jones, P. J. (2003a). Variation in morphology and flavonoid chemistry in *Hebe pimeleoides* (Scrophulariaceae), including a revised subspecific classification. *New Zealand Journal of Botany* 41: 233–53.

Kellow, A. V., Bayly, M. J., Mitchell, K. A., Markham, K. R. and Brownsey, P. J. (2003b). A taxonomic revision of *Hebe* informal group "Connatae" (Plantaginaceae), based on morphology and flavonoid chemistry. *New Zealand Journal of Botany* 41: 613–35.

Kellow, A. V., Bayly, M. J., Mitchell, K. A. and Markham, K. R. (2005). Geographic variation in the *H. albicans* complex (Plantaginaceae) – morphology and flavonoid chemistry. *New Zealand Journal of Botany* 43: 141–63.

Kinman, K. E. (1994). *The Kinman System: toward a stable cladisto-eclectic classification of organisms: living and extinct, 48 phyla, 269 classes, 1,719 orders.* Hays, Kansas: K. E. Kinman.

Kirk, T. (1878). Descriptions of new plants. *Transactions and Proceedings of the New Zealand Institute* 11: 463–6.

Kirk, T. (1896). Notes on certain Veronicas, and descriptions of new species. *Transactions and Proceedings of the New Zealand Institute* 28: 515–31.

Kirk, T. (1899). *The Students' Flora of New Zealand and the Outlying Islands.* Wellington: Government Printer.

Lloyd, D. G. (1985). Progress in understanding the natural history of New Zealand plants. *New Zealand Journal of Botany* 23: 707–22.

Lloyd, D. G. and Webb, C. J. (1986). The avoidance of interference between the presentation of pollen and stigmas in angiosperms I. Dichogamy. *New Zealand Journal of Botany* 24: 135–62.

Lois, R. and Buchanan, B. B. (1994). Severe sensitivity to ultraviolet radiation in an *Arabidopsis* mutant deficient in flavonoid accumulation. *Planta* 194: 504–9.

Lovis, J. D. (1990). *Hebe pimeleoides* var. *rupestris* in Canterbury. *Canterbury Botanical Society Journal* 24: 42–4.

McClure, J. W. and Alston, R. E. (1964). Patterns of selected chemical compounds of *Spirodela oligorrhiza* found under various conditions of axenic culture. *Nature* 201: 311–13.

McCusker, A. (1981). Glossary. *Flora of Australia. Vol. 1.* Canberra: AGPS. pp. 169–93.

Macdonald, A. D. (1980). Distribution maps of *Hebe* in Canterbury as an extension of the Canterbury checklist. *Canterbury Botanical Society Journal* 14: 40–5.

McGlone, M. S. (1985). Plant biogeography and the late Cenozoic history of New Zealand. *New Zealand Journal of Botany* 23: 723–49.

McGlone, M. S., Duncan, R. P. and Heenan, P. B. (2001). Endemism, species selection and the origin and distribution of the vascular plant flora of New Zealand. *Journal of Biogeography* 28: 199–216.

Malcolm, W. M. and Garnock-Jones, P. J. (2000). Photographing lichens without a camera. *Australasian Lichenology* 47: 17–22.

Mark, A. F. (1970). Floral initiation and development in New Zealand alpine plants. *New Zealand Journal of Botany* 8: 67–75.

Markham, K. R. (1982). *Techniques of Flavonoid Identification.* London: Academic Press.

Markham, K. R. (1988). The distribution of flavonoids in the lower plants and its evolutionary significance. In: Harborne, J. B., ed., *The Flavonoids – advances in research since 1980.* London: Chapman and Hall. pp. 426–64.

Markham, K. R., Hammett, K. R. W. and Ofman, D. J. (1992). Floral pigmentation in two yellow-flowered *Lathyrus* species and their hybrid. *Phytochemistry* 31: 549–54.

Markham, K.R., Mitchell, K. A., Bayly, M. J., Kellow, A. V., Brownsey, P. J. and Garnock-Jones, P. J. (2005). Composition and taxonomic distribution of leaf flavonoids in *Hebe* and *Leonohebe* (Plantaginaceae) in New Zealand – 1. "Buxifoliatae", "Flagriformes" and *Leonohebe*. *New Zealand Journal of Botany* 43: 165–203.

Martin, W. (1932). *The Vegetation of Marlborough.* Blenheim: [n.p.]. Reprinted from the *Marlborough Express.*

Meacham, C. A. and Duncan, T. (1987). The necessity of convex groups in biological classification. *Systematic Botany* 12: 78–90.

Melville, R. (1966). Continental drift, Mesozoic continents and the migration of the angiosperms. *Nature* 211: 116–20.

Metcalf, L. J. (1972). *The Cultivation of New Zealand Trees and Shrubs.* Auckland: Reed Methuen.

Metcalf, L. J. (1987). *The Cultivation of New Zealand Trees and Shrubs.* Auckland: Reed Methuen.

Metcalf, L. J. (2001). *International Register of* Hebe *Cultivars.* Lincoln: Royal New Zealand Institute of Horticulture (Inc.).

Meurk, C. D., Partridge, T. R. and Molloy, B. P. J. (1987). Botanical resources of Ryton Station, Lake Coleridge – based on a rapid autumn survey. Unpublished report, Botany Division, DSIR. [Copies are held in the libraries of Landcare Research, Lincoln, and the Museum of New Zealand Te Papa Tongarewa, Wellington.]

Meylan, B. A. and Butterfield, B. G. (1978). *The Structure of New Zealand Woods.* New Zealand Department of Scientific and Industrial Research Bulletin 222. Wellington: DSIR.

Mildenhall, D. C. (1980). New Zealand Late Cretaceous and Cenozoic plant biogeography: a contribution. *Palaeogeography, Palaeoclimatology, Palaeoecology* 31: 197–233.

Mitchell, K. A., Markham, K. R. and Bayly M. J. (1999). 6-Hydroxyluteolin-7-O-ß-D-[2-O-ß-D-xylosylxyloside]: a novel flavone xyloxyloside from *Hebe stenophylla*. *Phytochemistry* 52: 1165–7

Mitchell, K. A., Markham, K. R. and Bayly, M. J. (2001). Flavonoid characters contributing to the taxonomic revision of the *Hebe parviflora* complex. *Phytochemistry* 56: 453–61.

Moar, N. T. (1993). *Pollen Grains of New Zealand Dicotyledonous Plants*. Lincoln: Manaaki Whenua Press.

Moench, C. (1785). Verzeichniss ausländischer Bäume und Stauden des Lustschlosses Weissenstein bey Cassel. Frankfurt and Leipzig, in der J. G. Fleischerischen Buchhandlung.

Molloy, J., Bell, B., Clout, M., de Lange, P., Gibbs, G., Given, D., Norton, D., Smith, N. and Stephens, T. (2002). *Classifying Species According to Threat of Extinction – a system for New Zealand*. Wellington: Department of Conservation.

Moore, L. B. (1967). How to look at *Hebe*. *Tuatara* 15: 10–15.

Moore, L. B. and Edgar, E. (1970). *Flora of New Zealand. Vol. 2*. Wellington: Government Printer.

Moore, L. B. and Irwin, J. B. (1978). *The Oxford Book of New Zealand Plants*. Wellington: Oxford University Press.

Mueller, F. (1864). *The Vegetation of the Chatham Islands*. Melbourne: John Ferres, Government Printer.

Muller, J. (1981). Fossil pollen records of extant angiosperms. *Botanical Review* 47: 1–142.

Muñoz Pizarro, C. (1960). *Las Especies de Plantas Descritas por R. A. Philippi en el Siglo XIX*. Santiago de Chile: Ediciones de la Universidad de Chile.

Murray, B. G. and de Lange, P. J. (1999). Contributions to a chromosome atlas of the New Zealand Flora. *New Zealand Journal of Botany* 37: 511–21.

Murray, B. G., Braggins, J. E. and Newman, P. D. (1989). Intraspecific polyploidy in *Hebe diosmifolia* (Cunn.) Cockayne et Allan (Scrophulariaceae). *New Zealand Journal of Botany* 27: 587–9.

Nelson, G. J. and Platnick, N. (1981). *Systematics and Biogeography: cladistics and vicariance*. New York: Columbia University Press.

Newman, P. D. (1988). Analysis of variation in *Hebe diosmifolia* (A.Cunn.) Ckn. et Allan (Scrophulariaceae). Unpublished M.Sc. thesis, University of Auckland, New Zealand.

Nicol, E. R. (1997). *Common Names of Plants in New Zealand*. Lincoln: Manaaki Whenua Press.

Nicolson, D. H. and Fosberg, F. R. (2004). *The Forsters and the Botany of the Second Cook Expedition (1772–1775)*. Regnum Vegetabile 139. Ruggell, Liechtenstein: A. R. G. Gantner Verlag (for International Association of Plant Taxonomy).

Noack, L. E., Warrington, I. J., Plummer, J. A. and Andersen, A. S. (1996). Effect of low-temperature treatments on flowering in three cultivars of *Hebe* Comm. ex Juss. *Scientia Horticulturae* 66: 103–15.

Norton, D. A. (2000). Hebe cupressoides *Recovery Plan*. Wellington: Department of Conservation.

Norton, D. A. and de Lange, P. J. (1998). *Hebe paludosa* (Scrophulariaceae) – a new combination for an endemic wetland *Hebe* from Westland, South Island, New Zealand. *New Zealand Journal of Botany* 36: 531–8.

Oliver, W. R. B. (1910). The vegetation of the Kermadec Islands. *Transactions and Proceedings of the New Zealand Institute* 42: 118–75.

Oliver, W. R. B. (1925). Vegetation of the Poor Knights Islands. *New Zealand Journal of Science and Technology* 7: 376–84.

Oliver, W. R. B. (1944). The *Veronica*-like species of New Zealand. *Records of the Dominion Museum* 1: 228–31.

Ollier, C. D. (1986). The origin of alpine landforms in Australasia. In: Barlow, B. A., ed., *Flora and Fauna of Alpine Australasia: ages and origins*. Melbourne: CSIRO.

Olmstead, R. G. and Reeves, P. A. (1995). Evidence for the polyphyly of the Scrophulariaceae. *Annals of the Missouri Botanical Garden* 82: 176–93.

Olmstead, R. G., de Pamphilis, C. W., Wolfe, A. D., Young, N. D., Elisons, W. J. and Reeves, P. A. (2001). Disintegration of the Scrophulariaceae. *American Journal of Botany* 88: 348–61.

Parsons, M. J., Douglass, P. and Macmillan, B. H. (1998). *Current Names for Wild Plants in New Zealand*. Lincoln: Manaaki Whenua Press.

Pennell, F. W. (1921). "Veronica" in North and South America. *Rhodora* 23: 1–22, 29–41. [Reprinted in 1921 as *Contributions from the New York Botanical Garden* 230.]

Perry, N. B. and Foster, L. M. (1994). Antiviral and antifungal flavonoids plus a triterpene from *Hebe cupressoides*. *Planta Medica* 60: 491–2.

Petrie, D. (1926). Descriptions of new native plants. *Transactions and Proceedings of the New Zealand Institute* 56: 6–16.

Pickard, C. R. (1996a). New Zealand Map Grid Reference Conversion Utility for Windows. Version 1.0. Wellington: Department of Conservation.

Pickard, C. R. (1996b). Distribution Plotter for Windows. Version 1.0.1. Wellington: Department of Conservation.

Primack, R. B. (1978). Variability in New Zealand montane and alpine pollinator assemblages. *New Zealand Journal of Ecology* 1: 66–73.

Primack, R. B. (1983). Insect pollination in the New Zealand mountain flora. *New Zealand Journal of Botany* 21: 317–33.

Raven, P. H. (1973). Evolution of the subalpine and alpine plant groups in New Zealand. *New Zealand Journal of Botany* 11: 177–200.

Richard, M. A. (1832). *Essai d'une Flore de la Nouvelle-Zélande*. Paris: [n.p.].

Rozefelds, A. C. F., Cave, L., Morris, D. I. and Buchanan, A. M. (1999). The weed invasion in Tasmania since 1970. *Australian Journal of Botany* 47: 23–48.

Ryan, K. G., Markham, K. R., Bloor, S. J., Bradley, J. M., Mitchell, K. A. and Jordan, B. R. (1998). UVB radiation induced increase in quercetin:kaempferol ratio in wild-type and transgenic lines of *Petunia*. *Photochemistry and Photobiology* 68: 323–30.

Salmon, J. T. (1968). *A Field Guide to the Alpine Plants of New Zealand*. Wellington: A. H. & A. W. Reed.

Salmon, J. T. (1991). *Native New Zealand Flowering Plants*. Auckland: Reed Books.

Salmon, J. T. (1992). *A Field Guide to the Alpine Plants of New Zealand*. 3rd edn. Auckland: Godwit Publishing.

Sampson, F. B. and McLean, J. (1965). A note on the occurrence of domatia on the underside of leaves in New Zealand plants. *New Zealand Journal of Botany* 3: 104–12.

Saunders, E. R. (1934). A study of *Veronica* from the viewpoint of certain floral characters. *Botanical Journal of the Linnean Society* 49: 453–93.

Simonet, M. (1934). Contribution à l'étude caryologique des *Veronica*. *Comptes Rendus des séances de la Société de Biologie et de ser. filiales (Paris)* 117: 1153–6.

Simpson, G. (1945). Notes on some New Zealand plants and descriptions of new species (no. 4). *Transactions of the Royal Society of New Zealand* 75: 187–202.

Simpson, G. (1952). Notes on some New Zealand plants and descriptions of new species (no. 5). *Transactions of the Royal Society of New Zealand* 79: 419–35.

Simpson, G. and Thomson, J. S. (1942). Notes on some New Zealand plants and descriptions of new species (no. 2). *Transactions of the Royal Society of New Zealand* 72: 21–40.

Simpson, G. and Thomson, J. S. (1943). Notes on some New Zealand plants and descriptions of new species. *Transactions of the Royal Society of New Zealand* 73: 155–71.

Simpson, M. J. A. (1976). Seeds, seed ripening, germination and viability in some species of *Hebe*. *Proceedings of the New Zealand Ecological Society* 23: 99–108.

Skipworth, J. P. (1973). Continental drift and the New Zealand biota. *New Zealand Journal of Geography* 57: 1–13.

Smith-Dodsworth, J. C. (1991). *New Zealand Native Shrubs and Climbers*. Auckland: David Bateman.

Stafleu, F. A. and Cowan, R. S. (1976). *Taxonomic Literature. A selective guide to botanical publications and collections with dates, commentaries and types. Vol. 1: A–G*. 2nd edn. Regnum Vegetabile 94. Utrecht: Bohn, Scheltema & Holkema.

Stearn, W. T. (1973). *Botanical Latin*. Newton Abbot: David & Charles.

Swain, T. (1963). *Chemical Plant Taxonomy*. London: Academic Press.

Sykes, W. R. (1977). *Kermadec Islands Flora, an Annotated Check List*. New Zealand Department of Scientific and Industrial Research Bulletin 219. Wellington: DSIR.

Thieret, J. W. (1955). The seeds of *Veronica* and allied genera. *Lloydia* 18: 37–45.

Thomson, G. M. (1881). On the fertilisation etc. of New Zealand flowering plants. *Transactions and Proceedings of the New Zealand Institute* 13: 241–91.

Thomson, G. M. (1927). The pollination of New Zealand flowers by birds and insects. *Transactions and Proceedings of the New Zealand Institute* 57: 106–25.

Tiffney, B. H. (1985). Perspectives on the origin of the floristic similarity between Eastern Asia and Eastern North America. *Journal of the Arnold Arboretum* 66: 73–94.

Tootill, E. (1984). *The Penguin Dictionary of Botany*. London: Penguin Books.

Vahl, M. (1794). *Symbolae Botanicae. Vol. 3*. Hauniae [Copenhagen]: N. Möller and Son.

Van Royen, P. (1983). *The Alpine Flora of New Guinea. Vol. 4*. Vaduz: A. R. Gantner Verlag.

Wagstaff, S. J. and Garnock-Jones, P. J. (1998). Evolution and biogeography of the *Hebe* complex (Scrophulariaceae) inferred from ITS sequences. *New Zealand Journal of Botany* 36: 425–37.

Wagstaff, S. J. and Garnock-Jones, P. J. (2000). Patterns of diversification in *Chionohebe* and *Parahebe* (Scrophulariaceae) inferred from ITS sequences. *New Zealand Journal of Botany* 38: 389–407.

Wagstaff, S. J. and Wardle, P. (1999). Whipcord hebes – systematics, distribution, ecology and evolution. *New Zealand Journal of Botany* 37: 17–39.

Wagstaff, S. J., Bayly, M. J., Garnock-Jones, P. J. and Albach, D. C. (2002). Classification, origin, and diversification of the New Zealand hebes (Scrophulariaceae). *Annals of the Missouri Botanical Garden* 89: 38–63.

Walls, G., Baird, A., de Lange, P. and Sawyer, J. (2003). *Threatened Plants of the Chatham Islands*. Wellington: Department of Conservation.

Walsh, N. G. and Entwisle, T. J. (1999). *Flora of Victoria. Vol. 4*. Melbourne: Inkata Press.

Wardle, P. (1963). Evolution and distribution of the New Zealand flora, as affected by Quaternary climates. *New Zealand Journal of Botany* 1: 3–17.

Wardle, P. (1975). Vascular plants of Westland National Park (New Zealand) and neighbouring lowland and coastal areas. *New Zealand Journal of Botany* 13: 497–545.

Wardle, P. (1988). Effects of glacial climates on floristic distribution in New Zealand 1. A review of the evidence. *New Zealand Journal of Botany* 26: 541–55.

Wardle, P. (1991). *Vegetation of New Zealand*. Cambridge: Cambridge University Press.

Wardle, P., Ezcurra, C., Ramírez, C. and Wagstaff, S. (2001). Comparison of the flora and vegetation of the southern Andes and New Zealand. *New Zealand Journal of Botany* 39: 69–108.

Webb, C. J. and Lloyd, D. G. (1986). The avoidance of interference between the presentation of pollen and stigmas in angiosperms II. Herkogamy. *New Zealand Journal of Botany* 24: 163–78.

Webb, C. J. and Simpson, M. J. A. (2001). *Seeds of New Zealand Gymnosperms and Dicotyledons*. Christchurch: Manuka Press.

Webb, C. J., Lloyd, D. G. and Delph, L. F. (1999). Gender dimorphism in indigenous New Zealand seed plants. *New Zealand Journal of Botany* 37: 119–30.

Webb, D. A. (1972). Hebe. In: Tutin, T. G., Heywood, V. H., Burgess, N. A., Moore, D. M., Valentine, D. H., Walters, S. M. and Webb, D. A., eds, *Flora Europaea. Vol. 3, Diapensiaceae to Myoporaceae*. London: Cambridge University Press. pp. 251–2.

Wellman, H. W. (1979). An uplift map for the South Island of New Zealand, and a model for uplift of the Southern Alps. In: Walcott, R. I. and Cresswell, M. M., eds, *The Origin of the Southern Alps*. Royal Society of New Zealand Bulletin 18. Wellington: Royal Society of New Zealand.

Wettstein, R. von (1891). Scrophulariaceae. In: Engler, A. and Prantl, K., eds, *Die Natürlichen Planzenfamilien*. Vol. 4, Part 3a: 39–107. Leipzig: W. Engelmann.

Wheeler, C. and Wheeler, V. (2002). *Gardening with Hebes*. Lewes: Guild of Master Craftsman Publications.

Widyatmoko, D. and Norton, D. A. (1997). Conservation of the threatened shrub *Hebe cupressoides* (Scrophulariaceae), eastern South Island, New Zealand. *Biological Conservation* 82: 193–201.

Wiley, E. O. (1981). *Phylogenetics. The theory and practice of phylogenetic systematics*. New York: John Wiley & Sons.

Wilkinson, A. S. and Wilkinson, A. (1952). *Kapiti Island Bird Sanctuary: a natural history of the island*. Masterton: Masterton Printing Company.

Williams, C. A. and Harborne, J. B. (1994). Flavone and flavonol glycosides. In: Harborne, J. B., ed., *The Flavonoids – advances in research since 1986*. London: Chapman and Hall. pp. 337–70.

Williams, C. A. (2006). Flavone and flavonol glycosides. In: Anderson, O. M. and Markham, K. R., eds,

Flavonoids: Chemistry, Biochemistry and Applications. Boca Raton, Florida: CRC Press-Taylor and Francis.

Williams, P. A. (1993). The subalpine and alpine vegetation on the Central Sedimentary Belt of Paleozoic rocks in north-west Nelson, New Zealand. *New Zealand Journal of Botany* 31: 65–90.

Wilson, H. D. (1978). *Wild Plants of Mount Cook National Park.* Christchurch: Field Guide Publication.

Wilson, H. D. (1996). *Wild Plants of Mount Cook National Park.* 2nd edn. Christchurch: Manuka Press.

Wilson, H. D. and Galloway, T. (1993). *Small-leaved Shrubs of New Zealand.* Christchurch: Manuka Press.

Yamazaki, T. (1957). Taxonomical and phylogenetic studies of Scrophulariaceae – Veronicae with special reference to *Veronica* and *Veronicastrum* in Eastern Asia. *Journal of the Faculty of Science University of Tokyo. Section III Botany* 7: 91–162.

Picture Credits

In general, macro photos were taken by Bill Malcolm, photos of plant habit were taken by Mike Bayly, and scans of plant sprigs (the images in the top right-hand corner of the species plates) were made by Mike Bayly and/or Alison Kellow. Line drawings were by Tim Galloway. Listed below are the photos and illustrations contributed by others, as well as those that are exceptions to these general rules:

Mike Bayly: Figs 17, 43, 66I, 77, 82, 117I, 132, 166, 167.
Mike Bayly/Alison Kellow (digital scans): Figs 33, 50, 65J, 67I, 70H, 71H, 72H, 73I, 76H, 78E, 83H, 94E, 96D, 99E, 100E, 101E, 102G, 106E, 107E, 108E, 111E, 115H, 116I, 118I, 121H, 122E, 125F, 126G, 127H, 128E, 129E, 134F, 137E, 137H, 138I, 142E, 143E, 144C, 144D, 145C, 146G, 150E, 151G.
Ross Beever: Figs 138A, 138H.
Brian Butterfield: Fig. 19.
Gillian Crowcroft: Figs 58A, 105A, 107B, 108A.
Peter de Lange: Figs 106A, 145A.
Tim Galloway: Fig. 112B (adapted from a sketch provided by Phil Garnock-Jones), Appendix 4.
Phil Garnock-Jones: Figs 1A, 1D, 1E, 16, 68A, 68B, 92A, 125A.
Michael Hall: Fig. 35.
Alison Kellow: Figs 1C, 60.
Bill Malcolm: Figs 22, 23, 56A, 88A, 95A, 119A, 120A.
Jean-Yves Meyer[1]: Fig. 112A.
Vivienne Nicholls: Fig. 68I.
Rebecca Stanley: Fig. 146A.
Vanessa Thorn: Figs 68E, 68F.

---

1 Délégation à la Recherche, Tahiti.

Index

Accepted botanical names of wild species are shown in roman type. Synonyms, excluded names, botanical names applying to hybrids and horticultural forms, as well as those placed *incertae sedis*, are in *italic*. Informal names (tag names) are shown in quotation marks. **Bold face** type (for page numbers) indicates the principal reference for a taxon. Page numbers in *italic* indicate references to figures or tables. Entries for authors of botanical publications are not exhaustive.

2D-PC 40–1, *42–3*, 44
Adams, James 280
Adamson, R.S. 31
Agrostis capillaris 70–1
Aiton, W. 8, *9*
Albach, D.C. *12*, 13, 16, 17
Allan, H.H. *9*, 11, *12*
Alpine Fault 25–6
Andersen, J.C. 11, 302
androecium *see* anther; filament; stamen
aneuploidy 45
anther
　characters *84*, 86
　morphology and dehiscence *33*, 34, 61, 62, 64
Armstrong, J.B. *9*, *12*
Armstrong, J.F. *9*, 10, 108
arrangement of species 6, 18, 79
Astelia banksii 280
Atkinson, Esmond H. 182
Auckland Islands 19, 126, 216, 222, 262, 268
aute 331
Awaroa koromiko 330

Baptisia 44
Barker, Samuel D. 206
Barker's koromiko 330
Bartlett's hebe 330
beech forest hebe 330
Beever, J. 331
Bentham, George *9*, *12*, 126
biogeographic history 24–6
birds 63
Bishop, John J. 184
black-barked mountain hebe 330
blue-flowered mountain hebe 330
bog pine *see* Halocarpus bidwillii
Bohm, B.A. 41
Bollons, John P. (Capt.) 200
botanical nomenclature 5, 8, 79, 303
Boulder Lake hebe 330

boxwood 330
Brachyglottis huntii 70
bract
　characters 84, 88
　morphology and arrangement 27, 32–3, *112*
branches *see* stem
Breitwieser, I. 31
Briggs, B.G. 17
Brown, F. *9*, *213*, 331
Brownsey, Pat 5
brown-top 70–1
browsing animals 69–71
Buchanan, John *9*, 138
Butterfield, B.G. 28

calyx
　characters 84–5, 88
　morphology *27*, 33
Campbell Island 19, 126, *127*, 262
Canterbury hebe 330
Canterbury whipcords 330
capsule
　characters 87, 88, 89
　morphology and dehiscence 11, 35–6, 62, 64–6, 88
carpet grass *see* Chionochloa australis
Chalk, D. 3, 73
characters for descriptions 79, 81–7, 88
Chatham Island koromiko 330
Chatham Islands 69–70, 72, 262, 264
Cheeseman, Thomas *9*, 10, 11, *12*, 98, 296
Chionochloa
　australis 70, 220, 232, 319
　rubra subsp. occulta 71
Chionohebe 8, 15, 16, 28–9, 33–5, 37, 329
　ciliolata 323, 329
　densifolia 328, 329

glabra *7*, 16
chromosomes 45–60, 138
cladistic methods 13, 17
classification 5–6, *7–18*, 79
clubmoss hebe 330
clubmoss whipcord 330
Cockayne, Leonard *9*, 11, 37, 70, 248
Colenso, William *9*, 240
Colenso's hebe 330
common names 330–1
conservation 67–72
Cook, James 8, 10
corolla
　characters 85, 88
　morphology and development 33–4, 37, 61
Cretaceous 24, 26
Croizat, Leon C.M. 290
cultivation 69, 73, 77, 79, 80, 88
cuttings 27, 73
cypress koromiko 330
cypress whipcord 330
cypress-like hebe 330

de Lange, P.J. *9*, 67–72
Delph, Lynda 61–5
DELTA system 79, 81, 88
Dengler, N.G. 61
Department of Conservation (DoC) 72, 330
Derwentia 8, 15, 16
　formosa 328
　perfoliata *7*
Detzneria 8, 13, 35
developmental morphology 37
Dieffenbach, J.K. Ernst 208
Dieffenbach's koromiko 330
disc *see* nectarial disc
dish-leaved hebe 330
distribution
　Hebe 3, 19–26, 90
　Leonohebe 19–26, 290

381

distribution maps 79–80
DNA sequences 13–15, 17, 25–6
DoC 72, 330
Dracophyllum arboreum 70
Druce, A.P. 5, 216, 335
DSIR Botany Division 72
d'Urville, Dumont 10
D'Urville Island 150, 238, 258
dwarf whipcord 330

Eagle, A. 5, 216, 335
Elder, N.L. 240
endemism 22–4
English common names 330–1
Eocene 24
Erigeron karvinskianus 69
evolution 5, 7–18

Falkland Islands 8, 224, 262
field collections 77, *78*
filament
  characters *85*, 86, 88
  morphology *33*, 34
fire 70–1
flavonoid biochemistry 38–44, 230
Flora of New Zealand. Vol.1 (Moore & Ashwin) 3–6, 10–11
flower
  biology 61–4
  characters 83–7, 88
  morphology and development 32–5, 37
flowering times 62, 80, 88, 252
Forster, Georg 8, 10
Forster, Johann Reinhold 8, *9*, 10
fossils 24–5
Foundation for Research, Science and Technology (FRST) 5
Frankel, O.H. 64
French Polynesia 19, 90, 212
French polynesian names 331
fruit *see* capsule
fruiting times 62, 88

Garnock-Jones, P.J. 5, *9*, 12, 17, 27–37, 61–6, 80
gender 63, 64–5
generic limits 5–6, 11–13, 16–17
geological timescale *25*
germination 61, 66, 70
Gibbs, Frederick G. 134
glacial periods 25–6
Gmelin, J.F. 8, *9*, 11
Godley, E.J. 66
Gondwana 24–5
gynoecium *see* nectarial disc; ovary; stigma; style

habitats 3, 19–20, 69–70, 81
hair, morphology and distribution 27, 28, 30, 33–5, 37
Hair, J.B. 64

Halocarpus bidwillii 69, 70-1, 108
Hawai'i 284
Heads, M.J. 5, *9*, 10–11, *12*, 15, 18, 32, 35, 37
Hebe 3, 5–6, 11, 89, **90**, 302, 328–9
  sect. *Glaucae* 303
  sect. *Hebe* 303
  sect. *Subdistichae* 303
  ser. *Hebe* 33, 303
  ser. *Occlusae* 303
  "Apertae" 15, 18, 28, 44, 45, 64
    large-leaved 89, **266–85**
    small-leaved 89, **224–65**
  "Buxifoliatae" 15, 17–18, 30, 45, 64, 89, **214–23**
  "Connatae" 15, 32, 34, 37, 44, 45, 61–2, 64, 89, **112–27**
  "Flagriformes" 15, 17–18, 28–32, 34, 37, 45, 64, 81, 89, **91–111**, 331
  "Grandiflorae" 15, 17, 45, 64, 89, **286–7**
  "Occlusae" 15, 18, 28, 34, 45, 64–5, 89, **144–213**
  "Paniculatae" 5
  "Pauciflorae" 17–18, 45, 64, 89, **288–9**
  "Semiflagriformes" 12
  "Subcarnosae" 15, 45, 64, 89, **128–43**
  "Subdistichae" 15, 18, 44
acutiflora 20, *46*, *68*, 145, 182, 190, **192–3**, 194, 196, 198, 303–4, 340
adamsii 34, *46*, 61, *68*, 71–2, 200, 267, **280–1**, 304
"aff. bishopiana" 182
"aff. diosmifolia" 335
"aff. ligustrifolia" 198
"aff. pimeleoides" 335
"aff. pinguifolia" 335
"aff. rigidula" 335
×affinis 204, 325
albicans 18, 30, 44, 45, *46*, 63, *68*, 89, 129, 144, 146, 164, **168–9**, 268, 304, 331, 335, 341
albiflora 328
allanii 130, 304
amabilis, ×amabilis 325
  var. blanda 325
amplexicaulis 20, *46*, 77, 129, **130–1**, 132, *133*, 304
  f. amplexicaulis *68*, 304
  f. hirta 30, *46*, *68*, **130–1**, 304
  var. erecta 304
  var. suberecta 304
×andersonii 327
angustifolia 319, 325
×angustisala 325
angustissima 20, 34, *46–7*, 145, 182, 188, **190–1**, 192, 305, 335, 341

annulata 15, 20, 35, 45, *47*, *68*, 91, 98, 104, 106, 108, **110–11**, 305
anomala 315, 335
"arborea" 335
arganthera 32, 34, 36, *47*, *68*, 225, 248, **250–1**, 305, 335
armstrongii 15, 33, 45, *47*, *68*, 69–70, 72, 91, 104, 106, **108–9**, 110, 305, 310, 331
astonii 321
"Bald Knob Ridge" 335
balfouriana 327
barkeri 16, 23, 28, 30, 35, 45, *47*, *68*, 69–70, 72, 147, **206–7**, 208, 212, 305, 324, 330
"Bartlett" 202, 335
benthamii 15, 23, 32–5, *47*, 61, *68*, 112, 113, **126–7**, 305
bidwillii 328
biggarii 20, *47*, *68*, 129, **142–3**, 306
bishopiana *47*, 61, *68*, 69, 72, 145, 182, **184–5**, 186, 306, 331, 335
blanda 325–6
bollonsii 20, 31, 44, *47*, 145, **198–200**, 306
brachysiphon 20, *23*, *47*, 65, 218, 225, 234, 254, **256–7**, 258, 306, 341
brassii 328
brevifolia 20, 30, 34, 45, *47*, 61, *68*, 144, 202, **204–5**, 284, 306, 331, 335
breviracemosa 18, *23*, *47*, *68*, 69–70, 72, 267, **276–8**, 306, 330
brockiei 156, 158, 322
buchananii *23*, 28, 45, *48*, **128**, 129, 136, **138–9**, 306, 330–1
  var. exigua 136
  var. major 306
buxifolia 314
  var. odora 314
  var. pauciramosa 316
  var. prostrata 315
calcicola 20, 32, 44, *48*, *68*, 146, 160, **162–3**, 168, 268, 306, 335
canescens 328
canterburiensis *23*, *48*, 62, 225, 226, **228–9**, 232, 307, 330
carnea 327
carnosula 20, 45, *68*, 71, 136, 225, **234–5**, 307
carsei 326
carstensensis 328
cassinioides 326
catarractae 328
"cf. pinguifolia" 335
chathamica 20, *23*, 35, *48*, *68*, 147, 206, 208, **210–11**, 307, 330, 341

Hebe *continued*
  cheesemanii 323
  ciliata 328
  ciliolata 323
  coarctata 311
  cockayneana 27–8, *48*, 225, 244, 246, *248–9*, *250*, 307, 342
  colensoi 20, 33–4, *48*, 224, **240–1**, 242, 308, 330, 342
    var. *hillii* 308
  cookiana 320
  corriganii 30, 44, *48*, 204, 267, 274, **278–9**, 308, 313
  corymbosa 309
  coxiana 307
  crawii 230, 309
  crenulata 20, *48*, 225, **244–5**, 246, 248, 308, 342
  cryptomorpha 20, 32, *48*, 225, 244, **246–7**, 248, 308, 335, 342
  cupressoides 324
  dartonii 326
  darwiniana 324
  dasyphylla 328
  decumbens 18, *48*, 63, 136, 146, **158–9**, 308, 330
  demissa 311
  dieffenbachii 23, 35, *49*, *68*, 147, 206, **208–9**, 210, 212, 308, 324, 330, 342
  dilatata 33, 49, *68*, 224, **230–1**, 232, 248, 308–9
  diosmifolia 30, 32, 45, *49*, 66, *68*, 225, **252–3**, 258, 303, 309, 331, 335
    var. *trisepala* 309
    var. *vernalis* 309
  diosmoides 328
  divaricata 32, *49*, *63*, 134, 150, 225, 234, 252, 254, 256, **258–9**, 309, 331
  divergens 326
  dorrien-smithii 308
  "*egmontiana*" 335
  ×*ellipsala* 326
  elliptica 8, 11, 19–20, *23*, 25, 27, 30–31, 34-35, *49–50*, 66, 90, 150, 206, 224, **262–4**, *266*, 268, 302, 303, 309–10, 326, 327, 331, 343
    var. *crassifolia* *68*, 262, 310
  epacridea 15, 20, *23*, 30, 44, *50*, 62, 112, 113, **114–15**, 118, 120, 310, 331
  erecta 327
  evenosa 20, *50*, *68*, 144, 150, **152–3**, 154, 310, 331, 343
  ×*fairfieldii* 328
  flavida 20, 28, 30, *50*, 62, 145, 182, 192, **194–6**, 198, 206, 310, 343

fonkii 318
formosa 328
×*franciscana* 264, 327
fruticeti 321
gibbsii 20, 30, *50*, *68*, 129, 130, **134–5**, 310–11, 331
gigantea 324
glauca-caerulea 324
glaucophylla 20, *23*, *50*, 89, 129, 144, 146, 160, **164–5**, 166, 311, 335
  "NW Nelson" 168, 335
  "var. (i)" 335
godefroyana 327
gracillima 264, 270, 312
"Great Barrier" 204, 335
greyi 323
haastii 20, *23*, 35, *50*, 112, 113, 116, 118, **120–1**, 311, 331
  var. *humilis* 313
  var. *macrocalyx* 313
haustrata 315
hectorii 32, *50*, 91, 92, **94–8**, 98, 311–12, 336–7
  subsp. *coarctata* *50*, 62, **96–8**, 311, 331
  subsp. *demissa* *50*, **96–8**, 311–12, 330, 337
  subsp. *hectorii* *23*, *50*, *91*, **95–6**, 98, 311, 336–7
  subsp. *laingii* 96, 98, 311, 336
  subsp. *subsimilis* 321
  subsp. *subulata* 96, 312, 336
  var. *demissa* 311
hillii 308
hookeriana 328
hulkeana 328–9
imbricata *50*, *68*, 91, 98, 100, **102–3**, 312
  subsp. *poppelwellii* 312
imperialis 327
insularis 20, 23, *50*, *68*, 144, 225, **260–1**, 312
×*kirkii* 326
"Lady" 335
×*laevastoni* 326
laevis 322, 326
×*laevisala* 326
laingii 95, 311
lapidosa 318
"*latisepala*" 335
lavaudiana 329
leiophylla 28, *50*, 224, 258, **264–5**, 267, 270, 312, 326, 331, 344
  var. *strictissima* 320
×*leiosala* 326
lendenfeldii 329
lewisii 326
ligustrifolia 20, 34, *50–1*, *68*, 145, 182, 192, 194, **196–8**, 312–13, 335, 344
  "var. Surville" 198

linifolia 329
"Little Barrier" 335
loganioides 326–7
longiracemosa 325
lyallii 329
lycopodioides *23*, *51*, 91, **100–1**, 102, 313, 330–1, 338–9
  subsp. *patula* 100, 313, 339
  var. *patula* 100, 313
macrantha 15, 20, 29–30, 33–4, 35–6, 37, **286–7**, 313, 330
  var. *brachyphylla* *51*, **286–7**, 313, 344
  var. *macrantha* *23*, *51*, **286–7**, 344
macrocalyx 20, 30, 33, 44, 112, 113, **116–18**, 120, 240, 313, 331
  var. *humilis* *51*, **116–18**, 313, 344
  var. *macrocalyx* *51*, *68*, **116–18**, 344
macrocarpa 10, 30, 36, 45, *52*, 61–2, 145, 182, **202–4**, 274, 278, 303, 313–14, 327, 331, 335, 345
  var. *brevifolia* 306, 335
  var. *latisepala* 34, *51*, **202–3**, 314, 335, 345
  var. *macrocarpa* *51*, **202–3**
macrosala, ×*macrosala* 325, 327
macroura 320
magellanica 11, 302, 303, 310
"marble" 335
masoniae 20, 45, *52*, 215, 216, 218, **220–1**, 222, 314, 335
matthewsii *52*, *68*, 156, 325
menziesii 309
"Mokohinau" 335
montana 321
monticola 321
mooreae *23*, 25, 30, *52*, 215, 216, **218–19**, 222, 314, 335
murrellii 20, *52*, 61, 112, 113, 122, **124–5**, 314
myrtifolia 325
obovata 325
obtusata 20, *52*, *68*, 145, 184, **186–7**, 314, 331
ochracea 20, 45, *52*, *68*, 91, **104–5**, 106, 108, 110, 314, 331
odora 10, *23*, 32, 34, 44, 45, *52–3*, 62, 64, 92, *94*, 176, *214*, 215, **216–18**, 254, 256, 314–15, 326, 331, 335, 346
olsenii 329
paludosa 20, *23*, 54, 224, 264, 267, 268, **270–1**, 315, 320, 346
pareora 20, 28, 33, 35–6, *54*, *68*, 89, 129, 130, *131*, **132–3**, 240, 315, 335
parkinsoniana 319

INDEX 383

Hebe *continued*
  parviflora 28, 44, *54*, 64, 144, 147, 172, **174–6**, 188, 206, 315, 325, 331, 335
    var. *angustifolia* 319, 335
    var. *arborea* 315, 335
  pauciflora 18, 27, 30–1, 33, 35–6, *54*, *68*, **288–9**, 315
  pauciramosa 20, *23*, *54*, 71, 215, 216, 218, 220, **222–3**, 316, 331
    var. *masoniae* 314, 335
  perbella *54*, *68*, 69, 72, 145, 182, 198, **200–2**, 316, 330, 335
  petriei 15, 20, 34–5, *54*, 61, 112, 113, **122–3**, 124, 316
    var. *murrellii* 314
  pimeleoides 30, 32, 45, 89, 129, **140–2**, 316, 326, 327, 327, 330, 335
    subsp. faucicola 20, 44, *55*, 67, *68*, 71, **140–2**, 316, 335, 346
    subsp. pimeleoides 20, *23*, 44, *54–5*, **140–2**, 316, 335, 346
    var. *glauca-caerulea* *55*, 324
    var. *minor* 316
    var. *rupestris* 142, 316, 335
  "*pimeleoides agg.*" 335
  pinguifolia 20, *23*, 30, 36, 45, *55*, 62, 64, 66, *68*, 89, 129, 130, **136–8**, 158, 234, 303, 316
  polyphylla 329
  poppelwellii 102, 312
  propinqua *55*, 91, **98–9**, 316–17, 330–1
    var. *major* 317
  pubescens 20, 30, 44, 62, 224, 267, **272–6**, 317
    subsp. pubescens *55*, **272–6**, 346
    subsp. rehuarum *55*, *68*, 204, **272–6**, 317, 335, 346
    subsp. sejuncta 31, *55*, *68*, 200, **272–6**, 317, 335, 347
    "*form (i)*" 335
    "*var.*" 335
  rakaiensis 20, *23*, 34, *55*, 146, **160–1**, 162, 264, 317, 325, 326
  ramosissima 15, 20, *55*, *68*, 112, 113, **118–19**, 122, 317
  raoulii 329
    var. *maccaskillii* 329
    var. *pentasepala* 329
  rapensis 19, 35, 45, 90, 147, 206, **212–13**, 317, 331
  recurva 168, 304
  rigida 329
  rigidula 225, 236, **238–9**, 242, 246, 317–18, 335
    var. rigidula *56*, *68*, **238–9**
    var. sulcata *56*, *68*, 150, **238–9**, 318, 335
    "*form (i) & (ii)*" 335

"*var. (i) & (ii)*" 335
rotundata 328
rubra 329
"*rupestris*" 335
rupicola 20, *56*, 224, **236–7**, 238, 318
salicifolia 8, 19, *23*, 25, 28, 30, 36, 44, *56*, 62, 66, 90, 168, 176, 264, 266, 267, **268–9**, 270, 302, 318, 325–6, 331, 347
  var. *atkinsonii* 320
  var. *communis* 64, 318
  var. *longiracemosa* 319–20
  var. *paludosa* 315
  var. *stricta* 319
salicornioides 15, 20, 45, *56*, *68*, 69–71, 91, 104, **106–7**, 108, 110, 318, 331
scopulorum 28, *56*, *68*, 69, 72, 224, 238, **242–3**, 267, 318, 330, 335
scott-thomsonii 317
×*simmonsii* 327
societatis 20, 45, *56*, *68*, 69–70, 225, **232–3**, 234, 319
sp. (a) – (i) 335
sp. (l) 335
sp. (m) 198, 335
sp. (n) 188, 335
sp. (o) – (q) 335
sp. (t) – (v) 335
sp. (w) 204, 335
sp. (x) 202, 335
speciosa 20, 30, 33-4, *56*, 61–2, 63, *68*, 69, 80, 204, 267, **284–5**, 319, 327, 330–1
  var. *kermesina* 328
"*squalida*" 335
stenophylla 27, 30, 33, 61–2, 144, 146–7, 150, **170–2**, 174, 176, 319, 325, 327, 331
  var. *hesperia* *56*, *68*, **170–2**, 319
  var. *oliveri* *57*, *68*, 150, **170–2**, 319
  var. *stenophylla* *56*, *147*, 150, **170–2**, 319, 335
stricta 28, 32, 34–5, 45, *57–8*, 61, 65, 66, 145, 152, **178–83**, 190, 194, 268, 319–20, 326–7, 331
  var. *atkinsonii* *57*, 147, 150, **178–82**, 320, 325, 348
  var. *egmontiana* *57*, **178–83**, 188, 190, 320, 335, 348
  var. *lata* *57*, **178–83**, 320, 335, 348
  var. *macroura* *58*, **178–82**, 200, 320, 349
  var. *stricta* *57*, 61, **178–82**, 184, 186, 192, 194, 198, 204, 274, 276, 319–20, 326–7, 350
  "*form (i)*" 335
  "*var. (H. salicifolia var.*

angustissima)" 335
strictissima *58*, 61-2, 65, 146, 160, 162, 172, 174, **176–7**, 264, 268, 320, 350
subalpina *23*, 35, *58*, 61–5, 146, **148–9**, 150, 154, 156, 158, 160, 162, 320–1, 330–1, 350
subfulvida 309
subsimilis 321
  var. *astonii* 321
subulata 312
tairawhiti *58*, *68*, 145, 182, **188–9**, 190, 321, 335
"*takahe*" 335
tenuis 329
tetragona 91, **92–4**, 96, 321, 331
  subsp. subsimilis 44, *58*, **92–4**, 321
  subsp. tetragona *58*, *91*, **92–3**, 321
tetrasticha 292, 324
thymelaeoides 329
topiaria 20, 31, 45, *59*, 89, 129, 144, 146, 164, **166–7**, 311, 321, 351
townsonii 32, 34, *59*, 64, *68*, 225, 267, **282–3**, 321–2
traversii *23*, 30, *59*, 62–4, 147, 160, 162, **172–4**, 176, 256, 264, 322, 331, 351
treadwellii *23*, *59*, *68*, 146, **156–8**, 322, 330, 335, 351
truncatula 20, *59*, 144, 152, **154–5**, 162, 322, 351
tumida 324
uniflora 329
"*Unuwhao*" 280
urvilleana 30, *59*, *68*, 146, **150–1**, 322, 351
vanderwateri 329
venustula 20, 32, *59*, 218, 225, **254–5**, 256, 258, 322, 331
vernicosa 20, 29, 45, *60*, 225, **226–7**, 228, 232, 323, 327, 330–1
  var. *canterburiensis* 307
"*Wairoa*" 188, 335
"*Whangarei*" 198, 335
willcoxii 307
Hebe complex 7–8, 13–17, 24–5, 28, 35, 37, 45, 65
Hebe Society 73
Hedera 61
Heenan, P.B. 6
Helichrysum 31
Heliohebe 5, 8, 11, 15, 27, 29, 33, 37, 66, 327
×*fairfieldii* 328
hulkeana *7*, 328–9
  subsp. hulkeana 329
lavaudiana 329
pentasepala 329

Heliohebe *continued*
  raoulii
    subsp. maccaskillii 329
    subsp. raoulii 329
Hennig, W. 13
herbarium specimens 62, 77, 79, 80
high-pressure liquid chromatography *see* HPLC
hollow-leaved hebe 330
Hong, D. 35
Hooker, J.D. 9, 10, *12*, 62, 64, 218
horticultural forms 5–6, 327–8
HPLC 40–1, *42–3*, 44
Hutchins, G. 3, 6, 35, 73
hybrids 3, 5–6, 64, 90, 284, 324–5, 327
  identification 44
  possible wild hybrids 79, 325–7
  treatment of 79

ICBN 8, 17, 302
illustrations 352–6
implicit character states 88
inflorescence
  characters 83–4, 88, 112
  morphology and development 32–3, 62, 65, 112
informal names 5, 17, 89, 335
infrageneric classification 17–18
infraspecific ranks 77–9
insects 28, 29, 63
International Code of Botanical Nomenclature *see* ICBN
International Union for the Conservation of Nature (IUCN) 67
ivy 61

Jussieu, A.L. de 11
juvenile and reversion leaf
  characters 83
  morphology 37

Kampny, C.M. 61
Kermadec Islands 25, 276
Kermadec koromiko 330
Kirk, Thomas 9, 10, 37
kōkōmuka 331
kōkōmuka tāranga 331
kōkoromiko 331
kōkoromuka 331
korohiko 331
korokio 331
koromiko 330, 331
koromiko tāranga 331
koromuka 331

Lamiales 24, 27
Landcare Research 77, 138
large-flowered hebe 330
leaf
  characters 81–3, 88, 112

morphology and anatomy 29–32, 37, 112, 340–51
leaf bud
  characters 82
  morphology 29–30
leathery-leaved mountain hebe 331
Leonohebe 5–6, 11, *14*, 15, 16, 89, **290**, 302, 329
  sect. *Apiti* 303
  sect. *Aromaticae* 64, 89, **300**, 303
  sect. *Buxifoliatae* 303
  sect. *Connatae* 303
  sect. *Densifoliae* 329
  sect. *Flagriformes* 303
  sect. *Leonohebe* 15, 19, 61, 81, 89, 290, **291–9**, 302
  sect. *Salicornioides* 303
  annulata 305
  armstrongii 305
  benthamii 303, 305
  cheesemanii *23*, 35, *60*, 291, 292, 294, **296–7**, 298, 323, 330
  ciliolata *23*, 33–4, *60*, 62, 291, **292–3**, 294, 302, 323–4, 331
  coarctata 311
  cupressoides 15, 18, 19, *23*, 33, 35, *60*, *68*, 69–72, 80, 89, **300–1**, 303, 324, 330–1
  densifolia 329
  epacridea 303, 310
  haastii 311
    var. *humilis* 313
    var. *macrocalyx* 313
  hectorii 303, 311
    var. *demissa* 311
  imbricata 312
  laingii 311
  lycopodioides 313
    var. *patula* 313
  masoniae
    var. *masoniae* 314
    var. *rotundata* 314
  mooreae
    var. *mooreae* 314
    var. *telmata* 314
  ochracea 314
  odora 303, 314
  pauciflora 315
  pauciramosa 316
  petriei 316
    var. *murrellii* 314
  poppelwellii 312
  propinqua 316
  ramosissima 317
  salicornioides 303, 318
  subsimilis 321
    var. *astonii* 321
  subulata 312
  tetragona 321
  tetrasticha 27, 34, *60*, 291, 296, **298–9**, 324

tumida 15, *60*, 67, *68*, 291, 292, **294–5**, 296, 324
uniflora 329
Logania ciliolata 323

Malcolm, Bill 5, 80
male sterility 64
Māori names 330, 331
marble 20, 104, 162
Markham, K.R. 5, 38–44
Mason, Ruth 220
materials and methods 77–80, *81–7*
Metcalf, L.J. 6, 73
Mexican daisy 69
Meylan, B.A. 28
micropylar rim 87
Miocene 24–5
Mitrasacme
  cheesemanii 323
  hookeri 323
  petriei 316
Moar, N.T. 35
Moench, Conrad 8, *9*
Moore & Ashwin 3, *4*, 9, *12*, 13, 15, 17–18, 29, 35, 88, 240
Moore, Lucy B. 13, 218
morphology 5, 17, 27–37, 77, 79, *81–7*
mountain koromiko 330–1
Mt Arthur hebe 331
Mueller, F. 9, 10
Murrell, Robert 124
Myrsine chathamica 70

napuka 330, 331
nectar 62–3
nectarial disc
  character 86–7, 88
  structure and function 35, 62
Nelson Botanical Society 232
Nelson mountain hebe 331
New Guinea 13, *14*, 16
New Zealand Map Series (NZMS) 79
nodal joint of whipcords
  characters 81
  morphology 31–2
North Cape hebe 331
North Island whipcord 331
North-west Nelson hebe 331
Northern forest hebe 331
Northern koromiko 331

ochreous whipcord 331
Olearia chathamica 70
Oliver, W.R.B. 9, 11, 70, 172
Otari-Wilton's Bush 46, 58, 77, 148
ovary
  characters 86–7, 88
  morphology and function 35, 64

painaka 331
Panoxis 302
  macrocarpa 313
  salicifolia 318
Parahebe 8, 11, 15, 16, 27–9, 31, 33–5, 37, 61, 66
  albiflora 328
  ×*bidwillii* 328
  birleyi 114
  canescens 29, 328
  carstensensis 328
  catarractae 328
  ciliata 328
  diosmoides 328, 329
  giulianettii 328
  hookeriana *7*, 328, 329
  lendenfeldii 329
  linifolia 329
  lyallii 329
  planopetiolata 329
  polyphylla 329
  rigida 329
  rubra 329
  tenuis 329
  vanderwateri 329
pedicel
  characters 84, 88
  morphology 32–3, 66
Pennell, F.W. 11
Percy's Reserve 77
Petrie, Donald *9*, 122
photography and imaging 80, 352–6
phylogenetic trees 13–15
Plantaginaceae 7, 61
Plantago 28
Pleistocene 25
Pliocene 24–6
Podocarpus ? dieffenbachii 321
pōhutukawa 274
pollen
  morphology 35
  production and collection 34–5, 62, 63
pollination 63
Polynesian rat 70
polyploidy 45–6
Primack, R.B. 62
pumice whipcord 331
purple hebe 331
purple koromiko 331
Pygmea 8

Ranunculus grahamii 114
Raoul Island 70, 72, 276
Raoulia 31
Rapa 19, 24, 90, 144, 147, 212, *213*
Rattus exulans 70
recovery plans 72
recruitment failure 70–1
red tussock 71, 106

representative specimens *see* herbarium specimens
reproductive biology 61–6
reproductive morphology 32–7
reversion leaf *see* juvenile and reversion leaf
Richard, Achille *9*, 10
root, morphology 27–8
Ruapehu hebe 331

Salmon, J.T. 330
Saunders, E.R. 33, 34
scale bars 80
scent 61–2
scree 20, 114–24, 171–2, 230, 294–8
scree hebe 331
Scrophulariaceae 7, 24, 61
seed
  characters 87, 88
  morphology and biology 35, 36–7, 61, 64–6
seedlings 37, 69–70
semiwhipcords *see* Leonohebe sect. Leonohebe
shore hebe 331
shore koromiko 331
Simpson, M.J.A. 36, 66
sinus *see* leaf bud
Snares Islands 50, 262
snow tussock 142
Solander, Daniel 8
South America 11, 19, 24–5, 262, 266, 268
Southern Gap 24
species arrangement 6, 18, 79
species concepts 77–9
species richness 21
specimens examined *see* herbarium specimens
spiny whipcord 331
stamen
  characters *84*, 86, 88
  morphology *33*, 34
stem
  characters 81, 88
  morphology 28–9
Stenocarpon 35
sterility 64
Stewart Island 23, 77, *78*, 95, 96, *214*, 216, 222, 262, 268
stigma, morphology and receptivity *33*, 35, 61–3
style
  characters 86–7, 88
  morphology and elongation *33*, 35, 62
swamp hebe 331
synonyms 4, 10, 302–29
synopsis 79, 88–9

Takaka hebe 331
Tararua hebe 331
Tasmania 90, 224, 262
thick leaved hebe 331
Thieret, J.W. 37, 87
Thomson, G.M. 62
threat classification system 67
Three Kings Islands 260
tītīrangi 330, 331
Townson, William L. 202
trailing whipcord 331
Travers, William T.C. 174
Travers' (Traver's) hebe 331
Treadwell, Charles H. 158
tutaipainaka and tutai painaka 331
two-dimensional paper chromatography *see* 2D-PC

ultramafic substrates 20, 71, 204, 234, 258
University of Texas (Austin) 41

Vahl, Martin 8, *9*, 10
varnished koromiko 330–1
vegetative morphology 27–32
Veronica 7–8, 10–13, 15, 16–17, 28, 31, 34–5, 37, 61, 70
  subg. *Koromika* 302
  subg. *Pseudoveronica* 302
  sect. *Hebe* 302
    1. *Integrae* 303
    1. *Speciosae* 303
    2. *Decussatae* 303
    2. *Serratae* 303
  acutiflora 303
  adamsii 304
  albicans 304
  amabilis 325
    var. *blanda* 325
  amplexicaulis 304
  ×*andersonii* 327
  angustifolia 10, 319
    var. *abbreviata* 319
  annulata 305
  anomala 10, 315
  arborea 315
  armstrongii 305
    var. *annulata* 305
  astonii 321
  azurea 322
  balfouriana 327
  barkeri 305
  benthamii 303, 305
  biggarii 306
  ×*bishopiana* 306
  bollonsii 306
  brachysiphon 306
  breviracemosa 306
  buchananii 306
    var. *exigua* 306
    var. *major* 306

Veronica *continued*
　buxifolia 34, 62, 314
　　var. *odora* 314
　　var. *patens* 315
　　var. *prostrata* 315
　canterburiensis 307
　carnea 327
　carnosula 307
　carsei 326
　cassinioides 326
　chathamica 307
　　var. *coxiana* 307
　coarctata 311
　cockayneana 307
　colensoi 308
　cookiana 320
　coxiana 210, 307
　cupressoides 324
　　var. *variabilis* 316
　dartonii 326
　darwiniana 324
　decumbens 308
　decussata 8, 303, 309–10
　dieffenbachii 308
　diosmifolia 309, 327
　　var. *trisepala* 309
　divergens 326
　dorrien-smithii 308
　×*edinensis* 328
　elliptica 8, 62, 309–10
　　var. *odora* 314
　epacridea 310, 328
　erecta 327
　evenosa 310
　finaustrina 305
　fonkii 318
　forsteri 10, 309
　franciscana 327
　gibbsii 310
　gigantea 324
　gilliesiana 33–4, 323
　glauca-caerulea 324
　glaucophylla 311
　godefroyana 327
　gracillima 312
　greyi 323
　haastii 311
　　var. *macrocalyx* 313
　haustrata 315
　hectorii 311
　　var. *gracilior* 312
　hectorii × *V. pimeleoides* 328
　hendersonii 327
　hillii 308
　hulkeana 303
　imbricata 312
　insularis 312
　jasminoides 327
　kermesina 328
　kirkii 326
　laevis 254, 322
　　var. *carnosula* 307

　laingii 311
　latisepala 314
　leiophylla 312
　　var. *strictissima* 320
　lewisii 326
　ligustrifolia 312
　　var. *acutiflora* 304
　　var. *gracillima* 312
　lindleyana 328
　loganioides 326
　*longiracemosa*v325
　lycopodioides 302, 313
　macrantha 313
　　var. *brachyphylla* 313
　macrocalyx 313
　macrocarpa 8, 313
　　var. *affinis* 325
　　var. *crassifolia* 321–2
　　var. *latisepala* 314
　　var. *myrtifolia* 325
　macroura 320
　　var. *cookiana* 320
　　var. *dubia* 314
　marginata 310
　matthewsii 325
　menziesii 309
　　var. *divaricata* 309
　montana 320
　monticola 320–1
　myrtifolia, ×*myrtifolia* 325, 327
　obovata 325
　obtusata 184, 314
　odora 314
　parkinsoniana 319
　parviflora 8, 315
　　var. *angustifolia* 319
　　var. *arborea* 315
　　var. *obtusa* 325
　　var. *phillyreaefolia* 312
　　var. *strictissima* 320
　petriei 316
　pimeleoides 316
　　var. *glauca-caerulea* 324
　　var. *minor* 316
　pinguifolia 316
　plebeia 7, 16
　poppelwellii 312
　propinqua 316
　　var. *major* 317
　pubescens 302, 317
　quadrifaria 323
　rakaiensis 317
　rapensis 317
　rigidula 317
　rotundata 328
　rupicola 318
　salicifolia 8, 62, 184, 302, 318
　　var. *angustissima* 190, 305
　　var. *atkinsonii* 320
　　var. *communis* 318
　　var. *gigantea* 324
　　var. *gracilis* 325

　　var. *kirkii* 326
　　var. *longiracemosa* 320
　　var. *paludosa* 270, 315
　　var. *stricta* 319
　salicornioides 318
　serpyllifolia 7
　×*simmonsii* 327
　speciosa 303, 319, 328
　　var. *brevifolia* 306
　　var. *imperialis* 327
　squalida 319
　stenophylla 319
　stricta 319
　　var. *lindleyana* 328
　subalpina 320
　subsimilis 321
　tetragona 321
　tetrasticha 324
　townsonii 321–2
　traversii 62, 306, 322
　　var. *elegans* 322
　　var. *fallax* 311
　trisepala 252, 309
　truncatula 322
　tumida 324
　venustula 322
　vernicosa 323
　　var. *gracilis* 323
　　var. *multiflora* 323
　willcoxii 307
　Wrightii 322
Veroniceae 7
von Haast, Julius (Sir) 120
voucher specimens for photos and
　illustrations 80, 352–6

Wagstaff, S.J. 13–15
Waitakere City Council 72
Waitakere rock koromiko 331
Ward, J.M. 31
Webb, C.J. 36
weeds 69–70
Western mountain koromiko 331
Wettstein, R. von 12
Wheeler, C. and V. 6, 73
whipcord hebe (common name)
　335
　see also Hebe "Flagriformes"
wood anatomy 28

About the authors

Mike Bayly and Alison Kellow are botanists with extensive field and laboratory experience. They have co-authored many scientific papers on the classification and evolution of *Hebe*.

Both authors are Australian-trained: Mike has a Ph.D. in plant systematics from The University of Melbourne and Alison has a Ph.D. in Agricultural and Natural Resource Sciences from the University of Adelaide.

The pair researched and wrote *An Illustrated Guide to New Zealand Hebes* while employed as research scientists at the Museum of New Zealand Te Papa Tongarewa.

Currently, Mike is a Research Fellow at The University of Melbourne's School of Botany and Alison still works for Te Papa. They are married and live in Melbourne with their daughter.

Polar view of Southern Hemisphere

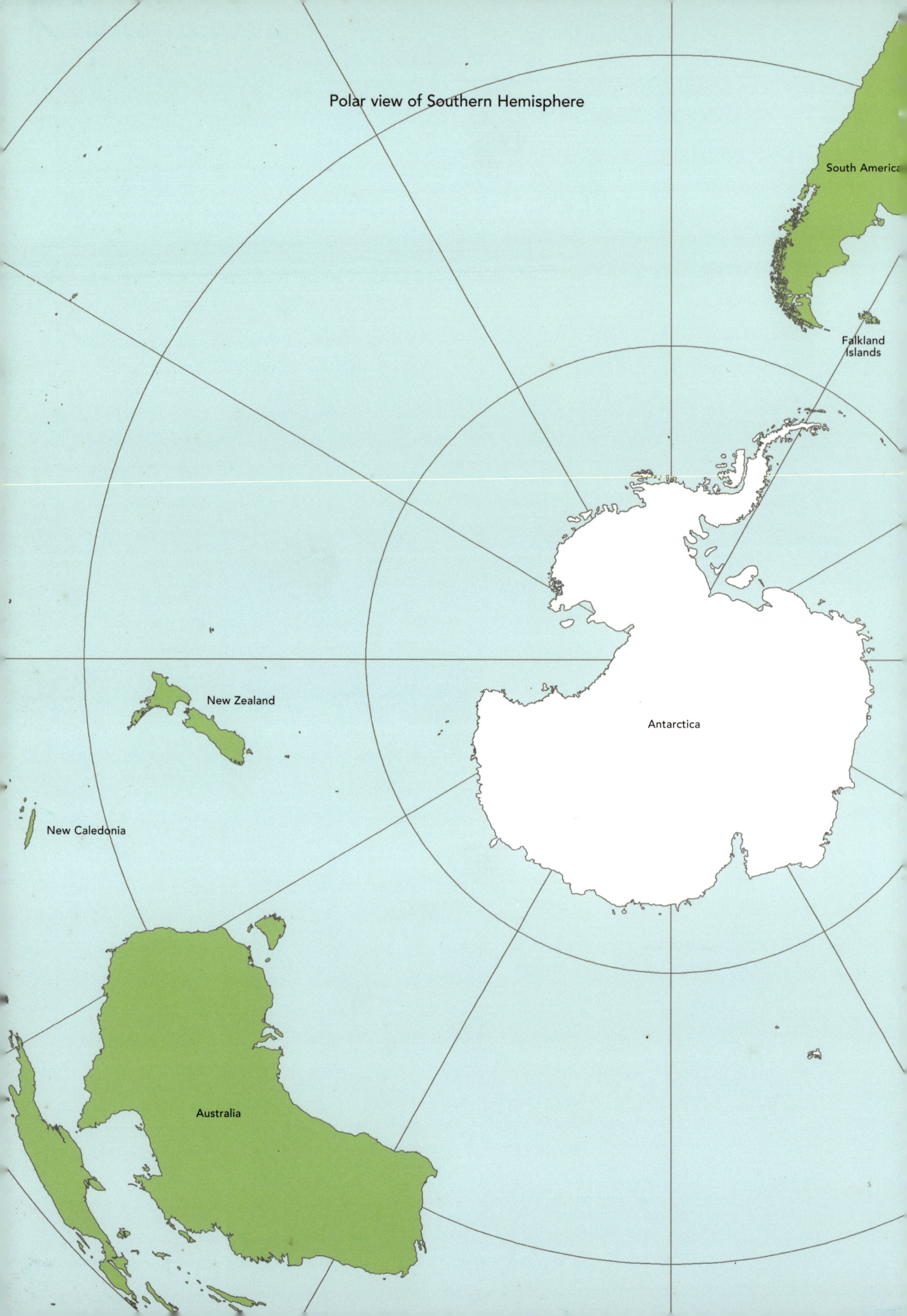

South America

Falkland Islands

New Zealand

New Caledonia

Antarctica

Australia